A Specialist Periodical Report

Organophosphorus Chemistry

Volume 1

A Review of the Literature Published up to June 1969

Senior Reporter
S. Trippett, *Department of Chemistry, The University, Leicester*

Reporters
R. S. Davidson, *The University, Leicester*
D. W. Hutchinson, *University of Warwick*
R. Keat, *Glasgow University*
R. A. Shaw, *Birkbeck College, London*
J. C. Tebby, *North Staffordshire College of Technology*

SBN: 85186 006 0

The Chemical Society
Burlington House, London, W1V 0BN

Organic formulae composed by Wright's Symbolset method

PRINTED IN GREAT BRITAIN BY JOHN WRIGHT AND SONS LTD., AT THE STONEBRIDGE PRESS, BRISTOL

Foreword

This Report was designed to cover all significant work in organophosphorus chemistry available to the Reporters in the year ending 30th June, 1969. As this is the first of an annual series most of the literature from the beginning of 1968 has been included. Hopefully, therefore, future Reports will be substantially smaller. The material has been organised on a compound type rather than a mechanistic basis and this may have tended to obscure the essential unity of the subject. Certainly the more theoretical aspects have not been adequately covered and a volunteer here would be welcome!

It is inevitable that next year's Report will have been organised before this one is published and that changes resulting from criticism may therefore be delayed. We should, however, be grateful for comments and suggestions for improvement.

Contents

Chapter 6 Quinquevalent Phosphorus Acids
By D. W. Hutchinson

Chapter 7 Phosphates and Phosphonates of Biochemical Interest
By D. W. Hutchinson

Abbreviations

The following abbreviations have been used

AIBN	bisazoisobutyrylnitrile
DCC	dicyclohexyl carbodi-imide
DMF	*NN*-dimethylformamide
DMSO	dimethyl sulphoxide
MO	molecular orbital
pp_i	inorganic pyrophosphate
n.q.r.	nuclear quadrupole resonance
THF	tetrahydrofuran
t.l.c.	thin-layer chromatography

1

Phosphines and Phosphonium Salts

BY S. TRIPPETT

PART I: Phosphines

1 Preparation

A. From Halogenophosphine and Organometallic Reagent.—For the preparation of tertiary phosphines this continues to be the method of choice when applicable. The lithioacetylide (1) with phosphorus trichloride gave the phosphine (2) [1] whose stability at 283° contrasted sharply with the thermal instability of triethynylphosphine. The silicon analogue (3) was prepared in a similar way [2] as well as from bis(trimethylsilyl)acetylene and phosphorus trichloride.

$$\underset{(1)}{\mathrm{Li{\cdot}C{:}C{\cdot}CMe_3}} \xrightarrow{\mathrm{PCl_3}} \underset{(2)\ 98\%}{\mathrm{P(C{:}C{\cdot}CMe_3)_3}}$$

$$\mathrm{Li{\cdot}C{:}C{\cdot}SiMe_3} \xrightarrow[49\%]{\mathrm{PCl_3}} \mathrm{P(C{:}C{\cdot}SiMe_3)_3}\ (3)$$

$$\mathrm{Me_3Si{\cdot}C{:}C{\cdot}SiMe_3} \xrightarrow[\mathrm{PCl_3}\ 74\%]{} \mathrm{P(C{:}C{\cdot}SiMe_3)_3}\ (3)$$

The previously described [3] preparation of tris(trifluorovinyl)phosphine from trifluorovinylmagnesium iodide and phosphorus trichloride is now reported [4] to give only polymeric material. Phosphorus tribromide gave the required phosphine.

Among other syntheses of this type, those of the phosphines (4) [5] and (5) [6] and of many fluoroalkylphosphines,[7] *e.g.* (6), may be mentioned.

[1] H. F. Reiff and B. C. Pant, *J. Organometallic Chem.*, 1969, **17**, 165.
[2] W. Siebert, W. E. Davidsohn, and M. C. Henry, *J. Organometallic Chem.*, 1969, **17**, 65.
[3] R. N. Sterlin, R. D. Yatsenko, L. N. Pinkina, and I. L. Knunyants, *Izvest. Akad. Nauk S.S.S.R., Ser. khim.*, 1960, 1991.
[4] A. H. Cowley and M. W. Taylor, *J. Amer. Chem. Soc.*, 1969, **91**, 1929.
[5] G. Marr and T. Hunt, *J. Chem. Soc.* (*C*), 1969, 1070.
[6] M. A. Bennett, W. R. Keen, and R. S. Nyholm, *Inorg. Chem.*, 1968, **7**, 552.
[7] K. Gosling, D. J. Holman, J. D. Smith, and B. N. Ghose, *J. Chem. Soc.* (*A*), 1968, 1909.

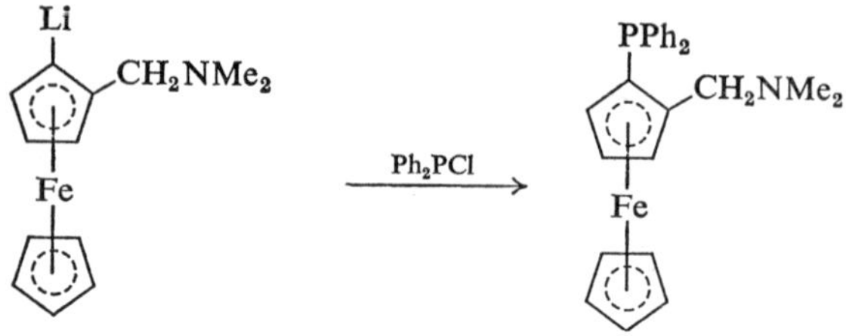

(4) 78%

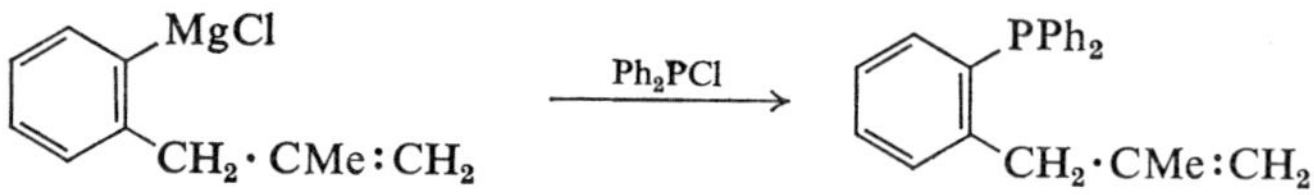

(5)

$$C_3F_7 \cdot Li + CF_3 \cdot PCl_2 \longrightarrow CF_3P(C_3F_7)_2$$

(6)

β-Ketoalkyltin compounds with halogenophosphines gave the corresponding β-ketoalkylphosphines[8–10] which are otherwise difficult to prepare, *e.g.*

$$Et_3Sn \cdot CH_2 \cdot CO_2 \cdot Et \xrightarrow{EtPBr_2} EtP(CH_2 \cdot CO_2Et)_2 \quad 66\%$$

$$Et_3Sn \cdot CH_2 \cdot CO_2 \cdot Et \xrightarrow{PhPCl_2} PhP(CH_2 \cdot CO_2Et)_2 \quad 70\%$$

$$Bu_3Sn \cdot CH_2 \cdot CO \cdot Me \xrightarrow{Ph_2PCl} Ph_2P \cdot CH_2 \cdot CO \cdot Me \quad 57\%$$

B. From Metallated Phosphines.—The synthesis of phosphiran from sodium phosphide and 1,2-dichloroethane in liquid ammonia[11] has been extended[12] to the preparation of both 1- and 2-substituted phosphirans. The 2-ethylphosphiran was a mixture of *cis*- and *trans*-isomers.

[8] M. V. Proskurnina, I. F. Lutsenko, Z. S. Novikova, and N. P. Voronova, *Khim. Org. Soedinenii Fosfora*, 1967, 8 (*Chem. Abs.*, 1968, **69**, 52217).
[9] G. M. Vinokurova, *Zhur. obshchei Khim.*, 1967, **37**, 1652.
[10] Z. S. Novikova, M. V. Proskurnina, L. I. Petrovskaya, I. V. Bogdanova, N. P. Galitskova, and I. F. Lutsenko, *Zhur. obshchei Khim.*, 1967, **37**, 2080.
[11] R. I. Wagner, Lev. D. Freeman, H. Goldwhite, and D. G. Rowsell, *J. Amer. Chem. Soc.*, 1967, **89**, 1102.
[12] S. Chan, H. Goldwhite, H. Keyzer, D. G. Rowsell, and R. Tang, *Tetrahedron*, 1969, **25**, 1097.

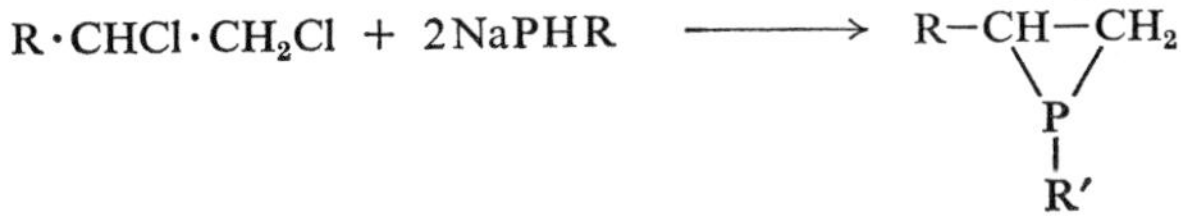

R	Me	Et	H	H
R′	H	H	Ph	Me
%	30	25	63	30

1-Deuteriophosphiran was obtained from 1,2-dichloroethane and sodium dideuteriophosphide prepared in tris(dimethylamino)phosphine oxide. Alkylphosphines (7) were similarly obtained, *e.g.* $EtPH_2$ (78%), $CH_2{:}CH{\cdot}CH_2{\cdot}PH_2$ (55%).

$$PH_3 \xrightarrow[(Me_2N)_3PO]{Na} NaPH_2 \xrightarrow{R.Hal} \underset{(7)}{RPH_2}$$

Convenient syntheses of methyl[13] and dimethylphosphine[14] have been described using dimethylsulphoxide as solvent. Other syntheses using metallated phosphines and alkyl halides include those of the amines (8)[15] and (9)[16] and of the diphosphine (10).[17] Whereas lithium diethyl-[18] and dicyclohexyl-phosphides[19] are stable in refluxing tetrahydrofuran, the corresponding dimethylphosphide[20] rapidly cleaves the solvent to give (11).

$$N(CH_2{\cdot}CH_2{\cdot}Cl)_3 + KPPh_2 \xrightarrow{THF} \underset{(8)}{N(CH_2{\cdot}CH_2{\cdot}PPh_2)_3}$$

$$EtNH{\cdot}CH_2{\cdot}CH_2{\cdot}Cl + NaPH_2 \xrightarrow{NH_3} \underset{(9)}{EtNH{\cdot}CH_2{\cdot}CH_2{\cdot}PH_2}$$

(10)

$$Me_2P)_2 \xrightarrow[THF]{Li} [Me_2PLi] \xrightarrow{THF} \underset{(11)}{Me_2P{\cdot}(CH_2)_4{\cdot}OLi} \xrightarrow{Me_3SiCl} \underset{91\%}{Me_2P{\cdot}(CH_2)_4{\cdot}O{\cdot}SiMe_3}$$

13 W. L. Jolly, *Inorg. Synth.*, 1968, **11**, 124.
14 W. L. Jolly, *Inorg. Synth.*, 1968, **11**, 126.
15 L. Sacconi and I. Bertini, *J. Amer. Chem. Soc.*, 1968, **90**, 5443.
16 K. Isslieb, R. Kümmel, H. Oehme, and I. Meissner, *Chem. Ber.*, 1968, **101**, 3613.
17 K. Isslieb and H. Weichmann, *Chem. Ber.*, 1968, **101**, 2197.
18 W. Hewertson and H. R. Watson, *J. Chem. Soc.*, 1962, 1490.
19 K. Isslieb and H.-R. Roloff, *Chem. Ber.*, 1965, **98**, 2091.
20 R. E. Goldsbury, D. E. Lewis, and K. Cohn, *J. Organometallic Chem.*, 1968, **15**, 491.

Typical of syntheses using vinyl halides were those of the diphosphine (12)[21] and of diphenyl-1-phenylvinylphosphine (13).[22] Perfluoroacyldiphenylphosphines have been obtained from the corresponding perfluoroacid halides or anhydrides.[23]

$$\underset{trans}{Bu_2PLi + Cl{\cdot}CH{:}CH{\cdot}Cl} \xrightarrow{\text{THF}} \underset{trans}{Bu_2P{\cdot}CH{:}CH{\cdot}PBu_2}$$
$$(12)\ 25\%$$

$$Ph_2PNa + PhCBr{:}CH_2 \xrightarrow{\text{THF}} Ph_2P{\cdot}CPh{:}CH_2$$
$$(13)\ 65\%$$

Aguiar showed[24] that the ready reaction of aryl halides with lithium diphenylphosphide does not involve an aryne. Isslieb has now shown[25] that such intermediates are involved in similar reactions with lithium di-t-butylphosphide (14) and aryl fluorides but not with the diethyl- or dibutyl-phosphides. While this difference was ascribed to the greater nucleophilicity of (14) it may be due to steric hindrance round the phosphorus. The reactions of lithium phosphides with aryl bromides are complicated by metal–halogen exchange. Thus (14) and *p*-bromotoluene gave only (15) together with the biphosphine (16).

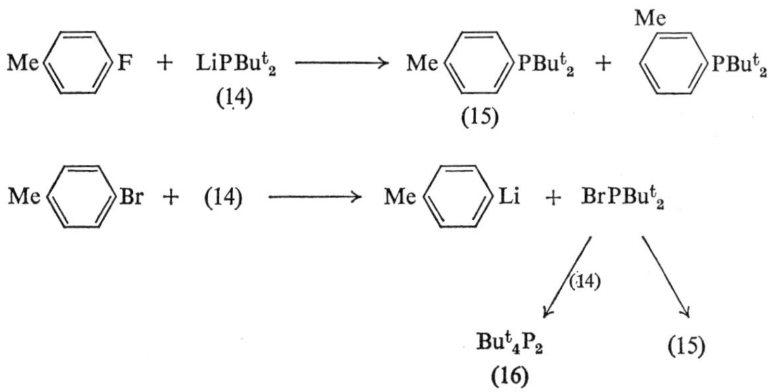

Metallated diphenylphosphine with carbon disulphide in tetrahydrofuran at −50° gave the pale orange-yellow salts (17) which formed stable red solutions in acetone and ethanol and did not react with nitrogen.[26] The corresponding reaction with the tetraphosphine (18) at 60° gave a rearranged salt (19) whose ochre solutions in polar solvents 'greedily' absorbed two

[21] M. A. Weiner and G. Pasternak, *J. Org. Chem.*, 1969, **34**, 1130.
[22] M. P. Savage and S. Trippett, *J. Chem. Soc.* (*C*), 1968, 591.
[23] E. Lindner and H. Kranz, *Chem. Ber.*, 1968, **101**, 3438.
[24] A. M. Aguiar, H. J. Greenberg, and K. E. Rubenstein, *J. Org. Chem.*, 1963, **28**, 2091.
[25] K. Isslieb, A. Tzschach, and H.-U. Block, *Chem. Ber.*, 1968, **101**, 2931.
[26] R. Kramolowsky, *Angew. Chem. Internat. Edn.*, 1969, **8**, 202.

molecules of nitrogen to give a species ($\nu_{N\equiv N}$ 2090 cm^{-1}) assigned a structure of which (20) is one of the contributing forms.[27]

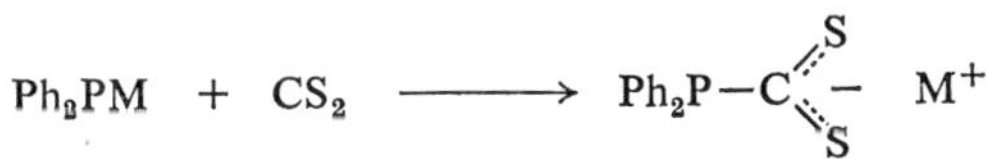

(17)

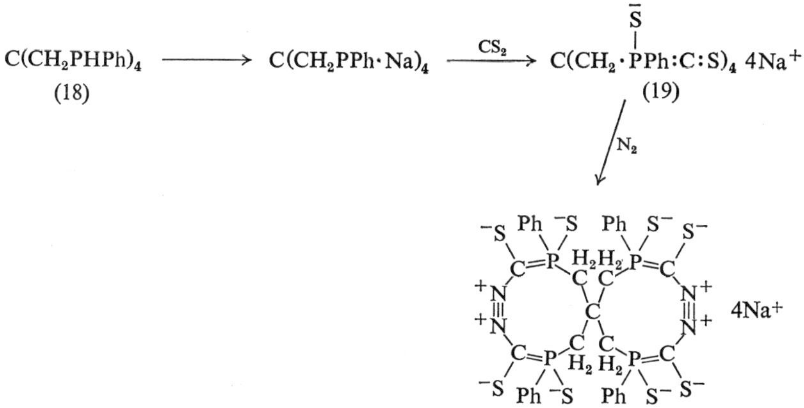

(20)

C. By Reduction.—Lithium aluminium hydride and trichlorosilane continue to be the reagents of choice. Among applications of the former are syntheses of the diphosphines (21)[28] and (22)[29] and of dimethylphosphine (70—81%) from tetramethyldiphosphine disulphide[30]

$$[(Pr^iO)RP(:O)\cdot CH_2]_2 \xrightarrow{LiAlH_4} RPH\cdot CH_2\cdot CH_2\cdot PHR$$

(21) R = Et, Ph, C_6H_{11}

$$[(EtO)_2P(:O)]_2(CH_2)_n \xrightarrow[Et_2O]{LiAlH_4} H_2P\cdot(CH_2)_n\cdot PH_2$$

(22) n = 1 (19%), 3 (56%), 4 (89%)

The triarylphosphines (23) containing functional groups sensitive to lithium aluminium hydride have been obtained[31] by the trichlorosilane reduction of the corresponding oxides. The use of hexachlorodisilane or octachlorotrisilane in refluxing benzene or in chloroform at room temperature has been recommended[32] for the reduction of optically active

[27] J. Ellerman, F. Poersch, R. Kunstmann, and R. Kramolowsky, *Angew. Chem. Internat. Edn.*, 1969, **8**, 203.
[28] K. Isslieb and H. Weichmann, *Chem. Ber.*, 1968, **101**, 2197.
[29] B.P. 1,130,487.
[30] G. W. Parshall, *Inorg. Synth.*, 1968, **11**, 157.
[31] G. P. Schiemenz and H.-U. Siebeneick, *Chem. Ber.*, 1969, **102**, 1883.
[32] K. Naumann, G. Zon, and K. Mislow, *J. Amer. Chem. Soc.*, 1969, **91**, 2788.

phosphine oxides. Almost complete inversion of configuration occurs and the mechanism shown has been suggested. The same reagents reduce acyclic phosphine sulphides and cyclic phosphine oxides with retention of configuration.

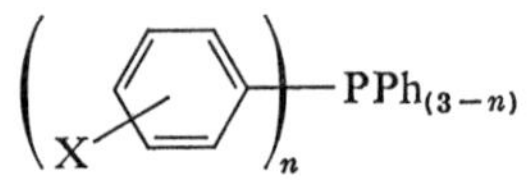

(23) X = CN, CO_2H, CO_2Me

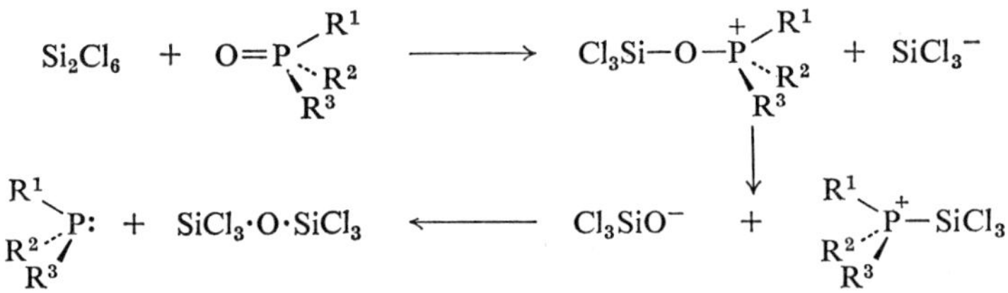

D. By the Radical Addition of P—H to Olefins.—Primary phosphines with allylamine in the presence of 2,2′-azobis-(2-methylpropionitrile) gave mixtures of the secondary (24) and tertiary (25) 3-aminopropylphosphines.[33]

$$RPH_2 + CH_2{:}CH{\cdot}CH_2{\cdot}NH_2 \longrightarrow \underset{(24)}{RPH{\cdot}CH_2{\cdot}CH_2{\cdot}CH_2{\cdot}NH_2} + \underset{(25)}{RP(CH_2{\cdot}CH_2{\cdot}CH_2{\cdot}NH_2)_2}$$

R = Ph, Bu

Similar addition of phenyl phosphine to the terminal dienes (26) gave the diphosphines (27).[34]

$$\underset{(26)}{PhPH_2 + CH_2{:}CH{\cdot}X{\cdot}CH{:}CH_2} \longrightarrow \underset{(27)}{(PhPH{\cdot}CH_2{\cdot}CH_2)_2X}$$

X	$—(CH_2)_6—$	$—CH_2{\cdot}O{\cdot}CH_2—$	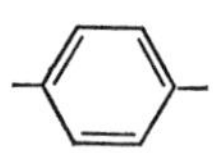	$—SiMe_2—O—SiMe_2—$
% (27)	27	19	32	52

Diallyl ether also gave 18% of the monophosphine $CH_2{:}CH{\cdot}CH_2{\cdot}O{\cdot}CH_2{\cdot}CH_2{\cdot}PHPh$. A series of additions of bicyclic secondary phosphines (28) to octa-1,7-diene has been described.[35a] The photochemical cyclisation of unsaturated secondary phosphines leads to cyclic tertiary phosphines (29).[35b]

[33] K. Issleib, H. Oehme, and E. Leissring, *Chem. Ber.*, 1968, **101**, 4032.
[34] V. V. Korshak, V. A. Zamyatina, A. I. Solomatina, E. I. Fedin, and P. V. Petrovskii, *J. Organometallic Chem.*, 1969, **17**, 193.
[35] (a) F.P. 1,502,250. (b) F.P. 1,488,936.

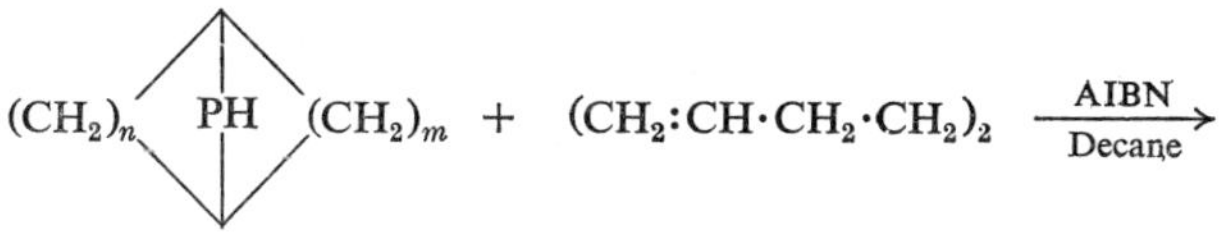

(28)

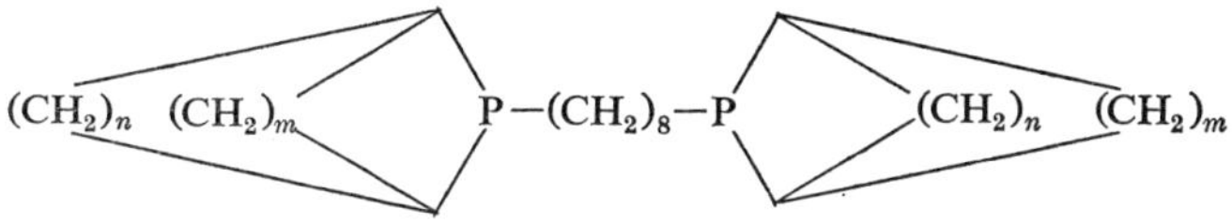

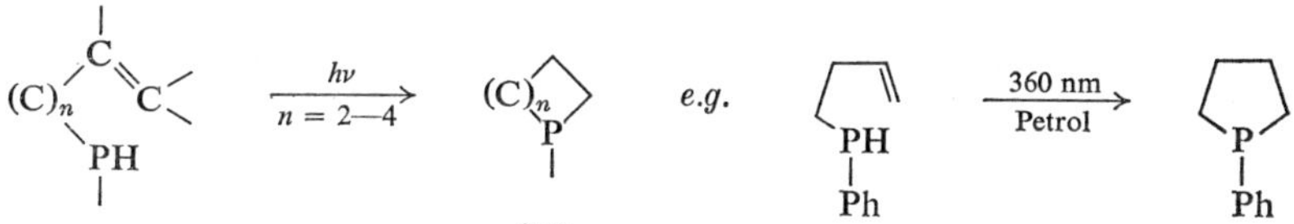

(29) 38%

E. Miscellaneous.—Tetraphenyldiphosphine on refluxing in aqueous ethanol with formaldehyde and diethylamine gave diethylaminomethyldiphenylphosphine (30) and the corresponding oxide.[36] A four-centre mechanism is proposed leading to diphenylphosphine and the phosphinite (31).

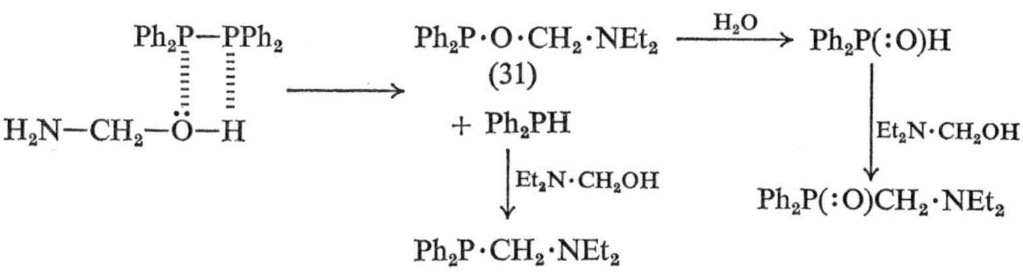

(30)

Treatment of tris-(hydroxymethyl)phosphine with phenacyl bromides followed by internal acetal formation and base-catalysed elimination of formaldehyde gave the interesting bicyclic phosphines (32).[37] Oxidation with hydrogen peroxide in methanol gave the acyclic oxides.

F. Optically Active Phosphines.—t-Butylmethylphenylphosphine has been resolved *via* the asymmetric platinum(II) complex (33) obtained from the binuclear compound (34) and (+)-deoxyephedrine.[38] Fractional crystallisation of (33) gave two diastereoisomers. Treatment of one of these with methanolic potassium cyanide liberated the optically active phosphine which was characterised as the oxide and as the optically active complex

[36] L. Maier, *Helv. Chim. Acta*, 1968, **51**, 1608.
[37] E. S. Kozlov, A. I. Sedlov, and A. V. Kirsanov, *Zhur. obshchei Khim.*, 1968, **38**, 1881.
[38] T. H. Chan, *Chem. Comm.*, 1968, 895.

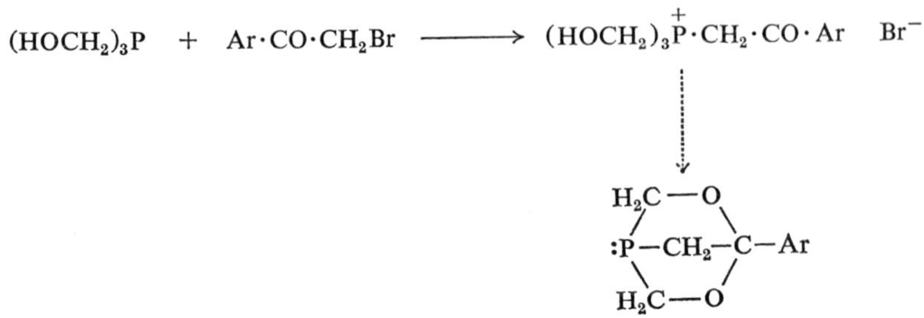

(32) 48—88%

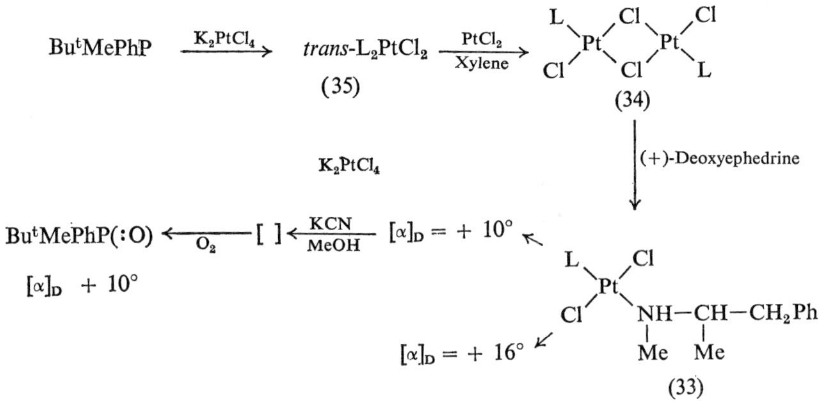

(35) having $[\alpha]_D = -11°$. The extension of this method to the resolution of other tervalent phosphorus compounds, *e.g.* phosphites, was proposed.

A method for determining the optical purity of phosphines has been described,[39] which involves quaternisation of the phosphine with the optically active bromide (36) and analysis of the ^1H n.m.r. spectrum of the resulting salt taking advantage of the chemical shift non-equivalence of the diastereotopic protons in the product mixture.

$$\underset{(36)}{PhCH(OMe)\cdot CH_2Br} + R^1R^2R^3P \longrightarrow R^1R^2R^3\overset{+}{P}\cdot CH_2\cdot CH(OMe)\cdot Ph\ \overline{Br}$$

Inversion of configuration at the phosphorus of the phosphetans (37) has been studied by n.m.r. techniques.[40] The methyl phosphetans did not invert at 162° for 4 days while the t-butyl and phenyl phosphetans inverted remarkably rapidly in view of the increased strain expected in the four-membered ring in the transition state.

[39] J. P. Casey, R. A. Lewis, and K. Mislow, *J. Amer. Chem. Soc.*, 1969, **91**, 2789.
[40] S. E. Cremer, R. J. Chorvat, C. H. Chang, and D. W. Davis, *Tetrahedron Letters*, 1968, 5799.

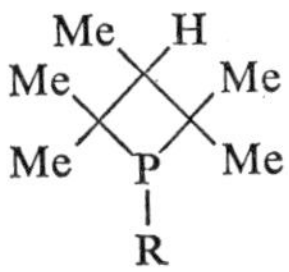

(37)

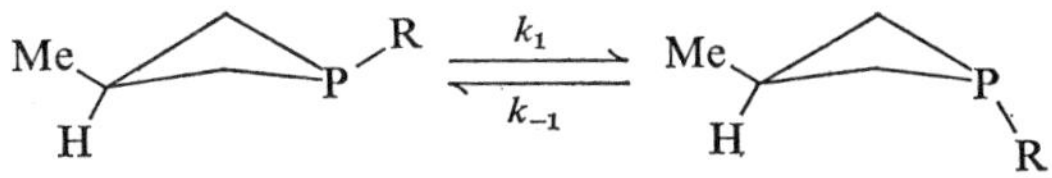

R	ΔH^* (kcal/mole)	ΔS^*(e.u.)	k_1/k_{-1}
Bu^t	28	−8	1·3
Ph	30	−8	1·5

2 Reactions

A. Nucleophilic Attack on Carbon.—(*i*) *Activated Olefins.* Tricyclohexylphosphine catalysed the addition of acrylonitrile and ethyl acrylate to aldehydes[41] to give the unsaturated alcohols (38), presumably *via* the betaines (39; R = C_6H_{11}). In contrast, the corresponding betaines from triphenylphosphine transfer a proton to give the ylides (40) before reacting with the aldehyde in a normal Wittig olefin synthesis.[42]

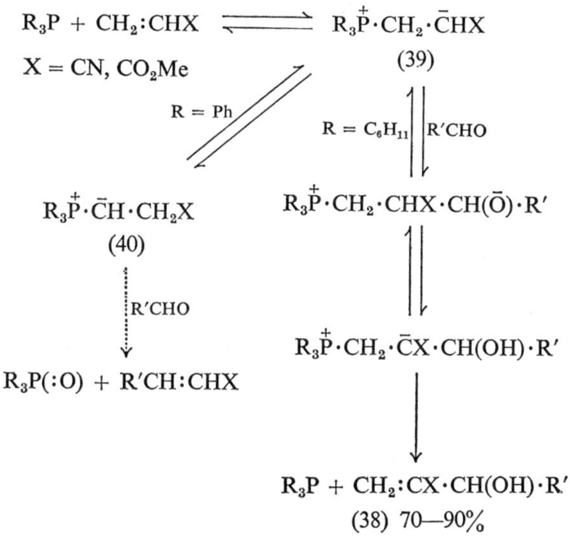

Triphenylphosphine and *N*-substituted maleimides in acetic acid gave the stable ylides (41).[43] The reaction is analogous to that previously described with maleic anhydride.[44] With either *cis*- or *trans*-β-haloacrylic acids, esters, or nitriles, tributyl- and triphenyl-phosphines in benzene at

[41] K. Morita, Z. Suzuki, and H. Hirose, *Bull. Chem. Soc. Japan*, 1968, **41**, 2815.
[42] R. Oda, T. Kawabata, and S. Tanimoto, *Tetrahedron Letters*, 1964, 1653.
[43] E. Hedaya and S. Theodoropulos, *Tetrahedron*, 1968, **24**, 2241.
[44] R. F. Hudson and P. A. Chopard, *Helv. Chim. Acta*, 1963, **46**, 2178.

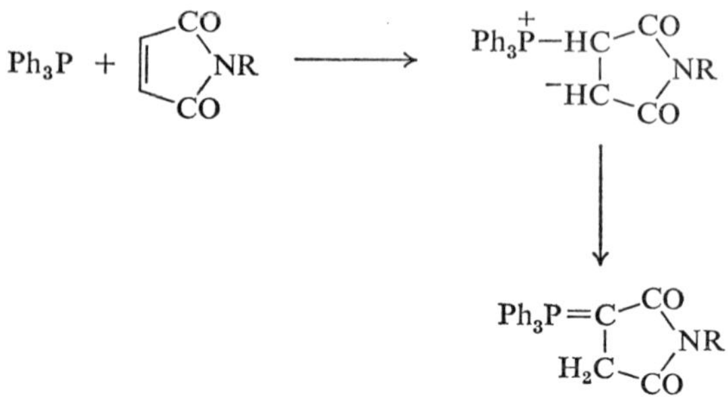

(41)

room temperature gave the *trans*-vinylphosphonium salts (42), probably by an addition–elimination mechanism.[45] No reaction occurred with the α- or β-methyl-β-haloacrylates. β-Bromoacrylic acid and triphenylphosphine also gave the bis-salt (43) which was formed exclusively at

$$\underset{cis \text{ or } trans}{R_3P + Hal\cdot CH{:}CH\cdot X} \longrightarrow R_3\overset{+}{P}\cdot CHHal\cdot \overset{-}{C}H\cdot X \longrightarrow \underset{trans}{R_3\overset{+}{P}\cdot CH{:}CHX\ Hal^-}$$

(42)

$X = CO_2H, CO_2Me, CN$

$Hal = Cl, Br$

higher temperatures. The salt (42; R = Ph, X = CO_2H) was not an intermediate in this reaction which may involve dehydrobromination of the β-bromoacrylic acid and addition of triphenylphosphonium bromide to the resulting propiolic acid. The last reaction is now reported to give a high yield of the bis-salt (43).

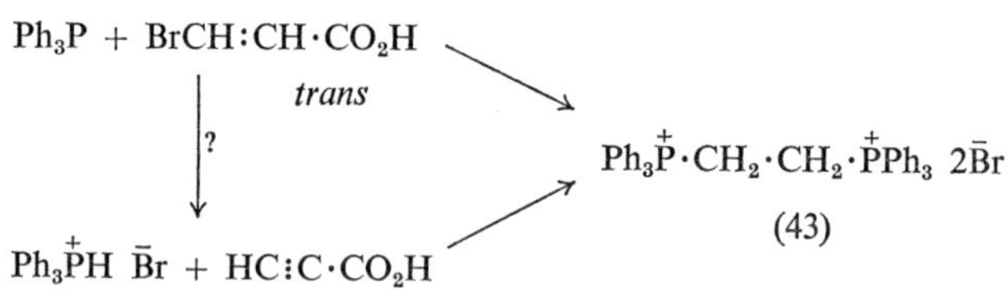

(43)

1,2-Dichloroperfluorocycloalkenes (44) and perfluorocycloalkenes with tertiary phosphines in wet acetic acid[46,47] gave the stable ylides (45) when $n = 1$ or 2 but not when $n = 3$, the major product in this case being the phosphine oxide together with tars and the 1-chlorocyclohexene (46).

[45] G. Pattenden and B. J. Walker, *J. Chem. Soc.* (*C*), 1969, 531.
[46] R. F. Stockel, F. Megson, and M. T. Beachem, *J. Org. Chem.*, 1968, **33**, 4395.
[47] S. E. Ellzey, *Canad. J. Chem.*, 1969, **47**, 1251.

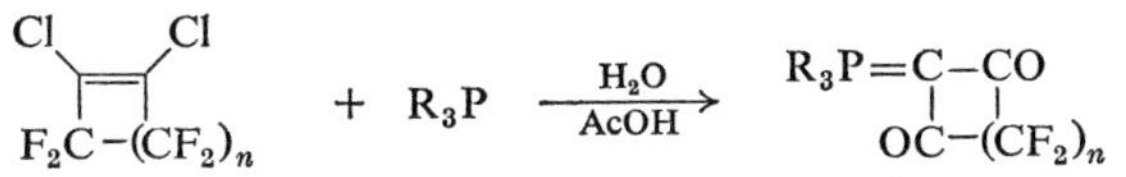

(44) $n = 1$ (29%), $n = 2$ (36%) (45)

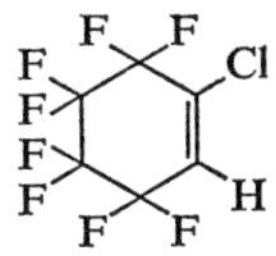

(46)

Triphenylphosphine and an excess of perfluorocyclobutene formed a 1 : 1-adduct,[46] which with water gave the ylide (45, R = Ph, n = 1), and for which, on the basis of ^{31}P and ^{19}F n.m.r. data, the unlikely looking structures (47) or (48) were suggested.

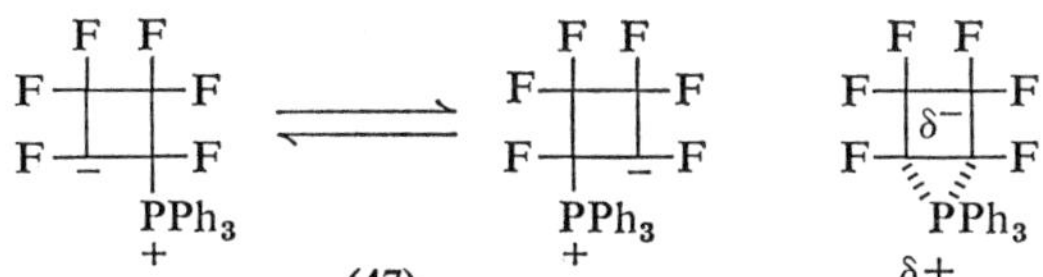

(47) (48)

Diphenylphosphine with 1,2-dichlorotetrafluorocyclobutene in dimethylformamide[48] gave the mono- (49, 46%) and di-phosphines (50, 20%) whereas in the absence of solvent only trifluorodiphenylphosphorane and diphenylphosphinyl fluoride had been identified.[49] The same phosphine with 1,2-dichlorohexafluorocyclopentene in dimethylformamide[50] gave only the monophosphine (51; R = Ph, 78%) while dicyclohexylphosphine also gave 8% of the diphosphine (52; R = C_6H_{11}). The diphosphine (50) had previously been obtained (11%) from diphenylphosphine and perfluorocyclobutene in the absence of solvent.[49]

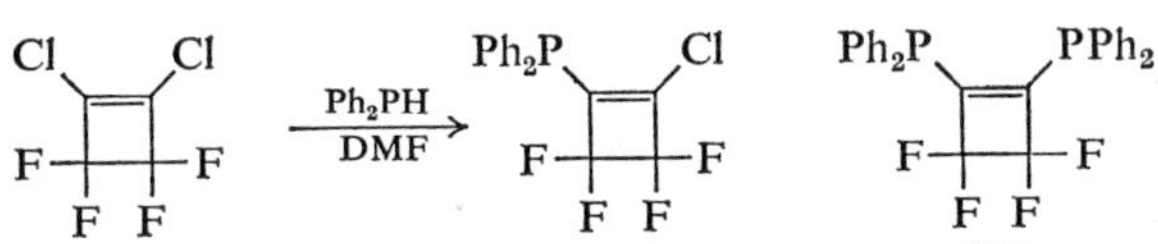

(49) (50)

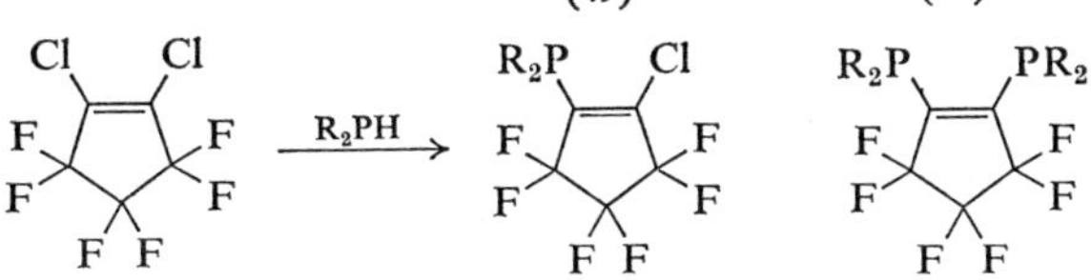

(51) (52)

[48] R. F. Stockel, *Canad. J. Chem.*, 1969, **47**, 867.
[49] W. R. Cullen, D. S. Dawson, and P. S. Dhahival, *Canad. J. Chem.*, 1967, **45**, 683.
[50] R. F. Stockel, *Canad. J. Chem.*, 1968, **48**, 2625.

Preparation of the phosphanone (53) has been improved[51] by catalysis with sodium alkoxides at 120—130°.

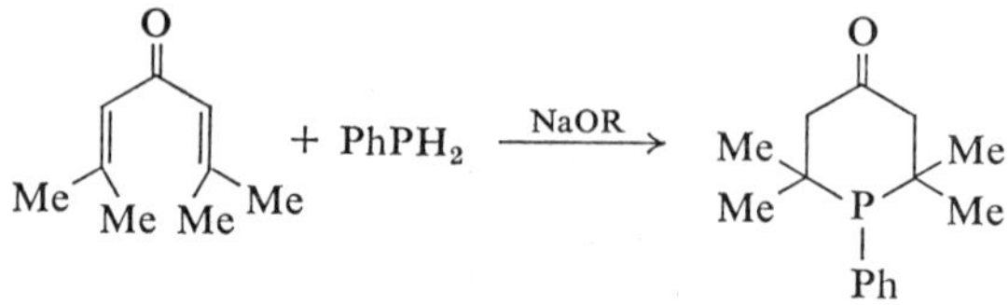

(53) 70%

The addition of dimethylphosphine to vinylsilanes is catalysed by lithium dimethylphosphide,[52] although with diphenyldivinylsilane vigorous polymerisation resulted.

$$Me_3Si{\cdot}CH{:}CH_2 + Me_2PH \xrightarrow{Me_2PLi} Me_3Si{\cdot}CH_2{\cdot}CH_2{\cdot}PMe_2 \quad 83\%$$

$$Me_2Si(CH{:}CH_2)_2 + Me_2PH \xrightarrow{Me_2PLi} Me_2Si(CH_2{\cdot}CH_2{\cdot}PMe_2)_2 \quad 85\%$$

Potassium diphenylphosphide added to 1,1-diphenylethylene gave a low yield of the phosphine (54).[53] Carbonation of the intermediate anion from the addition to stilbene resulted in the isolation of 6% of the acid (55).

$$Ph_2PK + Ph_2CH{:}CH_2 \longrightarrow Ph_2P{\cdot}CH_2{\cdot}CHPh_2 \quad (54)\ 20\%$$

$$Ph_2PK + Ph{\cdot}CH{:}CH{\cdot}Ph \longrightarrow [\ ] \xrightarrow{CO_2} Ph_2P({:}O){\cdot}CHPh{\cdot}CH(CO_2H){\cdot}Ph \quad (55)$$

(*ii*) *Activated Acetylenes.* The initial adducts (56) from the addition of triphenylphosphine to the acetylenic carboxylic esters (57) have been trapped[54] in the presence of sulphur dioxide and water as the betaines (58), also obtained, when R = Ph, CO_2Me, by the addition of bisulphite anion to the vinylphosphonium salts (59).

The yellow 1 : 2 adduct formed from triphenylphosphine and dimethyl acetylenedicarboxylate in refluxing ether[55] has now been shown[56] to be the stable ylide (60) formed by rearrangement of an intermediate phosphorane. Compound (60) gave a colourless perchlorate and reduction with zinc and acetic acid gave the oxide (61). It seems probable that the yellow 1 : 2-adduct formed from 1,2,5-triphenylphosphole and the same acetylene[57]

[51] F. Asinger, A. Saus, and E. Michel, *Monatsh.*, 1968, **99**, 1695.
[52] J. Grobe and U. Möller, *J. Organometallic Chem.*, 1969, **17**, 263.
[53] F. Rudolph, *Jenaer Jahrb.*, 1966, 221 (*Chem. Abs.*, 1968, **68**, 114,707).
[54] M. A. Shaw, J. C. Tebby, R. S. Ward, and D. H. Williams, *J. Chem. Soc.* (*C*), 1968, 2795.
[55] A. W. Johnson and J. C. Tebby, *J. Chem. Soc.*, 1961, 2126.
[56] N. E. Waite, J. C. Tebby, R. S. Ward, and D. H. Williams, *J. Chem. Soc.* (*C*), 1969, 1100.
[57] A. N. Hughes and S. Vaboonkul, *Tetrahedron*, 1968, **24**, 3437.

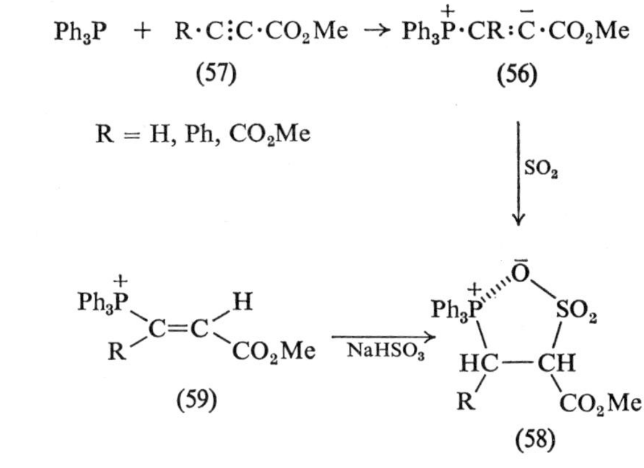

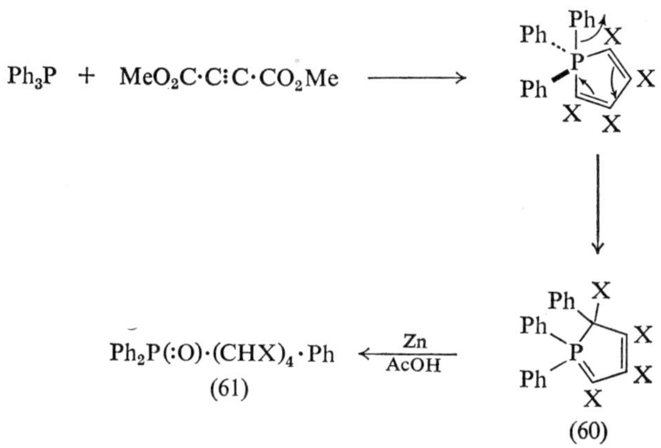

X = CO_2Me

was the result of a similar rearrangement and has the structure (62; X = CO_2Me). For the reactions of diphenyl-1-phenylvinylphosphine with this acetylene, with epoxides, and with activated olefins see Chapter 8, section 1A.

Dibutylphosphine and dibutylethynylphosphine oxide at 80° gave a low yield of the *trans*-oxide (63).[58a]

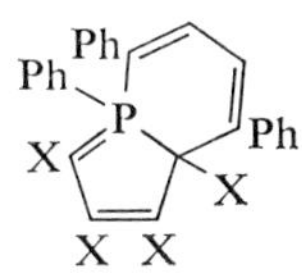

(62) X = CO_2Me

[58] [(a)] M. A. Weiner and G. Pasternack, *J. Org. Chem.*, 1969, **34**, 1130. [(b)] J. C. Kauer and H. E. Simons, *J. Org. Chem.*, 1968, **33**, 2720. [(c)] A. N. Hughes and M. Woods, *Tetrahedron*, 1967, **23**, 2973.

$$Bu_2P(:O)\cdot C:CH + Bu_2PH \longrightarrow \underset{(63)\ 20\%}{Bu_2P(:O)\cdot CH:CH\cdot PBu_2}$$

The tetramer of dimethyl acetylenedicarboxylate with triphenylphosphine gave[58b] the red ylide (63a), probably identical with the compound previously obtained from (impure) dimer.[58c]

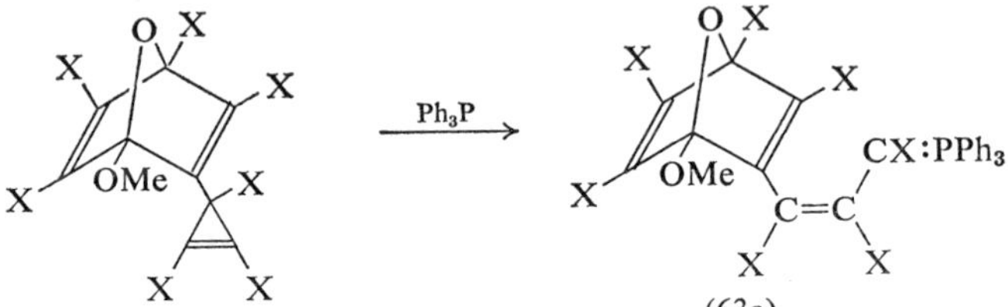

(63a)

(*iii*) *Carbonyls, etc.* Secondary phosphines added to keten[59] and to bis(trifluoromethyl)-keten[60] to give the acylphosphines (64). For the reaction of tris(dimethylamino)phosphine with dimethylketen see Chapter 2, section 6.

$$R_2PH + R'_2C:C:O \longrightarrow \underset{(64)}{R_2P\cdot CO\cdot CHR'_2}$$

Fluorosulphonyl isocyanate and triphenylphosphine in ether at room temperature gave a dipolar adduct (65).[61] *N*-Isothiocyanatodi-isopropylamine with trimethyl and triethylphosphine formed similar 1 : 1-adducts

$$Ph_3P + FSO_2\cdot NCO \longrightarrow \underset{(65)}{Ph_3\overset{+}{P}\cdot CO\cdot \overset{-}{N}\cdot SO_2F}$$

(66).[62] With secondary phosphines the products were di-isopropylamine thiocyanate and the diphosphine. A six-membered cyclic transition state is suggested.

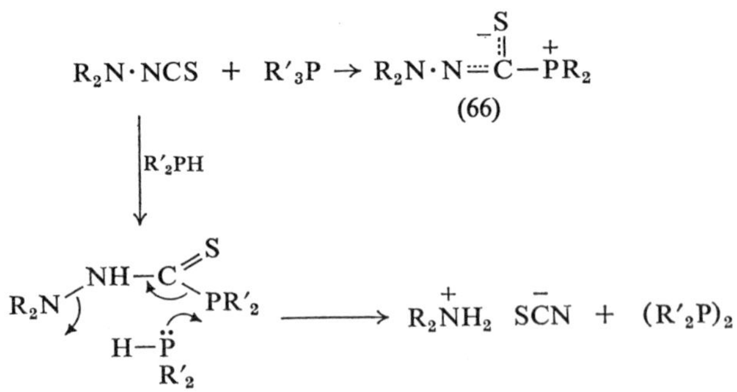

59 R. G. Kostyanovsky, V. V. Yakshin, and S. L. Zimont, *Tetrahedron*, 1968, **24**, 2995.
60 R. G. Kostyanovsky, V. V. Yakshin, S. L. Zimont, and I. I. Chervin, *Izvest. Akad. Nauk S.S.S.R., Ser. khim.*, 1968, 190.
61 H. Hoffmann, H. Förster, and G. Tor-Poghossian, *Monatsh.*, 1969, **100**, 311.
62 U. Anthoni, O. Dahl, C. Larsen, and P. H. Nielsen, *Acta Chem. Scand.*, 1969, **23**, 943.

Diphenylphosphine and cyanic acid gave the amide (67)[63] which with toluene-*p*-sulphonyl isocyanate did not form a urea but instead a low yield of a compound assigned the structure (68).

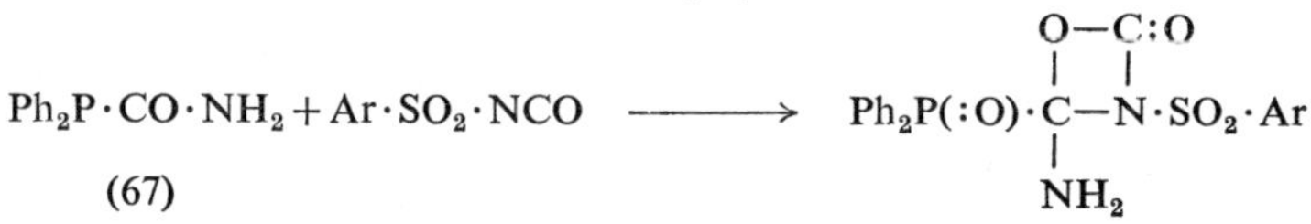

(67) (68) 3%

Addition of diphenylphosphine to the Schiff's bases (69) gave the phosphines (70).[64]

$$\underset{(69)}{\mathrm{R\cdot CH{:}N\cdot SO_2F + Ph_2PH}} \longrightarrow \underset{(70)}{\mathrm{Ph_2P\cdot CHR\cdot NH\cdot SO_2F}}$$

(*iv*) *Miscellaneous.* The mode of quaternisation of tertiary phosphines with triphenylmethyl chloride has been found to depend on the size of the phosphine.[65] Small phosphines, *e.g.* Et_3P, $PhPMe_2$, gave the expected triphenylmethylphosphonium salts (71) but more bulky phosphines, *e.g.* Ph_3P, Ph_2PMe, $PhBu^tPMe$, gave instead the 4-(diphenylmethyl)-phenylphosphonium salts (72).

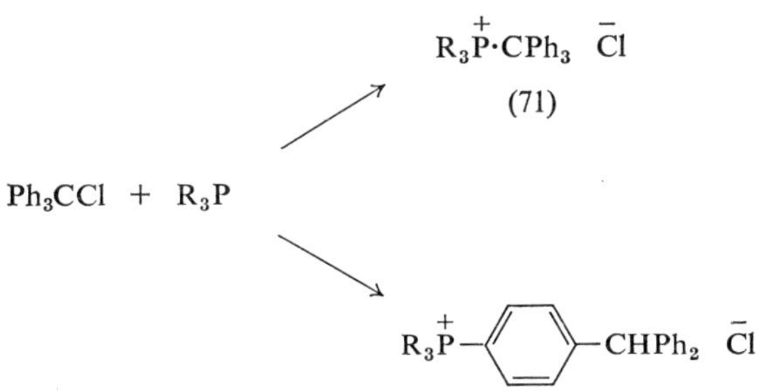

(71)

(72)

The quaternisation of triphenylphosphine with α-bromoketones is base catalysed.[66] Thus α-bromopropiophenone in acetonitrile at room temperature gave none of the salt (73) in the absence of base while in the presence of a catalytic amount of triethylamine 56% of (73) formed in 2 hr. Other

$$\underset{(73)}{\mathrm{Ph_3\overset{+}{P}\cdot CHMe\cdot CO\cdot Ph\ \overset{-}{Br}}}$$

effective catalysts included aqueous potassium cyanide and hydroxide. Phenacyl bromide and triphenylphosphine in refluxing benzene–methanol

[63] L. G. Vaughan and R. V. Lindsey, *J. Org. Chem.*, 1968, **33**, 3088.
[64] H. Hoffmann and H. Förster, *Monatsh*, 1968, **99**, 380.
[65] H. Hoffmann and P. Schellenbeck, *Chem. Ber.*, 1968, **101**, 2203.
[66] I. J. Borowitz, K. C. Kirby jun., P. E. Rusek, and R. Virkhaus, *J. Org. Chem.*, 1968, **33** 3687.

gave acetophenone (87%) *via* nucleophilic attack on halogen, while in the presence of triethylamine at room temperature in the same solvent mixture 92% of the phenacylphosphonium salt was produced. These remarkable effects were thought to involve addition of the base to the carbonyl group followed by attack of the phosphine, the transition state being 'stabilised by mesomeric electron release from the negatively charged oxygen atom.'

$$\mathrm{Ph{\cdot}CO{\cdot}CH_2Br} \xrightarrow{\mathrm{B:}} \mathrm{Ph{-}\underset{\underset{\bar{O}}{|}}{\overset{\overset{B}{|}}{C}}{-}CH_2Br} \xrightarrow{\mathrm{Ph_3P}} \mathrm{Ph{\cdot}CO{\cdot}CH_2{\cdot}\overset{+}{P}Ph_3\overset{-}{Br}} + \mathrm{B:}$$

The hindered isobutyrophenones (74; X = Cl, Br, O·SO$_2$·Me) with triphenylphosphine in aprotic solvents gave methacrylophenone in an elimination reaction.[67] Subsequent addition of triphenylphosphonium salt then gave the 3-ketophosphonium salts (75). In protic solvents the bromo-compound gave isobutyrophenone. Similar elimination–additions had previously been observed with α-halogenocyclohexanones.[68, 69]

$$\underset{(74)}{\mathrm{Ph{\cdot}CO{\cdot}CXMe_2}} + \mathrm{Ph_3P} \longrightarrow \mathrm{Ph{\cdot}CO{\cdot}CMe{:}CH_2} + \mathrm{Ph_3\overset{+}{P}H\ \overset{-}{X}}$$

$$\downarrow$$

$$\underset{(75)}{\mathrm{Ph{\cdot}CO{\cdot}CHMe{\cdot}CH_2{\cdot}\overset{+}{P}Ph_3\ \overset{-}{X}}}$$

B. Nucleophilic Attack on Halogen.—The conversion of alcohols into chlorides on treatment with a tertiary phosphine and carbon tetrachloride has been shown to involve inversion of configuration at the carbon and undoubtedly proceeds *via* the alkoxyphosphonium chloride (76).[70, 71] Thiols are similarly converted into chlorides with inversion[71] and carbon tetrabromide may be used for the preparation of alkyl bromides.[70] Trioctyl,

$$\mathrm{R_3\overset{+}{P}Cl} + \mathrm{R'OH} \longrightarrow \underset{(76)}{\mathrm{R_3\overset{+}{P}{\cdot}OR'\ \overset{-}{Cl}}} \longrightarrow \mathrm{R_3P({:}O)} + \mathrm{R'Cl}$$

triphenyl, and tris(dimethylamino)phosphine have been used, the last allowing particularly easy isolation of product.[72] The intermediate alkoxyphosphonium salt in the reaction of tris(dimethylamino)phosphine

[67] K. Fukui, R. Sudo, M. Masaki, and M. Ohta, *J. Org. Chem.*, 1968, **33**, 3504.
[68] P. A. Chopard and R. F. Hudson, *J. Chem. Soc.*, 1966, 1089.
[69] I. J. Borowitz, K. C. Kirby jun., and R. Virkhaus, *J. Org. Chem.*, 1966, **31**, 4031.
[70] J. Hooz and S. S. H. Gilani, *Canad. J. Chem.*, 1968, **46**, 86.
[71] R. G. Wells and E. I. Snyder, *Chem. Comm.*, 1968, 1358.
[72] I. M. Downie, J. B. Lee, and M. F. S. Matough, *Chem. Comm.*, 1968, 1350.

with pentan-1-ol and carbon tetrachloride has been trapped as the hexafluorophosphate.[72]

Benzotrichloride has been used in similar reactions,[73] the ease of reaction increasing with the nucleophilic character of the phosphine, and the sequence is suggested as a method for reducing suitable trichloromethyl to dichloromethyl compounds.

$$PhCCl_3 + EtOH + R_3P \longrightarrow PhCHCl_2 + EtCl + R_3P(:O)$$

With the trichloromethylcyclohexadienone (77) competitive reactions using tris(dibutylamino)phosphine showed that (77) was almost as reactive as benzotrichloride, and homoallylic stabilisation of the anion (78) was suggested.[74]

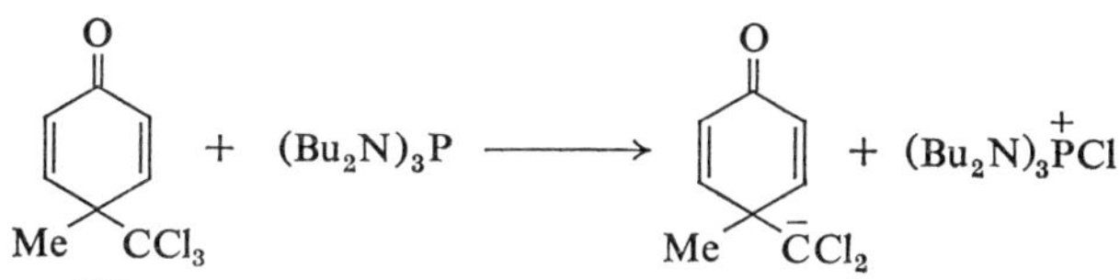

(77) (78)

The trichloromethyl anion formed from carbon tetrachloride and tris(dimethylamino)phosphine has been trapped[75] by addition to carbonyl compounds to give the alcohols (79).

$$(Me_2N)_3\overset{+}{P}Cl + \bar{C}Cl_3 \xrightarrow{R^1\cdot CO\cdot R^2} R^1-\underset{CCl_3}{\overset{\bar{O}}{C}}-R^2 \cdots\cdots\rightarrow CCl_3\cdot CR^1R^2\cdot OH$$

(79) ~ 50%

A similar sequence using esters or amides of trichloroacetic acid[76] gave the glycidic esters (80) or amides, while tributyltin trichloroacetate (81) and triphenylphosphine in the presence of benzaldehyde gave,[77] after treatment with aqueous sodium hydrogen carbonate, the dichloro-acid (82). The suggested intermediates here were dichloroketen and the β-lactone (83).

Debromination of α,α'-dibromodibenzylsulphone (84) with triphenylphosphine was stereospecific[78] involving inversion at both centres, the *meso*-form giving *cis*-stilbene and the ($\pm$)-form leading to *trans*-stilbene. The mechanism is illustrated for the former case.

The dechlorination of dichlorodiphenylmethane with tributylphosphine to give tetraphenylethylene (50%) was unaffected by the presence of butanol

[73] I. M. Downie and J. B. Lee, *Tetrahedron Letters*, 1968, 4951.
[74] B. Miller, *J. Amer. Chem. Soc.*, 1969, **91**, 751.
[75] B. Castro, R. Burgada, G. Lavielle, and J. Villiéras, *Compt. rend.*, 1969, **268**, *C*, 1067.
[76] J. Villiéras, G. Lavielle, R. Burgada, and B. Castro, *Compt. tend.*, 1969, **268**, *C*, 1164.
[77] M. Ohara, T. Okada, and R. Okawara, *Tetrahedron Letters*, 1968, 3489.
[78] F. G. Bordwell, R. B. Jarvis, and P. W. R. Corfield, *J. Amer. Chem. Soc.*, 1968, **90**, 5298.

$$(Me_2N)_3P + Cl_3C{\cdot}CO_2R \rightarrow (Me_2N)_3\overset{+}{P}Cl + Cl_2\bar{C}{\cdot}CO_2R$$

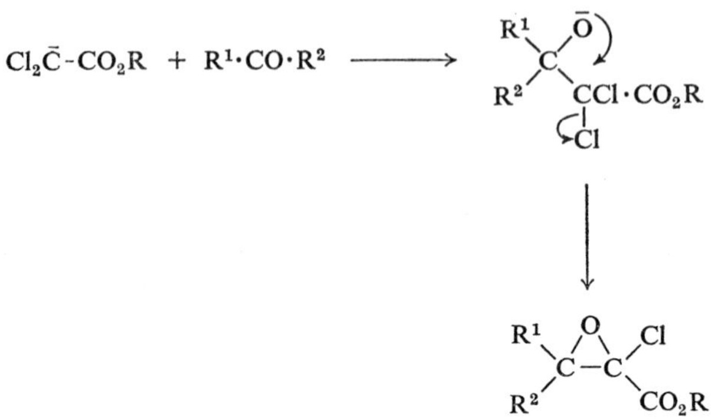

(80) 60—100%

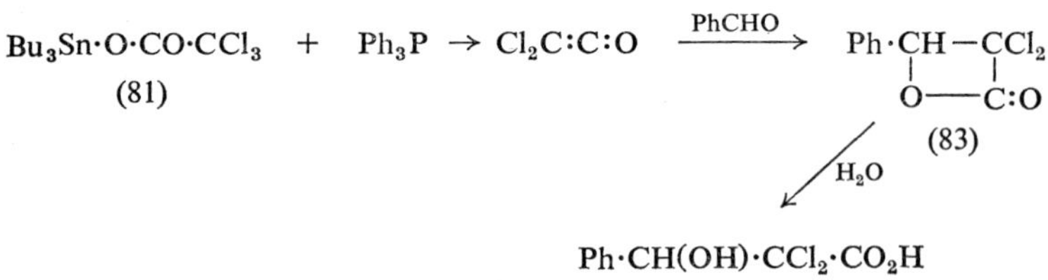

(81) (83)

(82) 33%

and was therefore held not to involve the formation of $Bu_3\overset{+}{P}Cl$.[79] Debromination of the dibromide (85) formed a convenient preparation of diphenylketen,[80] the dibromophosphorane being insoluble in the reaction mixture.

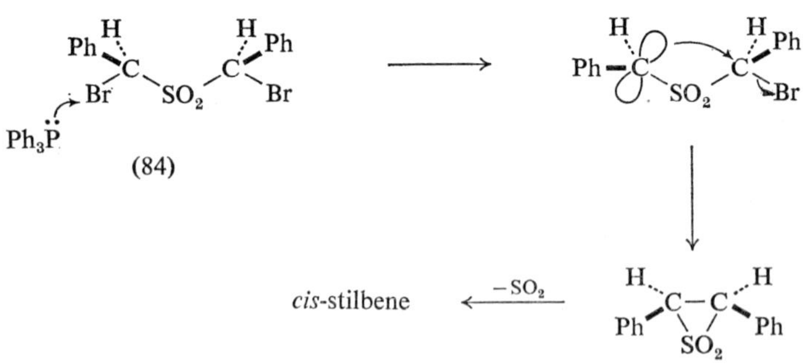

(84)

[79] K. L. Freeman and M. J. Gallagher, *Austral. J. Chem.*, 1968, **21**, 145.
[80] S. D. Darling and R. L. Kidwell, *J. Org. Chem.*, 1968, **33**, 3974.

$$Ph_2CBr\cdot CO\cdot Br + Ph_3P \xrightarrow[\text{or heptane}]{\text{benzene}} Ph_2C{:}C{:}O + Ph_3PBr_2$$

(85)

For the reactions of triphenylphosphine with 4-bromo-2,4,6-tri-t-butylcyclohexadienone see Chapter 10, section 7.

C. Nucleophilic Attack on Other Atoms.—A study of the reactions between triphenylphosphine and labelled ozonides [81] showed that attack of the phosphine was entirely on the peroxidic oxygen and was subject to steric interference, *e.g.*

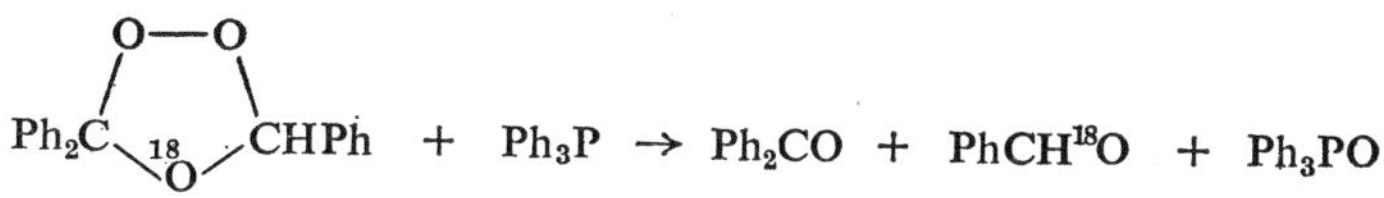

The cyclic peresters (86) with triarylphosphines gave mixtures of the olefins (87) and ketones (88) together with small amounts of the β-lactones (89).[82] The relative amounts of these varied; (87) was favoured by increasing polarity of solvent and by increasing nucleophilicity of the phosphine while increasing size of the group R led to a greater proportion of (88).

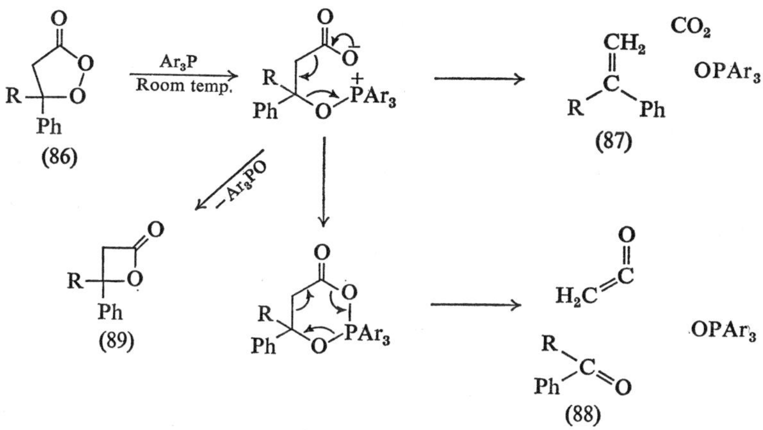

The conversion of cyclic carbonates into olefins on treatment with phosphines [83] has been applied [84] to the generation of benzyne from (90). The low yields of products are probably due to the reactions of benzyne and of benzyne-traps with phosphines.

Diphenyl and dicyclohexylphosphine with hexafluoroacetone gave 1 : 1-adducts [85] from which the phosphinate esters (91) were obtained on

[81] J. Carles and S. Fliszar, *Canad. J. Chem.*, 1969, **47**, 1113.
[82] W. Adam, R. J. Ramirez, and S.-C. Tsai, *J. Amer. Chem. Soc.*, 1969, **91**, 1254.
[83] P. T. Keough and M. Grayson, *J. Org. Chem.*, 1962, **27**, 1817.
[84] C. E. Griffin and D. C. Wysocki, *J. Org. Chem.*, 1969, **34**, 751.
[85] R. F. Stockel, *Chem. Comm.*, 1968, 1594.

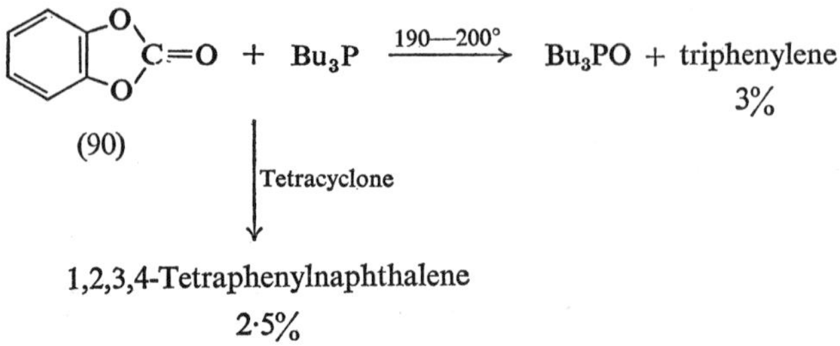

aerial oxidation showing that the initial attack of the phosphines was on oxygen. Proton transfer in the intermediate (92) successfully competed with the addition to a second molecule of ketone observed with tertiary tervalent phosphorus compounds.[86]

$$R_2PH + (CF_3)_2CO \longrightarrow (CF_3)_2\bar{C}\cdot O\cdot \overset{+}{P}HR_2 \quad (92)$$

$$(CF_3)_2CH\cdot O\cdot P(:O)R_2 \;(91) \xleftarrow{O_2} (CF_3)_2CH\cdot O\cdot PR_2$$

Cyclic, benzylic, aralkyl, and dialkyl but not diaryl disulphides were smoothly desulphurised to sulphides using tris(diethylamino)phosphine.[87] Aralkyl disulphides gave only the unsymmetrical sulphides but *p*-bromobenzyl benzyl disulphide led to a mixture of the three possible sulphides. The alicyclic thiolsulphonates (93; $n = 1, 2$) were smoothly desulphurised by the same phosphine to give the cyclic sulphinate esters (94).[88] The six-membered thiolsulphonate also gave a small amount of the sulphone (95).

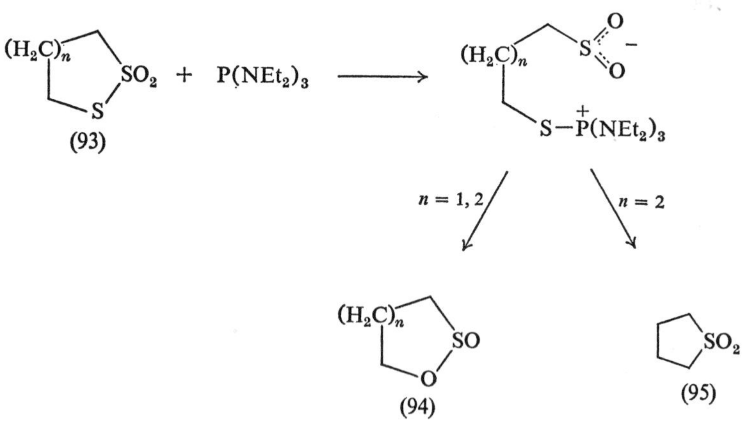

[86] F. Ramirez, S. B. Bhatia, and C. P. Smith, *Tetrahedron*, 1967, **23**, 2067.
[87] D. N. Harpp, J. G. Gleason, and J. P. Snyder, *J. Amer. Chem. Soc.*, 1968, **90**, 4181.
[88] D. N. Harpp and J. G. Gleason, *Tetrahedron Letters*, 1969, 1447.

D. Miscellaneous.—Karl Fischer reagent has been used to titrate tervalent phosphorus compounds.[89] It reacts quantitatively with substituted phosphines, metal complexes, and tertiary phosphonium salts, both of which presumably dissociate in solution, and with alkoxy-, aryloxy-, and chlorophosphines in the presence of an excess of pyridine. P^V compounds are unaffected.

The rates of dedeuteriation at the *o*-, *m*-, and *p*-positions of deuteriated tertiary phenylphosphines and their oxides, as well as the dedeuteriation of trideuteriomethyl and (trideuteriomethylphenyl)phosphines, have been studied both in liquid ammonia–potassium amide at 25° and in t-butanol–dimethylformamide catalysed by potassium t-butoxide at 120°.[90] The results indicate that the electron acceptor effect of tervalent phosphorus is similar to that of quinquevalent phosphorus in agreement with previous work on the effects of tervalent phosphorus attached to a benzene ring on substituents in that ring. The results were interpreted in terms of p_π–d_π conjugation with negligible p–π conjugation.

Part II: Phosphonium Salts

1 Preparation

The quaternisation of optically active tertiary phosphines *via* the complex salt method leads to complete retention of configuration at the phosphorus despite the high temperature (200°) involved.[1]

The addition of chlorophosphines to dienes to give five-membered phosphonium salts is a concerted disrotatory process ($\pi_s{}^4 + \omega_s{}^2$).[2] Thus *trans,trans*-hexa-2,4-diene with chlorodimethylphosphine gave the *cis*-phosphonium salt (1) and with dichloromethylphosphine gave after hydrolysis a mixture of the two *cis*-geometrical isomers (2) and (3).

Sodium diphenylphosphide and the trichloride (4) gave the interesting phosphetanium salt (5).[3] This synthesis conforms to the general principles laid down for the formation of azetidines,[4] the large group present on the β-carbon forcing the intermediate (6) into a conformation favouring cyclisation rather than polymerisation.

The synthesis of phosphonium salts from alcohols and triphenylphosphonium salts has been extended to include the use of ethers,[5] esters,[6] and the complexes obtained on reduction of aldehydes, ketones, acids, or esters with dialkylaluminium hydrides.[7] Examples are largely in the carotenoid

89 B. Hayton and B. C. Smith, *J. Inorg. Nuclear Chem.*, 1969, **31**, 1369.
90 E. A. Yakovleva, E. N. Tsvetkov, D. I. Lobanov, A. I. Shatenshtein, and M. I. Kabachnik, *Tetrahedron*, 1969, **25**, 1165.
1 L. Horner, R. Luckenbach, and W.-D. Balzer, *Tetrahedron Letters*, 1968, 3157.
2 A. Bond, M. Green, and S. C. Pearson, *J. Chem. Soc. (B)*, 1968, 929.
3 D. Berglund and D. W. Meek, *J. Amer. Chem. Soc.*, 1968, **90**, 518.
4 V. R. Gaertner, *Tetrahedron Letters*, 1966, 4691; B. J. Gaj and D. R. Moore, *ibid.*, 1967, 2155.
5 U.S.P. 3,347,932.
6 B.P. 1,102,064.
7 Neth. Appl. 6,606,916.

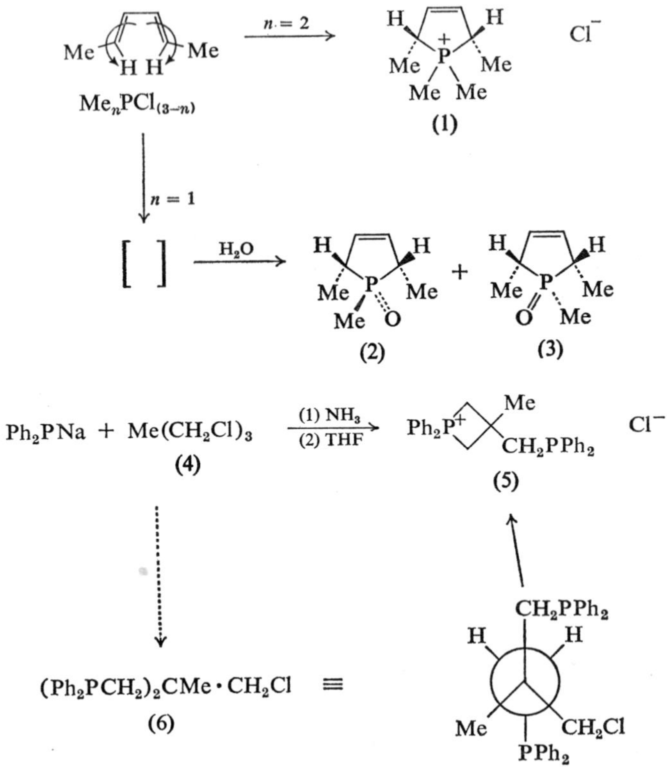

field or with allylic alcohols. Thus retinyl methyl ether with triphenylphosphine in methanolic hydrogen chloride gave the salt (7). Electron acceptors, *e.g.* BF_3, may also be used instead of a proton source.[8] Presumably all of these reactions involve formation of a carbonium ion which is captured by the phosphine.

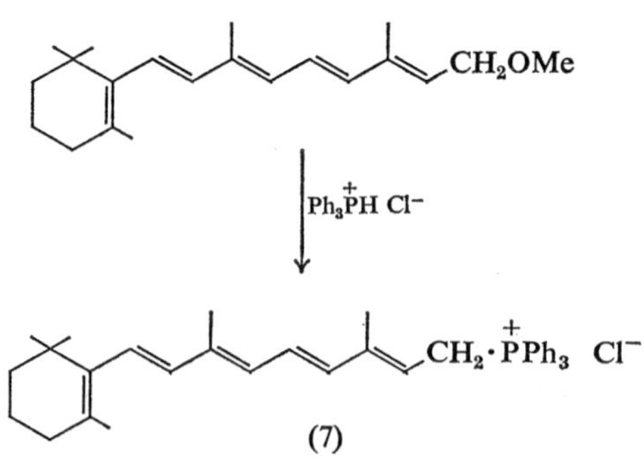

[8] Neth. Appl. 6,606,914.

High yields of tetramethylphosphonium salts have been obtained[9] from the action of white phosphorus on methyl chloride or bromide at 250°. Red phosphorus and 1,4-di-iodobutane at 200° in the presence of iodine gave the spiro-phosphonium salt (8) which on steam distillation gave the corresponding iodide.[10]

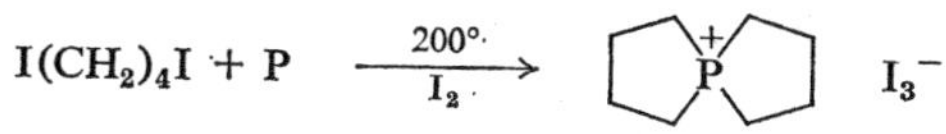

(8) 40·5%

2,6-Diphenylpyrylium perchlorate with triphenylphosphine in refluxing nitromethane (1 min) gave an almost quantitative yield of the salt (9).[11]

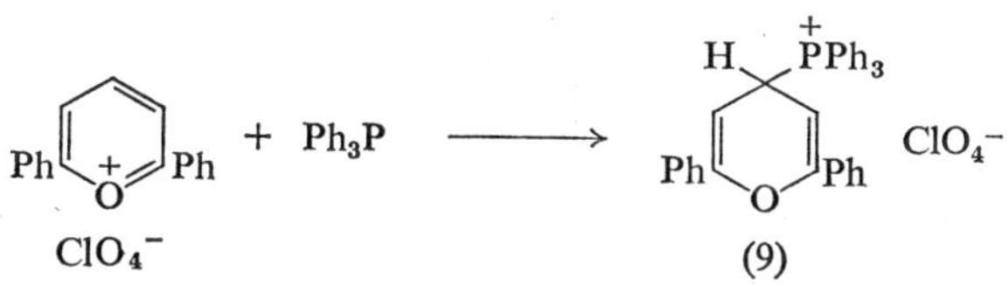

(9)

Quaternisation of tris(dimethylamino)phosphine with trifluoroiodomethane and perfluoroisopropyl iodide led to tetra(dimethylamino)phosphonium iodide and the aminophosphines (10).[12] A similar reaction had been previously observed using trimethylphosphine.[13]

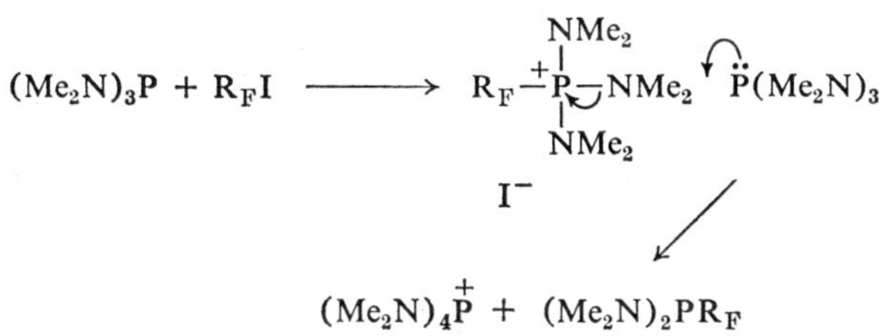

(10)

Quaternisation of a chloromethylated cross-linked polystyrene with (+)-methylphenylpropylphosphine gave an optically active anion exchange resin.[14a] However mandelic, atrolactic, and racemic tartaric acids could not be resolved on it.

Cyclopropenylphosphonium salts have been prepared from cyclopropenyl perchlorates and triphenylphosphine.[14b]

[9] L. Maier, *Helv. Chim. Acta*, 1968, **49**, 2458.
[10] N. Ya. Derkach and A. V. Kirsanov, *Zhur. obshchei Khim.*, 1968, **38**, 331.
[11] S. V. Krivun, *Doklady Akad. Nauk S.S.S.R.*, 1968, **182**, 347.
[12] H. G. Ang, G. Manoussakis, and Y. O. El-Nigumi, *J. Inorg. Nuclear Chem.*, 1968, **30**, 1715.
[13] R. N. Haszeldine and B. O. West, *J. Chem. Soc.*, 1956, 3631.
[14] (a) L. Horner and W.-D. Balzer, *Makromol. Chem.*, 1968, **115**, 245. (b) D. T. Longone and E. S. Alexander, *Tetrahedron Letters*, 1968, 5815.

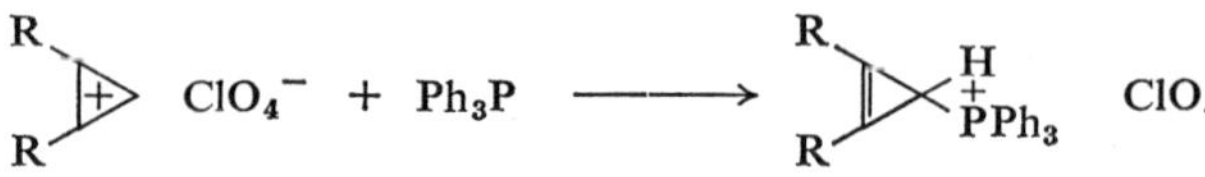

2 Reactions

A. Alkaline Hydrolysis.—A number of rearrangements are now known which conform to the general pattern of (11) in which a group occupying an apical position in an intermediate trigonal bipyramid migrates to an α-carbon bearing a substituent which can accommodate a negative charge.

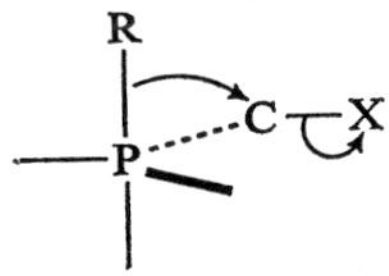

(11)

The alkaline hydrolysis of halogenomethylphosphonium salts is typical; the preference of four- and five-membered rings for an apical-equatorial position here leads to ring expansion. Thus the phosphetanium salt (12) gave the five-membered oxide (13),[15] while the five-membered phosphonium salts (14) gave the oxides (15).[16]

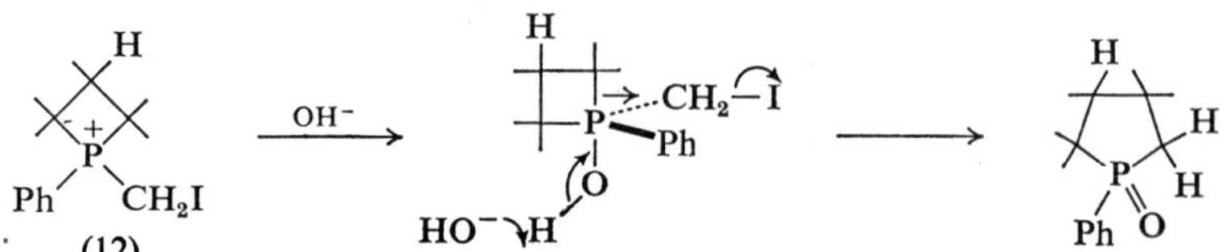

(12) (13)

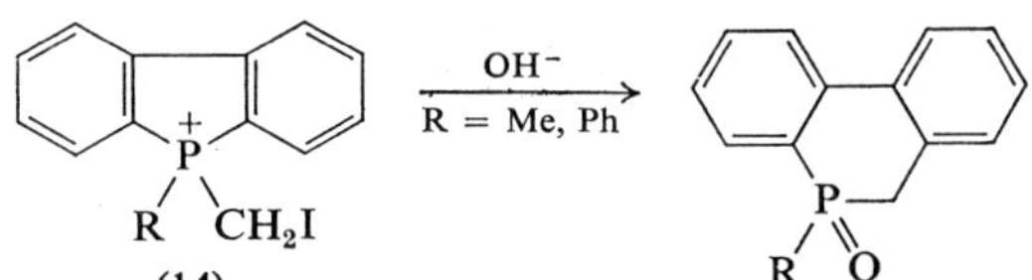

(14) (15)

A similar rearrangement was involved in the alkaline hydrolysis of styryltriphenylphosphonium bromide,[17, 18] and this salt is probably the intermediate in several previously described reactions which also lead to the oxide (16), among them the reactions of triphenylphosphine with styrene oxide[19] and with phenylacetylene in the presence of water,[20] the

[15] S. E. Fishwick, J. Flint, W. Hawes, and S. Trippett, *Chem. Comm.*, 1967, 1113.
[16] D. W. Allen and I. T. Millar, *J. Chem. Soc.* (*C*), 1969, 252.
[17] J. J. Brophy, K. L. Freeman, and M. J. Gallagher, *J. Chem. Soc.* (*C*), 1968, 2761.
[18] E. M. Richards and J. C. Tebby, *Chem. Comm.*, 1969, 494.
[19] S. Trippett and B. J. Walker, *J. Chem. Soc.* (*C*), 1966, 887.
[20] D. W. Allen and J. C. Tebby, *Tetrahedron*, 1967, **23**, 2795.

hydrolysis of the adduct (17) from methylenetriphenylphosphorane and the nitrone (18),[21] and the reaction of methyltriphenylphosphonium salts with benzaldehyde in ethanolic sodium ethoxide.[19]

Buta-1,3-dienyltriphenylphosphonium bromide on hydrolysis in the presence of a large excess of alkali gave the oxides (19) and (20) involving

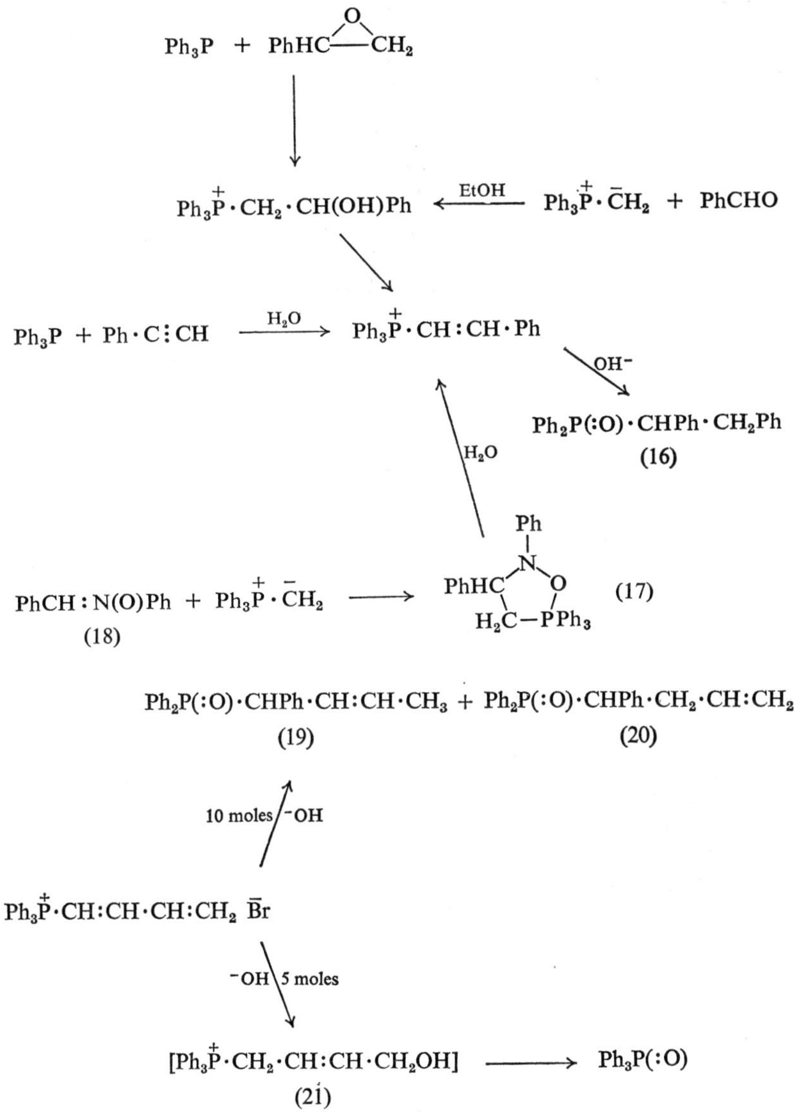

[21] R. Huisgen and J. W. Ulff, *Chem. Ber.*, 1969, **102**, 746.

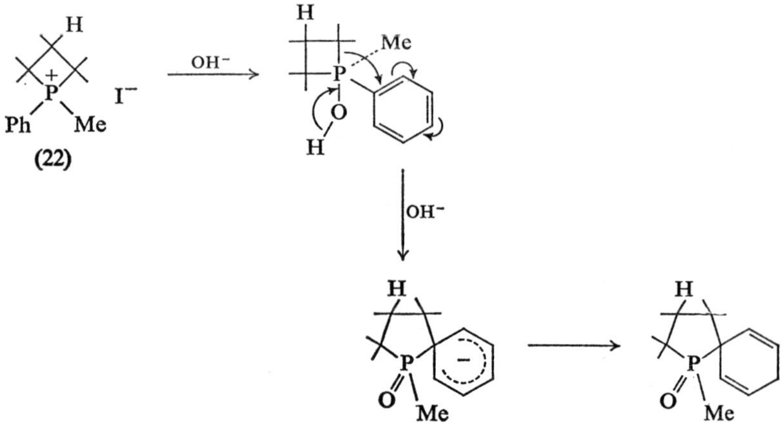

(22) (23) (24)

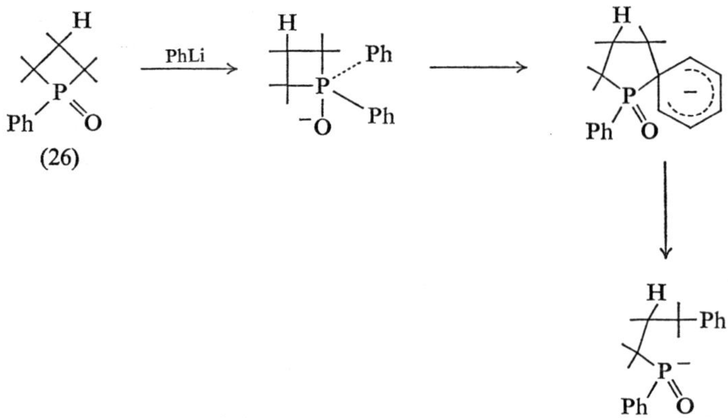

(26)

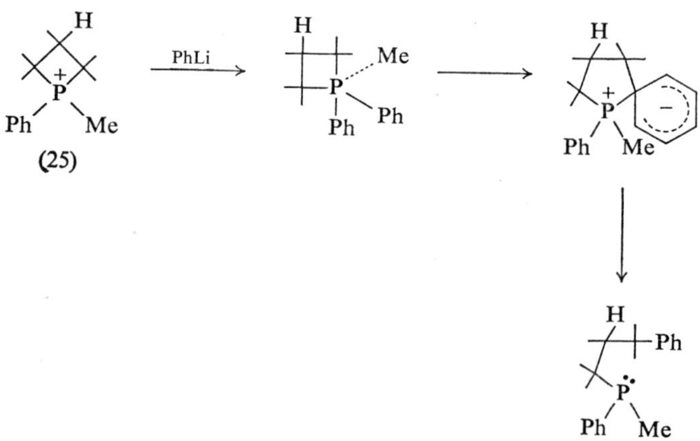

(25)

migration of phenyl from phosphorus,[17] but with less alkali triphenylphosphine oxide was the major product perhaps formed *via* hydration of the salt to give (21).

Also conforming to the general pattern of (11) was the alkaline hydrolysis of the 2,2,3,4,4-phosphetanium salt (22), the intermediate cyclohexadienyl anion (23) being rapidly protonated to give the non-conjugated isomer (24).[22] The structure (24) agrees with the results of experiments involving deuterium labelling in the phenyl ring.[23] Intermediates analogous to (23) are probably involved in the action of phenyl-lithium on phosphetanium salts, *e.g.* (25), and phosphetan oxides, *e.g.* (26).[24] Here, however, the cyclohexadienyl anions cannot be protonated and instead the five-membered rings open. The former reaction is reminiscent of the action of vinyl-lithium on tetraphenylphosphonium bromide to give styrene and triphenylphosphine[25] and this may be formulated in a similar way.

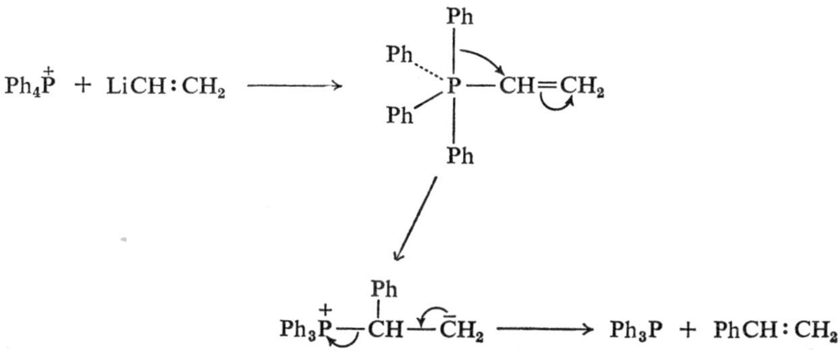

In contrast to the hydrolysis of (22), that of the 2,2,3-trimethylphosphetanium salt (27) gave the acyclic oxide (28) presumably due to the greater stability of the departing anion.[26]

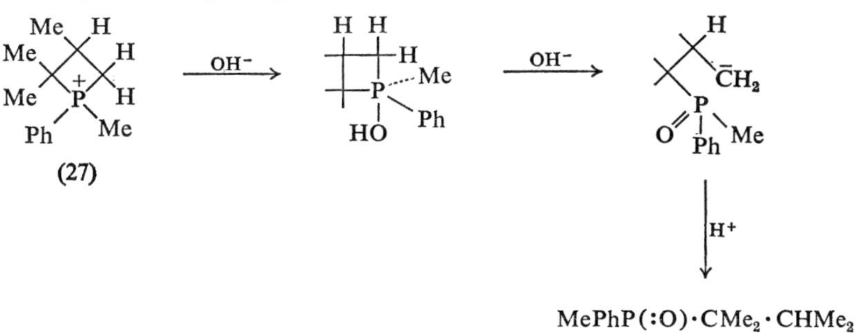

(27)

(28)

[22] S. E. Fishwick, J. Flint, W. Hawes, and S. Trippett, *Chem. Comm.*, 1967, 1113.
[23] S. E. Cremer, *Chem. Comm.*, 1968, 1132.
[24] S. E. Cremer and R. J. Chorvat, *Tetrahedron Letters*, 1968, 413.
[25] D. Seyferth, J. S. Fogel, and J. K. Heeren, *J. Amer. Chem. Soc.*, 1964, **86**, 307.
[26] S. E. Fishwick and J. Flint, *Chem. Comm.*, 1968, 182.

Alkaline hydrolysis of the benzylphosphetanium salt (29) gave the oxide (30) with predominant retention of configuration at phosphorus [27] and both isomers of the benzylphospholanium salt (31) also hydrolysed with retention,[28] a further consequence of the preference of four- and five-membered rings for the apical-equatorial position. The interesting point here is whether the benzyl anions left from equatorial positions of the first-formed trigonal bipyramids, *e.g.* (32), or whether pseudorotation was necessary in order to allow the anions to depart from apical positions. The latter seems unlikely as pseudorotation would move the only electronegative group from an apical to an equatorial position.

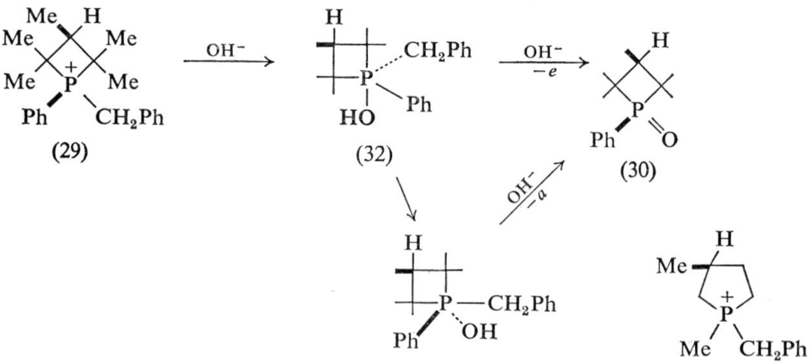

(29) (32) (30) (31)

Di-t-butyl phosphonium salts are extremely resistant to alkaline hydrolysis. Thus while the salt (33; R^1 = Ph, R^2 = Bu^t) was hydrolysed *ca.* 50 times more slowly than the salt (33; R^1 = Ph, R^2 = Me) to give the expected products, toluene and t-butyldiphenylphosphine oxide, 21% of the di-t-butylphosphonium salt (33; R^1 = R^2 = Bu^t) was unchanged after 11 days in 90% ethanolic N-NaOH at 100° and the major product was benzyl-t-butylphenylphosphine, the product of a normal Hofmann

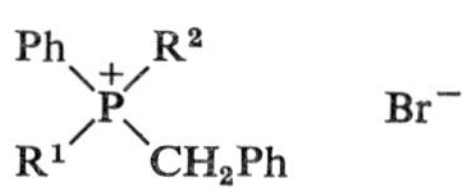

(33)

elimination.[29] To explain the different effects of one and two t-butyl groups it was proposed that severe hindrance to attack of a nucleophile is produced by a t-butyl group which is to occupy an equatorial position in the intermediate trigonal bipyramid. In agreement with the hypothesis alkaline hydrolysis of (−)-benzyl-t-butylmethylphenylphosphonium iodide

[27] W. Hawes and S. Trippett, *Chem. Comm.*, 1968, 295.
[28] K. L. Marsi, *Chem. Comm.*, 1968, 846.
[29] N. J. De'ath and S. Trippett, *Chem. Comm.*, 1969, 172.

was accompanied by predominant retention of configuration at the phosphorus.

A further example has appeared of the alkaline hydrolysis of 1,2-bisphosphonioethylenes (34) with elimination of acetylene. The salt (35) gave the phosphine (36), isolated as the dioxide.[30]

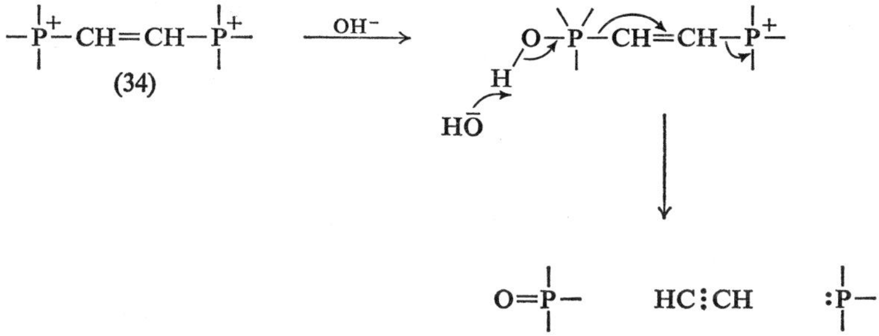

(34)

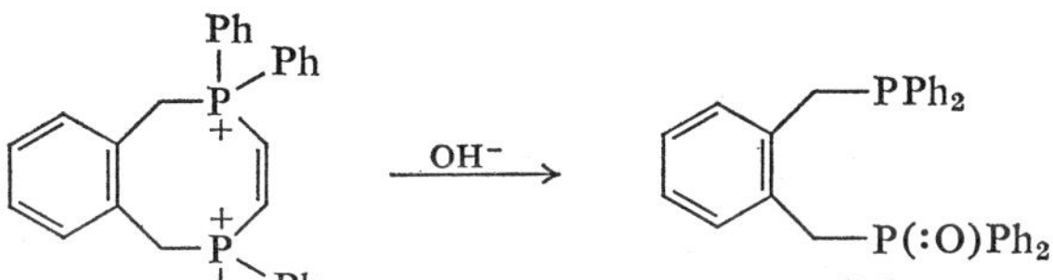

(36)

(35)

1-Phenylvinyl[31] and cymantrenyl (cyclopentadienylmanganesetricarbonyl)[32] proved to be better leaving groups than benzyl in the hydrolysis of benzylphosphonium salts. The hydrolyses of benzylcymantrenyldiphenylphosphonium bromide[32] and the cyclic salts (37) and (38)[33] were all third order, *i.e.* rate ∝ $[P^+][\bar{O}H]^2$

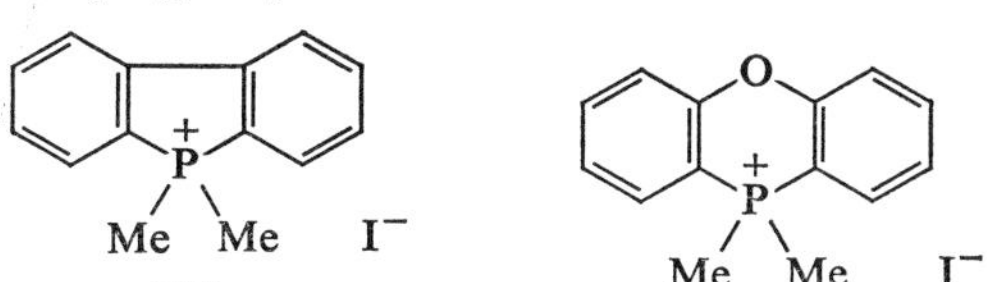

(37) (38)

B. Additions to Vinylphosphonium Salts.—The susceptibility of vinylphosphonium salts to Michael addition[34] has been exploited in a number

[30] A. M. Aguiar and M. G. Raghanan Nair, *J. Org. Chem.*, 1968, **33**, 579.
[31] M. P. Savage and S. Trippett, *J. Chem. Soc.* (*C*), 1968, 591.
[32] G. J. Reilly and W. E. McEwen, *Tetrahedron Letters*, 1968, 1231.
[33] D. W. Allen and I. T. Millar, *J. Chem. Soc.* (*B*), 1969, 263.
[34] P. T. Keough and M. Grayson, *J. Org. Chem.*, 1964, **29**, 631.

of ways. Those which involved trapping the ylide (39) in Wittig olefin syntheses are described in Chapter 8.

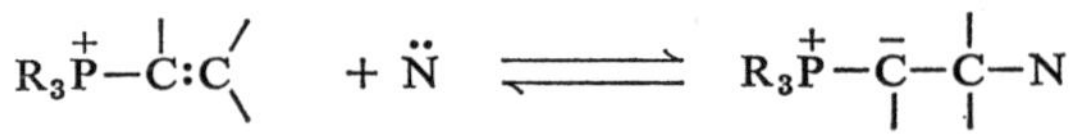

(39)

The β-acylvinylphosphonium salts (40) with aqueous sodium azide gave the 4-acyl-1,2,3-triazoles (41) in good yields as shown.[35]

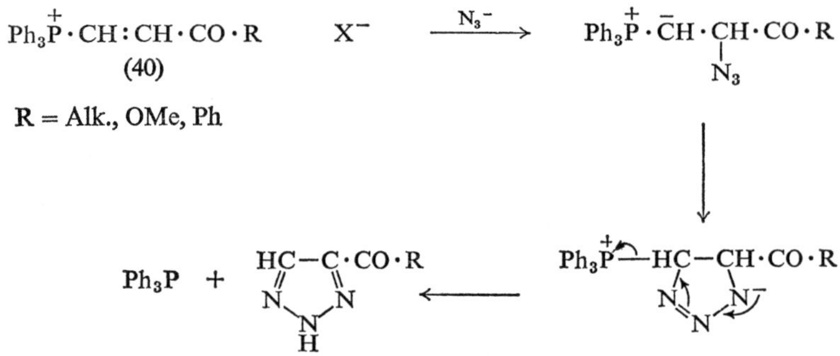

(40)

R = Alk., OMe, Ph

(41)

30—95%

Cyclisation of the initially formed vinylphosphonium salt occurred[36] when diphenylvinylphosphine was treated with (42). The structure of the product (43) was shown by hydrolysis. Similar cyclisations accompanied the quaternisations of the acetylenic phosphines (44) with α-bromoketones.[37]

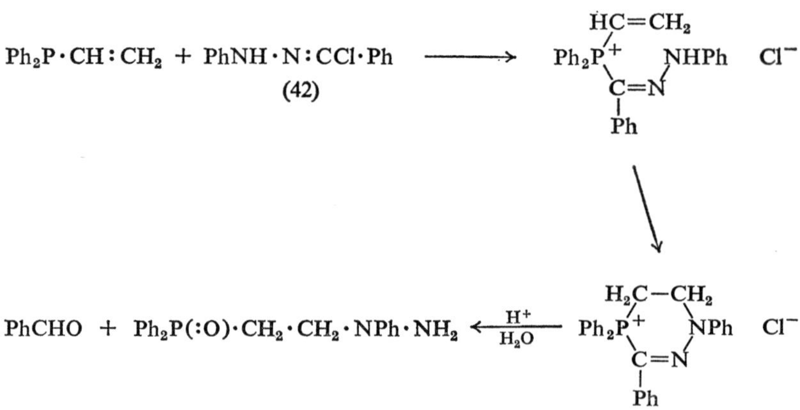

(42)

(43)

[35] M. Rasberger and E. Zbiral, *Monatsh.*, 1969, **100**, 64.
[36] I. G. Kolokol'tseva, V. N. Chistokletov, and A. A. Petrov, *Zhur. obshchei. Khim.*, 1968, **38**, 2819.
[37] M. Simalty and H. Chahine, *Compt. rend.*, 1968, **266**, *C*, 1098.

$Ph_2P\cdot C{\vdots}C\cdot R + R^1\cdot CO\cdot CH_2Br \longrightarrow$ 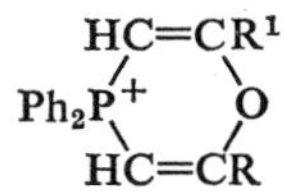 Br^-

(44)

R = Me, Ph R^1 = Me, Ph

When a dilute solution of dimethyl acetylenedicarboxylate was added to diphenylvinylphosphine in wet ether [38] a low yield of the 1 : 1 : 1-adduct (45) was obtained *via* cyclisation of the initial betaine (46). Reverse addition led to the 1 : 2 : 1-adduct (47), addition of the second molecule of acetylene to (48) being as shown in (49).

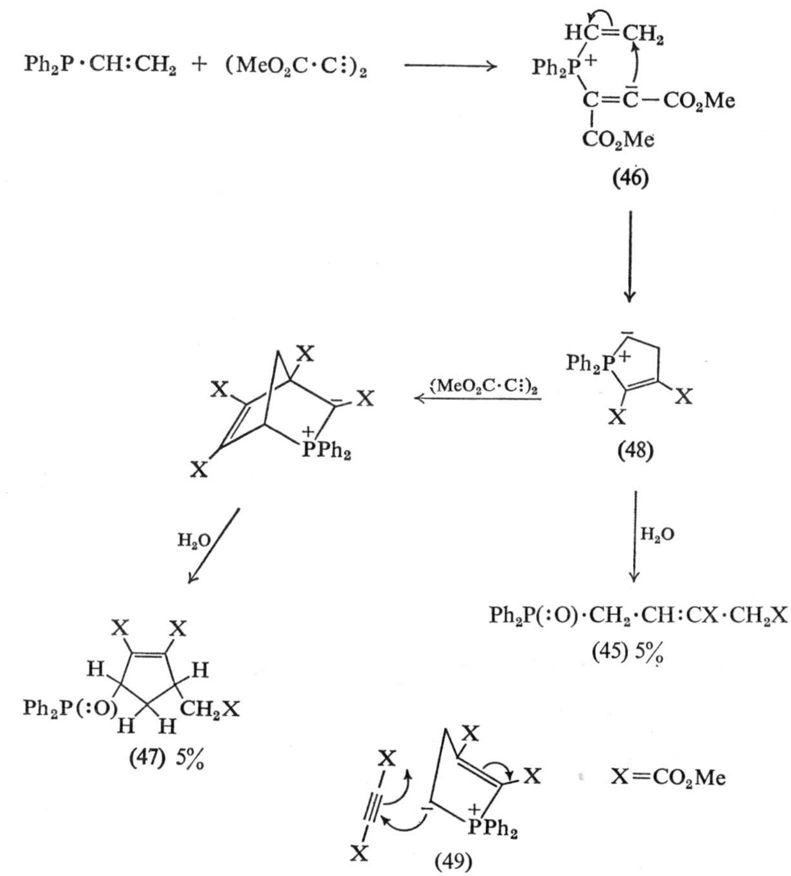

C. Miscellaneous Reactions.—The β-ketoalkylphosphonium isothiocyanates (50) with lead tetra-acetate [39] gave the allenes (51) and/or the acetylenes (52) the ratio depending on the nature of the groups R^1 and R^2.

[38] A. N. Hughes and M. Davis, *Chem. and Ind.*, 1969, 138.
[39] E. Zbiral and H. Hengstberger, *Monatsh.*, 1968, **99**, 412.

$$\mathrm{Ph_3\overset{+}{P}\cdot CHR\cdot CO\cdot CHR^1R^2\ \ S\overset{-}{C}N \xrightarrow{Pb(OAc)_4} R^1R^2C{:}C{:}CR\cdot SCN +}$$

(50) (51)

$$\mathrm{R^1R^2C(NCS)\cdot C{:}C\cdot R}$$

(52)

Thus when $R^1,R^2 = H$ the allene was formed but with $R^1,R^2 = Me$ the acetylene was isolated. The mechanism envisages the acetylene being produced by rearrangement of the allene. With the salts (50; R = OMe) the ketones (53) were produced by hydrolysis of the intermediates (54) followed by rearrangement.

$\mathrm{Ph_3\overset{+}{P}\cdot CHR\cdot CO\cdot CHR^1R^2}$ $\mathrm{S\bar{C}N}$ (50)

$\xrightarrow{Pb(OAc)_4}$ $\mathrm{Ph_3\overset{+}{P}\cdot CR(SCN)\cdot CO\cdot CHR^1R^2 + Pb(OAc)_2 + AcOH}$ $\mathrm{\bar{O}Ac}$ (54)

(54) $\longrightarrow$ $\mathrm{Ph_3\overset{+}{P}\cdot CR(SCN)\cdot C(\bar{O}){:}CR^1R^2}$ $\longrightarrow$ (51) + $\mathrm{Ph_3PO}$

(54) $\xrightarrow{R\,=\,OMe}$ $\mathrm{RHC(SCN)\cdot CO\cdot CHR^1R^2}$ $\longrightarrow$ $\mathrm{RHC(NCS)\cdot CO\cdot CHR^1R^2}$ (53)

Pyrolysis of the betaine (55) gave the phosphine (56) with migration of phenyl from phosphorus to oxygen.[40]

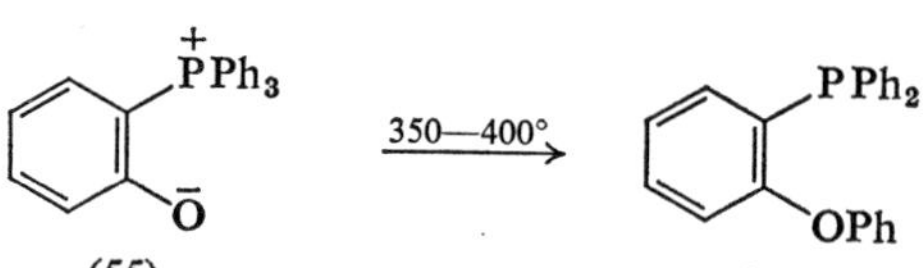

(55) (56)

Investigation of the electrolytic reduction of phosphonium salts[41] showed that with groups which do not interact with the phosphonium centre, *e.g.* Me, Ph, the percentage splitting off of a group depends only on the number and aptitude of that group, so that from data for one of the series $R^1{}_nR^2{}_{(4-n)}\overset{+}{P}$ the results for the other members of the series can be

[40] H. J. Bestmann and G. Hofmann, *Annalen*, 1968, **716**, 98.
[41] L. Horner and J. Haufe, *Chem. Ber.*, 1968, **101**, 2903.

predicted. This is not the case with groups such as *p*-anisyl; more anisole is produced when several such groups are attached to phosphorus than would be expected on the basis of the mono-*p*-anisylphosphonium salt.

Treatment of tetraphenylphosphonium bromide with potassium diarylphosphides in refluxing tetrahydrofuran gave triphenylphosphine and the diarylphenylphosphines (57), with transfer of a phenyl group.[42] This establishes part of the scheme originally proposed [43] to account for the products formed in the photolysis of triphenylphosphine.

$$\mathrm{Ph_4\overset{+}{P} + Ar_2\overset{-}{P} \longrightarrow Ph_3P + \underset{(57)}{Ar_2PhP}}$$

Ketones and triphenylphosphine were obtained from the action of Grignard reagents on α,α-disubstituted *β*-ketoalkyltriphenylphosphonium salts.[43b]

$$\mathrm{Ph_3\overset{+}{P}\cdot CR^1R^2\cdot CO\cdot R^3\ \overset{-}{X} + 2R^4MgBr \longrightarrow R^4{}_2 + Ph_3P + R^1R^2CH\cdot CO\cdot R^3}$$

3 Miscellaneous

The structures of the phosphonium salts (22) [44] and (58) [45] have been determined by *X*-ray analysis. The ring in (58) is planar and there is no evidence of delocalisation of charge.

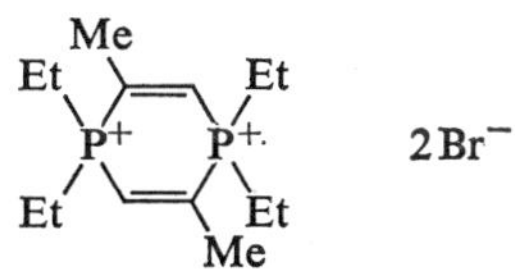

(58)

For a discussion of the magnetic non-equivalence of the methylene hydrogens in the ^{1}H n.m.r. spectra of benzylphosphonium salts see Chapter 11. Triarylmethylphosphonium salts have been shown [46] by n.m.r. measurements in chloroform to associate to contact ion pairs in which the anionic charge is localised on the phosphorus. Hammett constants determined in the same medium and by the same method have therefore been used to provide information on the non-inductive portion of the electronic effect of 'onium phosphorus. This indicates that R_3P^+ is a strong $-$M substituent. The reactivity of the salts $\mathrm{Ph\overset{+}{X}Me_3}$ (X = N, P, As, Sb) towards nitration in sulphuric acid has also been discussed in these terms.[47]

[42] L. Horner, P. Beck, and R. Luckenbach, *Chem. Ber.*, 1968, **101**, 2899.
[43] [(a)] L. Horner and J. Dörges, *Tetrahedron Letters*, 1965, 763. [(b)] T. Mukaiyama, R. Yoda, and I. Kuwajima, *Tetrahedron Letters*, 1969, 23.
[44] C. Moret and L. M. Trefonas, *J. Amer. Chem. Soc.*, 1969, **91**, 2255.
[45] R. Majeste and L. M. Trefonas, *J. Heterocyclic Chem.*, 1969, **6**, 269.
[46] G. P. Schiemenz, *Angew. Chem. Internat. Edn.*, 1968, **7**, 544.
[47] A. Gastaminza, T. O. Modro, J. H. Ridd, and J. H. P. Utley, *J. Chem. Soc.* (*B*), 1968, 534.

PART III: Phosphorins and Phospholes

1 Phosphorins

A. Preparation.—The preparation of phosphorins from pyrylium salts and tris(hydroxymethyl)phosphine [1] has been extended [2] to the synthesis of 2,4,6-tri-t-butylphosphorin (1). Oxidation of this with ethanolic hydrogen peroxide gave the furan (2).

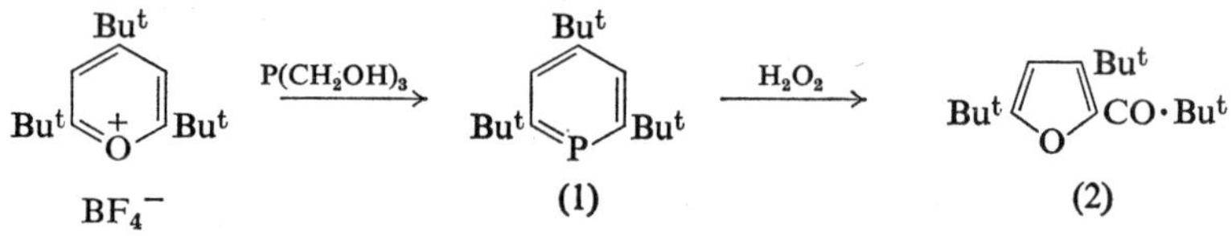

(1) (2)

9-Phospha-anthracene (3) [3] and 9-phosphaphenanthrene (4) [4] have been prepared in solution by dehydrochlorination of the relevant chlorophosphines with 1,5-diazabicyclo[5,4,0]undec-5-ene, and their u.v. spectra recorded. 10-Phenyl-9-phospha-anthracene [5] was much more stable than the parent (3) and was isolated as a crystalline solid which, however, reacted with oxygen much more rapidly than did monocyclic phosphorins.

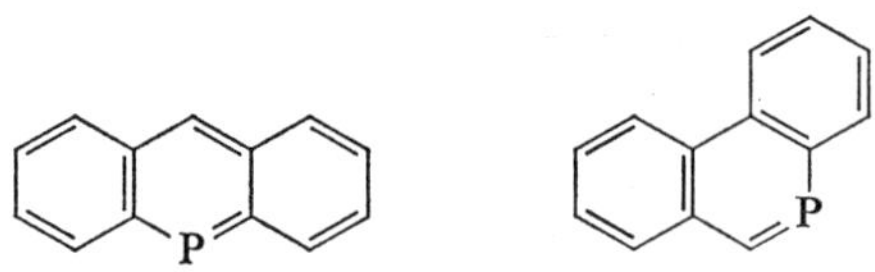

(3) (4)

A comparison [6] of the spectra of the phosphacyclohexadienone (5) with those of its oxide and methiodide did not entirely rule out a contribution from the 'phosphapyrylium' form (6). However, the alcohol (7), obtained from (5) and phenyl-lithium, protonated on phosphorus to give the salt (8) rather than form a phosphapyrylium salt.

The u.v. spectrum of the salt (9) did not support any contribution from the aromatic form (10).[7]

B. Structure.—*X*-Ray analysis [8] of 2,6-dimethyl-4-phenylphosphorin has supported the concept of a delocalised aromatic nucleus. The ring is planar and symmetrical with P—C bonds of 1·74 Å intermediate between the single and double bond values. Similar results have been obtained [9] for the phosphorins (11).

[1] G. Märkl, *Angew. Chem.*, 1966, **78**, 907.
[2] K. Dimroth and W. Mach, *Angew. Chem. Inernat. Edn.*, 1968, **7**, 460.
[3] P. de Koe and F. Bickelhaupt, *Angew. Chem. Internat. Edn.*, 1967, **6**, 567.
[4] P. de Koe, R. van Veen, and F. Bickelhaupt, *Angew. Chem. Internat. Edn.*, 1968, **7**, 465.
[5] P. de Koe and F. Bickelhaupt, *Angew. Chem. Internat. Edn.*, 1968, **7**, 889.
[6] G. Märkl and H. Olbrich, *Tetrahedron Letters*, 1968, 3813.
[7] M. Simalty and H. Chahine, *Bull. Soc. chim. France*, 1968, 4938.
[8] J. C. J. Bart and J. J. Daly, *Angew. Chem. Internat. Edn.*, 1968, **7**, 811.
[9] W. Fischer and E. Hellner, *Tetrahedron Letters*, 1968, 6227.

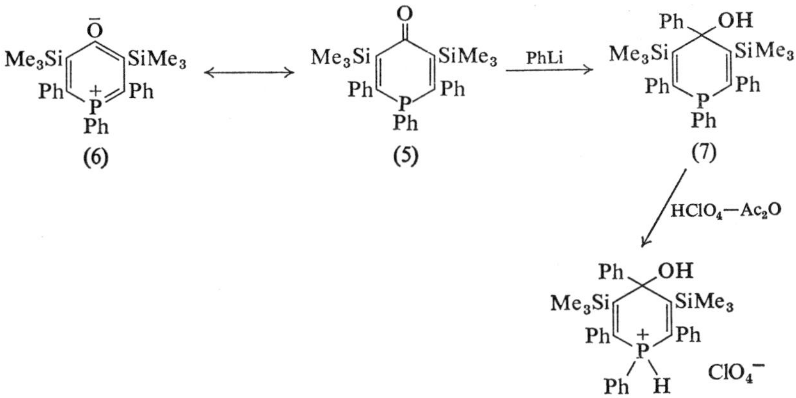

(8)

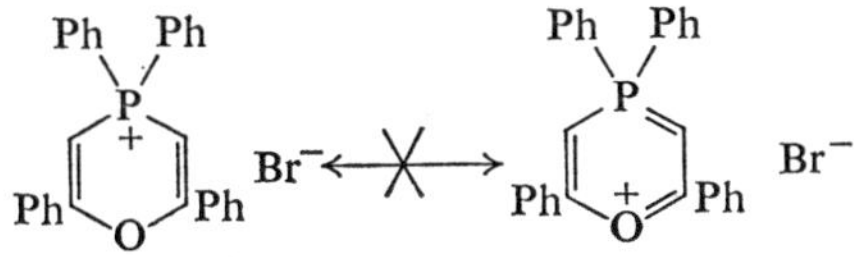

(9) (10)

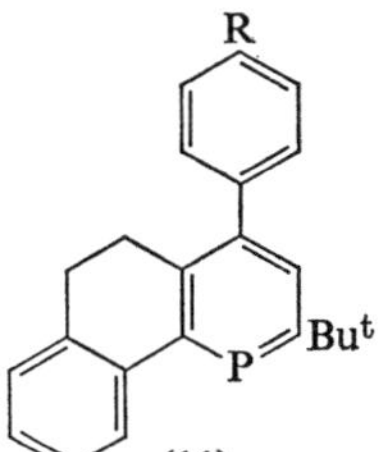

(11)

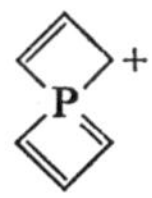

(12)

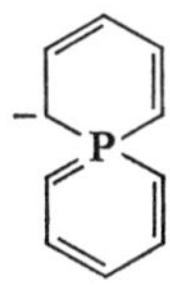

(13)

Molecular orbital calculations on phosphorins have been reported.[10] The models which agreed with the experimental data on electronic spectra and reactivity involved strong interaction between the π-orbital of the phosphorus and those of the neighbouring atoms. The stability and possible existence of $(n+1)$-phosphonia[n,n]spirarenes has been discussed.[11] These, *e.g.* (12) and (13), are aromatic systems, the two parts being at right angles but fully conjugated through $(d\text{–}p)\pi$ interaction.

C. Reactions.—Exposure to air of a solution of the phosphorin (14; R = Ph) in benzene gave [12] the phosphinic acid (15) and the anhydride (16). The phosphorin (14; R = Bu^t) under similar conditions formed the

[10] R. Vilceanu, A. Balint, and Z. Simon, *Rev. Roumaine Chim.*, 1968, **13**, 533.
[11] A. J. Ashe, *Tetrahedron Letters*, 1968, 359.
[12] K. Dimroth, K. Vogel, W. Mach, and U. Schoeler, *Angew. Chem. Internat. Edn.*, 1968, **7**, 371.

phosphonic acid (17) which was also obtained using other oxidising agents. Both (15) and (17) were hydrogen-bonded dimers.

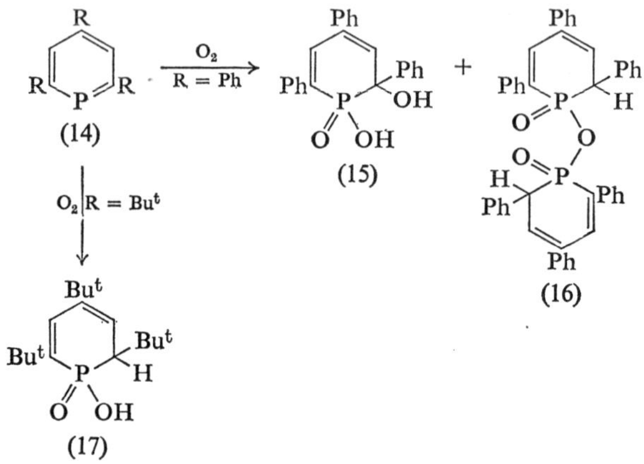

Oxidation of phosphorins with mercury(II) acetate in the presence of alcohols or phenols [13] proceeded *via* radical cations to give the 1,1-dialkoxy- or 1,1-diphenoxy-phosphorins (18). These were protonated by strong acids at the 2- or the 4-position. Two isomeric methyl phosphinates (19) and (20) were obtained on oxidation of the 1,1-dimethoxyphosphorin (21). Both isomers with trifluoracetic acid gave the deep blue cation (22). Similar cations may be responsible for the blue colorations obtained [12] when (15) and (16) were dissolved in trifluoracetic acid and when a solution of the phosphorin (14; R = Ph) in benzene was shaken with oxidising acids.

The ambident anion (23) formed from 2,4,6-triphenylphosphorin (24) and phenyl-lithium was alkylated on phosphorus with alkyl halides in tetrahydrofuran but at the 2-position in benzene.[14] The same anion (23) has also been obtained [15] by reduction of the pentaphenylphosphorin (25) with sodium–potassium. The phosphorin (24) did not react with maleic anhydride or with diethyl acetylenedicarboxylate.[16] However, with hexafluorobut-2-yne at 100° the 1-phosphabarralene (26) was formed.[16]

Arylation [15] of the phosphine (24) or of (27) with mercury diaryls at 240—270° gave the 1,1-diarylphosphorins (28) formed *via* the radicals (29). With the 2,4,6-triphenylphenoxy radical, (27) gave a stable yellow compound for which structure (30) was proposed.

The cation (31) was obtained from (27) and trityl perchlorate.[15] With phenyl-lithium it gave (25). The ambident nature of (31) was shown by its

[13] K. Dimroth and W. Städe, *Angew. Chem. Internat. Edn.*, 1968, **7**, 881.
[14] G. Märkl and A. Merz, *Tetrahedron Letters*, 1968, 3611.
[15] G. Märkl and A. Merz, *Tetrahedron Letters*, 1969, 1231.
[16] G. Märkl and F. Lieb, *Angew. Chem. Internat. Edn.*, 1968, **7**, 733.

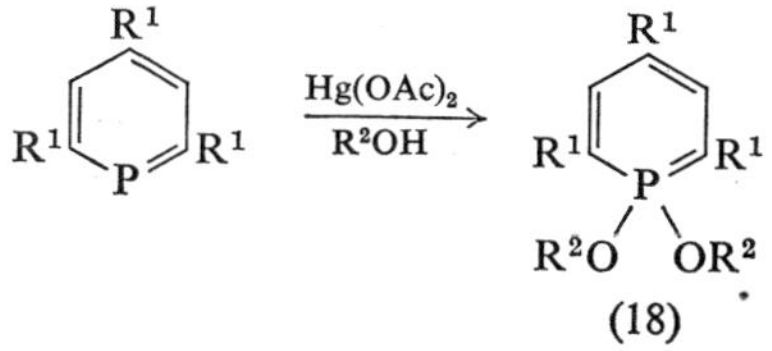

Hg(OAc)2
R2OH
(18)

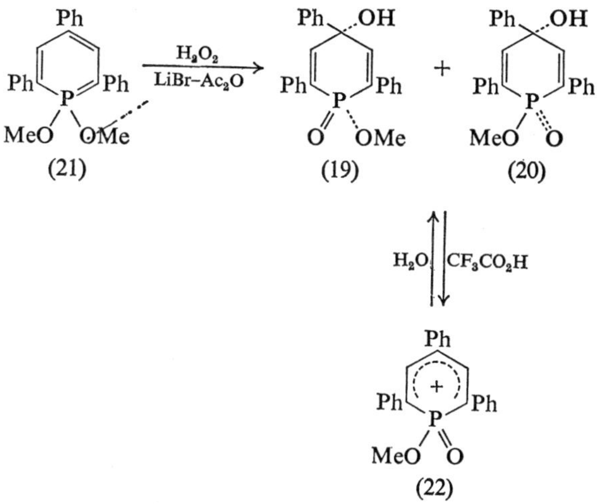

H2O2
LiBr–Ac2O
(21)
(19)
(20)
H2O
CF3CO2H
(22)

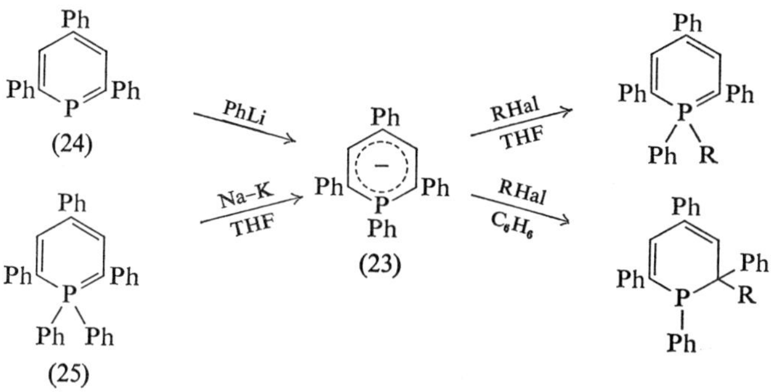

(24)
(25)
PhLi
Na–K
THF
(23)
RHal
THF
RHal
C6H6

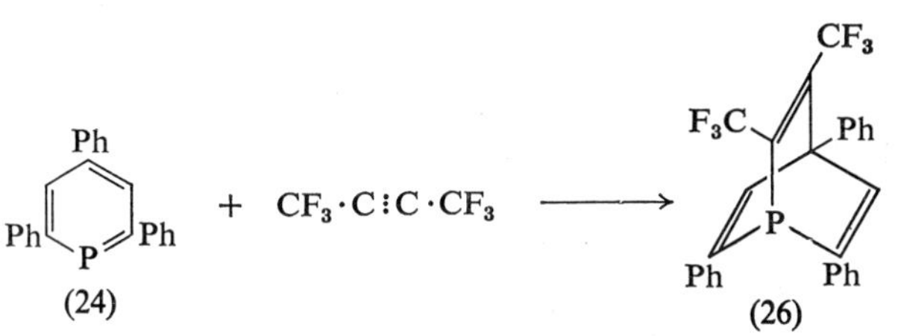

CF3·C⋮C·CF3
(24)
(26)

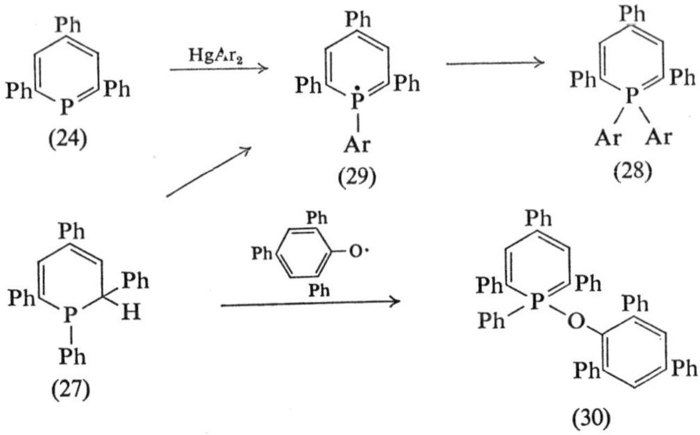

reaction with water from which the oxides (32) and (33) have been obtained.[15, 17]

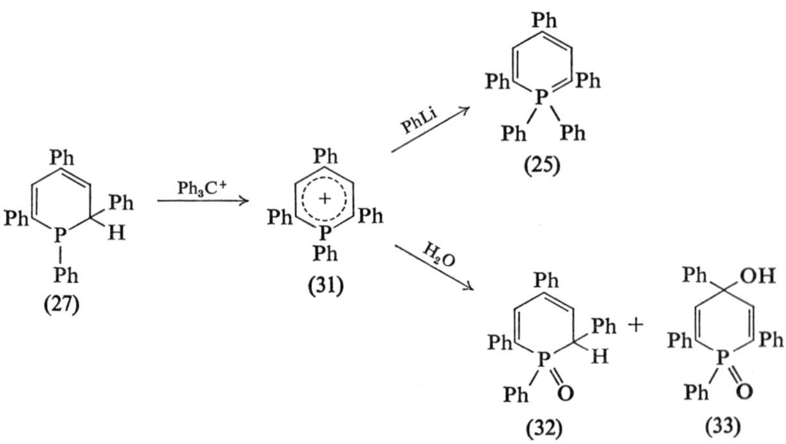

2 Phospholes

Similar syntheses of 1-methyl-[18] and 1-phenyl-phosphole[19] have been described. Their spectroscopic and chemical properties agreed with previous data for more highly substituted phospholes in indicating considerable aromatic character in the phosphole nucleus. 1-Phenylphosphole (34) was obtained as a distillable liquid which did not dimerise at room temperature in contrast to the oxide and sulphide (in which aromatic

[17] C. C. Price, T. Parasan, and T. V. Lakshminarayan, *J. Amer. Chem. Soc.*, 1966, **88,** 1034.
[18] L. D. Quin and J. G. Bryson, *J. Amer. Chem. Soc.*, 1967, **89**, 5948.
[19] G. Märkl and R. Potthast, *Tetrahedron Letters*, 1968, 1755.

stabilisation is not possible) which were isolated as the dimers (35). The sulphide of 2,5-dimethyl-1-phenylphosphole was monomeric possibly because of steric hindrance to dimerisation.

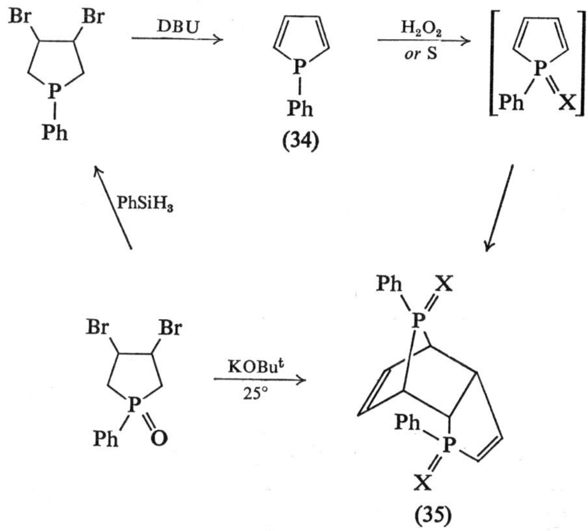

2
Quinquecovalent Phosphorus Compounds

BY S. TRIPPETT

The use of 3*d*-orbitals in bonding by phosphorus has been discussed[1, 2] and reviewed.[3]

1 Pseudorotation

Pseudorotation, that is the interconversion of trigonal bipyramids (1) and (3) by way of the square pyramid (2) which may be a transition state or an

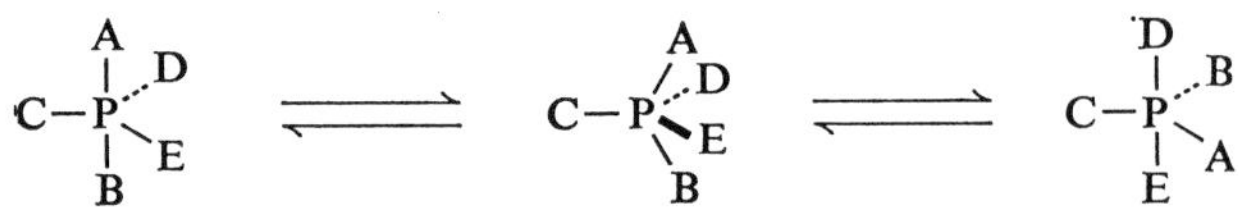

(1) (2) (3)

intermediate, has been reviewed by Westheimer[4] and the general question of ligand reorganisation in trigonal bipyramids[5–7] and in hexaco-ordinate structures[8] has been discussed. Peterson graphs are useful for following pseudorotations and for exploring the consequences of stereochemical restrictions which may be imposed on the process, for example by including in the molecule a small ring which will prefer to be apical-equatorial. One such graph[6] is given in (4); each vertex corresponds to one of the ten

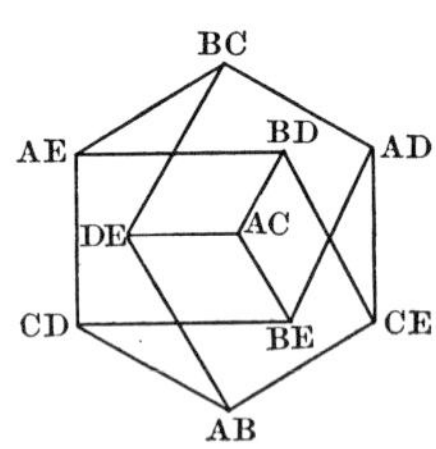

(4)

[1] K. A. R. Mitchell, *J. Chem. Soc.* (*A*), 1968, 2677.
[2] B. C. Webster, *J. Chem. Soc.* (*A*), 1968, 2909.
[3] K. A. R. Mitchell, *Chem. Rev.*, 1969, **69**, 157.
[4] F. H. Westheimer, *Accounts Chem. Res.*, 1968, **1**, 70.
[5] E. L. Muetterties, *Inorg. Chem.*, 1967, **6**, 635.
[6] J. D. Dunitz and V. Prelog, *Angew. Chem. Internat. Edn.*, 1968, **7**, 725.
[7] P. C. Lauterbur and F. Ramirez, *J. Amer. Chem. Soc.*, 1968, **90**, 6722.
[8] E. L. Muetterties, *J. Amer. Chem. Soc.*, 1968, **90**, 5097.

possible trigonal bipyramids, designated by the groups occupying apical positions, and each edge represents a possible pseudorotation. Conclusions which have been emphasised by several authors include: (*i*) a trigonal bipyramid may be converted into its mirror image by a series of five successive pseudorotations; and (*ii*) this process of racemisation is not possible if the restriction is imposed that two groups cannot both be equatorial at the same time. It follows from the first that a stereospecific substitution at phosphorus cannot involve a trigonal bipyramidal intermediate which is sufficiently long lived to pseudorotate without restriction.

The outstanding need in this area of organophosphorus chemistry is for quantitative data on the preference of electronegative groups for apical positions, and on the energy barrier to pseudorotation and how it varies with substituents. Such information is beginning to appear. Calculation[9] from spectroscopic data of the energy barrier to pseudorotation in phosphorus halides gave values, for example, of 7·6 for PF_5 and 15·0 kcal/mole for $MePF_4$. Molecular orbital calculations[10] on the cyclic phosphate ester (5) and the intermediates in its hydrolysis gave an upper value to the barrier between (6) and (7) of 12—15 kcal/mole and fully supported the mechanism for the hydrolysis of (5) proposed by Westheimer.[4]

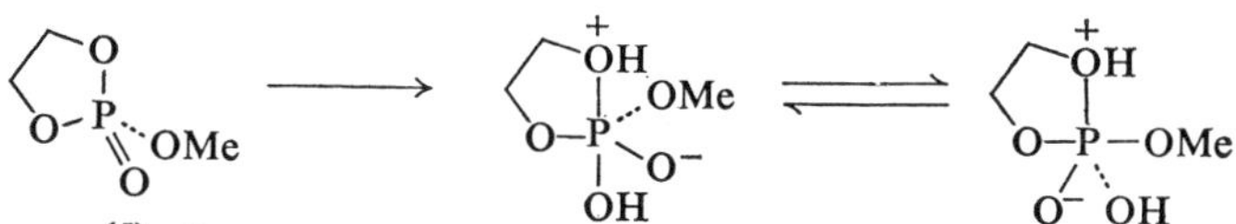

(5) (6) (7)

The temperature-dependent ^1H n.m.r. spectrum of the bis(2,2′-biphenylylene)phosphorane (8; R = *o*-isopropylphenyl) was interpreted[11] in

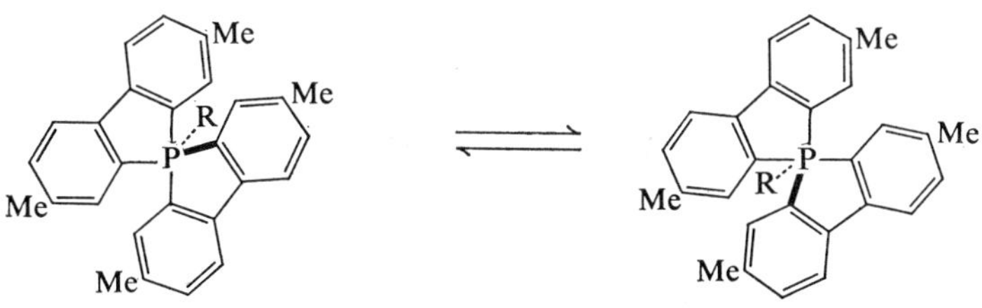

(8)

terms of pseudorotation in which the *o*-isopropylphenyl group remained equatorial, *i.e.* occupied the apical position of the intermediate square pyramid, and with activation parameters of $E = 20{\cdot}8 \pm 0{\cdot}4$ kcal/mole and log A = $15{\cdot}0 \pm 0{\cdot}3$. Similar studies[12] on a wide variety of phosphoranes

[9] R. R. Holmes and Sr. R. M. Deiters, *J. Amer. Chem. Soc.*, 1968, **90**, 5021.
[10] D. B. Boyd, *J. Amer. Chem. Soc.*, 1969, **91**, 1200.
[11] G. M. Whitesides and W. M. Bunting, *J. Amer. Chem. Soc.*, 1967, **89**, 6801.
[12] D. Hellwinkel, *Chimia (Switz.)*, 1968, **22**, 488.

(8; R = alkyl, aryl) gave free energies of activation for pseudorotation (at the coalescence temperature of the signals due to the biphenylylene methyls) varying from 11·9 (R = β-naphthyl) to 17·1 kcal/mole (R = α-naphthyl).

A study [13] of the temperature-dependent ^{1}H n.m.r. spectra of (9) and (10) has given data for the energy requirements for forcing the 1,3,2-dioxaphospholan ring into a diequatorial position. Two processes of pseudorotation were identified: those [(11) ⇌ (12)] in which the rings remained apical-

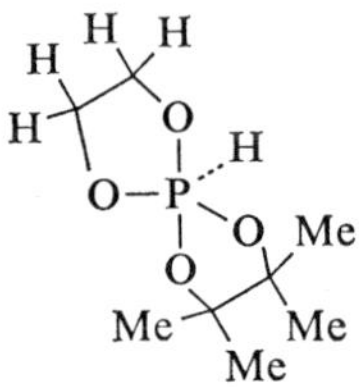

(9)

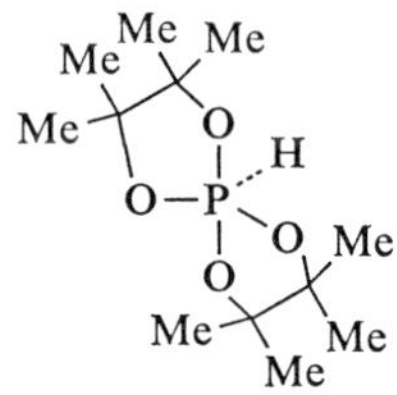

(10)

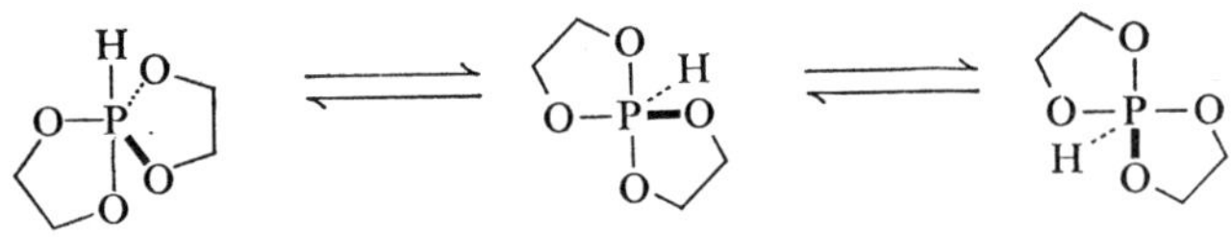

(13) (11) (12)

equatorial were rapid at $-70°$ whereas the pseudorotations [(11) ⇌ (13)] which involved placing a ring diequatorial were slow on the n.m.r. time-scale. The observed spectral changes led to values for the free energy difference between (11) and (13) of 15·6 kcal/mole in the case of (9) and of 18·4 kcal/mole in the case of (10).

A major development in the chemistry of quinquecovalent phosphoranes has been the appreciation that small rings stabilise such compounds. This may be because the small ring partly offsets crowding difficulties in the quinquecovalent state, or because the ring is less strained when spanning an apical-equatorial position of a trigonal bipyramid than when it contains a tetrahedral phosphorus. Several examples of this stabilisation are contained in the following account.

2 2,2′-Biphenylylenephosphoranes

Details have appeared [14] of the preparation of bis(2,2′-diphenylylene)-phosphorane (14) by reduction of the salt (15). A benzene solution of (14) in the dark became deep violet due to the radical (16) which decayed over a period of weeks to give mainly the phosphines (17) and (18) with small

[13] D. Houalla, R. Wolf, D. Gagnaire, and K. B. Robert, *Chem. Comm.*, 1969, 443.
[14] D. Hellwinkel, *Chem. Ber.*, 1969, **102**, 528.

quantities of (19) and (20). In benzene–methanol the sole product was compound (18) and deuterium labelling showed that the transfer of hydrogen involved was from (14) to (16) and did not involve the solvent.

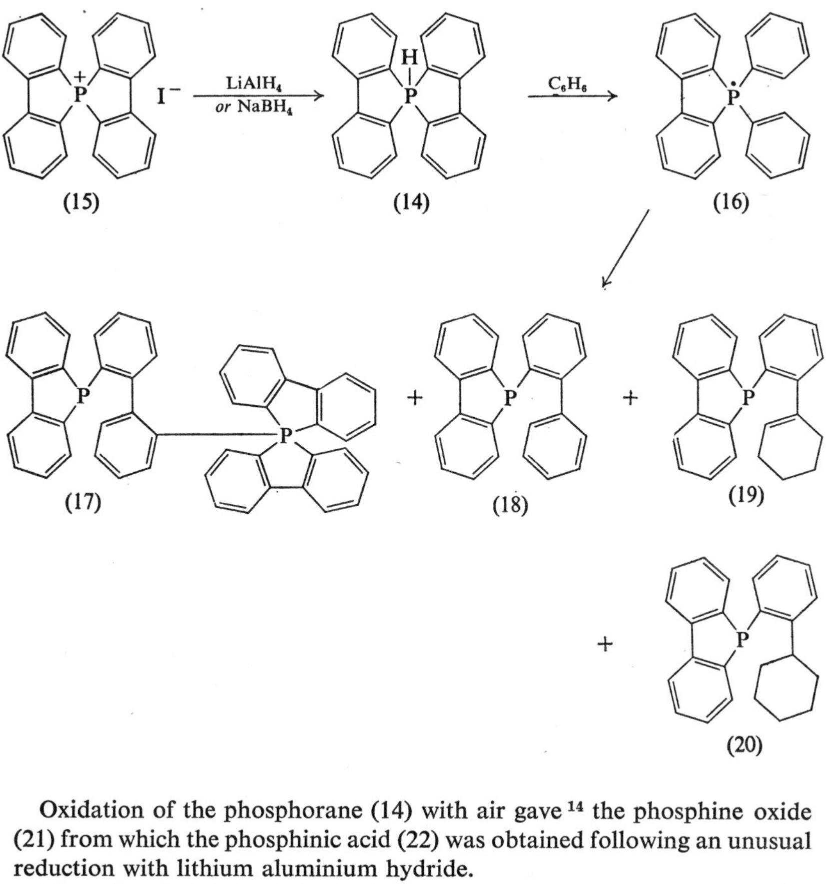

Oxidation of the phosphorane (14) with air gave [14] the phosphine oxide (21) from which the phosphinic acid (22) was obtained following an unusual reduction with lithium aluminium hydride.

The anion (23) from (14) was in equilibrium with the phosphine anion (24), an equilibrium which could be established from either side.[14] When compound (14) was treated with t-butyl-lithium in benzene or ether and the solution quenched with deuterium oxide the resulting phosphine (18) was predominantly undeuteriated, *i.e.* it was present in the reaction mixture before hydrolysis. A catalytic conversion of (14) into (18) *via* the radical anions (25) and (26) was suggested.[14]

The phosphoranes (27; R = Me, CH_2Ph) with bromine gave the phosphonium salts (28).[15]

[15] D. Hellwinkel, *Chem. Ber.*, 1969, **102**, 548.

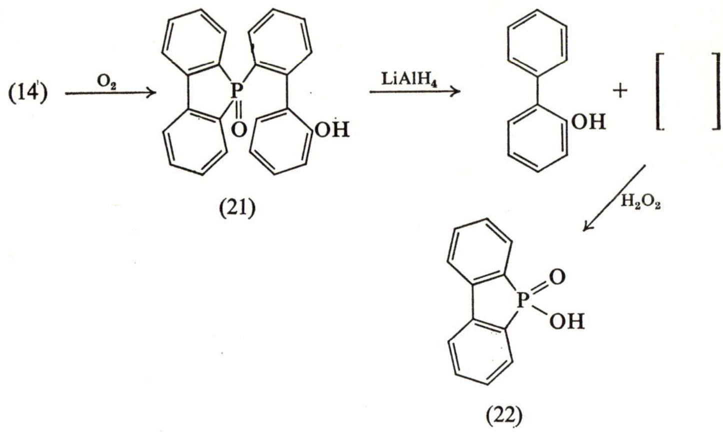
(14')
O2
P
O
OH
LiAlH4
OH
+
H2O2
P
O
OH
(21)
(22)

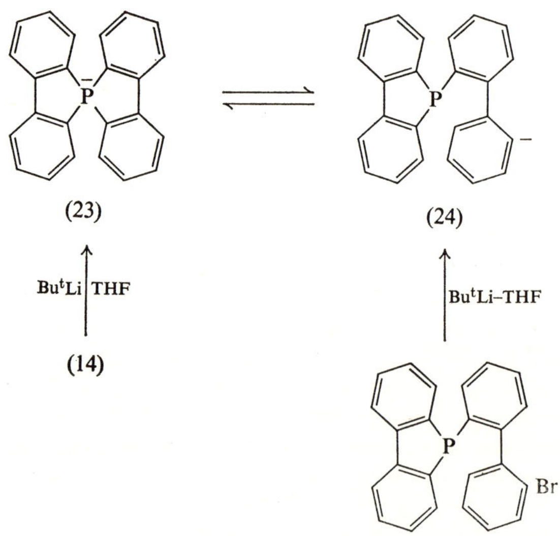
P
P
(23)
(24)
Bu^tLi THF
(14)
Bu^tLi–THF
P
Br

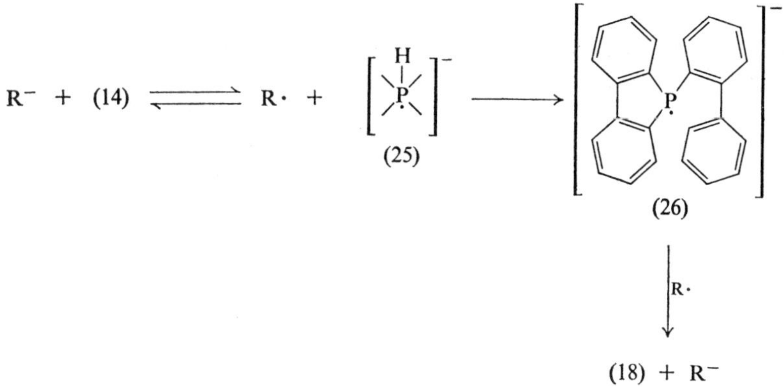

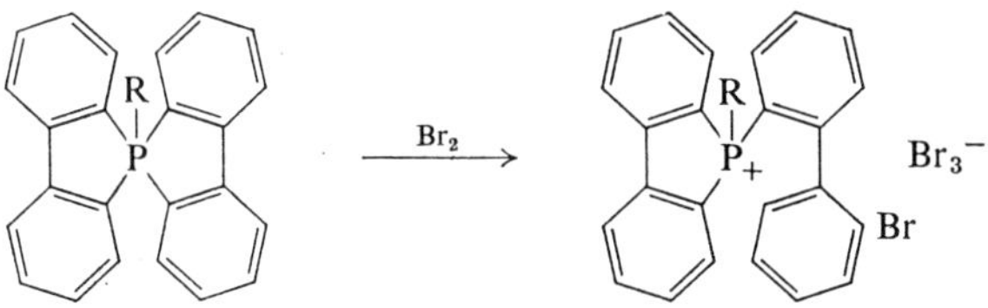

(27) (28)

3 1,3,2-Dioxaphospholens

The synthesis of 1,3,2-dioxaphospholens from α-diketones and phosphites has been extended to the use of glyoxal[16] and of cyclic phosphites and phosphoramidites.[17] The glyoxal–trimethyl phosphite adduct (29) was much less sensitive to water than the corresponding biacetyl adduct and has been used in the synthesis of various sugar-like phosphates, *e.g.* (30). The bicyclic adducts, *e.g.* (31), had remarkable thermal stability.

Biacetyl with the aminofluorophosphine (32) gave the dioxaphospholen (33).[18]

Successive exchange of the methoxy-groups of the biacetyl–trimethyl phosphite adduct (34) for benzyloxy-groups occurred when (34) was heated[19] with benzyl alcohol at 100°. With acyl chlorides (34) gave the phosphate esters (35) of α-hydroxy-β-diketones.[20] This is an example of the nucleophilic addition of phospholens to carbonyl compounds, a wide variety of which has been investigated by Ramirez and his co-workers. Among

[16] F. Ramirez, S. L. Glaser, A. J. Bigler, and J. F. Pilot, *J. Amer. Chem. Soc.*, 1969, **91**, 496.
[17] F. Ramirez, M. Nagabhushanam, and C. P. Smith, *Tetrahedron*, 1968, **24**, 1785.
[18] G. I. Drozd, S. Z. Ivin, and V. V. Sheluchenko, *Zhur. obshchei Khim.*, 1968, **38**, 1906.
[19] F. Ramirez, K. Tasaka, N. B. Desai, and C. P. Smith, *J. Amer. Chem. Soc.*, 1968, **90**, 751.
[20] F. Ramirez, S. B. Bhatia, A. J. Bigler, and C. P. Smith, *J. Org. Chem.*, 1968, **33**, 1192.

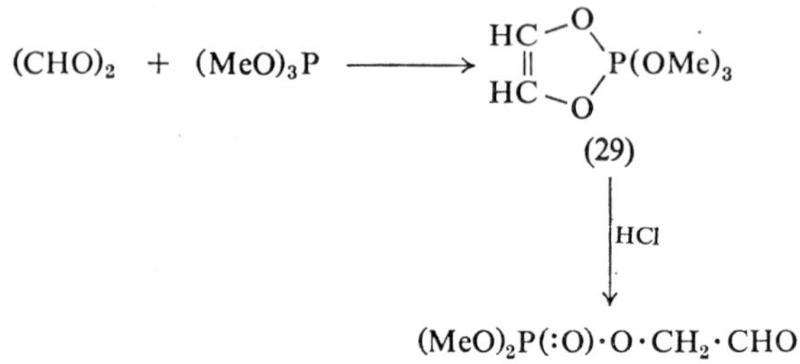

(29) (30)

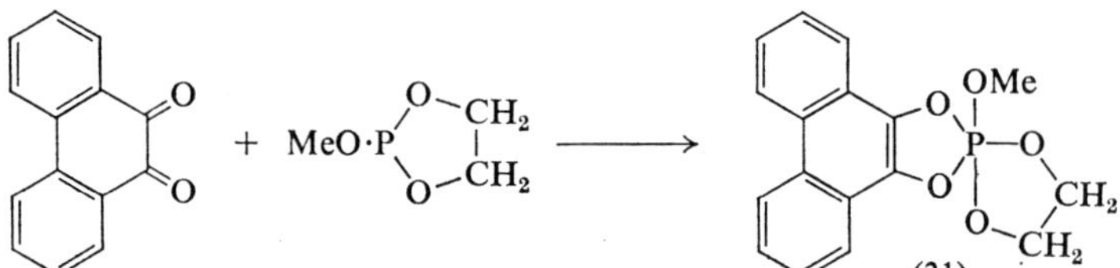

(31)

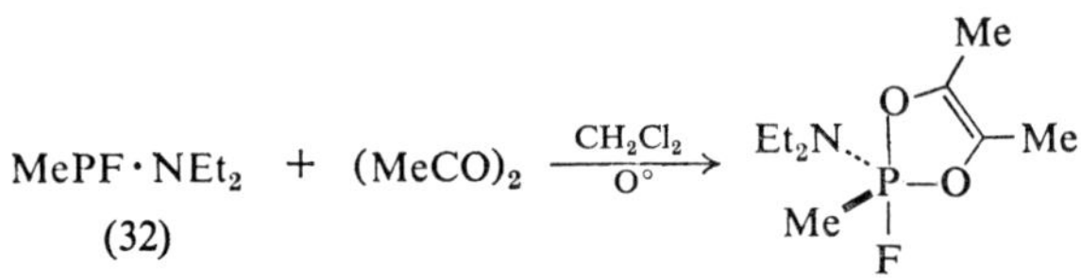

(32) (33)

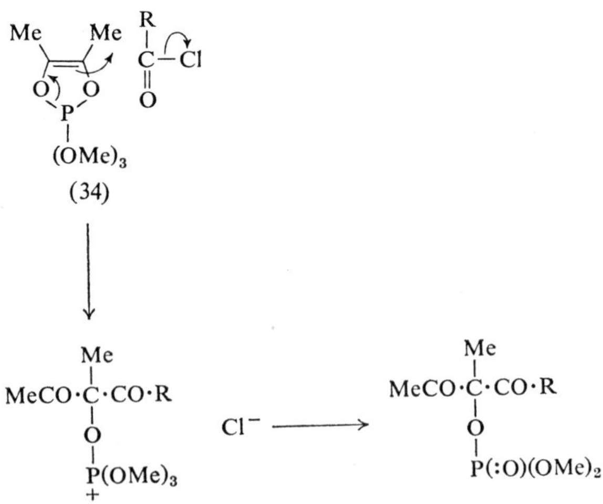

(34) (35)

recent examples are additions to carbon suboxide,[21] $\alpha\beta$-unsaturated aldehydes,[22, 23] acyl isocyanates,[24] and sulphonyl isocyanates.[24]

Carbon suboxide [21] gave the lactones (36); the suggested mechanism is as shown. Acraldehyde [22] and crotonaldehyde [23] with (34) gave the phospholans (37); the complex hydrolyses and methanolyses of these compounds have been described.

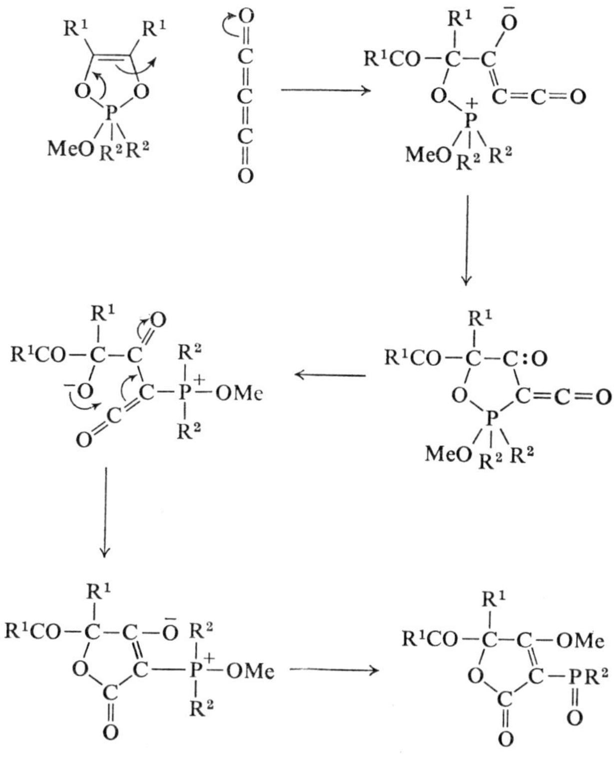

(36)

The initial betaines (38) formed from dioxaphospholens and acyl isocyanates [24] cyclised with expulsion of trialkyl phosphate to form the oxazolinones (39). Similar cyclisation of the intermediate from the reaction of (34) with two molecules of benzenesulphonyl isocyanate gave the *NN′*-ditosylhydantoin (40).[24]

[21] F. Ramirez and G. V. Loewengart, *J. Amer. Chem. Soc.*, 1969, **91**, 2293.
[22] F. Ramirez, H. J. Kugler, A. V. Patwardhan, and C. P. Smith, *J. Org. Chem.*, 1968, **33**, 1185.
[23] F. Ramirez, H. J. Kugler, and C. P. Smith, *Tetrahedron*, 1968, **24**, 3153.
[24] F. Ramirez and C. D. Telefus, *J. Org. Chem.*, 1969, **34**, 376.

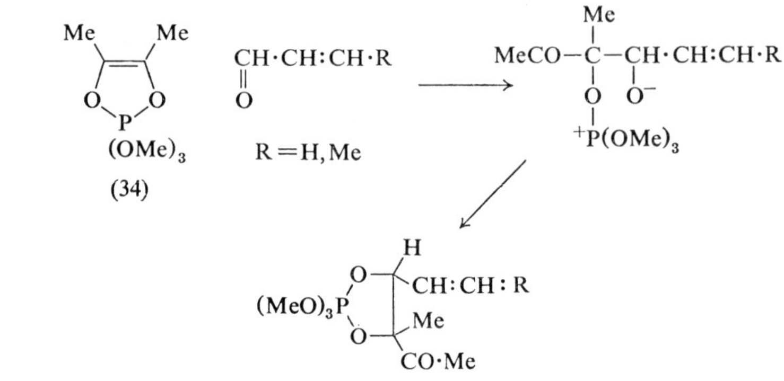

(34) (37)

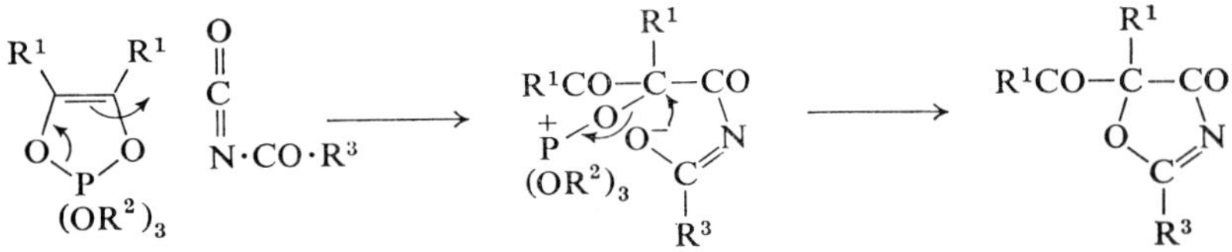

(38) (39)

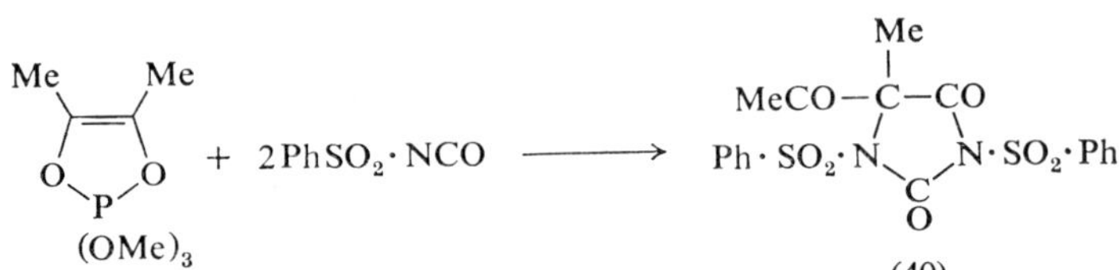

(34) (40)

4 1,3,2-Dioxaphospholans

Tertiary phosphines and hexafluoroacetone formed the 1,3,2-dioxaphospholans (41).[25] That from trimethylphosphine had only one doublet in its ^{1}H n.m.r. spectrum at $-62°$ showing that pseudorotation was rapid on the n.m.r. time-scale even at this temperature. The 1,3,2,-dioxaphospholans rearranged on heating to give the 1,2-oxaphosphetans (42) which at higher temperatures formed phosphinates and olefins.[26] The rates of rearrangement were in the order $Et_3P > Et_2PhP > EtPh_2P > Me_3P$ and the rates of olefin formation in the order $Me_3P > Et_2PhP > Et_3P > EtPh_2P$. The oxaphosphetan from diethylphenylphosphine was obtained as the two diastereoisomers (43) and (44) which equilibrated on heating. This equilibration could be by a process of pseudorotation only if the four-membered ring became diequatorial in an intermediate trigonal bipyramid.

[25] F. Ramirez, C. P. Smith, J. F. Pilot, and A. S. Gulati, *J. Org. Chem.*, 1968, **33**, 3787.
[26] F. Ramirez, C. P. Smith, and J. F. Pilot, *J. Amer. Chem. Soc.*, 1968, **90**, 6726.

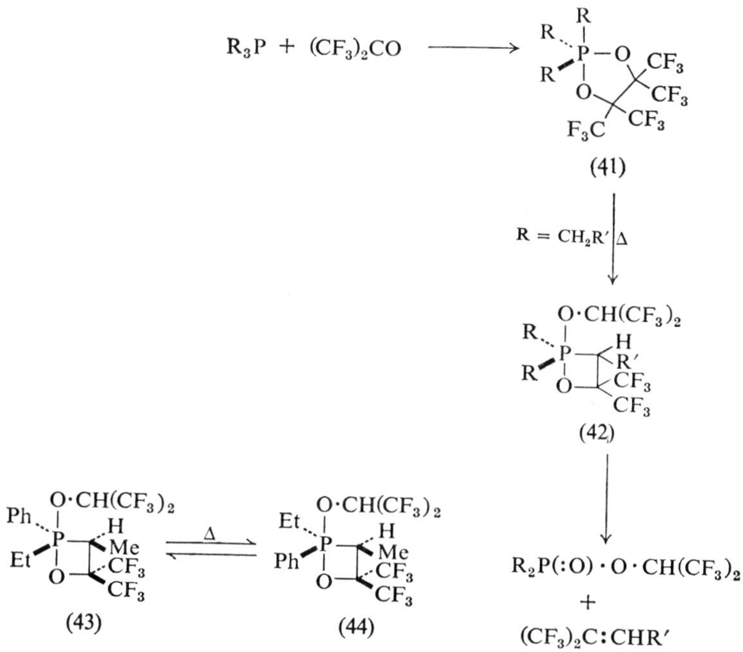

(41) (42) (43) (44)

The 1,3,2-dioxaphospholan (45) was formed from ninhydrin and trialkyl phosphites but with triphenylphosphine the stable ylide (46) and triphenylphosphine oxide were obtained.[27]

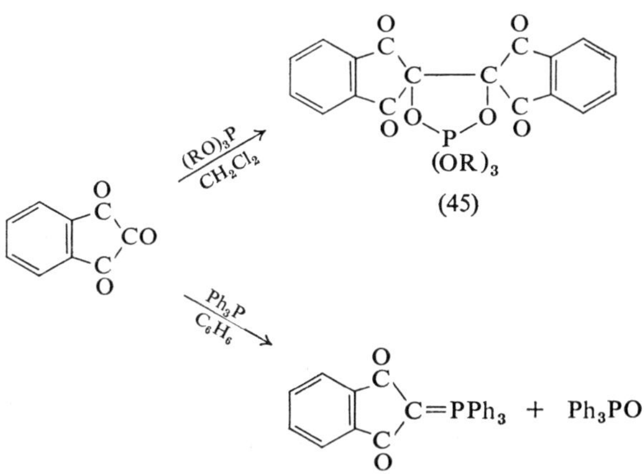

(45) (46)

[27] A. Mustafa, M. M. Sidky, S. M. A. D. Zayed, and M. R. Mahran, *Annalen*, 1968, **712**, 116.

The 1,3,2-dioxaphospholans (47), formed from triethyl phosphite and the fluorenones (48), when dissolved in acetonitrile at 43—46° for 20 hr. gave[28]

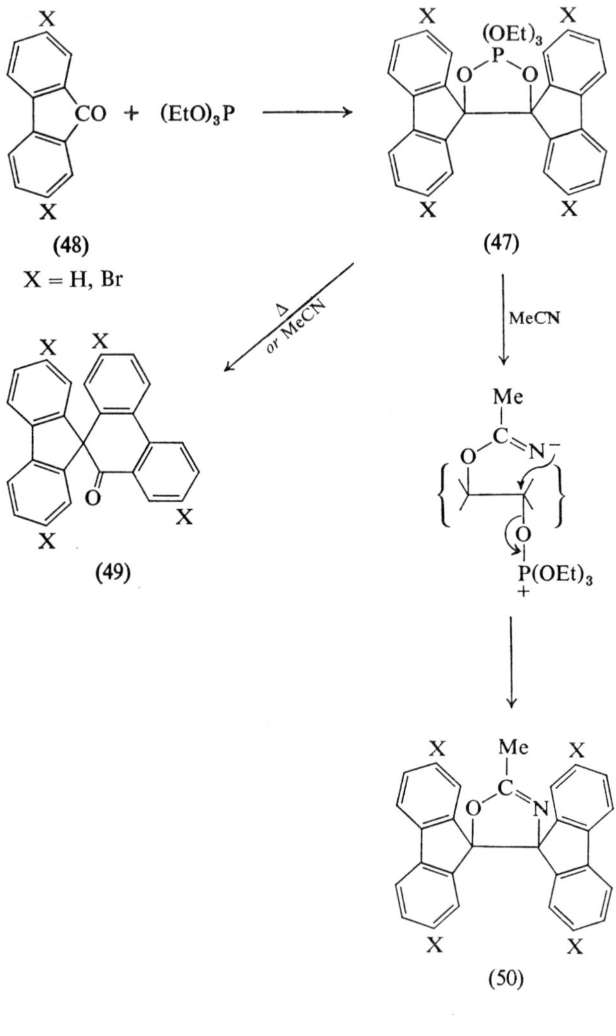

the spiro-ketones (49)—also formed[29, 30] on heating the adducts to 160°—and the oxazolines (50).

[28] I. J. Borowitz, P. D. Readio, and P. Rusek, *Chem. Comm.*, 1968, 240.
[29] F. Ramirez and C. P. Smith, *Chem. Comm.*, 1967, 662.
[30] I. J. Borowitz and M. Anschel, *Tetrahedron Letters*, 1967, 1517.

The tetraoxyphosphorane (51) has been condensed with benzaldehyde and with trichloroacetaldehyde to give the alcohols (52).[31] The corresponding alcohol from benzaldehyde and the tetraoxyphosphorane derived

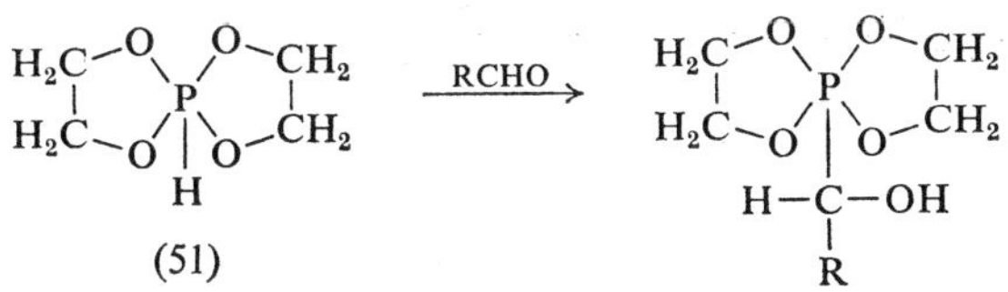

(51) (52)

from *cis*-cyclohexan-1,2-diol was shown from its ^1H n.m.r. spectrum to contain three diastereoisomers. See Chapter 10, section 1 for the photolysis of 1,3,2-dioxaphospholans.

5 1,2-Oxaphospholens

The addition of trimethyl phosphite to 3-benzylidene-2,4-pentandione to form a 1,2-oxaphospholen[32] has been extended to the use of phosphonite and phosphinite esters.[33] A study of the ^1H n.m.r. spectrum[33,34] of the dimethyl phenylphosphonite adduct showed that at low temperatures the two isomers (53) and (54) were stable but that at higher temperatures (> 52°) they were in rapid equilibrium by a process involving four successive

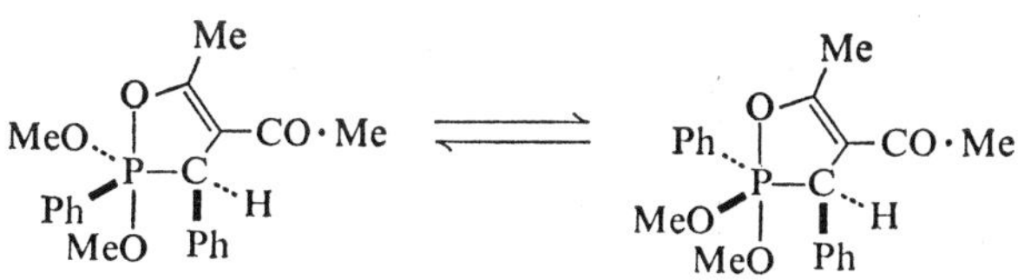

(53) (54)

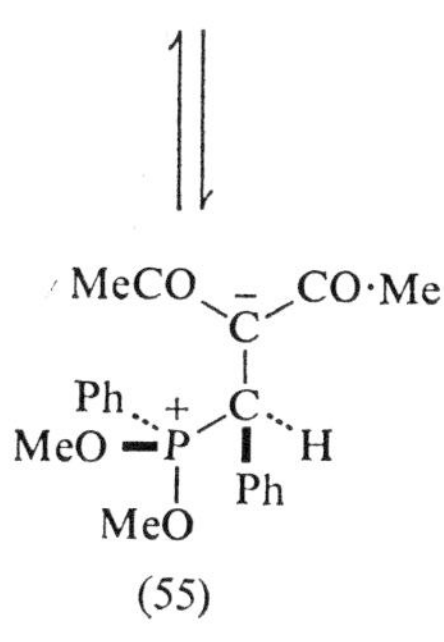

(55)

[31] R. Burgada and H. Germa, *Compt. rend.*, 1968, **267**, *C*, 270.
[32] F. Ramirez, O. P. Madan, and S. R. Heller, *J. Amer. Chem. Soc.*, 1965, **87**, 731.
[33] F. Ramirez, J. F. Pilot, O. P. Madan, and C. P. Smith, *J. Amer. Chem. Soc.*, 1968, **90**, 1275.
[34] D. Gorenstein and F. H. Westheimer, *Proc. Nat. Acad. Sci. U.S.A.*, 1967, **58**, 1747.

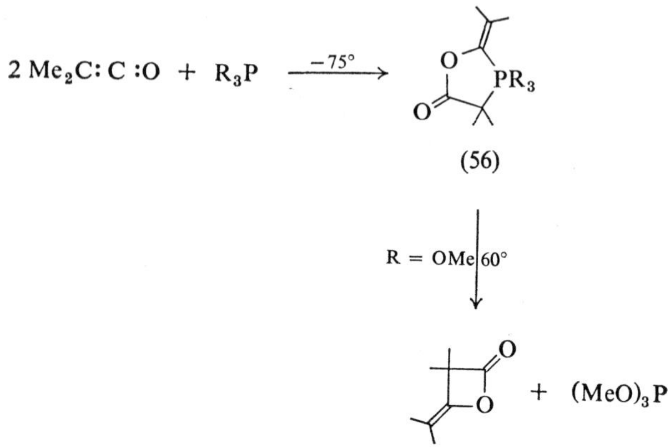
2 Me₂C: C :O + R₃P
−75°
O
PR₃
O
R = OMe 60°
O
O
+ (MeO)₃P

(56)

(58)

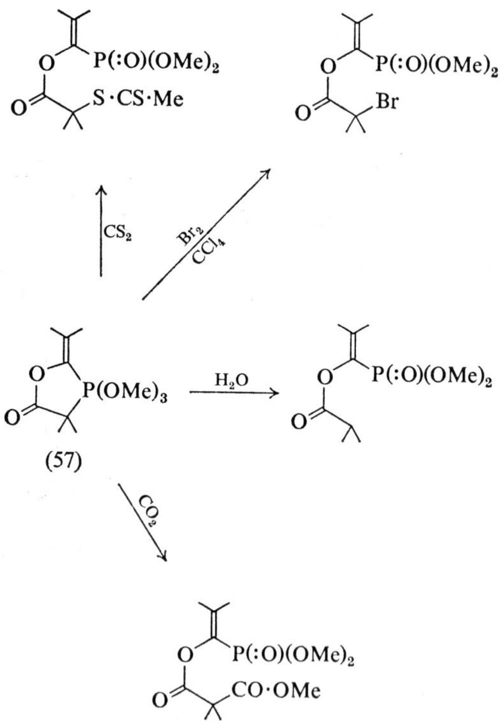
O
P(:O)(OMe)₂
O
S·CS·Me
O
P(:O)(OMe)₂
O
Br
CS₂
Br₂
CCl₄
O
P(OMe)₃
O
H₂O
O
P(:O)(OMe)₂
O
CO₂
O
P(:O)(OMe)₂
O
CO·OMe

(57)

pseudorotations. Above 70° opening of the ring began and at 125° the adduct was entirely the dipolar form (55).

6 1,3-Oxaphospholans

The 1,3-oxaphospholans (56) were the initial products from the reactions[35] of phosphites and aminophosphines with dimethylketen at −75°. The trimethyl phosphite adduct (57) gave the keten dimer (58) quantitatively at 60°. Dimerisation of dimethylketen to (58) in the presence of phosphites had previously been observed.[36] The reactions of the adduct (57) are indicative of a highly polarised carbon–phosphorus bond.

7 Penta(aryloxy)phosphoranes

Pentaphenoxyphosphorane has been prepared[37] by the addition of phenol (5 moles) in benzene to a hexane solution of γ-collidine (5 moles) at 0° containing phosphorus pentachloride (1 mole). None of the previously reported syntheses was successful. With catechol in benzene the phosphorane readily exchanged two phenoxy-groups to give (59) while a further

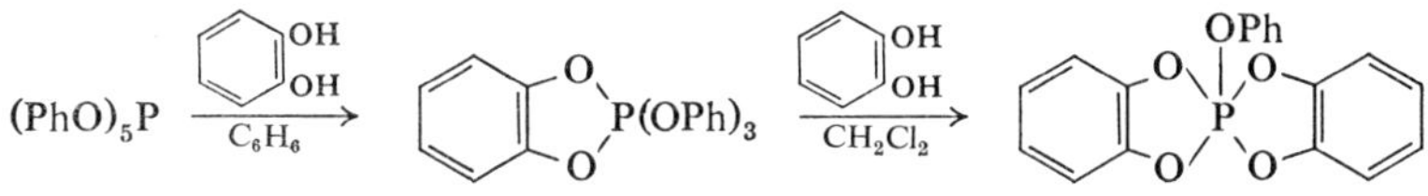

(59) (60)

exchange occurred in methylene chloride at 25° to give the bicyclic phosphorane (60).[38] This is a striking demonstration of the stabilising effect of five-membered rings on phosphoranes.

8 Miscellaneous

The minor products in the preparations of the phosphinites (61) from catechol and dichlorophosphines in the presence of triethylamine have been identified[39] as the phosphoranes (62). These were also formed from the phosphinites (61) on prolonged heating at 100° in the presence of an acid catalyst or on treatment with *o*-benzoquinone.[39]

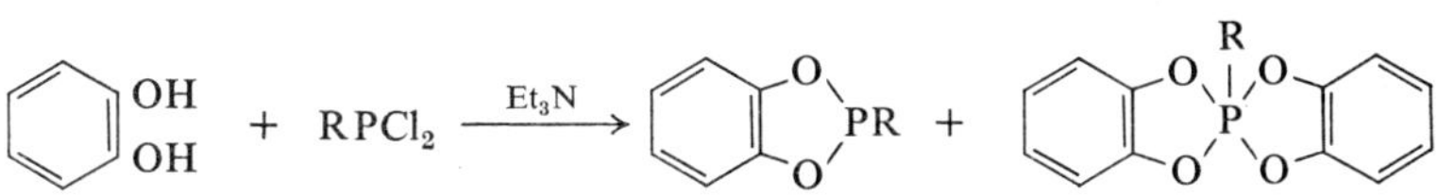

R = Me,Ph (61) (62)

[35] W. G. Bentrude and W. D. Johnson, *J. Amer. Chem. Soc.*, 1968, **90**, 5924.
[36] E. U. Ullam, *J. Org. Chem.*, 1967, **32**, 215.
[37] F. Ramirez, A. J. Bigler, and C. P. Smith, *J. Amer. Chem. Soc.*, 1968, **90**, 3507.
[38] F. Ramirez, A. J. Bigler, and C. P. Smith, *Tetrahedron*, 1968, **24**, 5041.
[39] M. Wieber and W. R. Hoos, *Tetrahedron Letters*, 1968, 5333.

The phosphorane (63) was the major product formed from *o*-aminophenol and various halogenocyclophosphazenes in refluxing xylene.[40] While undoubtedly existing as the phosphorane in solution there was some i.r. evidence that in the solid state the hexaco-ordinate betaine structure (64) might be present.

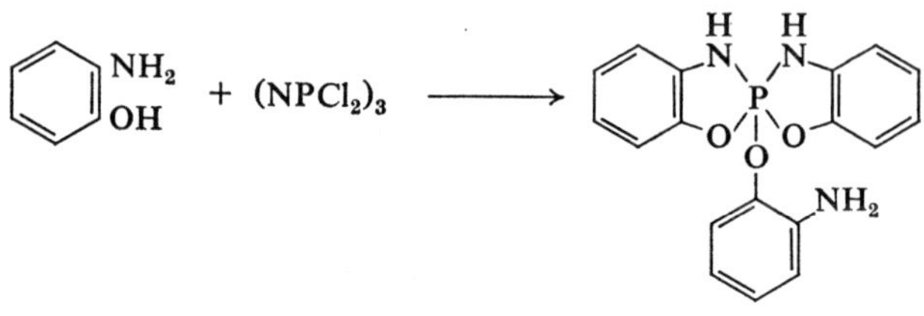

(63)

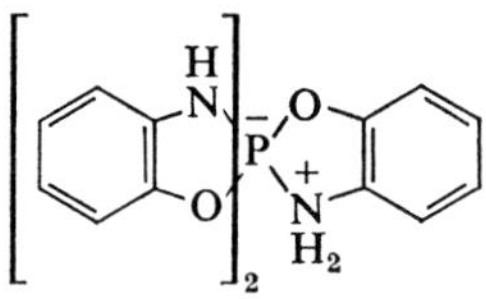

(64)

The phosphorane (65) was obtained [41] from bis(trifluoromethyl)phosphine and the stable radical (66).

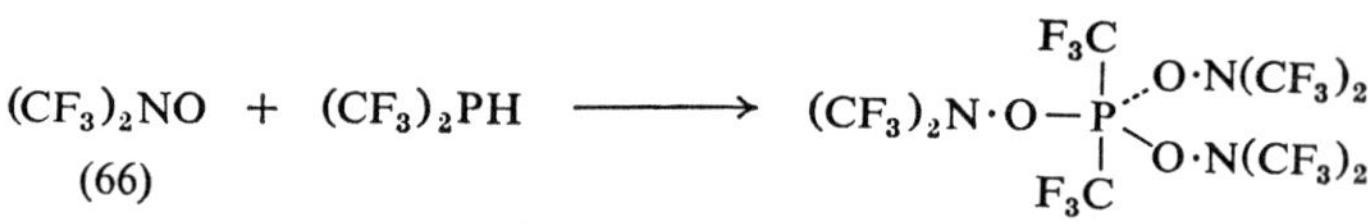

(66)

(65)

The slow exchange of ligands on treatment of pentaphenylphosphorane with phenyl-lithium has been followed kinetically [42] using tritium labelling, and compared with similar exchange in the pentaphenyl derivatives of arsenic, antimony, and bismuth.

[40] H. R. Allcock and R. L. Kugel, *Chem. Comm.*, 1968, 1606.
[41] H. G. Ang, *Chem. Comm.*, 1968, 1320.
[42] H. Daniel and J. Paetsch, *Chem. Ber.*, 1968, **101**, 1451.

3

Halogenophosphines and Related Compounds

BY S. TRIPPETT

1 Halogenophosphines

A. Preparation.—*p*-Dihalogenobenzenes with white phosphorus and phosphorus trichloride at 340° in the presence of iodine [1] gave *p*-bis(dichlorophosphino)-benzene (38%). Under similar conditions the chlorophosphine (1) was obtained.[2]

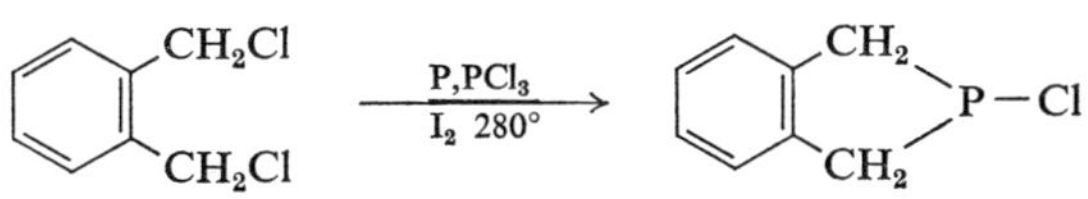

(1)

Methane, ethane, and benzene passed over phosphorus pentachloride at 200°, and the resulting vapour then passed over heated quartz chips, gave the corresponding dichlorophosphine.[3] Terminal olefins with phosphorus pentachloride in benzene or phosphorus trichloride followed by reduction with dichloromethoxyphosphine gave the dichlorovinylphosphines (2) in 38—87% yield.[4]

$$R^1R^2C{:}CH_2 + PCl_5 \longrightarrow [R^1R^2C{:}CH{\cdot}PCl_4] \xrightarrow{MeO\cdot PCl_2} R^1R^2C{:}CH{\cdot}PCl_2 + MeCl + OPCl_3$$

(2)

Difluoropentafluorophenylphosphine has been prepared by the action of sodium fluoride in acetonitrile (not sulpholan [5]) or of antimony trifluoride on the corresponding dibromo- or dichloro-compound.[6] The same reagents gave fluorodi(pentafluorophenyl)phosphine from the chloride or bromide.

1,2-Dibromo-1,2-diphenyldiphosphine (3) has been obtained in a variety of ways as shown.[7]

1 Yu. I. Baranov, O. F. Filippov, S. L. Varshavskii, and M. I. Kabachnik, *Doklady Akad. Nauk S.S.S.R.*, 1968, **182**, 337.
2 U.S.S.R.P. 210,155.
3 E. A. Smirnov, Yu. M. Zinov'ev, and V. A. Petrunin, *Zhur. obshchei Khim.*, 1968, **38,** 1551.
4 Ya. A. Levin, V. S. Galeev, and E. K. Trutneva, *Zhur. obshchei Khim.*, 1967, **37**, 1872.
5 J. M. Miller, *J. Chem. Soc.* (*A*), 1967, 828.
6 M. Fild and R. Schmutzler, *J. Chem. Soc.* (*A*), 1969, 840.
7 M. Baudler, O. Gehlen, K. Kipker, and P. Backes, *Z. Naturforsch.*, 1967, **22b**, 1354.

$$(PhP)_5 \xrightarrow[CS_2 \cdot R \cdot T]{PhPBr_2} PhPBr \cdot PPh \cdot Br \xleftarrow{Br_2-CS_2} (PhP)_5$$

73% (3) 92%

$$Ph_2P_2I_2 \xrightarrow{AgBr,\ (CH_2Cl)_2} (3) \quad 47\%$$

B. Reactions.—(*i*) *Nucleophilic Attack on Phosphorus.* Sodium cyanate and dichlorophenylphosphine gave successively the mono- (4) and di-(iso-cyanato)phosphine (5).[8] The latter polymerised slowly in an inert atmosphere.

$$PhPCl_2 + NaNCO \xrightarrow[Reflux]{MeCN} \underset{57\%}{\underset{(4)}{PhPCl \cdot NCO}} \longrightarrow \underset{67\%}{\underset{(5)}{PhP(NCO)_2}}$$

The carbodi-imide (6) was obtained from the reaction of chlorodiphenylphosphine with silver cyanamide as a complex with silver chloride and solvent.[9] Addition of sulphur gave the disulphide (7), and the urea (8) was obtained on oxidation with aqueous hydrogen peroxide. The same chlorophosphine with calcium carbide gave carbon and tetraphenyldiphosphine isolated as the dioxide or disulphide in 50% yield.[10]

$$\underset{(6)}{Ph_2P \cdot N{:}C{:}N \cdot PPh_2} \xrightarrow{S} \underset{(7)}{Ph_2P({:}S) \cdot N{:}C{:}N \cdot P({:}S)Ph_2}$$

$$\underset{(6)}{Ph_2P \cdot N{:}C{:}N \cdot PPh_2} \xrightarrow[H_2O]{H_2O_2} \underset{(8)}{Ph_2P({:}O) \cdot NH \cdot CO \cdot NH \cdot P({:}O)Ph_2}$$

Whereas hydrazine gave no recognisable products, the expected hydrazinophosphines were obtained from 1,1-dimethylhydrazine and trimethylhydrazine on treatment with trifluoromethyldi-iodophosphine and with di(trifluoromethyl)iodophosphine,[11, 12] *e.g.*,

$$(CF_3)_2PI + Me_2N \cdot NHMe \xrightarrow{Ether} \underset{83\%}{(CF_3)_2P \cdot NMe \cdot NMe_2}$$

The first diaminophosphine (9) was obtained[13] from t-butyldichlorophosphine and ammonia in ether at $-50°$. Compound (9) was extremely sensitive to oxygen and water and with trimethylchlorosilane gave the bis(trimethylsilyl) derivative (10).

8 J. J. Pitts, M. A. Robinson, and S. I. Trotz, *Inorg. Nuclear Chem. Letters*, 1968, **4**, 483.
9 A. Weisz and K. Utvary, *Monatsh.*, 1968, **99**, 2498.
10 E. J. Spanier and F. E. Caropreso, *Tetrahedron Letters*, 1969, 199.
11 L. K. Peterson and G. L. Wilson, *Canad. J. Chem.*, 1968, **46**, 685.
12 L. K. Peterson, G. L. Wilson, and K. I. Thé, *Canad. J. Chem.*, 1969, **47**, 1025.
13 O. J. Scherer and P. Klusman, *Angew. Chem. Internat. Edn.*, 1968, **7**, 541.

$$\mathrm{Bu^tPCl_2} \xrightarrow{\mathrm{NH_3}} \underset{(9)\ 60\%}{\mathrm{Bu^tP(NH_2)_2}} \xrightarrow{\mathrm{Me_3SiCl}} \underset{(10)\ 70\%}{\mathrm{Bu^tP(NH{\cdot}SiMe_3)_2}}$$

Sodium arylsulphinates with chlorodiphenylphosphine in dimethylformamide gave the aryl diphenylphosphinothiolates (11).[14] There was no reaction with diphenylphosphinic chloride, and aliphatic chlorophosphines gave only diphenyl disulphide and the phosphinic acids. Sodium arylsulphonates did not react with chlorophosphines.

$$\mathrm{Ph_2PCl + Ar{\cdot}SO_2Na} \longrightarrow \underset{(11)}{\mathrm{Ph_2P(:O){\cdot}S{\cdot}Ar}}$$

Phenyl diphenylphosphinoselenolate (12) has been obtained in a variety of ways as shown.[15] The route from the dioxide was thought to involve the intermediate $Ph_2P{\cdot}O{\cdot}Se{\cdot}Ph$ which rearranged to (12).

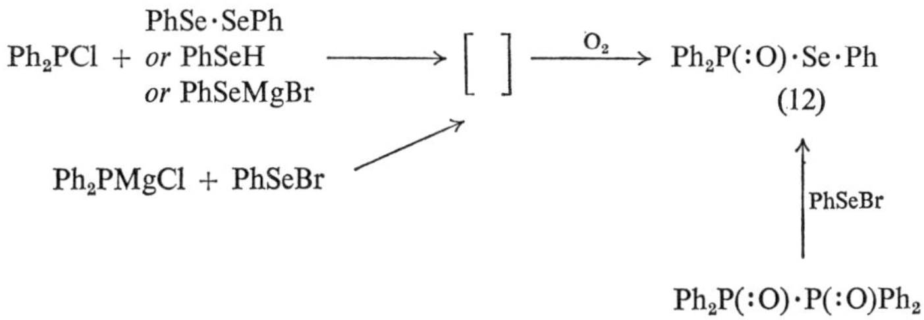

Chlorodimethylphosphine with metallated pentaborane(9) gave μ-dimethylphosphinopentaborane(9) (13) containing a B—P—B three-centre two-electron bond.[16] The n.m.r. spectrum of (13) showed the two methyl groups to be in different environments and chloro(methyl)trifluoromethylphosphine gave rise to two similar isomeric compounds. However, chlorobis(trifluoromethyl)phosphine gave 1-bis(trifluoromethyl)phosphinopentaborane(9) with no evidence of bridging phosphorus.

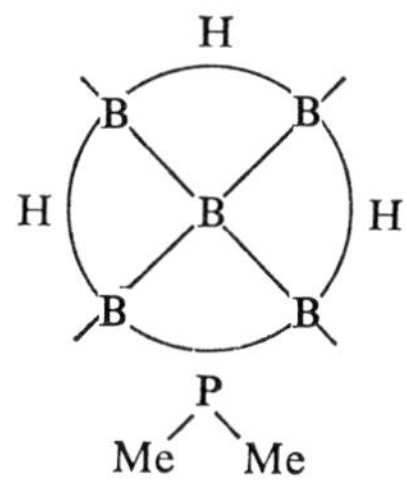

(13)

(*ii*) *Nucleophilic Attack by Phosphorus.* The formation of benzoyl and alkyl chlorides together with the phosphine oxides (14) from the reactions of

[14] H. Schindlbauer and W. Prikoszovich, *Monatsh.*, 1968, **99**, 1792.
[15] N. Petragnani, V. G. Toscano, and M. de Moura Campos, *Chem. Ber.*, 1968, **101**, 3070.
[16] A. B. Burg and H. Heinen, *Inorg. Chem.*, 1968, **7**, 1021.

chlorodiphenylphosphine with benzoate esters has been rationalised in terms of competing nucleophilic attacks of the chlorophosphine on the carbonyl carbon or on the alkyl oxygen as shown.[17]

$$PhCO_2^- + RPh_2\overset{+}{P}Cl \longrightarrow Ph{\cdot}CO{\cdot}O{\cdot}\overset{+}{P}Ph_2{\cdot}R\ \bar{C}l$$

$$\uparrow \qquad\qquad\qquad\qquad \downarrow$$

$$Ph{\cdot}CO{\cdot}OR + Ph_2PCl \qquad\qquad Ph{\cdot}CO{\cdot}Cl + RPh_2P(:O)$$

(14)

$$\downarrow$$

$$Ph{\cdot}CO{\cdot}\overset{+}{P}Ph_2{\cdot}Cl\ \bar{O}R \longrightarrow Ph{\cdot}CO{\cdot}\overset{+}{P}Ph_2{\cdot}OR\ \bar{C}l$$

$$\downarrow$$

$$[Ph{\cdot}CO{\cdot}P(:O){\cdot}Ph_2] + RCl$$

Aliphatic carboxylic anhydrides with diethylchlorophosphine gave the vinylphosphine oxides (15) presumably *via* the highly reactive acylphosphine oxides (16).[18] For other reactions of these oxides see Chapter 4, section 1D.

$$Et_2PCl + (RCH_2{\cdot}CO)_2O \longrightarrow R{\cdot}CH_2{\cdot}CO{\cdot}\overset{+}{P}Et_2{\cdot}Cl + RCH_2{\cdot}C\bar{O}_2$$

$$R{\cdot}CH_2{\cdot}CO{\cdot}P(:O)Et_2 + RCH_2{\cdot}CO{\cdot}Cl$$

(16)

$$\downarrow (RCH_2{\cdot}CO)_2O$$

$$R{\cdot}CH{:}C(O{\cdot}CO{\cdot}CH_2R){\cdot}P(:O)Et_2 \longleftarrow R{\cdot}CH_2{\cdot}C(O{\cdot}CO{\cdot}CH_2R)_2{\cdot}P(:O)Et_2$$

(15)

2-Cyanoethylphosphinic chlorides (17) resulted from the addition of dichlorophosphines to acrylamide, perhaps *via* the cyclic intermediates (18).[19, 20] Dichloro(methyl)phosphine with methacrylic acid gave the acid

[17] S. T. McNeilly and J. A. Miller, *Chem. Comm.*, 1969, 620.

[18] V. S. Tsivunin, Yu. N. Afanas'ev, R. G. Ivanova, T. A. Zyablikova, and G. Kh. Kamai, *Zhur. obshchei Khim.*, 1968, **38**, 1523.

[19] V. K. Khairullin, R. M. Kondrat'eva, and A. N. Pudovik, *Izvest. Akad. Nauk S.S.S.R., Ser. khim.*, 1967, 2097.

[20] V. K. Khairullin, T. I. Sobchuk, and A. N. Pudovik, *Zhur. obshchei Khim.*, 1968, **38**, 584.

chloride (19) and with propiolic acid followed by ethanol gave the ester (20).[21]

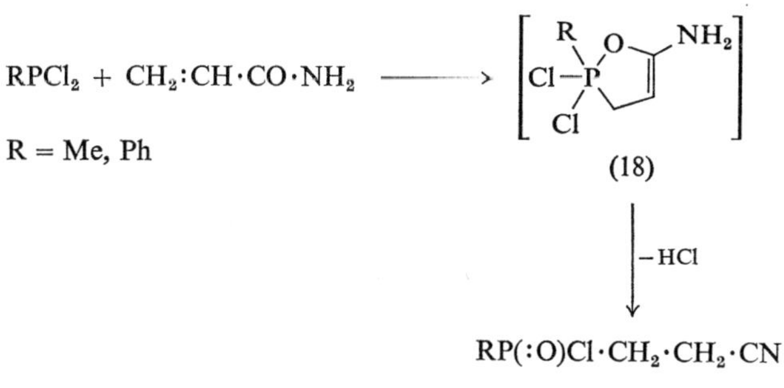

(17) 37—56%

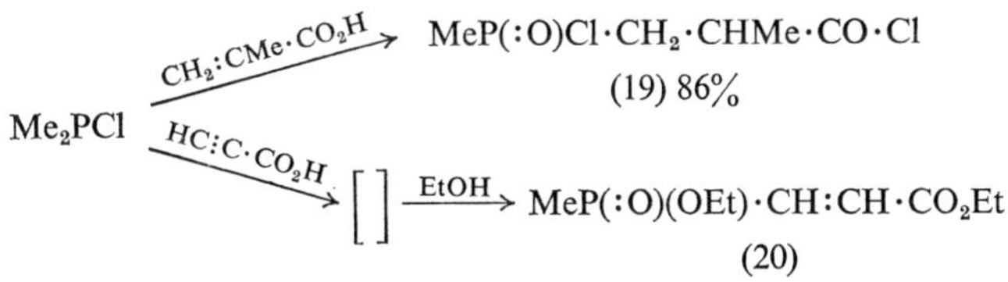

Chlorodiphenylphosphine and triethyl orthoformate reacted exothermically[22] to give the acetal (21). Acetals, dithioacetals, and ethoxymethylamines similarly gave the oxides (22) and (23). The acetals (24) were obtained from triethyl orthoformate and dichlorophosphines.

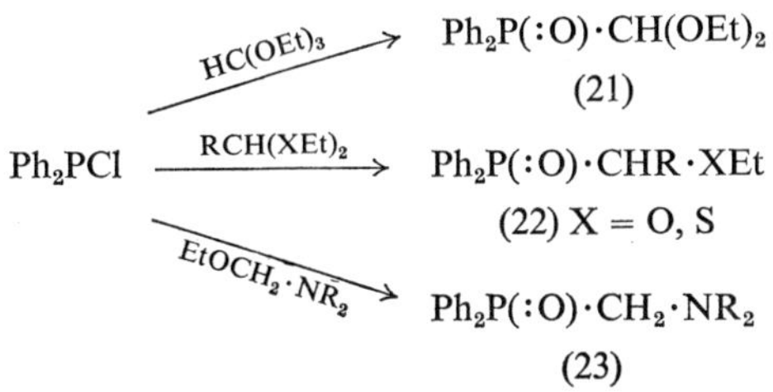

$$RPCl_2 + HC(OEt)_3^- \longrightarrow RP(:O)(OEt)\cdot CH(OEt)_2$$

(24)

Details have appeared[23] of the reaction between chlorodiphenylphosphine and benzoyl peroxide which gave benzoyl chloride, benzoic anhydride, and diphenylphosphinic anhydride. The intermediate mixed anhydride (25)

[21] V. K. Khairullin, R. M. Kondrat'eva, and A. N. Pudovik, *Zhur. obshchei Khim.*, 1968, **38**, 288.
[22] W. Dietsche, *Annalen*, 1968, **712**, 21.
[23] G. Sosnovsky and D. J. Rawlinson, *J. Org. Chem.*, 1968, **33**, 2325.

was detected by its characteristic carbonyl absorption at 1740 cm^{-1}. Authentic (25) in refluxing benzene rapidly formed an equilibrium mixture of the two symmetrical anhydrides and an appreciable proportion of (25).

$$Ph_2PCl + (PhCO)_2O_2 \longrightarrow Ph_2P(:O)Cl + (PhCO)_2O$$

$$\downarrow$$

$$\begin{array}{ccc} Ph_2P(:O)\cdot O\cdot P(:O)Ph_2 & \longleftarrow & Ph_2P(:O)\cdot O\cdot CO\cdot Ph + PhCO\cdot Cl \\ + (PhCO)_2O & & (25) \end{array}$$

(*iii*) *Miscellaneous.* Attempts to extend to other ketones the reaction of benzophenone with dichlorophenylphosphine in the presence of partially hydrated aluminium chloride to give the phosphinic chloride (26) met with limited success.[24] Alkyl aryl and acyclic dialkyl ketones gave complex mixtures of neutral products but cyclohexanone gave the phosphinic acid (27). Biphthalyl (34%) was formed from phthalic anhydride and dichlorophenylphosphine in the absence of catalyst at 100° in a reaction analogous to that using triethyl phosphite.[25]

$$Ph_2CO + PhPCl_2 \xrightarrow[H_2O]{AlCl_3} Ph_2CCl\cdot P(:O)\cdot Ph\cdot Cl$$

(26)

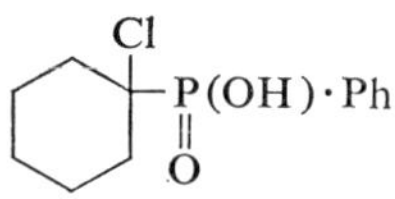

(27)

A study[5, 26] of the disproportionation of difluorophosphines (28) to give tetrafluorophosphoranes and cyclopolyphosphines has shown that the rates of disproportionation are in the order R = Me > Ph > C_6F_5 > CF_3.

$$\underset{(28)}{RPF_2} \longrightarrow RPF_4 + (RP)_n$$

2 Halogenophosphoranes

A. Dihalogenophosphoranes.—(*i*) *Structure.* Examination of the ^{31}P n.m.r. spectra of the chlorine adducts of the phosphines Ph_nPCl_{3-n} (n = 3, 2, or 1) showed[27] that in this solvent they exist as pentaco-ordinate species but that an excess of chlorine promotes dissociation to the ion pairs $R_4\overset{+}{P}\,\overset{-}{Cl_3}$. This ionisation is less ready as the number of chlorines attached to the phosphorus increases. Hydrogen chloride also promotes the dissociation of dichlorotriphenylphosphorane to give $Ph_3\overset{+}{P}Cl\,\overset{-}{HCl_2}$.

[24] K. L. Freeman and M. J. Gallagher, *Austral. J. Chem.*, 1968, **21**, 2297; *Tetrahedron Letters*, 1966, 121.
[25] F. Ramirez, H. Yamanaka, and O. H. Basedow, *J. Amer. Chem. Soc.*, 1961, **83**, 173.
[26] H. G. Ang and R. Schmutzler, *J. Chem. Soc. (A)*, 1969, 702.
[27] D. B. Denney, D. Z. Denney, and B. C. Chang, *J. Amer. Chem. Soc.*, 1968, **90**, 6332.

The tetra(dihalogenophosphoranes) (29) are covalent in methylene chloride but in strongly polar solvents such as dimethylformamide and acetonitrile the chlorine adduct forms a 1 : 2, the bromine adduct a 1 : 3, and the iodine adduct a 1 : 4 electrolyte.[28] These were formulated as the spirocyclic (30), bicyclic (31), and bridge-free (32) structures.

$C[CH_2PX_2 \cdot Ph_2]_4$
(29)

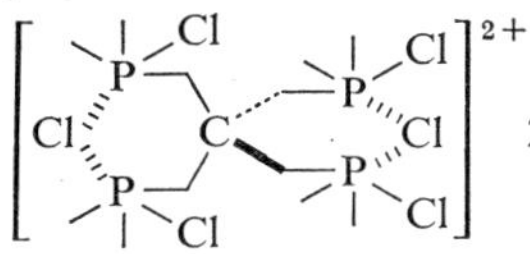
$2Cl^-$
(30)

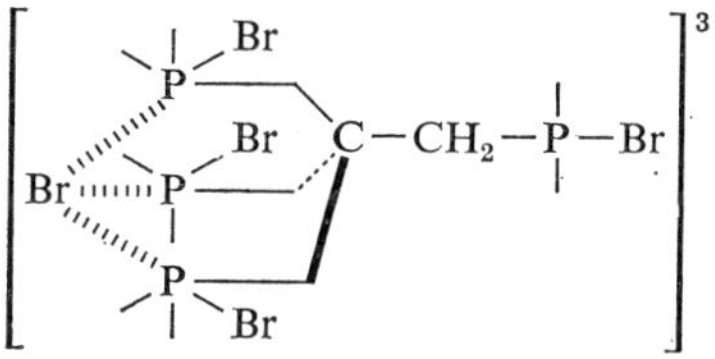
$3Br^-$
(31)

$C[CH_2\overset{+}{P}I \cdot Ph_2]_4\ 4\bar{I}$
(32)

Spectroscopic evidence has shown [29] that dichlorotriphenylphosphorane in chloroform or bromoform solutions forms hydrogen-bonded nonionic dimeric species (33) with chlorine bridges between the phosphorus atoms. The solvates are crystalline and stable under anhydrous conditions.

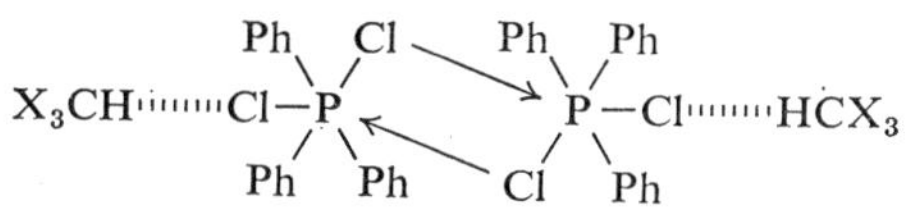
(33)

(*ii*) *Reactions.* Dichlorophosphoranes have been reduced [30] to phosphines on heating with yellow phosphorus at 150—180°. In contrast to the behaviour of the arsenic analogue, dibromotriphenylphosphorane with silver nitrate, N_2O_4, or N_2O_5 gave only triphenylphosphine oxide with no evidence of a nitrato-phosphorus compound.[31]

The dehydration of amides to nitriles using dibromotriphenylphosphorane (34) in the presence of triethylamine [32] has been extended [33] to the synthesis of aminonitriles, *e.g.* (35). With the same reagent *N*-alkylformamides gave isocyanides, *NN'*-dialkylureas gave carbodi-imides, *e.g.* (36), and the substituted amides (37) gave ketenimines.

[28] J. Ellermann and D. Schirmacher, *Chem. Ber.*, 1969, **102**, 289.
[29] G. G. Arzoumanidis, *Chem. Comm.*, 1969, 217.
[30] G.P. 1,247,310.
[31] G. C. Tranter, C. C. Addison, and D. B. Sowerby, *J. Organometallic Chem.*, 1968, **12**, 369.
[32] L. Horner, H. Oediger, and H. Hoffmann, *Annalen*, 1959, **626**, 26.
[33] H. J. Bestmann, J. Lienert, and L. Mott, *Annalen*, 1968, **718**, 24.

$$\underset{(34)}{Me_2N\cdot CO\cdot NH_2 + Ph_3PBr_2} \xrightarrow{Et_3N} \underset{(35)\ 67\%}{Me_2N\cdot CN + Ph_3PO}$$

$$R\cdot NH\cdot CHO + (34) \xrightarrow{Et_3N} R\cdot NC$$

$$Ph\cdot NH\cdot CO\cdot NH\cdot Bu + (34) \xrightarrow{Et_3N} \underset{(36)\ 72\%}{Ph\cdot N{:}C{:}N\cdot Bu}$$

$$\underset{(37)}{R^1R^2CH\cdot CO\cdot NH\cdot R^3} + (34) \xrightarrow[CH_2Cl_2]{Et_3N} R^1R^2C{:}C{:}N\cdot R^3$$

Vilsmeier formylation has been carried out[33] with (34) and dimethylformamide presumably *via* the intermediate (38). Indole gave the 3-aldehyde (78%) and phenylacetylene gave β-bromocinnamaldehyde (60%).

$$Ph_3PBr_2 + Me_2N\cdot CHO \longrightarrow \underset{(38)}{Ph_3\overset{+}{P}Br\cdot O\cdot \overset{+}{C}H\cdot NMe_2\ \bar{B}r}$$

$$(38) + R^1R^2C{:}CHR^3 \xrightarrow{H_2O} R^1R^2C{:}CR^3\cdot CHO + Ph_3PO + Me_2NH$$

o-t-Butylphenols on heating with (34) gave t-butyl bromide and steam distillation of the residues gave triphenylphosphine oxide and the dealkylated phenols.[34] Thus *o*-t-butylphenol was obtained from 2,6-di-t-butylphenol at 200°, 4-methyl-2,6-di-t-butylphenol gave 4-methyl-2-t-butylphenol, and *o*-t-butylphenol at 240° gave phenol. The suggested intermediate in the last case (39) could not be isolated. It is, however, a known compound[35] and is thermally stable up to 350° (see Chapter 1, Part II, section 2C).

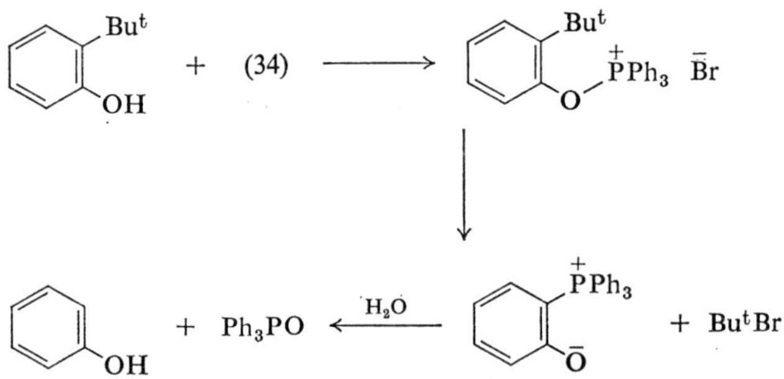

(39)

A series of perfluoroalkylfluorophosphoranes has been prepared.[36] ^{19}F N.m.r. studies on the trifluorophosphoranes R_2PF_3 ($R = CF_3$, C_2F_5) have shown that the fluorines attached to phosphorus are in apparently equivalent environments and it was suggested that the perfluoroalkyl groups are sufficiently electronegative to prefer apical positions.

[34] D. G. Lee, *Chem. Comm.*, 1968, 1554.
[35] H. J. Bestmann and G. Hofmann, *Annalen*, 1968, **716**, 98.
[36] J. F. Nixon, *J. Inorg. Nuclear Chem.*, 1969, **31**, 1615.

B. Hydridofluorophosphoranes.—Interest in hydridofluorophosphoranes continues. Difluorohydridodimethylphosphorane was obtained by the action of hydrogen fluoride on both fluorodimethylphosphine and tetramethyldiphosphine.[37] The same reagent with alkyl and aryldichlorophosphines at −20° gave the hydridophosphoranes (40) which on chlorination followed by treatment with primary or secondary amines gave the aminotrifluorophosphoranes (41).[38] These with the same amines at 40—60° gave the diaminodifluorophosphoranes (42).[39]

$$\mathrm{RPCl_2} \xrightarrow{\mathrm{HF}} \underset{(40)}{\mathrm{RPHF_3}} \xrightarrow[-10^\circ]{\mathrm{Cl_2}} [\ ] \xrightarrow{\mathrm{R^1R^2NH}} \underset{(41)}{\mathrm{RPF_3(NR^1R^2)}}$$

$$\xrightarrow[40-60^\circ]{\mathrm{R^1R^2NH}} \underset{(42)}{\mathrm{RPF_2(NR^1R^2)_2}}$$

Aminodifluorohydridophosphoranes (43) have been obtained from primary and secondary amines and alkyldifluorophosphines as stable distillable liquids.[40] Ethylenediamine gave the diphosphoranes (44) with interesting possibilities of isomerism.[41] In all cases the apical positions were occupied by the two fluorine atoms.

$$\underset{(43)}{\mathrm{(R^1{}_2N)R^2PHF_2}}$$

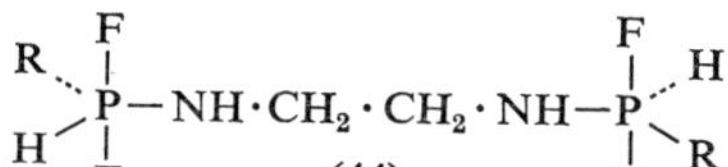

(44)

C. Hydridofluorophosphate Anions.—Potassium difluoride added to fluorobis(trifluoromethyl)phosphine and to difluoro(trifluoromethyl)phosphine either at 60–100° without solvent or in acetonitrile at room temperature to form hexaco-ordinate phosphorus salts, *e.g.* (45), containing phosphorus–hydrogen bonds.[42] The ^{19}F n.m.r. spectrum of (45) favours the structure (46).

$$\mathrm{(CF_3)_2PF + H\bar{F}_2} \longrightarrow \underset{(45)}{\mathrm{(CF_3)_2\bar{P}F_3H}}$$

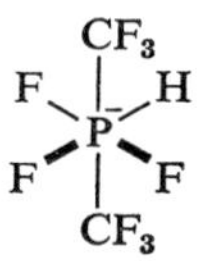

(46)

[37] F. Seel and K. Rudolph, *Z. anorg. Chem.*, 1968, **363**, 233.
[38] G. I. Drozd, S. Z. Ivin, V. V. Sheluchenko, B. I. Tetel'baum, and A. D. Varshavskii, *Zhur. obshchei Khim.*, 1968, **38**, 567.
[39] G. I. Drozd, S. Z. Ivin, M. A. Landau, and V. V. Sheluchenko, *Zhur. obshchei Khim.*, 1968, **38**, 1654.
[40] G. I. Drozd, S. Z. Ivin, V. V. Sheluchenko, B. I. Tetel'baum, G. M. Luganskii, and A. D. Varshavskii, *Zhur. obshchei Khim.*, 1967, **37**, 1631.
[41] G. I. Drozd, S. Z. Ivin, and V. V. Sheluchenko, *Zhur. obshchei Khim.*, 1968, **38**, 1655.
[42] J. F. Nixon and J. R. Swain, *Chem. Comm.*, 1968, 997.

3 Phosphines Containing a P—X (X = Si, Ge, Sn, Pb) Bond

A. Phosphorus–Silicon.—Tris(trimethylsilyl)phosphine treated with stoicheiometric amounts of water, methanol, or deuterium oxide in tetrahydrofuran is a convenient source of trimethylsilylphosphine and bis(trimethylsilyl)phosphine and their deuterium analogues.[43] The ^{1}H n.m.r. and i.r. spectra of these silylphosphines suggested pyramidal structures and did not support p_π–d_π interaction in the phosphorus–silicon bonds.

$$(Me_3Si)_3P \xrightarrow{0\cdot5\text{ mole }H_2O} (Me_3Si)_2PH$$

$$(Me_3Si)_3P \xrightarrow[H_2O]{1\text{ mole}} (Me_3Si)PH_2$$

The polyphosphines and phosphino-arsines (47) and (48), all extremely sensitive to oxygen, have been obtained from trimethylsilylphosphines and arsines as shown.[44]

$$(Me_3Si)_3M^1 + 3Ph_2M^2Cl \xrightarrow[\text{Room temp.}]{\text{Ether}} \underset{(47)}{(Ph_2M^2)_3M^1}$$

$$(Me_3Si)_2M^1Ph + 2Ph_2M^2Cl \longrightarrow \underset{(48)}{(Ph_2M^2)_2M^1Ph}$$

$$M^1, M^2 = P, As$$

Bis(trifluoromethyl)trimethylsilylphosphine (49) was obtained as indicated.[45] With hydrogen bromide it gave bis(trifluoromethyl)phosphine and with methyl iodide the phosphine (50).

$$(CF_3)_2PH \text{ or } (CF_3)_2PI \xrightarrow{Hg(SiMe_3)_2} \underset{(49)}{(CF_3)_2P\cdot SiMe_3}$$

$$(CF_3)_2PH + (Me_3Si)_2PH \longrightarrow (CF_3)_2P\cdot SiMe_3$$

$$(CF_3)_2P\cdot SiMe_3 \xrightarrow{HBr} Me_3SiBr + (CF_3)_2PH$$

$$(CF_3)_2P\cdot SiMe_3 \xrightarrow{MeI} Me_3SiI + \underset{(50)}{(CF_3)_2PMe}$$

The insertion reactions of diphenyl(trimethylsilyl)phosphine have been explored.[46] While definite structural assignments were not possible in all

[43] H. Bürger and U. Goetze, *J. Organometallic Chem.*, 1968, **12**, 451.
[44] H. Schumann, A. Roth, and O. Stelzer, *Angew. Chem. Internat. Edn.*, 1968, **7**, 218.
[45] J. Grobe, *Z. Naturforsch*, 1968, **23b**, 1609.
[46] E. W. Abel and I. H. Sabherwal, *J. Chem. Soc.* (*A*), 1968, 1105.

cases, the available evidence indicated that in the products from CO_2, CS_2, phenyl isothiocyanate, and hexafluoroacetone the phosphorus was attached to carbon. A minor product in the last case was the oxide (51). Phenyl isocyanate gave only its dimer whereas with keten, insertion occurred in the carbon–carbon double bond to give the phosphine (52). The product from sulphur dioxide decomposed on attempted distillation and gave a crystalline product formulated as the sulphoxide (53). Other reported reactions of this silylphosphine include those with silver halides and chlorodiphenylphosphine.[47]

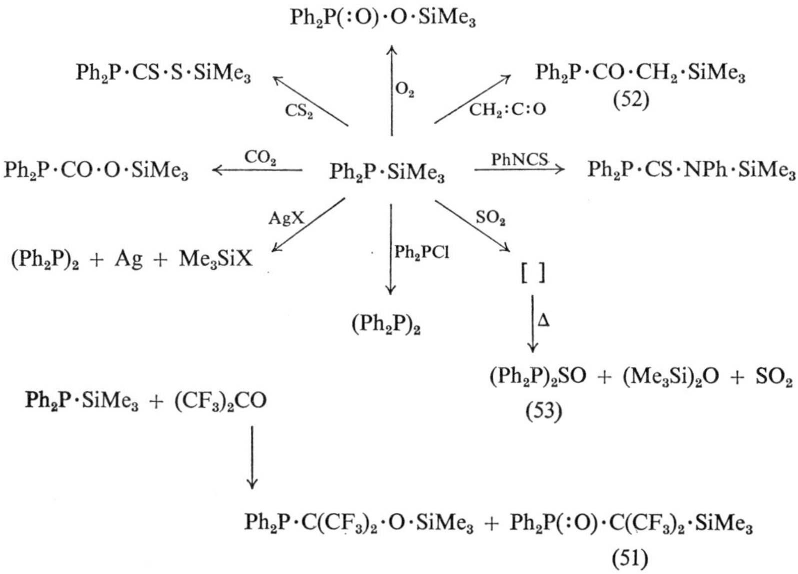

B. Phosphorus–Germanium.—Diethyl(triethylgermyl)phosphine (54) with keten[48] and diphenylketen[49] gave the phosphines (55) formed by insertion into the carbonyl groups of the keten (contrast the behaviour of the silylphosphine above). Hydrolysis of (55; R = H) gave acetyldiethyl-

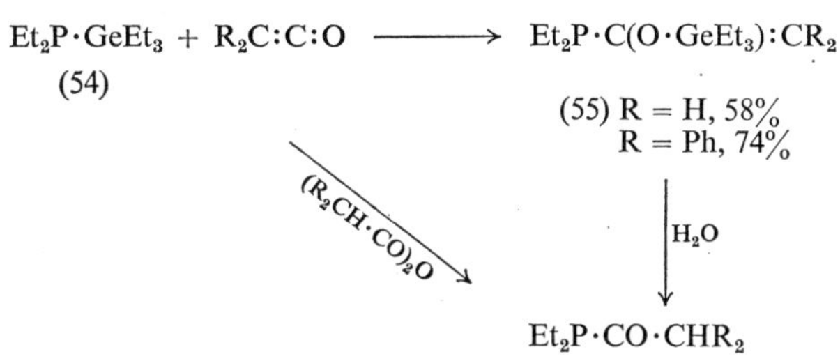

47 E. W. Abel, R. A. N. McLean, and I. H. Sabherwal, *J. Chem. Soc. (A)*, 1968, 2371.
48 J. Satgé and C. Couret, *Compt. rend.*, 1968, **267**, *C*, 173.
49 J. Satgé and C. Couret, *Bull. Soc. chim. France*, 1969, 333.

phosphine (79%) whereas hydrolysis of (55; R = Ph) led to a mixture of acylphosphine (80%) and diethylphosphine (20%). The acylphosphines were also obtained from (54) on treatment with the corresponding anhydrides.[49] Similar additions of (54) to the carbonyls of aldehydes produced the phosphines (55) while with $\alpha\beta$-unsaturated aldehydes 1,4-addition gave the phosphines (56) from which the aldehydes (57) were obtained on hydrolysis.[48]

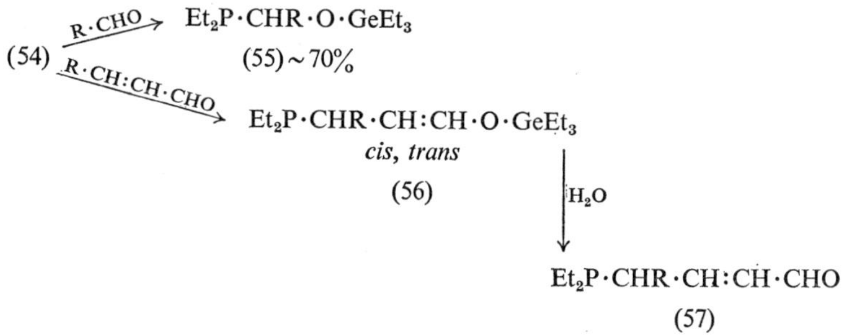

C. Phosphorus–Tin and Phosphorus–Lead.—The six-membered heterocyclic compounds (58; R = Me, Bu, Ph) were obtained [50] from phenylphosphine and the tin dichlorides (59). Pyrolysis of (58; R = Me) gave phenylbis(trimethylstannyl)phosphine, tin, and the cyclophosphine (60).

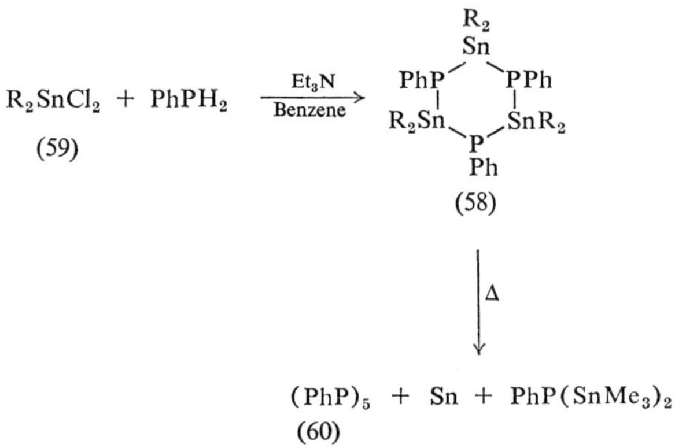

The insertion reactions of the (triphenylstannyl)phosphine (61) into carbonyl and thiocarbonyl groups have been investigated.[51] They were very similar to those of phosphorus–germanium compounds, the phosphorus always adding to carbon.

Migration of the organometallic groups from phosphorus to nitrogen occurred [52] when (61) and the lead analogue were treated with phenyl azide,

[50] H. Schumann and H. Benda, *Angew. Chem. Internat. Edn.*, 1968, **7**, 812.
[51] H. Schumann and P. Jutzl, *Chem. Ber.*, 1968, **101**, 24.
[52] H. Schumann and A. Roth, *J. Organometallic Chem.*, 1968, **11**, 125.

the product from (61) being the phosphine-imine (62). Successive migrations occurred starting with the bis(triphenylstannyl)- or bis(triphenylplumbyl)-phosphines (63), but with phenylbis(trimethylstannyl)phosphine the only organometallic compound isolated was phenylbis(trimethylstannyl)amine. No reaction occurred between triphenylsilylazide or triphenylstannylazide and organotin or organolead phosphines in refluxing benzene or between phenyl azide and tris(triphenylstannyl)phosphine or tris(triphenylplumbyl)phosphine at 100°.

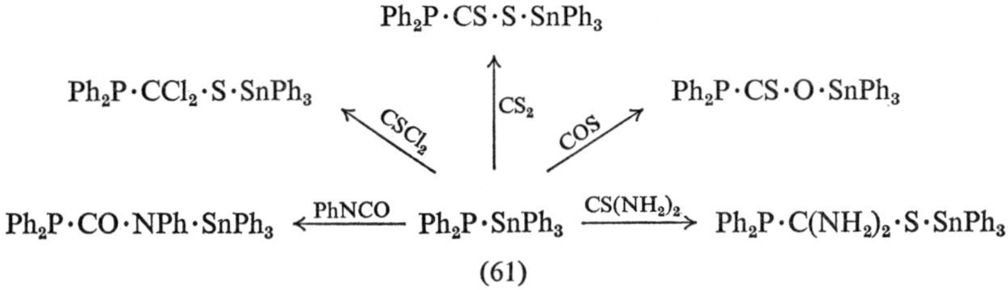

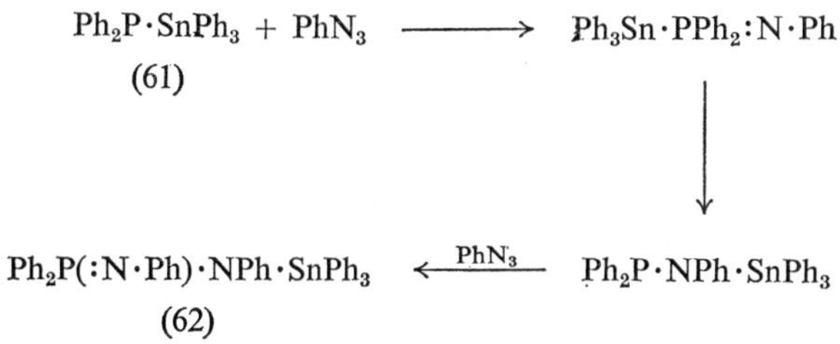

$$\underset{(63)}{PhP(MPh_3)_2} + 3\,PhN_3 \dashrightarrow Ph_3M{\cdot}NPh{\cdot}PPh(:NPh){\cdot}NPh{\cdot}MPh_3$$

4

Phosphine Oxides

BY S. TRIPPETT

1 Preparation

A. Using Organometallic Reagents.—Diethyl phosphite with alkyl magnesium halides[1] gave initially the tervalent magnesium salts (1) which reacted with further reagent to form the phosphinous acid salts (2). These with aqueous potassium carbonate gave the corresponding secondary phosphine oxides. This method had previously failed with lower alkyl Grignard reagents. The salts (2) are useful intermediates for the preparation of phosphine oxides of various types.

$$(EtO)_2PH({:}O) + R{\cdot}MgX \longrightarrow \underset{(1)}{(EtO)_2P{\cdot}OMgX} \rightarrow \underset{(2)}{R_2P{\cdot}OMgX}$$

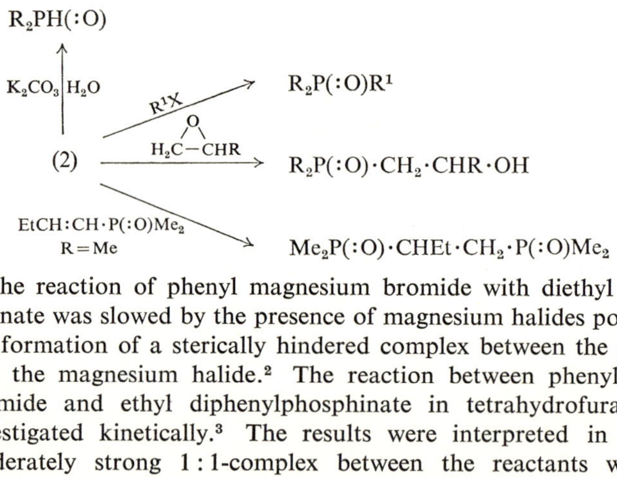

The reaction of phenyl magnesium bromide with diethyl phenylphosphonate was slowed by the presence of magnesium halides possibly due to the formation of a sterically hindered complex between the phosphonate and the magnesium halide.[2] The reaction between phenyl magnesium bromide and ethyl diphenylphosphinate in tetrahydrofuran has been investigated kinetically.[3] The results were interpreted in terms of a moderately strong 1 : 1-complex between the reactants which slowly rearranged to give a complex of triphenylphosphine oxide with magnesium salt.

A careful reinvestigation[4] of the formation of diphosphine disulphides from thiophosphoryl halides and phosphonothioic dihalides with Grignard

[1] H. R. Hays, *J. Org. Chem.*, 1968, **33**, 3690.
[2] H. R. Hays, *J. Org. Chem.*, 1968, **33**, 4201.
[3] H. R. Hays, *J. Amer. Chem. Soc.*, 1969, **91**, 2736.
[4] P. C. Crofts and I. S. Fox, *J. Chem. Soc.* (*B*), 1968, 1416.

reagents has supported a mechanism in which a complex (3) between the halide and the Grignard reagent leads to a phosphinidine sulphide (4). This then gives the diphosphine disulphide either *via* insertion into a phosphorus halogen bond or by reaction with the Grignard reagent to give an intermediate phosphinothioyl magnesium halide (5).

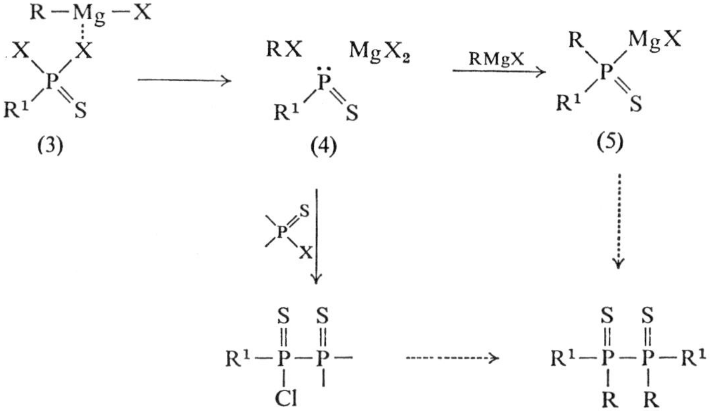

(3) (4) (5)

Of several buta-1,3-dienylphosphonic dichlorides only (6) gave an acceptable yield of phosphine oxide with Grignard reagents.[5] Other buta-1,3-dienylphosphine oxides were therefore prepared *via* the corresponding 4-chlorobut-2-enylphosphine oxides (7).

$$\underset{(6)}{\mathrm{Me\cdot(CH{:}CH)_2\cdot P({:}O)Cl_2}} \xrightarrow{\mathrm{RMgX}} \mathrm{Me\cdot(CH{:}CH)_2P({:}O)R_2}$$

R = Et (57%)

$$\mathrm{ClCH_2\cdot CR{:}CH\cdot CH_2\cdot P({:}O)Cl_2} \xrightarrow{\mathrm{RMgX}} \underset{(7)}{\mathrm{ClCH_2\cdot CR{:}CH\cdot CH_2\cdot P({:}O)R_2}}$$

R = H (22%) | EtOH—KOH ↓

$$\mathrm{CH_2{:}CR\cdot CH{:}CH\cdot P({:}O)R_2}$$

Tertiary di(chloromethyl)phosphine oxides have been prepared from alkyl and aryl magnesium halides and di(chloromethyl)phosphinyl chloride.[6] Dialkylphosphinic chlorides with ethynyl magnesium bromide gave the ethynylphosphine oxides (8),[7]

$$\mathrm{R_2P({:}O)Cl + HC{:}C\cdot MgBr} \xrightarrow[-25^\circ]{\mathrm{THF}} \underset{(8)}{\mathrm{R_2P({:}O)C{:}CH}}$$

[5] L. N. Mashlyakovskii, B. I. Ionin, I. S. Okhrimenko, and A. A. Petrov, *Zhur. obshchei Khim.*, 1968, **38**, 2124.
[6] L. Maier, *Helv. Chim. Acta*, 1969, **52**, 845.
[7] W. Hagens, H. J. T. Bos, W. Voskuil, and J. F. Arens, *Rec. Trav. chim.*, 1969, **88**, 71.

Full details have appeared[8] of the synthesis of optically active tertiary phosphine oxides by the action of Grignard reagents on diastereoisomeric menthyl phosphinates. The menthyl esters are in general highly crystalline and readily separated into diastereoisomers. Reaction with the Grignard reagent proceeds with inversion of configuration at the phosphorus.

B. From Olefins and Dienes.—Full details have appeared[9] of the addition of dihalophosphines to 1,3-dienes to give, after treatment with water, phospholene oxides. Methyldichlorophosphine and phenyldibromophosphine with isoprene or penta-1,3-diene gave the 3-phospholene oxides (9) while phenyldichlorophosphine with the same dienes gave the 2-phospholene

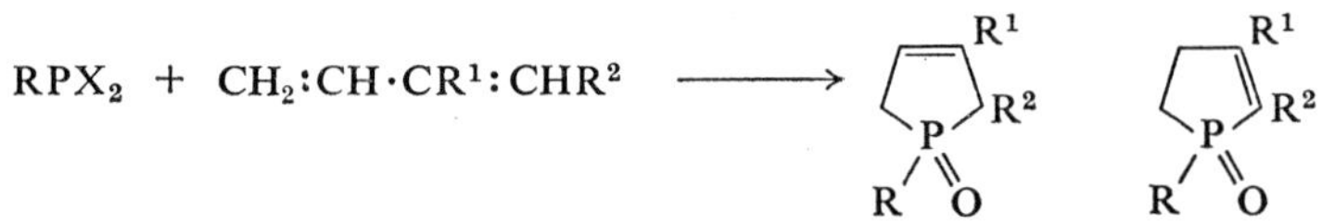

(9) (10)

oxides (10). 2-Chlorobuta-1,3-diene with methyldichlorophosphine gave a 9 : 1 mixture of the 3- and 2-phospholene oxides[10] from which the 3-methoxy-phospholene oxide (11) was obtained on treatment with sodium methoxide. Acidic hydrolysis of this gave the ketone (12) which is in tautomeric equilibrium with the enol form (13). Crystalline material is largely the enol form, which is probably a strongly hydrogen-bonded dimer, while dilute solutions in chloroform contain predominantly the ketone.

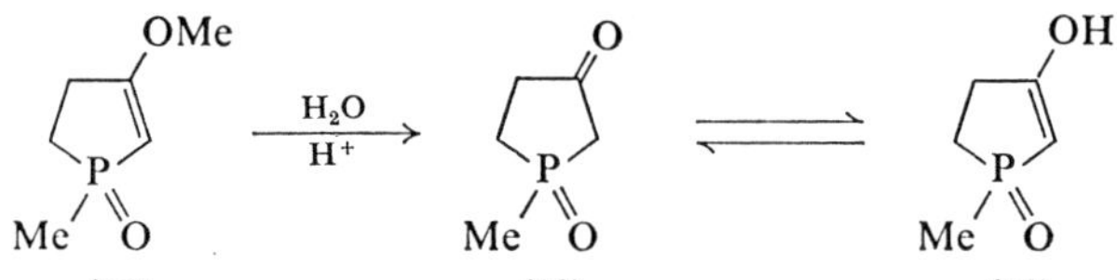

(11) (12) (13)

The synthesis of a phosphetan by the action of phosphorus trichloride and aluminium chloride on 2,4,4-trimethylpent-2-ene[11] has been extended[12] to include the use of alkyl and aryl dichlorophosphines and of various methyl substituted t-butylethylenes. Some of the resulting oxides (14; $R^1 = H$) exist as geometrical isomers; the ratio of isomers produced in a given case depended on the mode of decomposition of the intermediate. Thus when R^1, R^2, R^3 = H and R^4 = Ph slow addition of water to the

[8] O. Korpiun, R. A. Lewis, J. Chickos, and K. Mislow, *J. Amer. Chem. Soc.*, 1968, **90**, 4842.
[9] L. D. Quin, J. P. Gratz, and T. P. Barket, *J. Org. Chem.*, 1968, **33**, 1034.
[10] L. D. Quin and J. A. Caputo, *Chem. Comm.*, 1968, 1463.
[11] J. J. McBride, E. Jungermann, J. V. Killhefer, and R. J. Clutter, *J. Org. Chem.*, 1962, **27**, 1833.
[12] S. E. Cremer and R. J. Chorvat, *J. Org. Chem.*, 1967, **32**, 4066.

reaction mixture gave a 19 : 1 ratio of isomeric oxides while slow addition of the reaction mixture to an excess of water gave a 7 : 3 ratio of oxides.

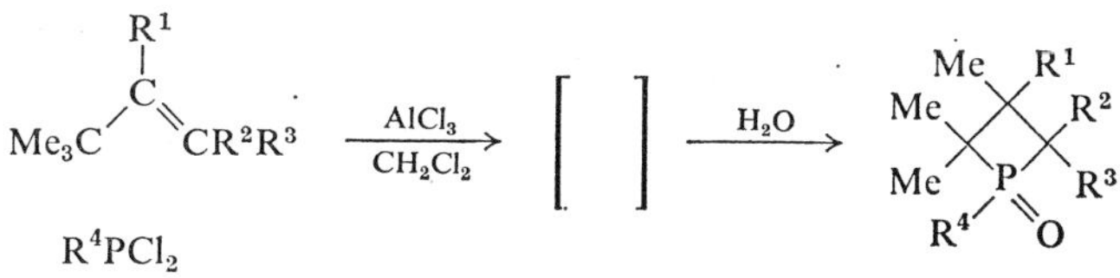

(14)

Optimum conditions have been described[13] for the preparation of phosphine oxides, *e.g.* (15) and (16), by the addition of phosphorus pentachloride or trichlorodiphenylphosphorane to styrene and phenylacetylene.

$$\mathrm{PhCH{:}CH_2 + Ph_2PCl_3 \xrightarrow[\text{NaI}]{110^\circ} Ph_2P({:}O){\cdot}CH{:}CHPh}$$

(15) 22%

$$\mathrm{PhC{\vdots}CH + PCl_5 \longrightarrow (PhCCl{:}CH)_3P({:}O)}$$

(16) 25%

C. From Alkyl Phosphinites.—Arbusov reactions using alkyl phosphinites and chloromethylphosphine oxides have been used[14–16] to prepare a wide range of di-, tri-, and tetra-phosphine oxides, *e.g.* (17) and (18).

$$\mathrm{(ClCH_2)_3P({:}O) + Ph_2POEt \longrightarrow [Ph_2P({:}O){\cdot}CH_2]_3P({:}O)}$$

(17) 68%

$$\mathrm{ClCH_2{\cdot}P({:}O)Ph_2 + Ph_2POEt \longrightarrow [Ph_2P({:}O)]_2CH_2}$$

(18) 58%

D. Miscellaneous.—The rearrangement of alk-2-enyl diphenylphosphinites *via* a five-membered cyclic transition state[17] has been shown[18] by deuterium labelling to be >95% specific and the activation parameters for the rearrangement of (19) to (20) have been determined.

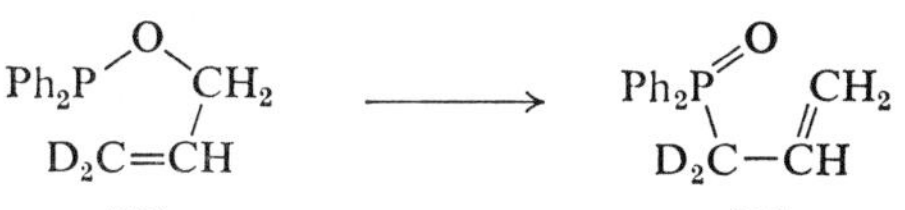

(19) (20)

ΔH^* 23 kcal./mole
ΔS^* − 14 e.u.

[13] G. K. Fedorova, Ya. P. Shaturskii, L. S. Moskalevskaya, Yu. S. Grushin, and A. V. Kirsanov, *Zhur. obshchei Khim.*, 1967, **37**, 2686.

[14] M. I. Kabachnik, T. Y. Medved, Y. M. Polikarpov, and K. S. Yudina, *Izvest. Akad. Nauk S.S.S.R., Ser. khim.*, 1967, 591.

[15] L. Maier, *Angew. Chem. Internat. Edn.*, 1968, **7**, 384, 385.

[16] T. Y. Medved, Y. M. Polikarpov, S. A. Pisareva, E. I. Matrosov, and M. I. Kabachnik, *Izvest. Akad. Nauk S.S.S.R., Ser. khim.*, 1968, 2062.

[17] M. P. Savage and S. Trippett, *J. Chem. Soc.* (*C*), 1966, 1842.

[18] A. W. Herriott and K. Mislow, *Tetrahedron Letters*, 1968, 3013.

Tris(chloromethyl)phosphine on oxidation with hydrogen peroxide[19] gave the methylphosphine oxide (21) while treatment with methanolic sodium methoxide[20] gave the methylphosphine oxide (22). Both reactions are thought to involve intermediate ylides (23).

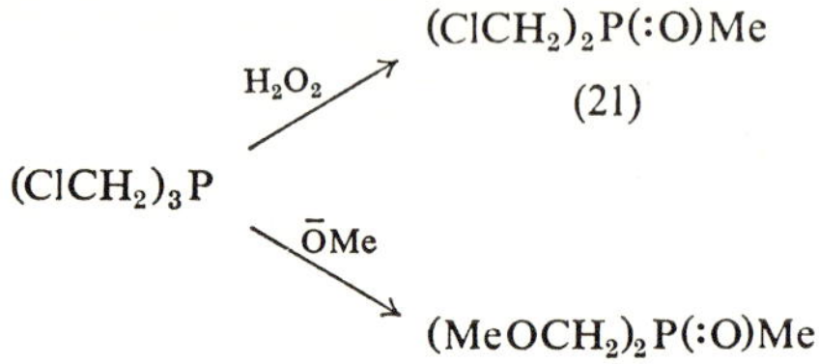

(21)

(22)

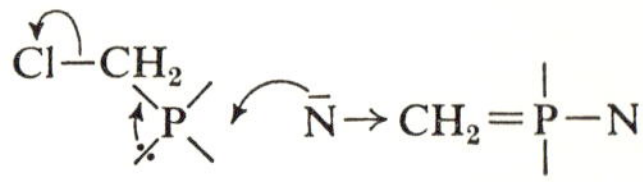

(23)

Oxidation of 1-diphenylphosphinoanthracene or its oxide[21] with sodium chlorate and vanadium pentoxide in glacial acetic acid gave the anthraquinone (24). Oxidation of acetyldiphenylphosphine with oxygen has now

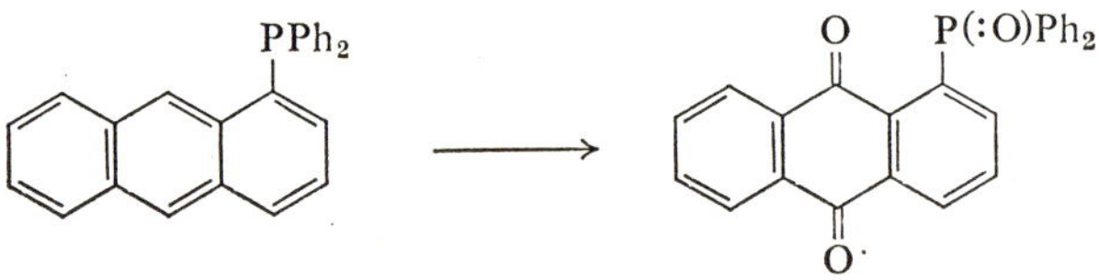

(24) 70%

been reported[22] to give the α-hydroxydiphosphine dioxide (25) formed by the addition of diphenylphosphine oxide to the very reactive acetyldiphenylphosphine oxide.

$$CH_3\cdot CO\cdot PPh_2 \xrightarrow{O_2} CH_3\cdot CO\cdot P(:O)Ph_2 \xrightarrow{Ph_2PHO} CH_3\cdot C(OH)[P(:O)Ph_2]_2$$

(25)

Similar acylphosphine oxides were intermediates in the reactions of tetraphenyldiphosphine with carboxylic acids.[22] The resulting α-hydroxydiphosphine dioxides (26) rearranged rapidly (R = aryl) or slowly (R = alkyl) to the corresponding phosphinyloxyalkylphosphine oxides (27).

The alkylations of red phosphorus and of P_2I_4 by alkyl iodides at 250° were unaffected by the presence of dibenzoyl peroxide or on irradiation with u.v. light.[23] Treatment of the reaction mixtures with water gave high yields of tertiary phosphine oxides.

[19] L. Maier, *Helv. Chim. Acta*, 1969, **52**, 858.
[20] I. O. Shapiro, E. N. Tsvetkov, A. I. Shatenshtein, and M. I. Kabachnik, *Doklady Akad. Nauk S.S.S.R.*, 1968, **179**, 888.
[21] H. Schindlbauer, *Monatsh.*, 1969, **100**, 567.
[22] R. S. Davidson, R. A. Sheldon, and S. Trippett, *J. Chem. Soc.* (*C*), 1968, 1700.
[23] N. G. Feshenko, T. I. Alekseeva, and A. V. Kirsanov, *Zhur. obshchei Khim.*, 1968, **38**, 122.

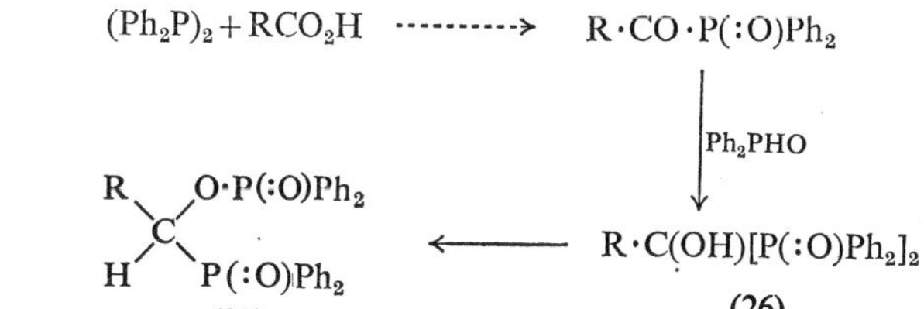

(27) (26)

Secondary phosphine oxides have been obtained[24] by the reduction of phosphinate esters with lithium aluminium hydride and the synthesis has been extended to that of the optically active secondary phosphine oxide (28) by reduction of the high-melting diastereoisomeric menthyl ester (29).

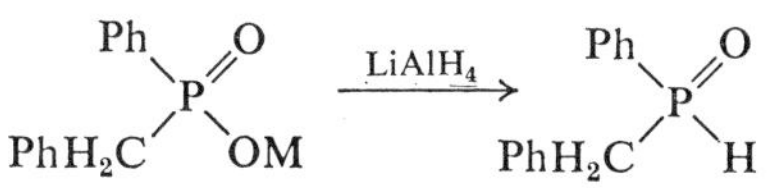

(29) (28)

$[\alpha]_D = -65{\cdot}4°$ $[\alpha]_D = -0{\cdot}52°$

Optically active (28) exchanged its proton rapidly in deuteriomethanol with no loss of activity except in the presence of acid or base. Exchange with retention of activity, and configuration, was suggested to occur *via* the deuterioxyphosphine (30), and racemisation in the presence of base to

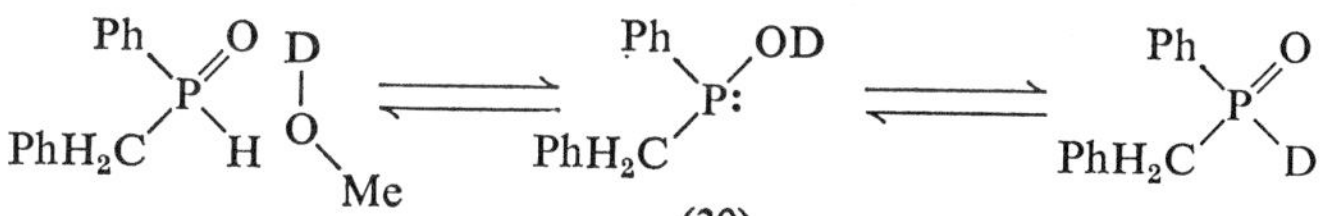

(30)

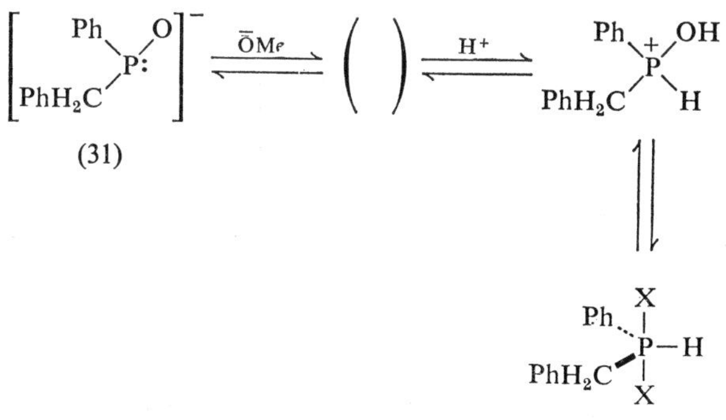

(31)

(32)

[24] T. L. Emmick and R. L. Letsinger, *J. Amer. Chem. Soc.*, 1968, **90**, 3459.

be *via* the anion (31). Acid-catalysed racemisation may have involved the symmetrical phosphorane (32).

Secondary phosphine oxides have also been obtained[24] by the reaction of Grignard reagents with monosubstituted phosphinate esters (33).

$$\underset{(33)}{R^1HP(:O)\cdot OEt} + R^2MgBr \longrightarrow \underset{25\text{—}75\%}{R^1R^2HP(:O)}$$

R^1 = Ph, Bu R^2 = aryl, aralkyl

2 Reactions

A. Metallated Phosphine Oxides.—Metallated phosphine oxides with *p*-toluenesulphonyl azide gave the α-diazophosphine oxides (34).[25, 26] The base used to generate the anion was usually butyl-lithium or potassium t-butoxide. With the α-formylphosphine oxide (35) cyclisation of the intermediate and elimination of the diphenylphosphinate anion gave the triazine (36).

$$R_2P(:O)\cdot\bar{C}H\cdot R^1 + N_3\cdot SO_2\cdot Ar \longrightarrow R_2P(:O)\cdot CHR^1\cdot N{:}N\cdot\bar{N}\cdot SO_2\cdot Ar$$

$$\downarrow$$

$$R_2P(:O)\cdot CN_2\cdot R^1 + \bar{N}H\cdot SO_2\cdot Ar \longleftarrow R_2P(:O)\cdot\bar{C}R^1\cdot N{:}N\cdot NH\cdot SO_2\cdot Ar$$

(34) 14—83%

$Ph_2P(:O)\cdot CH(CHO)\cdot Ph + N_3\cdot SO_2\cdot Ar$

(35)

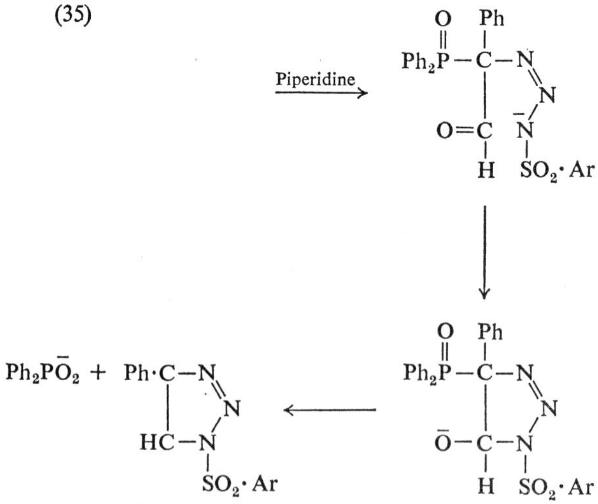

(36)

[25] G. Petzold and H. G. Henning, *Naturwiss.*, 1967, **54**, 469.

[26] M. Regitz, W. Anschutz, W. Bartz, and A. Liedhegener, *Tetrahedron Letters*, 1968, 3171.

Photochemical decomposition of the oxide (37) in aqueous dioxan[26] gave as the major product the α-hydroxyphosphine oxide (38) together with a small amount of the phosphinic acid (39) formed by a Wolff rearrangement of the intermediate carbene. The oxide (37) had previously been obtained[27] from nitrous acid and the corresponding amine.

$$\underset{(37)}{Ph_2P(:O)\cdot CN_2\cdot Ph} \xrightarrow{h\nu} Ph_2P(:O)\cdot \ddot{C}\cdot Ph \begin{cases} \xrightarrow{H_2O} \underset{(38)}{Ph_2P(:O)\cdot CH(OH)\cdot Ph} \\ \longrightarrow PhP(:O):CPh_2 \xrightarrow{H_2O} \underset{(39)}{(Ph_2CH)PhP(:O)\cdot OH} \end{cases}$$

Methyldiphenylphosphine oxide with butyl-lithium and trimethylchlorosilane gave the oxide (40)[28] which was stable at 180° in contrast to the corresponding sulphoxide. The same lithiomethyl oxide with cuprous chloride in tetrahydrofuran at −40° followed by oxygenation at 20° gave the diphosphine dioxide (41).[29]

$$Ph_2P(:O)\cdot CH_2Li \begin{cases} \xrightarrow{Me_3SiCl} \underset{(40)}{Ph_2P(:O)\cdot CH_2\cdot SiMe_3} \\ \xrightarrow[O_2]{Cu_2Cl_2} \underset{(41)\ 43\%}{Ph_2P(:O)\cdot CH_2\cdot CH_2\cdot P(:O)Ph_2} \end{cases}$$

The product previously obtained from triphenylphosphine oxide and phenyl-lithium was not a tetraphenylphosphonium salt but probably the adduct $(Ph_3P:O)_4\cdot LiI$.[30] Phenyl-lithium metallated the oxide at the meta position(s); treatment of the reaction mixture with benzophenone gave the oxides (42; 56%) and (43; 19%). Ligand exchange occurs prior to metallation. Thus biphenylyl-α-naphthylphenylphosphine oxide gave naphthalene (27%) and biphenyl (2%) on treatment with *p*-tolyl-lithium in ether at room temperature. The intramolecular version of the aryl-lithium plus phosphine oxide reaction has given phosphonium salt,[31] the oxide (44) when treated successively with butyl-lithium and acid giving 24% of the spirocyclic salt (45).

[27] L. Horner, H. Hoffmann, H. Ertel, and G. Klahre, *Tetrahedron Letters*, 1961, 9.
[28] A. G. Brook and D. G. Anderson, *Canad. J. Chem.*, 1968, **46**, 2115.
[29] T. Kauffmann, G. Beissner, H. Berg, E. Köppelmann, J. Legler, and M. Schonfelder, *Angew. Chem. Internat. Edn.*, 1968, **7**, 540.
[30] G. Wittig and H.-J. Cristan, *Bull. Soc. chim. France*, 1969, 1293.
[31] M. A. Weiner and G. Pasternack, *J. Org. Chem.*, 1969, **34**, 1130.

(42)

(43)

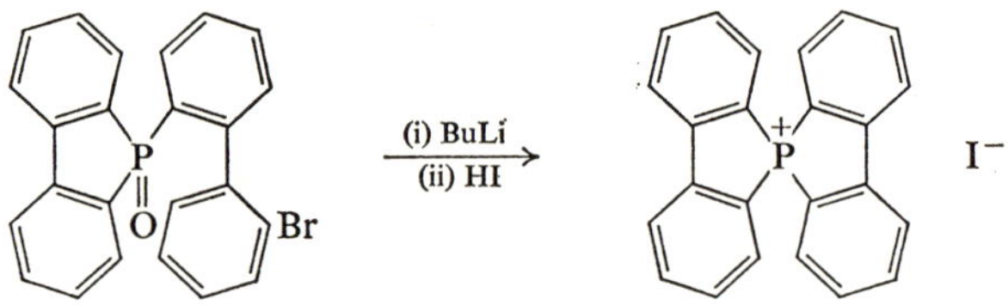

(44) (45)

Phenyl-lithium with triphenylphosphine sulphide[30] gave triphenylphosphine (56%) and lithium thiophenoxide (41%), a reaction formally analogous to that of phenyl-lithium with the methylthiophosphonium salt (46) which gave triphenylphosphine and methyl phenyl sulphide, the suggested intermediate here being (47).

$$\underset{(46)}{Ph_3\overset{+}{P}SMe\ \overset{-}{B}F_4} \xrightarrow{PhLi} Ph_3P{=}SMe\cdot Ph \longrightarrow \underset{62\cdot5\%}{Ph_3P} + \underset{55\%}{Ph\cdot S\cdot Me}$$

$$\updownarrow$$

$$\underset{(47)}{Ph_3\overset{+}{P}{-}\overset{-}{S}Me\cdot Ph}$$

B. Additions to Unsaturated Phosphine Oxides.—Addition of dibutylphosphine to the ethynylphosphine oxide (48; R = Bu) at 80° gave a low yield of the *trans*-oxide (49).[32] However, addition of HX (X = R^1_2N, R^1O, R^1S, or halogen) to the oxides (48; R = Me or Bu) led initially to the *cis*-oxides (50)[7] which isomerised on distillation to the *trans*-oxides, the ease of isomerisation being in the order X = $R^1_2N > R^1O$, $R^1S >$ halogen. In contrast ethynylphosphines are not susceptible to nucleophilic addition.[33]

$$\underset{(48)}{R_2P(:O)\cdot C{\vdots}CH} \xrightarrow{Bu_2PH} \underset{(49)\ 20\%\ trans}{Bu_2P\cdot CH{:}CH\cdot P(:O)Bu_2}$$

$$\underset{(48)}{R_2P(:O)\cdot C{\vdots}CH} \xrightarrow[\bar{X}]{HX} \underset{(50)\ cis}{R_2P(:O)\cdot CH{:}CHX}$$

[32] D. Hellwinkel, *Chem. Ber.*, 1969, **102**, 548.
[33] W. Voskuil and J. F. Arens, *Rec. Trav. chim.*, 1962, **81**, 993.

1 : 3-Dipoles with diphenylvinylphosphine oxide and sulphide gave heterocyclic oxides,[34] 2-pyrazolines (51) from nitrile-imines, 2-isoxazolines from nitrile oxides, and isoxazolidines from nitrones.

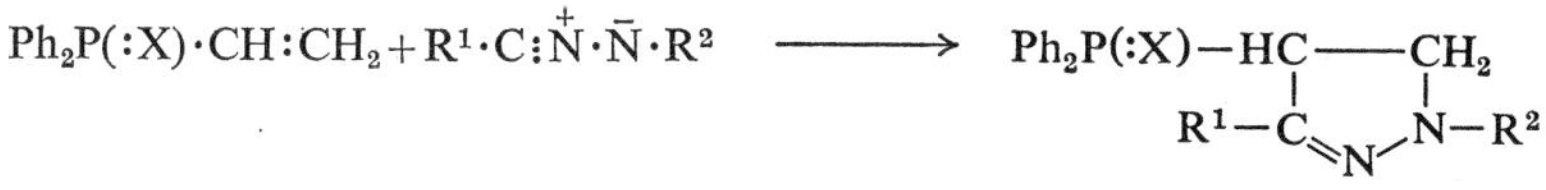

(51)

C. Miscellaneous.—Baeyer–Villiger oxidations occurred[35] when tetraphenyldiphosphine dioxide and disulphide were treated with perbenzoic acid in methylene chloride, the high migratory aptitudes of the phosphinyl and thiophosphinyl groups leading to the anhydrides (52). Triphenylphosphine sulphide with perbenzoic acid gave the phosphine oxide (70%).

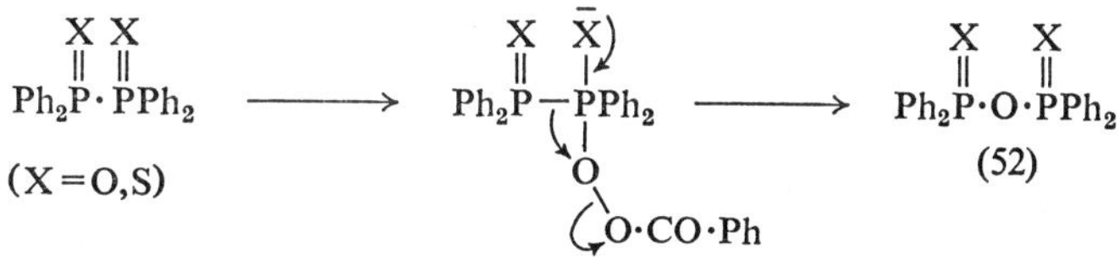

(X=O,S) (52)

Addition of the secondary phosphine oxide (53) to diethyl maleate[36] led eventually to the oxide (54). However, all attempts to resolve both this and the oxide (55) were unsuccessful.

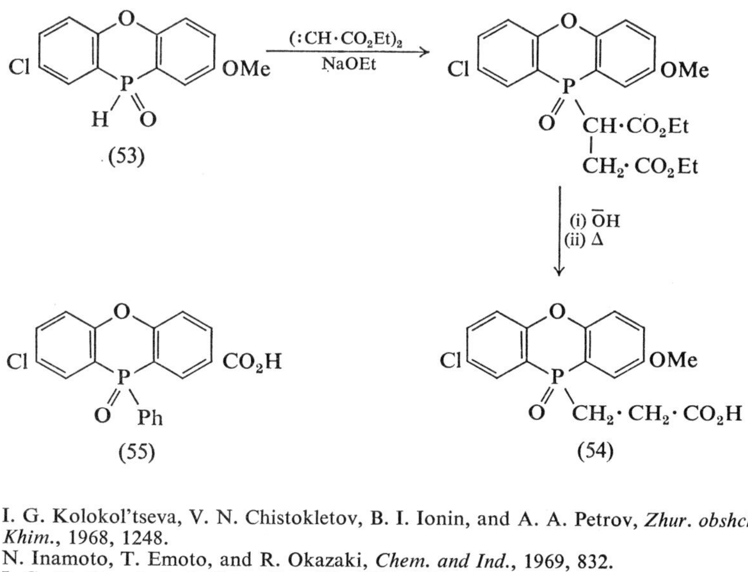

(53) (55) (54)

[34] I. G. Kolokol'tseva, V. N. Chistokletov, B. I. Ionin, and A. A. Petrov, *Zhur. obshchei Khim.*, 1968, 1248.
[35] N. Inamoto, T. Emoto, and R. Okazaki, *Chem. and Ind.*, 1969, 832.
[36] I. G. M. Campbell, *J. Chem. Soc. (C)*, 1968, 3026.

Debenzylation and decarboxylation occurred when the phosphinylformate ester (56) was treated with iodide ion.[37] The resulting diphenylphosphinyl anion has been trapped with carbonyl compounds and with iodine.

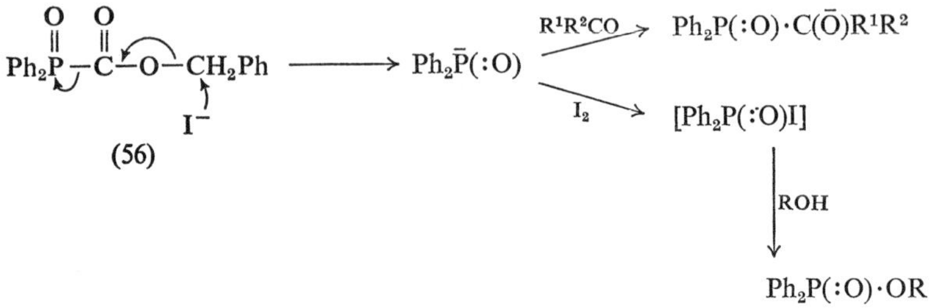

(56)

Tris(chloromethyl)phosphine oxide with phosphorus pentachloride gave as the major product a yellow crystalline compound from which the trichlorophosphorane (57) was obtained on crystallisation from chloroform.[38] This remarkable compound was stable to boiling water but

$$(ClCH_2)_3P(:O) + PCl_5 \longrightarrow [\ ] \xrightarrow{CHCl_3} (Cl_3C)_2PCl_3 \quad (57) \qquad + CCl_4$$

hydrolysis with aqueous ethanolic sodium hydroxide gave bis(trichloromethyl)phosphinic acid. Pyrolysis gave hexachloroethane and phosphorus trichloride. Similar ready fissions of phosphorus-carbon bonds occurred when chloromethylphosphonic dichloride and di(chloromethyl)phosphinic chloride were treated with phosphorus pentachloride.

The conversion of phenyl isocyanate into diphenylcarbodi-imide in the presence of the phospholene oxides (58) has been investigated kinetically.[39] The catalytic (second-order) rate constants obeyed a linear $\sigma^0\rho$ relationship ($\rho = -0{\cdot}75$), where σ^0 is the 'normal' substituent constant in agreement with the lack of conjugation between the aromatic ring and the phosphoryl group.

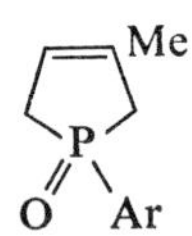

(58)

Phosphine oxides form 1 : 1 hydrogen-bonded adducts with phenols.[40, 41] With triphenylphosphine oxide the aromatic rings could be substituted by

[37] S. Warren and M. R. Williams, *Chem. Comm.*, 1969, 180.
[38] A. W. Franks, *Canad. J. Chem.*, 1968, **46**, 3573.
[39] G. Ostrogovich, F. Kerek, A. Buzás, and N. Doca, *Tetrahedron*, 1969, **25**, 1875.
[40] H. Schindlbauer and H. Stezenberger, *Monatsh.*, 1968, **99**, 2468, 2474.
[41] G. Aksnes and P. Albriktsen, *Acta Chem. Scand.*, 1968, **22**, 1866.

Me but not by Cl, NH_2, NO_2, or COOH. The product from triphenylphosphine and tetrachloro-*o*-benzoquinone was the adduct of triphenylphosphine oxide with tetrachlorocatechol.[40] The phospholan oxide (59) was particularly efficient at intermolecular hydrogen bonding with phenols and it was suggested that this may be due to the ease with which the five-membered ring can be accommodated in an adduct (60) having the geometry of a trigonal bipyramid.[41] The poor hydrogen-bonding ability of the phosphetan oxide (61) was probably due to steric hindrance.

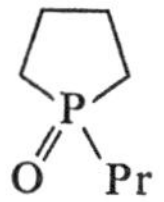

(59)

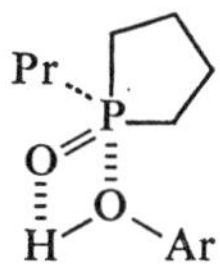

(60)

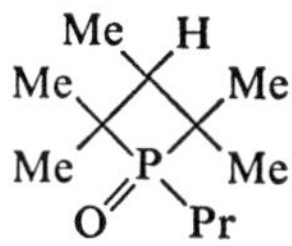

(61)

The optical rotatory dispersions and circular dichroisms of several tertiary phosphine oxides and sulphides have been recorded.[42] There are indications that compounds of absolute structure (62), where A and B are alkyl or aralkyl groups and A is larger than B, show a negative Cotton effect in the region of the *p*-band (200—225 nm) and a positive effect in the region of the α-band (260—280 nm).

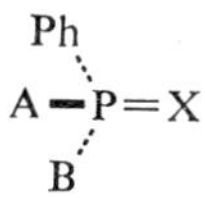

(62)

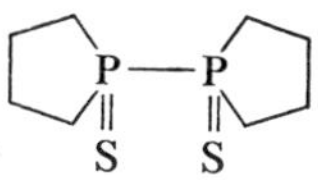

(63)

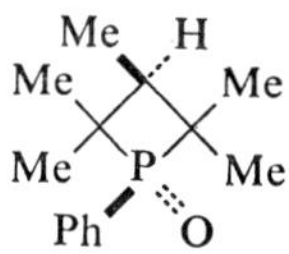

(64)

The structures of the disulphide (63)[43] and the oxide (64)[44] have been determined by *X*-ray crystallography.

[42] W.-D. Balzer, *Tetrahedron Letters*, 1968, 1189.
[43] J. D. Lee and G. W. Goodacre, *Naturwiss.*, 1968, **55**, 543.
[44] Mazhar-Ul-Haque and C. N. Caughlan, *Chem. Comm.*, 1968, 1228.

5

Tervalent Phosphorus Acids

BY D. W. HUTCHINSON

1 Introduction

The oxyacids of phosphorus have been the subject of several reviews [1] which describe the field up to the present, and as a very large number of papers on tervalent phosphorus acids appear annually, a considerable amount of selection has been exercised in the preparation of this Chapter. In general, papers which contribute to the understanding of the reactions and properties of phosphorus oxyacids have been included, while papers of a purely preparative nature (in particular many papers from the Russian literature) have been omitted owing to lack of space.

2 Phosphorous Acid and its Derivatives

A. Nucleophilic Reactions.—(*i*) *Attack on Saturated Carbon.* In a recent example of the Arbusov reaction,[2] 1-chloro-2-hydroxy-3-alkoxypropanes (1) when heated with triethyl phosphite at 120—150° give rise to the expected 2-hydroxyphosphonates (2).[3] Ester exchange can also occur under these

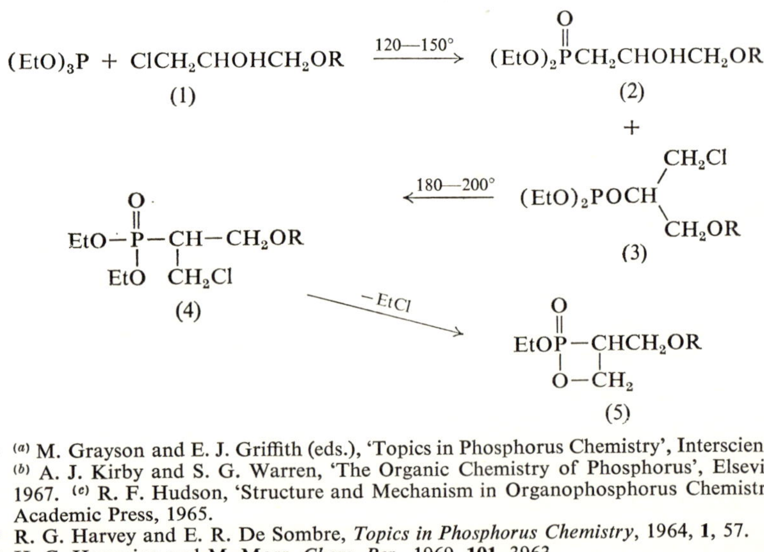

[1] (*a*) M. Grayson and E. J. Griffith (eds.), 'Topics in Phosphorus Chemistry', Interscience. (*b*) A. J. Kirby and S. G. Warren, 'The Organic Chemistry of Phosphorus', Elsevier, 1967. (*c*) R. F. Hudson, 'Structure and Mechanism in Organophosphorus Chemistry', Academic Press, 1965.

[2] R. G. Harvey and E. R. De Sombre, *Topics in Phosphorus Chemistry*, 1964, **1**, 57.

[3] H. G. Hemming and M. Morr, *Chem. Ber.*, 1969, **101**, 3963.

conditions producing the phosphite (3) which rearranges to the phosphonate (4) among other products at higher temperatures. Elimination of ethyl chloride from (4) can then take place to generate the β-phostone (5).

(*ii*) *Attack on Unsaturated Carbon.* The synthesis of aryl phosphonates can be achieved if the Arbusov reaction is initiated either by ^{60}Co–γ rays [4] or by u.v. irradiation,[5] and in both cases the reaction presumably proceeds *via* free radical intermediates. In general, however, ready displacement of halogen from an unsaturated carbon atom by tervalent phosphorus only occurs when the carbanion intermediate is stabilised by delocalisation or by d–π orbital overlap. Thus, 1,2-dichloroperfluorocyclopentene (6, R = Cl) and 1-chloro-2-ethylthioperfluorocyclopentene (6, R = SEt), both of which can give rise to stabilised carbanions, undergo substitution of chlorine by alkyl phosphites to give the phosphonates (7) and (8).[6] With (6, R = Cl) only the diphosphonate (8) is produced, and it has been suggested that (7, R = Cl) must react with the phosphite more rapidly than (6, R = Cl). One reason for this increased reactivity may be because the phosphonate group can stabilise carbanions very effectively [7] and hence it is easier to form a carbanion from (7, R = Cl) than from (6, R = Cl). Both *trans*-β-bromoviny phenyl sulphone (9) and *gem*-α-bromovinyl phenyl sulphone (10) undergo

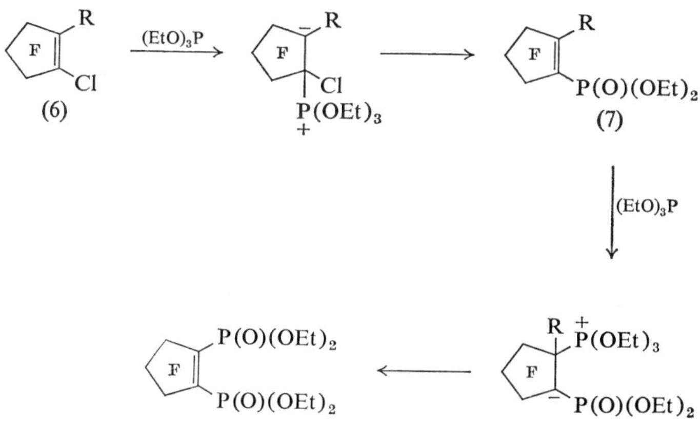

(8)

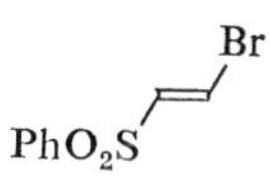

(9)

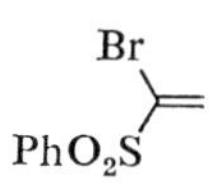

(10)

[4] G. Caspari, H. Drawe, and A. Henglein, *Radiochim. Acta*, 1967, **8**, 102 (*Chem. Abs.*, 1968, **68**, 105,304).
[5] P. Obrycki and C. E. Griffin, *J. Org. Chem.*, 1968, **33**, 632.
[6] J. D. Park and O. K. Furuta, *Tetrahedron Letters*, 1969, 393.
[7] W. S. Wadsworth jun. and W. D. Emmons, *J. Amer. Chem. Soc.*, 1961, **83**, 1733.

displacement reactions with phosphite esters;[8] here the carbanion intermediates are stabilised by the sulphonyl group.

Numerous papers on the addition of tervalent phosphorus compounds to Mannich phenol bases (*e.g.* 11) have appeared recently[9] and a kinetic investigation[10] reveals that the reaction is first order with respect to (11), and the rate of decomposition of (11) to the quinone methide (12) is some 20 times slower than the reaction of (12) with triethyl phosphite. A similar addition reaction can take place with 2-benzylidene-3(2*H*)-thianaphthenone-1,1-dioxide (13) when (14),[11a] (15),[11b] or (16)[11a] can be formed depending on the reaction conditions. Pentaco-ordinate products are also formed by the addition of trialkyl phosphites to ethyl ethylidene acetoacetate (17)[12] or 1,3-dienes,[13] although in the last instance tetraco-ordinate products can also be obtained. The formation of pentaco-ordinate products from α-diketones and phosphite esters is reviewed in Chapter 2.

Triethyl orthoformate will react with tervalent phosphorus halides giving rise to phosphonates,[14] presumably by nucleophilic attack by the phosphorus on (18) and ethanolysis of the intermediate.

When tris(triethylsilyl)phosphite is added to acetyl chloride the products include triethylsilyl chloride and bis(triethylsilyl)acetylphosphonate.[15] Chloride ion will attack the electrophilic organosilyl group in the phosphonium intermediate by either an intra- or inter-molecular process in preference to an ethyl group.

The Perkow reaction has continued to attract attention, and evidence has been presented[16] that the reaction can take place with attack by phosphorus on positive halogen. In trichloroacetylthiourea (19) the halogen

8 E. G. Kataev, F. R. Tantesheva, E. G. Yarkova, and E. A. Berdnikov, *Doklady Akad. Nauk S.S.S.R.*, 1968, **179**, 862 (*Chem. Abs.*, 1968, **69**, 77,350).

9 *E.g.* [a] B. E. Ivanov and L. A. Valitova, *Izvest. Akad. Nauk S.S.S.R., Ser. khim.*, 1967, 1087, 1090 (*Chem. Abs.*, 1968, **68**, 39,723, 39,724). [b] B. E. Ivanov, A. B. Ageeva, and R. R. Shagidullin, *Izvest. Akad. Nauk S.S.S.R., Ser. khim.*, 1967, 1994 (*Chem. Abs.*, 1968, **68**, 69,079).

10 B. E. Ivanov, L. A. Valitova, and T. G. Vavilova, *Izvest. Akad. Nauk S.S.S.R., Ser. khim.*, 1968, 768 (*Chem. Abs.*, 1968, **69**, 76,330).

11 [a] A. Mustafa, M. M. Sidky, S. M. A. D. Zayed, and W. M. Abdo, *Tetrahedron*, 1968, **24**, 4725. [b] B. A. Arbusov, T. D. Sorokina, and V. S. Vinogradova, *Doklady Akad. Nauk S.S.S.R.*, 1967, **177**, 1337 (*Chem. Abs.*, 1968, **68**, 105,301).

12 B. A. Arbusov, N. A. Polezhaeva, and V. S. Vinogradova, *Izvest. Akad. Nauk S.S.S.R., Ser. khim.*, 1967, 2281 (*Chem. Abs.*, 1968, **68**, 4984).

13 [a] Z. L. Evtikhov, N. A. Razumova, and A. A. Petrov, *Zhur. obshchei Khim.*, 1968, **38**, 2341 (*Chem. Abs.*, 1969, **70**, 37,879). [b] *Doklady Akad. Nauk S.S.S.R.*, 1968, **181**, 877 (*Chem. Abs.*, 1969, **70**, 78,084). [c] N. A. Razumova, Z. L. Evtikhov, L. I. Zubtsova, and A. A. Petrov, *Zhur. obshchei Khim.*, 1968, **38**, 2342 (*Chem. Abs.*, 1969, **70**, 47,546). [d] K. I. Novitskii, N. A. Razumova, and A. A. Petrov., *Khim. Org. Soedin. Fosfora, Akad. Nauk S.S.S.R. Otd. obshchei Tekh. Khim.*, 1967, 248 (*Chem. Abs.*, 1968, **69**, 43,981).

14 [a] S. S. Krokhina, R. I. Pyrkin, Y. A. Levin, and B. E. Ivanov, *Izvest. Akad. Nauk S.S.S.R.*, 1968, 1420 (*Chem. Abs.*, 1968, **69**, 67,478). [b] W. Dietsche, *Annalen*, 1968, **712**, 21.

15 N. F. Orlov and B. L. Kaufman, *Zhur. obshchei Khim.*, 1968, **38**, 1842 (*Chem. Abs.*, 1969, **70**, 4193).

16 J. S. Ayres and G. O. Osborne, *Chem. Comm.*, 1968, 195.

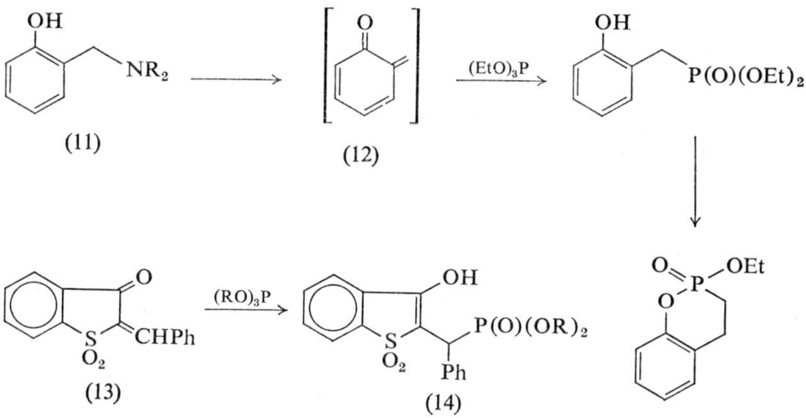

(11) (12) (13) (14)

(15) (16)

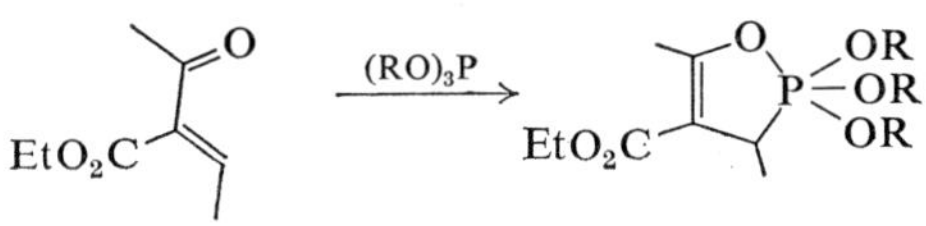

(17)

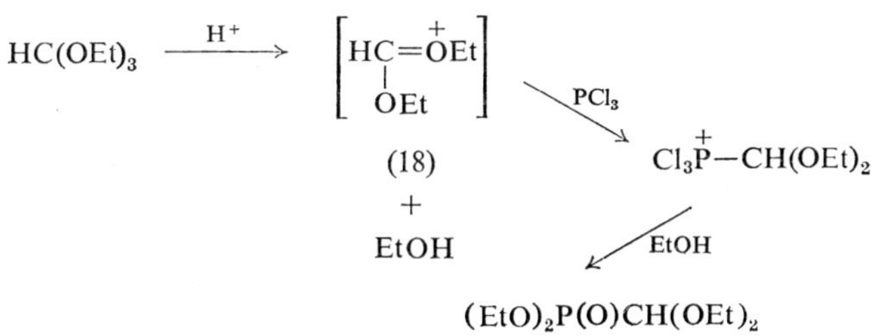

(18)

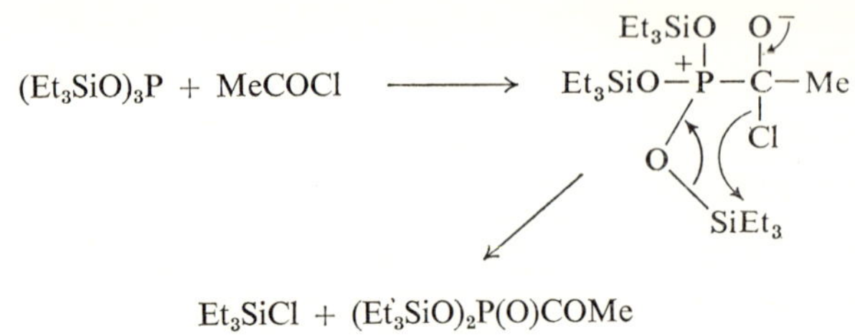

atoms have considerable positive character and, as attack on the carbonyl group is sterically hindered, attack at halogen is a favoured process. The reaction products, dichloroacetylthiourea (20), *S*-ethyltrichloroacetylisothiourea (21), and diethyl phosphorochloridate, support the proposed mechanism, (21) arising from the dealkylation of the phosphonium intermediate by the isothiouronium ion. This example of the Perkow reaction appears to be an exception to the more usual case when initial attack by

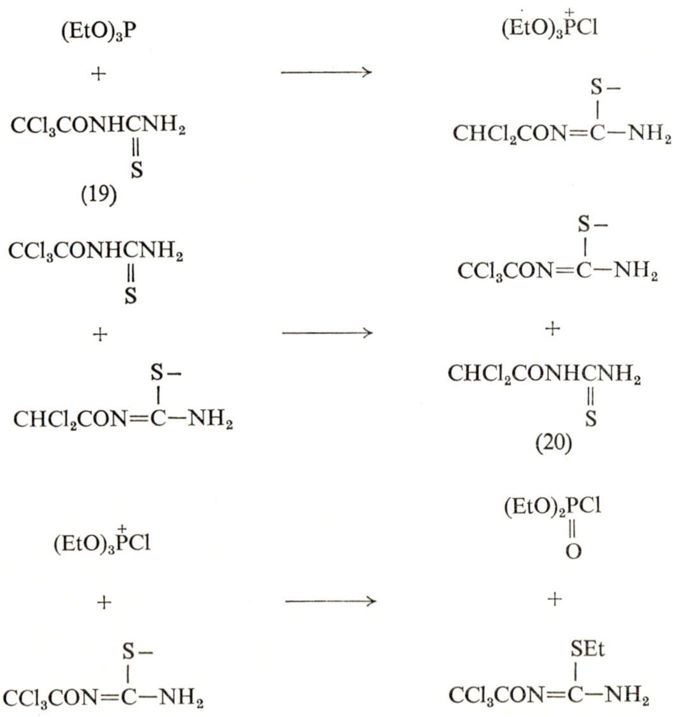

(19) (20) (21)

(22)

phosphorus occurs at carbonyl carbon followed by rearrangement of the intermediate.[17a]

Sulphur can expand its outer valence shell by the use of *d* orbitals and this could reduce the electron availability on the carbonyl oxygen atom of chlorothiolacetates (22). The synthesis of 1-thiovinyl phosphates from (22) and trialkyl phosphites has been cited [17b] as an example of the Perkow reaction in which nucleophilic attack by phosphorus occurs on carbonyl oxygen. There appears, however, to be little evidence for this mode of attack and the reaction probably proceeds in the normal manner. The relative yield of enol phosphate formed by the action of triethyl phosphite and *p*-substituted phenacyl halides (23) increases when Y is electron releasing and decreases with alteration of X in the order Cl > Br > I.[18]

Irradiation of a mixture of chloroacetone (24) and triethyl phosphite gives the ketophosphonate (25), the enol phosphate (26), and other products.[19] The yield of (25) is decreased on addition of a radical scavenger while the yield of (26) is unaffected. It is suggested that the primary process of this reaction is the $n \rightarrow \pi^*$ excitation of the carbonyl group, and addition of methanol or acetic acid which inhibit such excitation suppresses the entire reaction. The byproducts in the photo-Perkow reaction could arise from the photoisomerisation of triethyl phosphite to diethyl ethylphosphonate (27). E.s.r. studies of this isomerisation show [20] that the radical (28) is formed initially and then rapidly decays to (29). Combination of the latter with an ethyl radical leads to (27).

When α-bromo-α,β-dicyanopropionates (30, X = CO_2Et) are treated [21] with trimethyl phosphite, a reaction analogous to the Perkow reaction takes place to produce a ketenimine (31). On the other hand, when α-bromo-α,β-dicyanopropionitriles (30, X = CN) are used,[22] (32) is formed rather than a ketenimine. It has been proposed [22] that attack by the phosphite on (30, X = CN) occurs at the comparatively positive halogen atom to generate the bromophosphonium cation (33) together with the mesomeric anion (34), and that further interaction of (33) and (34) gives rise to the observed product.

N-Fluorosulphonyl isocyanates (35) will form zwitterionic, crystalline adducts (36) with triphenyl phosphine, phosphorous trisdimethylamide and triethyl phosphite.[23] (36, R = OEt) does not dealkylate to (37, R = Et) at room temperature but is converted into (37, R = H) by the action of hydrogen chloride. *N*-Fluorosulphonyl carbodi-imides are not formed

[17] (a) P. A. Chopard, V. M. Clark, R. F. Hudson, and A. J. Kirby, *Tetrahedron*, 1965, **21**, 1961. (b) L. F. Ward jun., R. R. Whetstone, G. E. Pollard, and D. D. Phillips, *J. Org. Chem.*, 1968, **33**, 4470.
[18] A. Arcoria and S. Fisichella, *Ann. Chim.* (*Italy*), 1967, **57**, 1228.
[19] H. Tomoika, Y. Izawa, and Y. Ogata, *Tetrahedron*, 1968, **24**, 5739.
[20] K. Terauchi and H. Sakurai, *Bull. Chem. Soc. Japan*, 1968, **41**, 1736.
[21] A. Foucaud and R. Leblanc, *Tetrahedron Letters*, 1969, 509.
[22] R. Leblanc and A. Foucaud, *Tetrahedron Letters*, 1969, 2441.
[23] H. Hoffmann, H. Forster, and G. Tor-Poghossian, *Monatsh.*, 1969, **100**, 311.

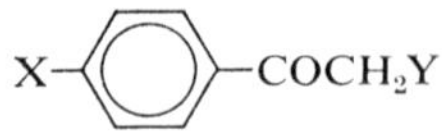

(23)

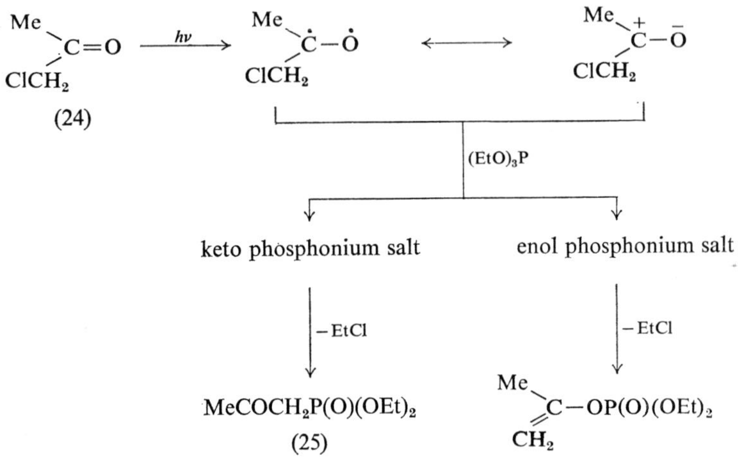

(24) (25) (26)

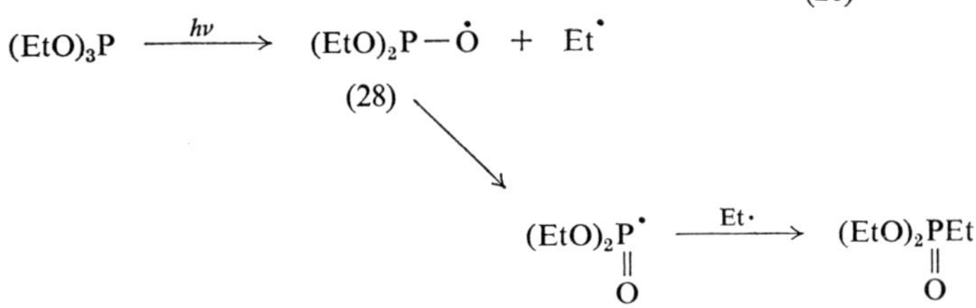

(28) (29) (27)

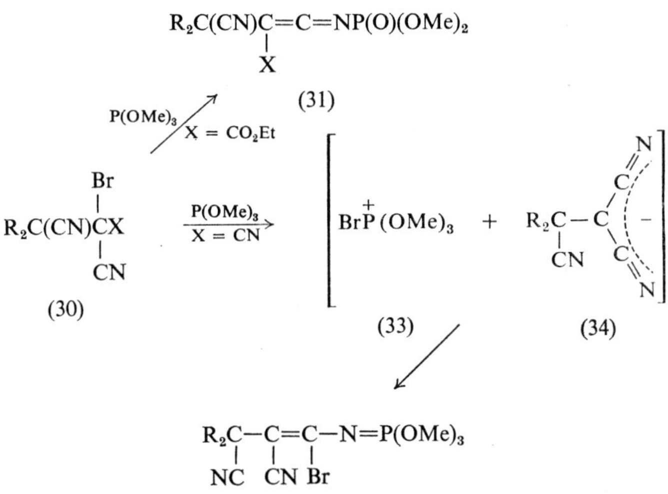

(31) (30) (33) (34) (32)

from (35) and the phospholene oxide (38),[24] as the reaction does not proceed beyond the formation of the *N*-fluorosulphonyl phospholene imine (39). Carbodi-imides are, however, postulated as intermediates in the reaction between monoaryl amidophosphites (40) and aromatic isocyanates [25] when imidoyl phosphonates (41) are produced.

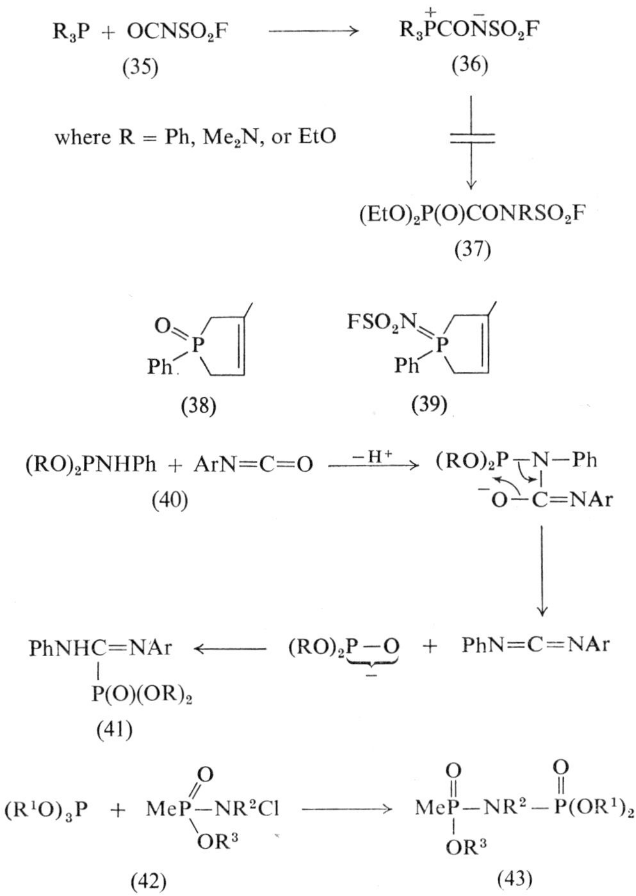

(*iii*) *Attack on Nitrogen*. *N*-Alkyl-*N*-chloro-methylphosphonamides (42) undergo nucleophilic attack by trialkyl phosphites to give aza-analogues of mixed anhydrides (43) [26] which can also be synthesised by the action of

[24] T. W. Campbell, I. I. Monagle, and V. S. Foldi, *J. Amer. Chem. Soc.*, 1962, **84**, 3673.
[25] A. N. Pudovik and E. S. Betyeva, *Zhur. obshchei Khim.*, 1968, **38**, 285 (*Chem. Abs.*, **69**, 52,231).
[26] V. A. Shokol, G. A. Golik, and G. I. Derkach, *Khim. Org. Soedin. Fosfora, Akad. Nauk S.S.S.R., Otd. obshchei Tekh. Khim.*, 1967, 96 (*Chem. Abs.*, 1968, **69**, 10,511).

alkyl phosphorochloridates on the corresponding *N*-alkyl-methylphosphonamides.

(*iv*) *Attack on Oxygen.* The oxidation of trialkyl phosphites to the corresponding phosphates by the action of *p*-benzoquinone and benzyl alcohol [27] can be regarded as an example of nucleophilic attack on oxygen by a tervalent phosphorus derivative. The quinone is reduced in the initial reaction to the hydroquinone anion which can then debenzylate the phosphonium cation to produce hydroquinone benzyl ethers which can be isolated from the reaction.

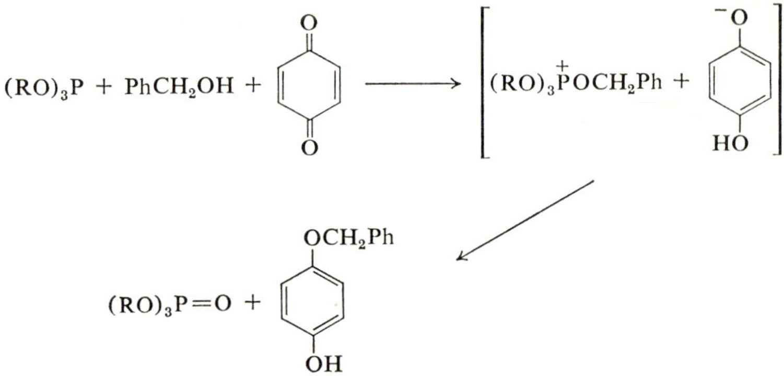

(*v*) *Attack on Halogen.* Alkyl chlorides can be synthesised under mild conditions by treating the alcohol with carbon tetrachloride and a tertiary phosphine [28a] or trialkyl phosphite.[28b] Phosphorous trisdimethylamide can be used [29] in place of the phosphine and the alkyl chlorides can be readily separated from the phosphoric trisdimethylamide by extraction with water. The isolation of chloroform and the phosphonium intermediate (44) from the reaction supports an earlier suggestion [30] for the mechanism of this reaction. The reaction between trialkyl phosphites, alkyl thiols, and bromotrichloromethane which leads to *OO*-dialkyl *S*-alkyl phosphonothiolates is a complex one,[31, 32] and appears to be ionic when carried out in hexane; whereas in the presence of azobis-isobutyronitrile, radical intermediates are involved.

The anion of diethyl phosphite will displace halide ion from pentachlorobenzene (45) in the presence of ethanol to generate 1,2,4,5-tetrachlorobenzene and triethyl phosphate.[33] This reaction could take place by attack by

[27] O. Mitsonobu, K. Kodera, and T. Mukaiyama, *Bull. Chem. Soc. Japan*, 1968, **41**, 461.
[28] [(a)] J. Hooz and S. S. H. Gilani, *Canad. J. Chem.*, 1968, **46**, 86. [(b)] H. R. Hudson, *J. Chem. Soc.* (*B*), 1968, 664.
[29] I. M. Downie, J. B. Lee, and M. F. S. Matough, *Chem. Comm.*, 1968, 1350.
[30] B. Miller, *Topics in Phosphorus Chemistry*, 1965, **2**, 133.
[31] L. L. Murdock and T. L. Hopkins, *J. Org. Chem.*, 1968, **33**, 907.
[32] R. E. Atkinson, J. I. G. Cadogan, and J. T. Sharp, *J. Chem. Soc.* (*B*), 1969, 138.
[33] M. C. Demarcq, *Bull. Soc. Chim. France*, 1969, 1716.

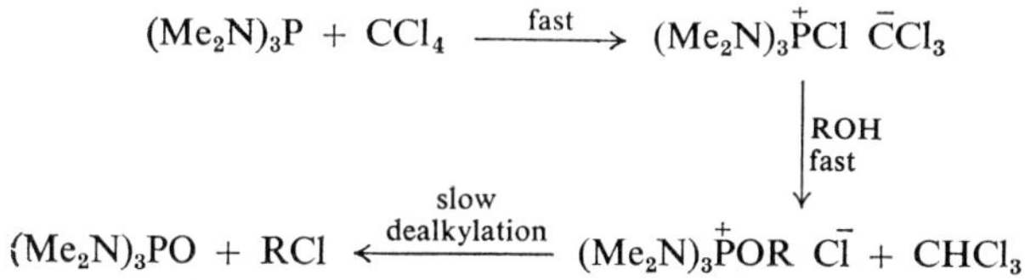

(44)

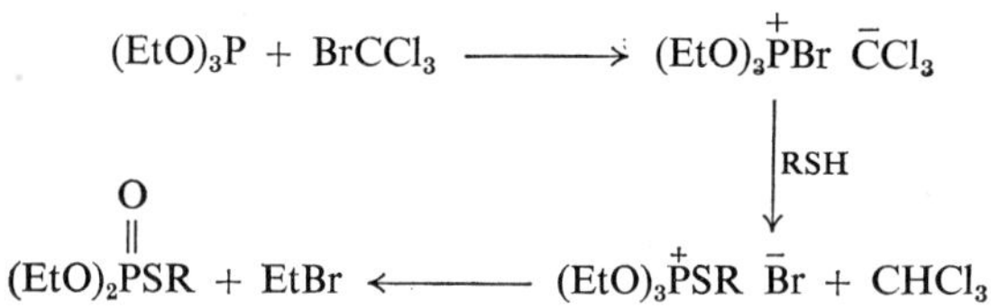

phosphorus on halogen to form diethyl phosphorochloridate and the relatively stable anion of 1,2,4,5-tetrachlorobenzene; these would then react with ethanol to give the observed products. In the absence of ethanol a mixture of products is formed including the diethyl ester of 2,3,5,6-tetrachlorobenzenephosphonic acid.

Diethyl bromomalonate reacts with sodium diethyl phosphite to give 1,1,2,2-ethanetetracarboxylate;[34] here again the initial reaction is attack

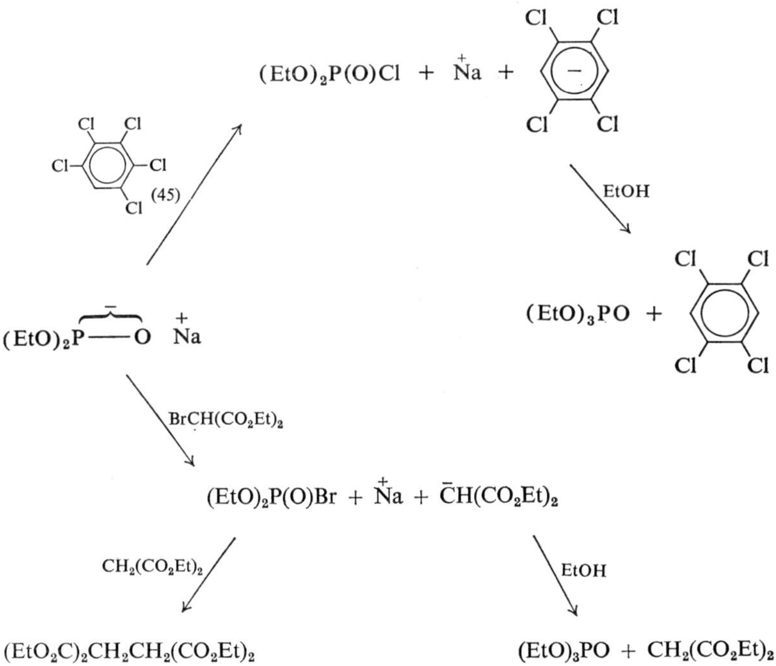

[34] K. H. Takemura and D. J. Tuma, *J. Org. Chem.*, 1969, **34**, 252.

on positive halogen. When ethanol is added to the reaction only triethyl phosphate and diethyl malonate are observed.

(*vi*) *Attack on Hydrogen.* The acetolysis of trialkyl phosphites with acetic acid involves protonation of the phosphorus atom followed by dealkylation of the intermediate to produce acetate esters and dialkyl phosphites.[35] The use of trialkyl phosphites in the synthesis [36a] and alkylation of *OO*-dialkyl dithiophosphates,[36b] together with the synthesis of mixed anhydrides by the action of carboxylic acids on acetyl phosphites,[37] may proceed by a similar mechanism.

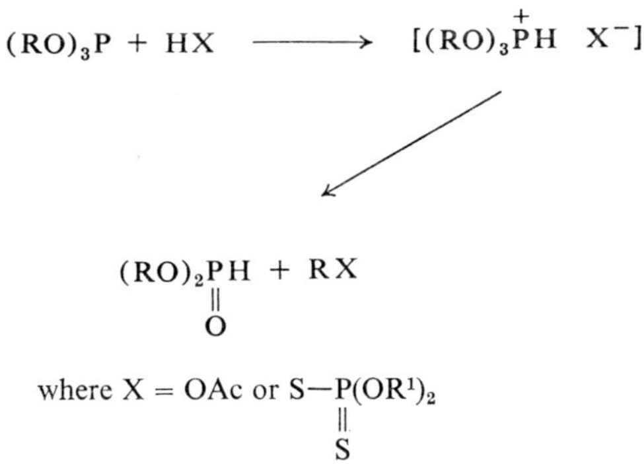

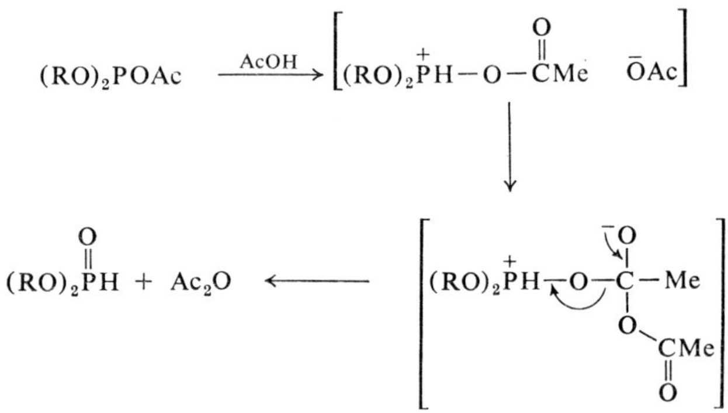

B. Electrophilic Reactions.—Transesterification reactions of phosphorous esters appear to involve a four-centre aggregate (46) in which the phosphorus atom acts as an electrophile.[38] Phosphorous trisdialkylamides undergo ready exchange with imidoyl ethers, possibly by a similar mechanism.[39]

[35] E. S. Huyser and J. A. Deiter, *J. Org. Chem.*, 1968, **33**, 4205.

[36] [(a)] I. S. Akhmetzhanov, *Zhur. obshchei Khim.*, 1968, **38**, 1090 (*Chem. Abs.*, 1968, **69**, 66,829). [(b)] A. N. Pudovik and V. K. Krupnov, *Zhur. obshchei Khim.*, 1968, **38**, 304 (*Chem. Abs.*, 1968, **69**, 77,369).

[37] E. E. Nifant'ev and I. V. Fursenko, *Zhur. obshchei Khim.*, 1968, **38**, 1295 (*Chem. Abs.*, 1968, **69**, 58,786).

[38] A. N. Pudovik and G. I. Evstaf'ev, *Doklady Akad. Nauk S.S.S.R.*, 1968, **183**, 842 (*Chem. Abs.*, 1969, **70**, 67,280).

[39] Y. Charbonnel, R. Burgada, and J. Barrans, *Compt. rend.*, 1968, **266**, *C* 1241.

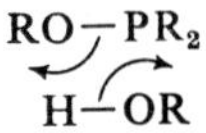

(46)

When acyclic *OO*-dialkyl thiophosphites are treated with nitrosyl chloride, *OO*-tetra-alkyl monothiopyrophosphates (47) are formed.[40] Cyclic thiophosphites [*e.g.* 2-thiono-5,5-dimethyl-1,3,2-dioxaphosphorinan (48)], however, give rise[41] to a mixture of the asymmetric (49, X = Y = S, Z = O) and the symmetric[42] (49, X = Z = S, Y = O) dithiopyrophosphates under the same conditions. Studies with ^{18}O-labelled nitrosyl

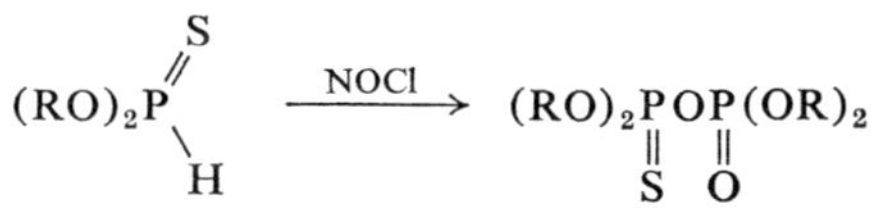

(47)

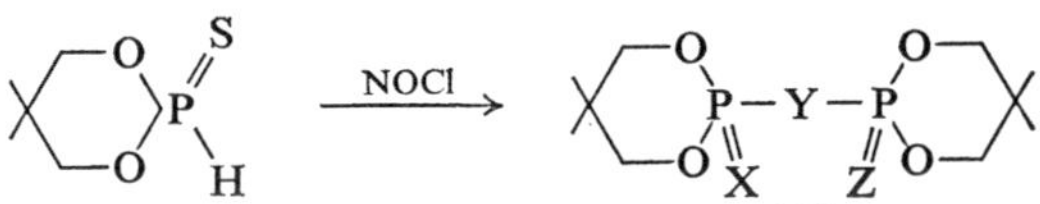

(48) (49)

chloride and dialkyl phosphites show[43] that the bridge oxygen atom of the tetra-alkyl pyrophosphate produced in this reaction originates from the nitrosyl chloride which is behaving as an ambident nucleophile. Electrophilic attack on nitrosyl chloride can, therefore, occur on nitrogen or oxygen and the difference in products when acyclic or cyclic thiophosphites are used may be due to a difference in the mode of electrophilic attack by the two thiophosphites. There appears to be little reason for such a difference except for steric reasons. From a comparison of the rate of reaction of dialkyl phosphites with nitrosyl chloride and with halogens,[44] it is considered unlikely that tautomerism of the dialkyl phosphite occurs to give the tercovalent phosphite in the reaction with nitrosyl chloride, though this appears to be the rate-determining step in the reaction with halogens.

Electrophilic substitution at phosphorus has been reviewed[45] and the reactions classified according to mechanism and to the co-ordination number of the phosphorus atom.

[40] L. Almasi and L. Paskucz, *Chem. Ber.*, 1963, **96**, 2024.
[41] A. Zwierzak and J. Michalski, *Canad. J. Chem.*, 1969, **47**, 1163.
[42] J. Michalski, M. Mikołajczyk, and B. Młotkowska, *Chem. Ber.*, 1969, **102**, 90.
[43] D. Samuel and B. Silver, *J. Chem. Soc.*, 1963, 3582.
[44] Z. Luz and B. Silver, *J. Amer. Chem. Soc.*, 1962, **84**, 1091, 1095.
[45] S. G. Warren, *Angew. Chem. Internat. Edn.*, 1968, **7**, 606.

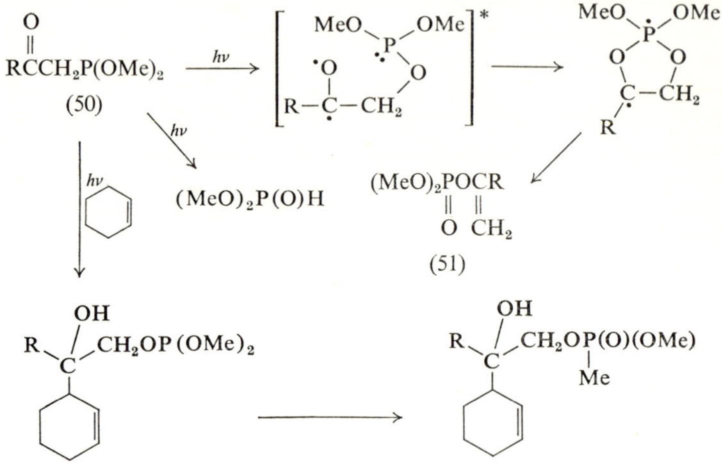

(50) (51) (52) (53)

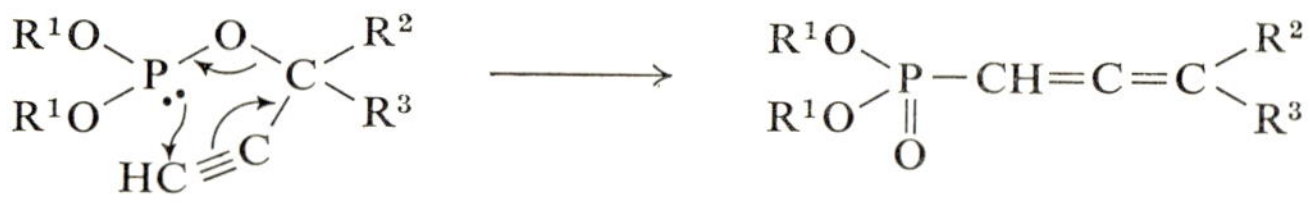

(54)

(a) R^1 = 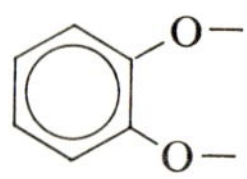 , $R^2 = R^3 = H$

(b) $R^1 = -CMe(CO_2Et)C{\equiv}CH$, $R^2 = Me$, $R^3 = CO_2Et$

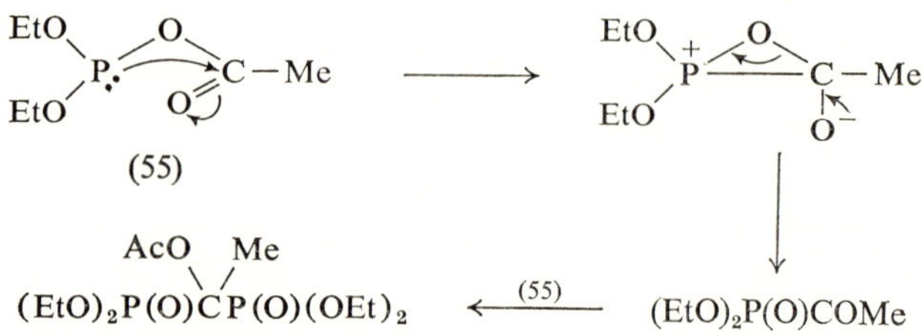

(55) (56) (57)

C. Rearrangements.—Dimethyl vinyl phosphates (51) can be synthesised by the photorearrangement of dimethyl β-ketoethyl phosphites (50).[46] Dimethyl phosphite is also formed in this reaction and the product ratio depends on the availability of readily abstractable hydrogen atoms in the solvent. For example, when the reaction is carried out in cyclohexene, little (51) is obtained and (52) is the major product. Presumably (53) which is formed initially, undergoes a photo-Arbusov rearrangement [47] to yield (52).

Several papers concerned with the Arbusov rearrangement of phosphite esters have appeared in the past year.[48, 49] These include a kinetic investigation of the rearrangement of propargyl catechol phosphite (54a) and the first-order rate constant of this rearrangement has been determined over a range of temperatures.[48a] Another acetylene–allene rearrangement [that following treatment of ethyl 2-hydroxy-2-methylbut-3-ynoate (54b) with phosphorus trihalides] has been studied both in the presence and absence of pyridine.[50] As the rearrangement occurs with ease in the absence of pyridine, an S_Ni type of mechanism is suggested for this transformation rather than an intermolecular two-step process.

Diethyl acetyl phosphite (55) rearranges on heating to the α-keto-phosphonate (56) [51] which can then break down to keten and diethyl phosphite; alternatively (56) can react with a second molecule of (55) to form (57).

D. Cyclic Esters of Phosphorous Acid.*—It has been deduced [52] from dipole moment and ^{1}H n.m.r. measurements that the cyclic phosphites (58a) and (58b) together with their borine adducts have chair conformations in which the methoxy-groups attached to phosphorus are axial. Two signals, which can be ascribed to the ring protons, are observed in the ^{1}H n.m.r. spectrum, the relative areas of which vary with temperature due to interconversion of (58a) and (58b) by inversion at phosphorus. This interconversion is a

[46] C. E. Griffin, W. G. Bentrude, and G. M. Johnson, *Tetrahedron Letters*, 1969, 969.
[47] R. B. LaCount and C. E. Griffin, *Tetrahedron Letters*, 1965, 3071.
[48] [(a)] Y. A. Kondrat'ev, V. V. Tarasov, A. S. Vasil'ev, N. M. Ivakina, and S. Z. Ivin, *Zhur. obshchei Khim.*, 1968, **38**, 1791 (*Chem. Abs.*, 1968, **69**, 106,074). [(b)] S. Z. Ivin V. N. Pastushkov, Y. A. Kondrat'ev, K. F. Oglobin, and V. V. Tarasov, *Zhur. obshchei Khim.*, 1968, **38**, 2069 (*Chem. Abs.*, 1969, **70**, 11,759). [(c)] Y. A. Kondrat'ev, V. V. Tarasov, A. S. Vasil'ev, N. M. Ivakina, S. Z. Ivin, and V. M. Pastushkov, *Zhur. obshchei Khim.*, 1969, **38**, 2590 (*Chem. Abs.*, 1969, **70**, 47,031). [(d)] L. V. Nesterov and R. I. Mutalapova, *Zhur. obshchei Khim.*, 1967, **37**, 1843, 1847 (*Chem. Abs.*, 1968, **68**, 38,804, 38,895).
[49] N. I. Rizpolozhenskii and F. S. Mukhametov, *Izvest. Akad. Nauk S.S.S.R., Ser. khim.*, 1968, 2163 (*Chem. Abs.*, 1969, **70**, 78,092).
[50] M. Verny and R. Vessière, *Bull. Soc. chim. France*, 1968, 3004.
[51] [(a)] A. N. Pudovik, T. K. Gazizov, and A. P. Pashinkin, *Khim. Org. Soedin. Fosfora, Akad. Nauk S.S.S.R., Otd. obshchei Tekh. Khim.*, 1967, 25 (*Chem. Abs.*, 1968, **69**, 59,333). [(b)] A. N. Pudovik and T. K. Gazizov, *Zhur. obshchei Khim.*, 1968, **38**, 140 (*Chem. Abs.*, 1968, **69**, 96,836).
[52] D. W. White, G. K. McEwen, and J. G. Verkade, *Tetrahedron Letters*, 1968, 5369.

* Although Ring Index nomenclature can be used for these compounds, it is simpler to name them as esters of the appropriate oxyacids of phosphorus when derivatives can be recognised without difficulty.

relatively slow process and attainment of equilibrium can take several days; moreover, the methyl resonances do not broaden over a wide temperature range, which may explain reports [53] that such cyclic phosphites are conformationally immobile. Cyclic phosphates appear to possess conformational mobility.[54]

Analysis of the ^1H n.m.r. spectra of ethylene phosphites suggests [55] that the ring has the twisted conformation (60) rather than the simple envelope (59) as the P—O—C—H_A and P—O—C—H_B couplings indicate markedly different dihedral angles in the two P—O—C—H systems.

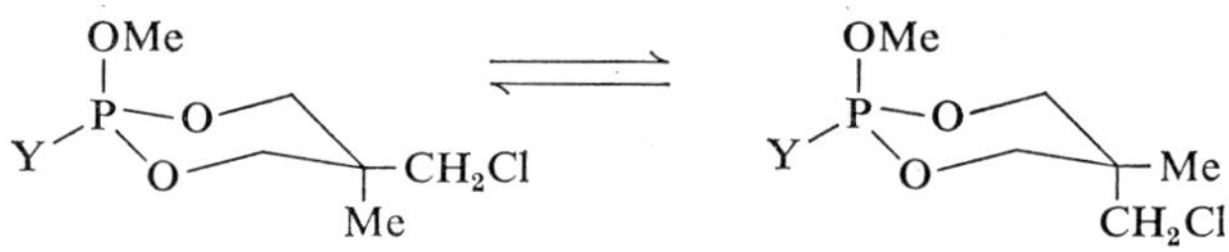

(58a) (58b)

where Y = electron pair or BH_3

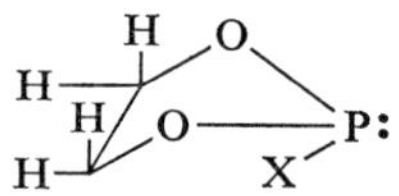

(59)

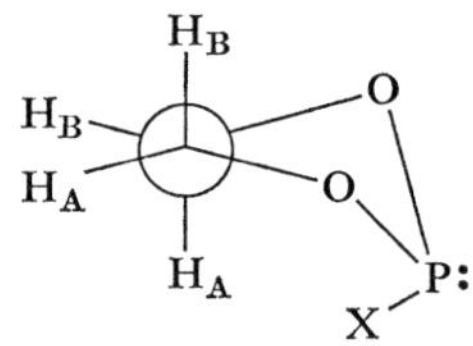

(60)

where X = OMe, OPh or Cl

The enhanced reactivity of cyclic phosphoramidites towards benzaldehyde has been attributed,[56] by analogy with the hydrolysis of cyclic phosphates,[57] to the release of steric strain on formation of the transition state. The heats of hydrolysis of some cyclic and acyclic phosphoramidites have been determined [58] and are very similar, suggesting that release of ring strain may not be the major factor in determining the relative reactivity of these compounds. However, the rate of reaction of (61, R = Ph) with phenyl isocyanate [59] to give the insertion product [60] is considerably greater than the rate of reaction of (61, R = OMe). This could mean that release of ring strain still does have some influence on the relative reactivities of these compounds.

[53] D. Gagnaire, J. B. Robert, and J. Verrier, *Bull. Soc. chim. France*, 1968, 2392.
[54] R. S. Edmundson, *Tetrahedron Letters*, 1969, 1905.
[55] P. Haake, J. P. McNeal, and E. J. Goldsmith, *J. Amer. Chem. Soc.*, 1968, **90**, 715.
[56] R. Greenhalgh and R. F. Hudson, *Chem. Comm.*, 1968, 1300.
[57] F. H. Westheimer, *Accounts Chem. Res.*, 1968, **1**, 70.
[58] R. Greenhalgh, J. E. Newberry, R. Woodcock, and R. F. Hudson, *Chem. Comm.*, 1969, 22.
[59] R. F. Hudson and R. J. G. Searle, *J. Chem. Soc.* (*B*), 1968, 1349.
[60] R. F. Hudson and A. Mancuso, *Chem. Comm.*, 1969, 522.

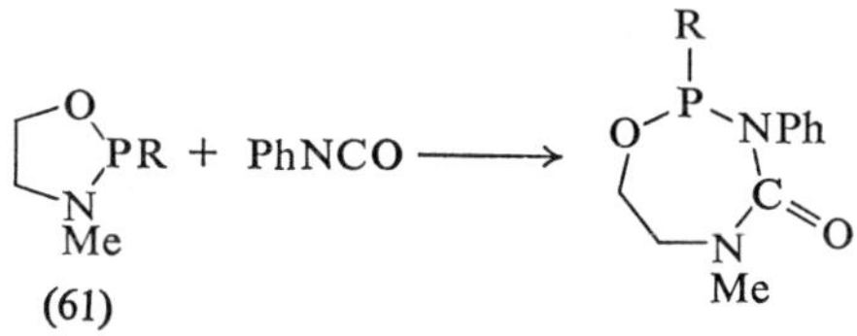

(61)

E. Miscellaneous Reactions.—A new method for the synthesis of dialkyl phosphites consists of treating a mixture of the alcohol and methanol with phosphorus trichloride. Elimination of methyl chloride from the intermediate occurs smoothly and removal of the volatile products *in vacuo* leaves the phosphite ester.[61] The synthesis of phosphonites and phosphinites by the action of trimethyl phosphite on alkyl chlorides is a similar process.[62]

Phosphorus fluorides react with trialkyl phosphines or aminophosphines to give 1 : 1-adducts with P—P bonds.[63] When trimethyl phosphite is treated with phosphorus pentafluoride, a 1 : 1-adduct is formed in the first place which breaks down to tetrafluoro(methoxy)phosphorane (62) and fluoro(dimethoxy)phosphine (63).[64] Interaction of (62) and (63) gives a complex mixture of products. A range of products is also formed when phosphorus trifluoride reacts with trimethyl phosphite.

$$2ROH + 1MeOH + PCl_3 \longrightarrow [(RO)_2POMe] \xrightarrow{HCl} (RO)_2P(O)H + MeCl$$

$$PF_5 + (MeO)_3P \longrightarrow \underset{(62)}{PF_4OMe} + \underset{(63)}{(MeO)_2PF}$$

$$\downarrow$$

$$(MeO)_3\overset{+}{P}Me\ \overset{-}{P}F_6 + MeOP(O)F_2 + MeP(O)F_2 + MeOPF_2 + (MeO)_2PF$$

$$PF_3 + (MeO)_3P$$

$$\downarrow$$

$$MeOPF_2 + MeP(O)(F)(OMe) + MeP(O)(OMe)_2 + MeP(O)F_2$$

Esters of pyrophosphoric and pyrothiophosphoric acids are formed when sodium diethyl phosphite is treated with sulphur dioxide.[65] The mechanism

[61] Y. A. Mandel'baum, A. L. Itskova, and N. N. Mel'nikov, *Khim. Org. Soedin. Fosfora, Akad. Nauk S.S.S.R., Otd. obshchei Tekh. Khim.*, 1967, 288 (*Chem. Abs.*, 1968, **69**, 43,338).

[62] I. Hechenbleikner and K. R. Molt, U.S.P. 3,316,333 (*Chem. Abs.*, 1968, **68**, 49,765).

[63] D. H. Brown, K. D. Crosbie, G. W. Fraser, and D. W. A. Sharp, *J. Chem. Soc.* (*A*), 1969, 551.

[64] D. H. Brown, K. D. Crosbie, G. W. Fraser, and D. W. A. Sharp, *J. Chem. Soc.* (*A*), 1969, 872.

[65] W. Stec, A. Zwierzak, and J. Michalski, *Tetrahedron Letters*, 1968, 5873.

of this reaction is uncertain, but since dialkyl phosphites can function as ambident nucleophiles, two of the intermediates may be (64) and (65) which must interact with dialkyl phosphite and/or sulphur dioxide to generate the anhydrides. Tervalent phosphorus compounds can be determined by means of the Karl Fischer reagent[66] when the sulphur dioxide again acts as an oxidising agent. Phosphonium salts and quinquevalent phosphorus derivatives do not interfere with the determination.

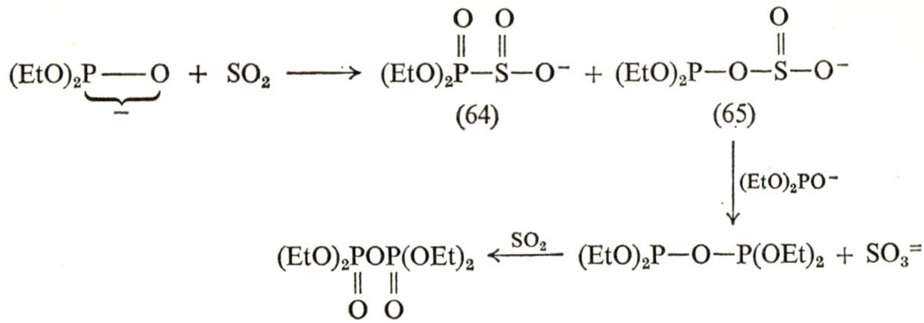

(64) (65)

Diethyl phosphite has been used in a study of the interaction of thiamine with nucleophiles.[67] Nucleophilic attack by the phosphorus atom occurs to produce phosphonates.

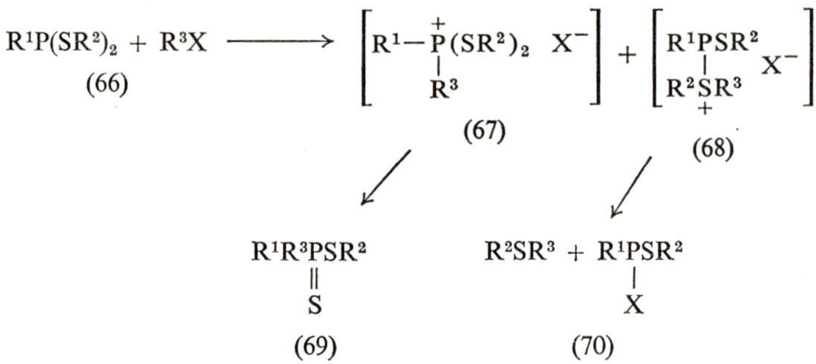

(66) (67) (68) (69) (70)

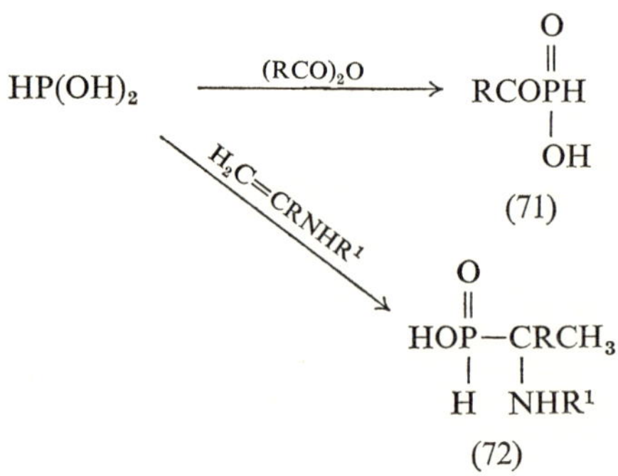

(71) (72)

[66] B. Hayton and B. C. Smith, *J. Inorg. Nuclear Chem.*, 1969, **31**, 1369.
[67] A. Takamizawa, K. Hirai, Y. Hamashima, Y. Matsumoto, and S. Tanaka, *Chem. and Pharm. Bull* (*Japan*), 1968, **16**, 1764.

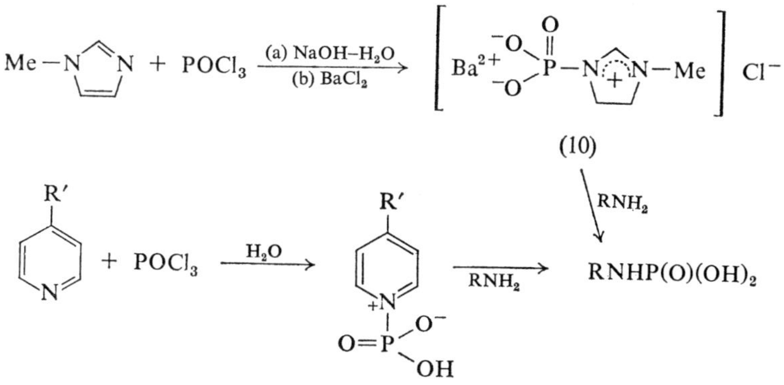

(10)

(11)

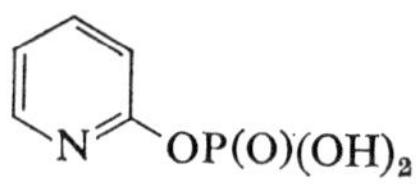

(12)

an important part in the hydrolysis of monoesters of polyphosphoric acids.[14]

Phosphorylated triazines (*e.g.* 14) can be isolated from the reaction between cyanuric chloride and salts of *OO*-diethyl phosphorodithioate.[16] These are P—XYZ systems and when *OO*-diethyl phosphorodithioate ion is

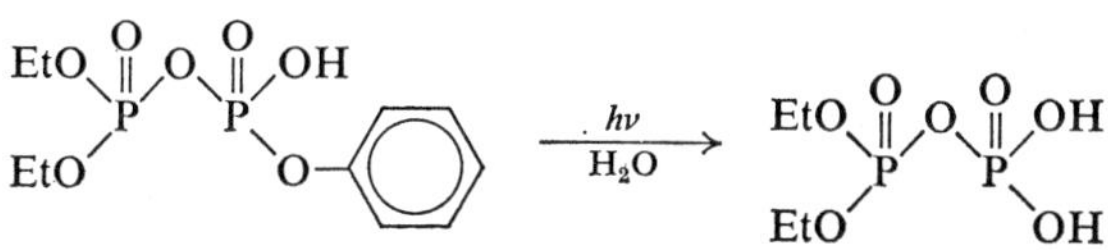

(13)

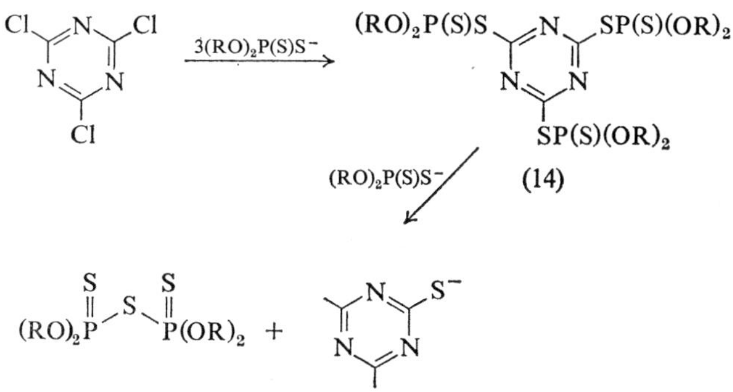

(14)

[16] D. Harding and G. O. Osborne, *Austral. J. Chem.*, 1968, **21**, 1093.

present in a large excess, *OO*-tetraethyl trithiopyrophosphate is the major phosphorus-containing product. Similar phosphorylated triazines have previously been obtained from cyanuric chloride and salts of *OO*-dialkyl phosphorothioic acid.[17]

When phosphorus oxychloride is heated with *NN*-dimethyl-4-methylaniline (15), the major product of the reaction is the eight-membered heterocycle (16) whose structure has been proved by *X*-ray crystallography.[18] The individual stages in the formation of (16) are not known; however, in one possible scheme, (15) functions as an ambident nucleophile which is capable of dealkylating the quaternary compound (17). An oxidation step must occur during the formation of (16) and either the *NN*-dimethyl-2,4-dimethylaniline undergoes aerial oxidation to (18), or phosphorus

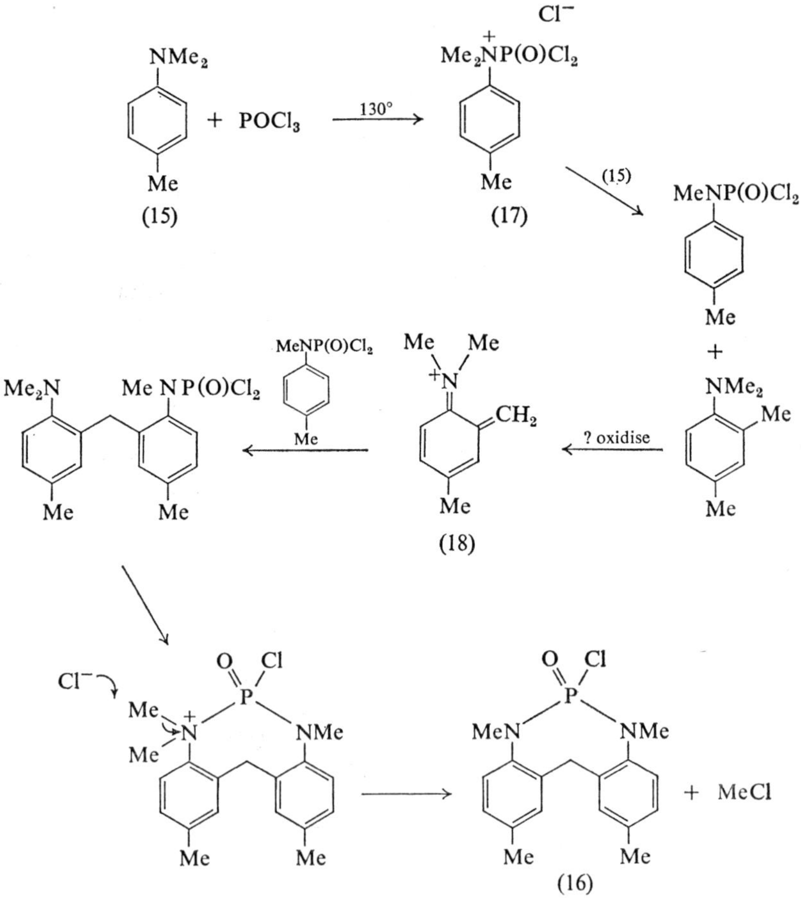

[17] G. O. Osborne and G. Page, *J. Chem. Soc.* (*C*), 1967, 1192.
[18] C. Y. Cheng, R. A. Shaw, T. S. Cameron, and C. K. Prout, *Chem. Comm.*, 1968, 616.

oxychloride itself functions as an oxidising agent to produce (18) and phosphorus trichloride. The former possibility being the more likely, (18) could then undergo Michael-type addition followed by cyclisation to give (16).

Steric hindrance has little effect on the reaction between 2,6-dialkylphenols and phosphorus oxychloride unless the alkyl group is very bulky (*e.g.* t-butyl).[19] In the latter instance reaction only takes place in the presence of a Friedel–Crafts catalyst when either rearrangement occurs to give the 4-substituted product, or t-butyl chloride is eliminated. No phosphonates can be detected under the conditions employed.

Acrylonitrile can be used instead of pyridine to remove hydrogen chloride in the synthesis of phosphate esters from phosphorus oxychloride.[20] The technique could be useful in the synthesis of compounds which are subject to attack by pyridine *e.g.* benzyl phosphates.

A review has appeared on the preparation and properties of phosphorus isocyanates,[21] which are most readily prepared by the action of cyanate or thiocyanate ion on a phosphorus chloride[22] or by treating a phosphoramidate with oxalyl chloride.[23] The isocyanate group in phosphorus isocyanates reacts readily with a range of nucleophiles; for example, chlorination of dialkyl phosphoric isothiocyanates yields sulphur dichloride and *NN*-dichloromethylene phosphoramidates.[24]

Addition of ethylene oxide to a slurry of phosphoric oxide in refluxing chloroform gives a clear solution, which on cooling precipitates bis-ethylene pyrophosphate (19). Removal of solvent *in vacuo* from the filtrate leaves ethylene bis[ethylene phosphate] (20).[25] (19) reacts violently with water with evolution of heat, and addition of one equivalent of water to a solution of (19) results in the rapid formation of ethylene phosphate. Alcohols will cleave the pyrophosphate bond of (19) leading to ethylene phosphate and its alkyl esters. (20) is also sensitive to water and tends to polymerise on standing.

The expected products of the reaction of phosphorus pentaselenide with alcohols, *viz. OO*-dialkyl diselenophosphoric acids, are unstable and have to be isolated rapidly as their potassium salts or as chromium(III) complexes.[26] If the salts are not isolated from the reaction, derivatives having bridging selenium atoms are obtained.

19 G. M. Kosolapoff, C. K. Arpke, R. W. Lamb, and H. Reich, *J. Chem. Soc. (C)*, 1968, 815.
20 N. K. Bliznyuk, P. S. Khokhlov, R. V. Strel'tsov, Z. N. Kvasha, and A. F. Kolomiets, *Zhur. obshchei Khim.*, 1967, **37**, 1119 (*Chem. Abs.*, 1968, **68**, 105,296).
21 G. I. Derkatsch, *Angew. Chem. Internat. Edn.*, 1969, **8**, 421.
22 G. I. Derkach and N. I. Liptuga, *Zhur. obshchei Khim.*, 1968, **38**, 1779 (*Chem. Abs.*, 1969, **70**, 4231).
23 L. I. Samaray, O. I. Kolodjaznij, and G. I. Derkatsch, *Angew. Chem. Internat. Edn.*, 1968, **7**, 618.
24 P. I. Alimov and M. P. Alimov, *Izvest. Akad. Nauk S.S.S.R., Ser. khim.*, 1967, 1344 (*Chem. Abs.*, 1968, **68**, 48,958).
25 R. D. Wilcox, G. H. Harris, and R. S. Olson, *Tetrahedron Letters*, 1968, 6001.
26 M. V. Kudchadker, R. A. Zingaro, and K. J. Irgolic, *Canad. J. Chem.*, 1968, **46**, 1415.

$$(RO)_2P(O)NH_2 + (COCl)_2 \longrightarrow (RO)_2P(O)NCO$$

$$(RO)_2P(O)Cl + KSCN \longrightarrow (RO)_2P(O)NCS$$

$$(RO)_2P(O)NCS \xrightarrow{Cl_2} (RO)_2P(O)N{=}CCl_2 + SCl_2$$

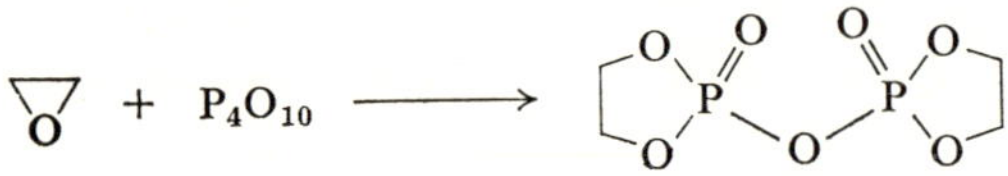

(19)

+

(20)

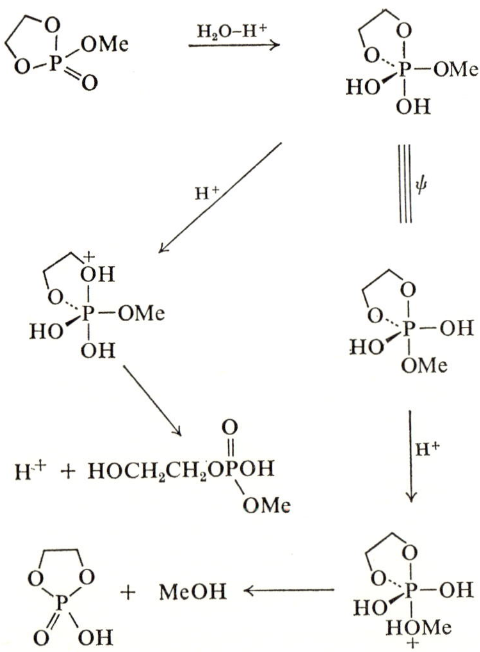

B. Hydrolysis of Phosphate Esters.—Molecular orbital calculations have been carried out [27] on cyclic and acyclic phosphate esters, and corroboration has been found for the main features of the Westheimer mechanism [28] for the facile hydrolysis and oxygen exchange of cyclic phosphate esters. The calculations show that ring strain leads to a larger net atomic charge on phosphorus in cyclic than in acyclic esters. This is due to a lowered occupation of the 3*d* orbitals on phosphorus and favours nucleophilic attack at phosphorus leading to a trigonal bipyramidal pentaoxyphosphorane with the entering and leaving groups in apical positions. Pseudorotation of the methyl ethylene phosphate adduct can occur [29] without undue difficulty when the phosphoryl oxygen remains in the basal plane. The apical oxygen atom in the pentaoxyphosphorane containing the five-membered ring is quite negative and protonation of the methoxy group can occur in this position even though the phosphoryl oxygen is also quite negative. Hydrolysis with ring-opening does not require pseudorotation and can proceed either by an S_N2 displacement or by the formation of a pentaco-ordinate adduct in which the ring opens without rotation.

Dimethyl phosphoacetoin (21) undergoes rapid base-catalysed hydrolysis to dimethyl phosphate and acetoin.[30] When the hydrolysis is carried out in $H_2{}^{18}O$, 90—100 atom % incorporation of oxygen from the solvent into the dimethyl phosphate occurs,[31] and control experiments indicate that oxygen incorporation and hydrolysis occur in the same stage of the reaction. The isotopic experiments, together with the fact that the rates of hydrolysis are 10^5—10^6 times greater than the rate for trimethyl phosphate, suggest that the hydrolysis involves a pentaco-ordinate intermediate.[32] The rate of hydrolysis of 4,5-dimethyl-2-hydroxy-2-oxo-1,3,2-phospholene (22) to the ring-opened product occurs at a much faster rate than the hydrolysis of the corresponding saturated phosphate.[33] Presumably relief of ring strain in the pentaco-ordinate intermediate is the important factor in this reaction.

The rates of alkaline hydrolysis and reaction with n-butylamine have been measured [34] for a series of phosphorylating agents [23, where X = CN, Cl, F, and $(Pr^iO)_2P(O)O$]. It was observed that (23, X = CN) reacts faster than the other phosphorylating agents; and (23, X = F) did not react with butylamine. The differences in reactivity have been explained in terms of the relative leaving-group tendencies of X, water, and butylamine from a pentaco-ordinate adduct (24).

The neutral hydrolysis of methyl *N*-cyclohexylphosphoramidothioic chloride (25) in aqueous dimethoxyethane is stereospecific and presumably

[27] D. B. Boyd, *J. Amer. Chem. Soc.*, 1969, **91**, 1200.
[28] F. H. Westheimer, *Accounts Chem. Res.*, 1968, **1**, 70.
[29] J. D. Dunitz and V. Prelog, *Angew. Chem. Internat. Edn.*, 1968, **9**, 725.
[30] F. B. Ramirez, B. Hansen, and N. B. Desai, *J. Amer. Chem. Soc.*, 1962, **84**, 4588.
[31] D. M. Brown and M. J. Frearson, *Chem. Comm.*, 1968, 1342.
[32] D. S. Frank and D. A. Usher, *J. Amer. Chem. Soc.*, 1967, **89**, 6360.
[33] V. E. Bel'skii, N. N. Bezzubova, and I. P. Gozman, *Zhur. obshchei Khim.*, 1968, **38**, 1330 (*Chem. Abs.*, 1968, **69**, 85,866).
[34] R. F. Hudson and R. Greenhalgh, *J. Chem. Soc.* (*B*), 1969, 325.

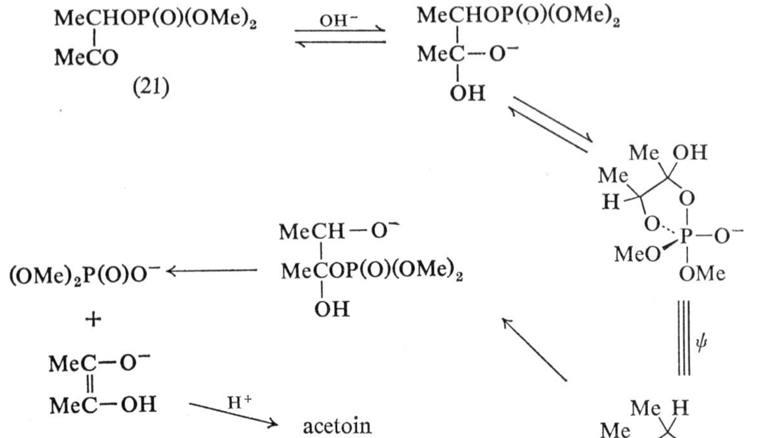

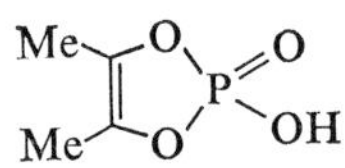

(22)

takes place by an S_N2 displacement at phosphorus.[35] The rate of alkaline hydrolysis is 10^4—10^5 times faster and gives a racemic product, hence by analogy with carbon chemistry the intermediate could be planar and could be an analogue (26) of monomeric metaphosphate. Alkaline hydrolysis of the 4-nitrophenyl ester corresponding to (25), on the other hand, gives extensive but not complete racemisation.[36] The hydrolysis of methyl

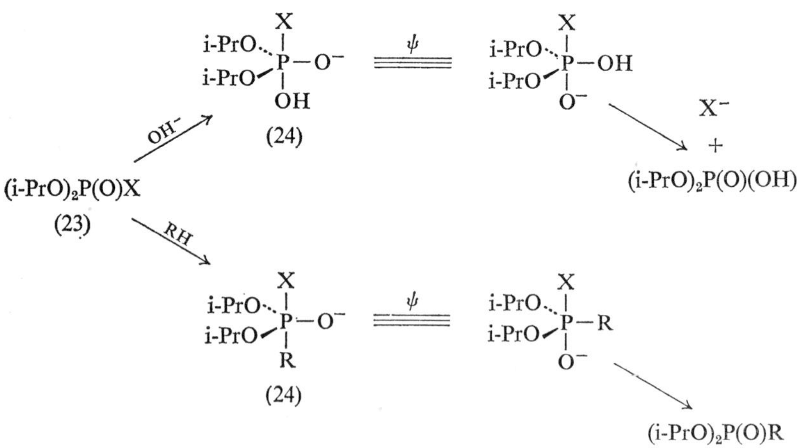

where R = OH or *n*-BuNH

[35] A. F. Gerrard and N. K. Hamer, *J. Chem. Soc.* (*B*), 1968, 539.
[36] A. F. Gerrard and N. K. Hamer, *J. Chem. Soc.* (*B*), 1967, 1122.

N-cyclohexylphosphoramidic chloride has also been investigated[37] and although the hydrolysis proceeds *via* the anion, a planar intermediate similar to (26) is apparently not involved.

Alkaline hydrolysis of diesters of phosphoramidic acids leads principally to P—O rather than P—N bond cleavage;[38] however, the acid hydrolysis of (27, R^1 = Me, R^2 = 2,4-$Cl_2C_6H_3$) takes place with negligible P—O bond cleavage.[39] The kinetic evidence suggests that this is direct substitution by water when the rate-determining step involves the breaking of the P—N bond in a one-step reaction rather than a two-stage process involving a pentaco-ordinate intermediate. Hydrolysis of *O*-methyl *O*-(2,2-dichloro-vinyl)*NN*-dimethylphosphoramidate (28) can take place either by cleavage of the P—N bond or by cleavage of the dichlorovinyl ester bond. At low temperatures P—N bond fission predominates, while at high temperatures and in the presence of alkali P—O bond fission is the major reaction.[40]

The hydrolysis of bis(2,4-dinitrophenyl) phosphate to the monoester occurs with P—O fission,[41] and the rate is independent of pH in the range

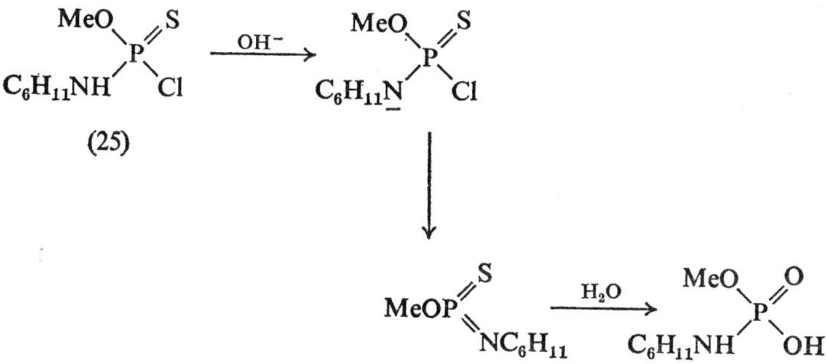

(25)

(26)

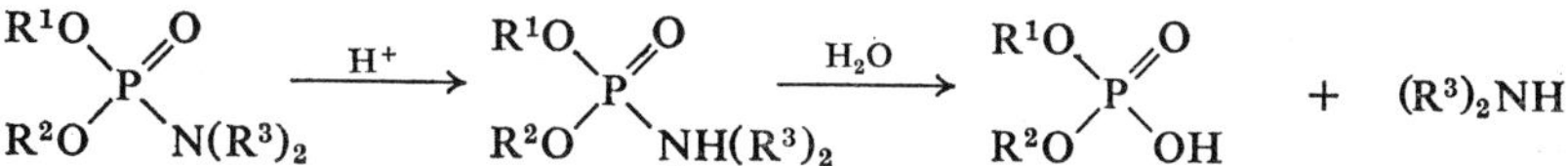

(27)

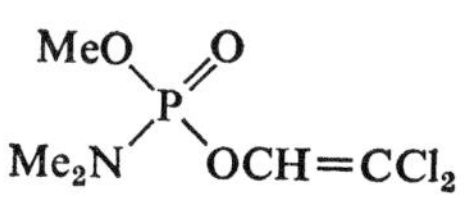

(28)

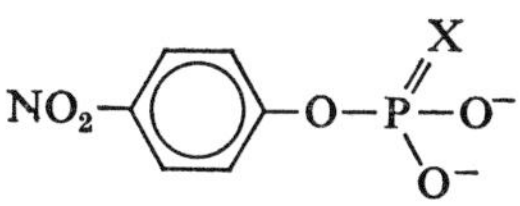

(29)

[37] A. F. Gerrard and N. K. Hamer, *J. Chem. Soc. (B)*, 1969, 369.
[38] J. E. Berger and E. Wittner, *J. Phys. Chem.*, 1966, **70**, 1025.
[39] A. W. Harrison and C. E. Boozer, *J. Amer. Chem. Soc.*, 1968, **90**, 3486.
[40] V. E. Bel'skii and Z. V. Lustina, *Doklady Akad. Nauk S.S.S.R.*, 1968, **178**, 1077 (*Chem. Abs.*, 1968, **69**, 66,588).
[41] C. A. Bunton and S. J. Farber, *J. Org. Chem.*, 1969, **34**, 767.

of pH 4—7 when the diester is present wholly as the monoanion. In this case, it appears that the rate-limiting step is attack by water rather than elimination of a metaphosphate derivative as occurs during the hydrolysis of the monoester. The reactivity of dianions of phospho-monoesters towards nucleophiles depends strongly on the leaving group [42a] but only weakly on the basicity of the nucleophile,[42b] and the displacement appears to be a bimolecular process. Hindered tertiary bases (*e.g.* 2-picoline and 2,6-lutidine) catalyse the hydrolysis of the dianion of 2,4-dinitrophenyl phosphate. This is not general base catalysis [42c] as there is no significant solvent deuterium isotope effect. However, most organic solvents catalyse the hydrolysis of 2,4-nitrophenyl phosphate [42a] and so the catalysis by hindered tertiary bases may be merely a solvent effect. The reactivity of monoanions of phospho-monoesters is similar to that of the dianions.[42a]

The preparation of a number of aryl phosphorochloridates has been described and their hydrolysis rates measured.[43] In general the rates of hydrolysis are the same as the corresponding phosphate esters. It has been suggested [44] that steric factors may influence the rates of hydrolysis of phosphorodihalidates, as phosphorofluorochloridates undergo a much more rapid elimination of chloride than do phosphorodichloridates. However, the high electronegativity of fluorine would render the phosphorus more susceptible to nucleophilic attack and this should be an important factor in hydrolytic reactions.

Cetyltrimethylammonium salts catalyse the hydrolysis of the dianions of 2,4- and 2,6-dinitrophenyl phosphates but do not affect the rate of attack of hydroxide ion upon them, or the hydrolysis of the monoanion of 4-nitrophenyl phosphate.[45] It is proposed that the presence of cationic micelles in the hydrolysis medium assists in the generation of two monoanions (*viz.* phenoxide and metaphosphate) from one dianion. It has also been observed that the rates of hydrolysis of dianions of 2,4- and 2,6-dinitrophenyl phosphates in an aqueous organic medium increase with decreasing water content of the medium.[42a] Hence, the binding of a dianion to a non-aqueous micelle would be expected to increase its rate of hydrolysis in an aqueous medium. A similar catalytic effect by cationic micelles on the hydrolysis of 4-nitrophenyl esters of carboxylic acids,[46] 2,4-dinitrofluorobenzene,[47] and 4-nitrophenyl diphenyl phosphate [48] has also been observed.

[42] [(a)] A. J. Kirby and A. G. Varvoglis, *J. Amer. Chem. Soc.*, 1967, **89**, 425. [(b)] A. J. Kirby and A. G. Varvoglis, *J. Chem. Soc.* (*B*), 1968, 135. [(c)] G. O. Dudek and F. H. Westheimer, *J. Amer. Chem. Soc.*, 1959, **81**, 2641.

[43] A. G. Varvoglis, *Chim. Chron. A*, 1968, **33**, 54 (*Chem. Abs.*, 1968, **69**, 106,075).

[44] A. A. Neimysheva and I. L. Knunyants, *Zhur. obshchei Khim.*, 1968, **38**, 593 (*Chem. Abs.*, 1968, **69**, 66,629).

[45] C. A. Bunton, E. J. Fendler, L. Sepulveda, and K. U. Yang, *J. Amer. Chem. Soc.*, 1968, **90**, 5512.

[46] C. Gittler and A. Ochoa-Solano, *J. Amer. Chem. Soc.*, 1968, **90**, 5004.

[47] C. A. Bunton and L. Robinson, *J. Org. Chem.*, 1969, **34**, 780.

[48] C. A. Bunton and L. Robinson, *J. Org. Chem.*, 1969, **34**, 773.

The aqueous hydrolysis of (29, X = S) proceeds some 50 times faster than the hydrolysis of (29, X = O),[49] which is consistent with the participation of a metaphosphate-like intermediate in these reactions. When a metaphosphate derivative is formed from a phosphomonoester, the P—O (P—S) bond order increases and the charge on oxygen (sulphur) decreases; since sulphur is less electronegative than oxygen this is a more favourable process for (29, X = S) than for (29, X = O). On the other hand, (29, X = S) is hydrolysed by alkaline phosphatase much more slowly than (29, X = O) and the rate ratio k_S/k_O is similar to that for the non-enzymic hydrolysis of phosphotriesters.[50] It has been suggested that pentaco-ordinate intermediates are involved in this hydrolysis and hence pentaco-ordinate intermediates are probably involved in the enzymic reaction. The participation of pentaco-ordinate intermediates in the hydrolysis of nucleoside phosphorothioates is discussed in Chapter 7.

The acid-catalysed and spontaneous hydrolyses of α-D-ribofuranose 1-phosphate (30) proceed at several hundred times the rates of analogous hydrolyses of α-D-glucopyranose 1-phosphate (31).[51] These differences in rates are considerably greater than the rate differences between cyclopentyl and cyclohexyl compounds for S_N1 reactions.[52] The rate-limiting step in the hydrolysis of (31) in moderately concentrated acid (when the phosphate is present in the undissociated form) is the generation of the glucosyl 1-cation (32).[53] This cation can be stabilised by overlap of its empty *p* orbital with the lone pair of electrons on the ethereal oxygen and this overlap would be greater with the ribose phosphate (30) than with the glucose phosphate (31);[54] thus (30) might be expected to have an enhanced reactivity.

Glucose 6-phosphate decomposes in aqueous alkali in a series of parallel, consecutive reactions.[55] The first common step is the setting up of an equilibrium between the dianion of glucose 6-phosphate and the enediolate anion (33). The latter can then give either 6-phosphoglucometasaccharinic acid (34) or fructose 6-phosphate (35). Further degradation of (35) leads to glyceraldehyde 3-phosphate and ultimately to lactic acid and orthophosphate.[56] The alkaline degradation of glucose 6-phosphate shows many similarities to the enzymic anaerobic metabolism of carbohydrates.

The hydrolysis of phosphate esters of 2-hydroxy acids is catalysed by molybdate ions,[57a] and 2,3-diphosphoglyceric acid is hydrolysed to

[49] R. Breslow and I. Katz, *J. Amer. Chem. Soc.*, 1968, **90**, 7376.
[50] J. Ketelaar, H. Gersmann, and K. Koopmans, *Rec. Trav. Chim.*, 1952, **71**, 1253.
[51] C. A. Bunton and E. Humeres, *J. Org. Chem.*, 1969, **34**, 572.
[52] [a] C. S. Foote, *J. Amer. Chem. Soc.*, 1964, **86**, 1853. [b] P. v. R. Schleyer, *J. Amer. Chem. Soc.*, 1964, **86**, 1854.
[53] C. A. Bunton, D. R. Llewellyn, K. G. Oldham, and C. A. Vernon, *J. Chem. Soc.*, 1958, 3588.
[54] J. N. Bemiller, *Adv. Carbohydrate Chem.*, 1967, **22**, 25.
[55] C. Degani and M. Halmann, *J. Amer. Chem. Soc.*, 1968, **90**, 1313.
[56] N. Y. Kozlova, I. V. Mel'nichenko, and A. A. Yasnikov, *Ukrain. khim. Zhur.*, 1968, **34**, 1041 (*Chem. Abs.*, 1969, **70**, 86,743).
[57] [a] Z. B. Rose and L. I. Pizer, *J. Biol. Chem.*, 1968, **243**, 4806. [b] R. Parvin and R. A. Smith, *Analyt. Biochem.*, 1969, **27**, 65.

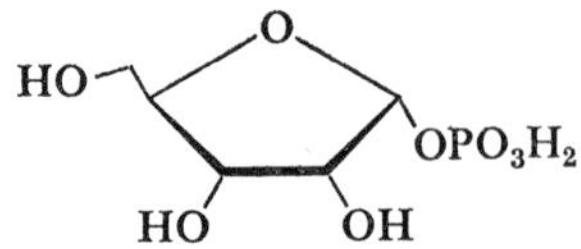

(30)

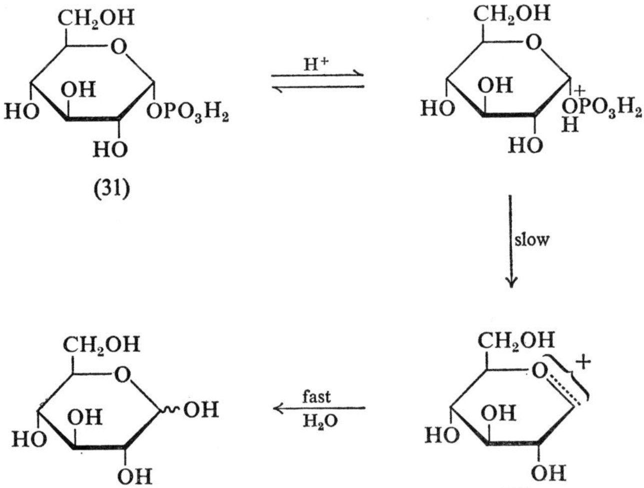

(31) (32)

3-phosphoglyceric acid under these conditions. A rapid procedure for the determination of orthophosphate in the presence of labile organic phosphates has been developed.[57b] This consists of the formation of a phosphomolybdovanadate complex which can be extracted from an aqueous medium into butanol.

The phosphorylation of a phosphorothioate to generate a pyrophosphorothionate occurs by direct attack of phosphorus on the oxygen atom of the ambident thioate anion and not by the isomerisation of a P(O)SP(O) intermediate.[58] Thiophosphorylation appears to proceed in a similar manner as the stability of the P(S)SP(O) system renders its rearrangement unlikely.[59] Thus, phosphorylation and thiophosphorylation of optically active phosphorothioates should lead to optically active pyrophosphorothionates with the same configuration and optical purity as the original thioacid. A number of optically active pyrophosphonothionates have been

[58] J. Michalski, M. Mikołajczyk, and A. Ratajczak, *Bull. Acad. polon. Sci., Sér. Sci. chim.*, 1965, **13**, 277.

[59] L. Almasi and L. Paskucz, *Monatsh.*, 1968, **99**, 187.

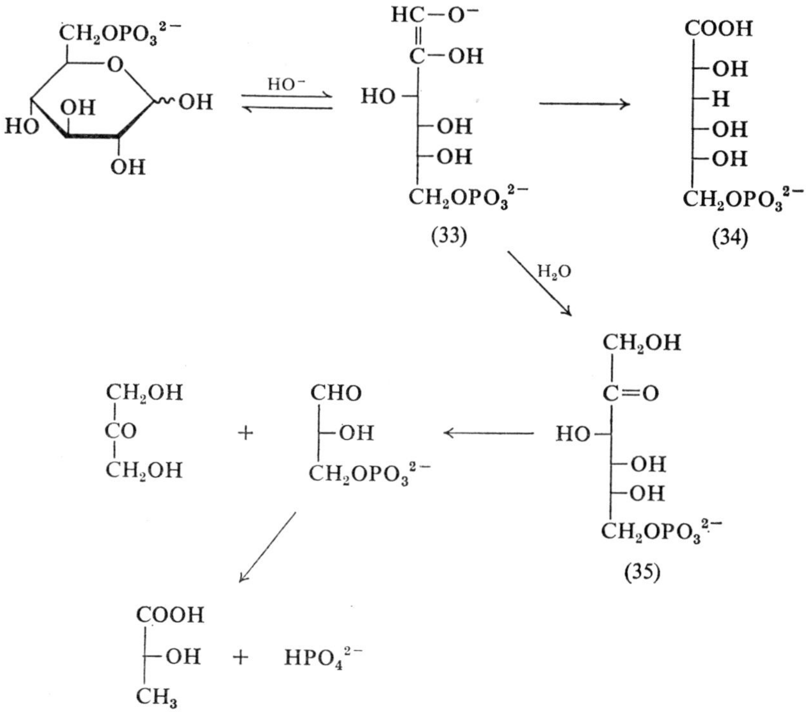

prepared[60] and inversion of configuration occurs during alkaline hydrolysis. This observation appears to exclude pseudorotation of a pentaco-ordinate transition state in these reactions.

The activation energy of the aqueous hydrolysis of *OO*-diethyl dithiopyrophosphate[61a] is much larger than that of its monothio analogue.[61b] It has been claimed[61c] that d_π–p_π bonding has little effect in the transition state during nucleophilic attack on phosphoryl or thiophosphoryl compounds and that reactivities are determined by the inductive effect of substituents.

The rate of acid-catalysed hydrolysis of aliphatic acyl phosphates depends on the size of the acyl group,[62] and the acid-catalysed hydrolysis occurs by attack by water on carbonyl carbon followed by expulsion of inorganic phosphate. This behaviour is in direct contrast to the mode of reaction at higher pH values.[63]

60 J. Michalski, M. Mikołajczyk, B. Młotkowska, and J. Omelanczuk, *Tetrahedron*, 1969, **25**, 1743.

61 (a) V. E. Bel'skii and M. V. Efremova, *Izvest. Akad. Nauk S.S.S.R., Ser. khim.*, 1968, 409 (*Chem. Abs.*, 1968, **69**, 66,596). (b) V. E. Bel'skii, M. V. Efremova, and Z. V. Lustina, *Izvest. Akad. Nauk S.S.S.R., Ser. khim.*, 1967, 1236 (*Chem. Abs.*, 1968, **68**, 21,342). (c) A. A. Neimysheva, V. I. Savchuk, M. V. Ermolaeva, and I. L. Knunyants, *Izvest. Akad. Nauk. S.S.S.R., Ser. khim.*, 1968, 2222 (*Chem. Abs.*, 1969, **70**, 56,923).

62 D. R. Phillips and T. H. Fife, *J. Amer. Chem. Soc.*, 1968, **90**, 6803.

63 G. Di Sabato and W. P. Jencks, *J. Amer. Chem. Soc.*, 1961, **83**, 4400.

C. Reactions of Derivatives of Phosphoric Acid.—(*i*) *Dephosphorylation Reactions.* The synthesis of alkynes from enol phosphates and other enol esters has been reviewed.[64] Both *cis*- and *trans*-elimination can occur and the highest yields of alkynes are obtained when the reaction is carried out in

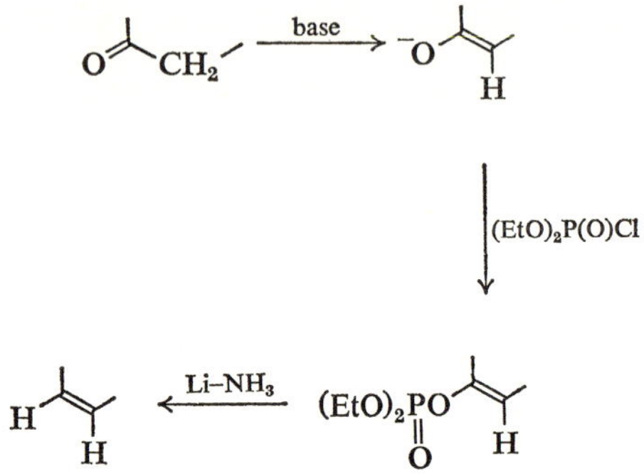

a non-aqueous medium. A structurally specific olefin synthesis has been developed[65, 66] which consists of elimination of a diethyl phosphate residue from an enol phosphate by reduction with lithium in ammonia. The enol phosphates can be prepared by treatment of the enolate anion with diethyl phosphorochloridate[65] or by the Perkow reaction.[66]

The acidolysis and halogenation of α-alkoxy enol phosphates takes place with fragmentation.[67] The acid (or halogen) adds on to the carbon–carbon double bond of the enol phosphate to give (36) which can then undergo unimolecular decomposition with C—O bond fission.

The reaction between *N*-bromophthalimide and dimethyl esters of enol phosphates is complex as α-bromoketones, trimethyl phosphate, and

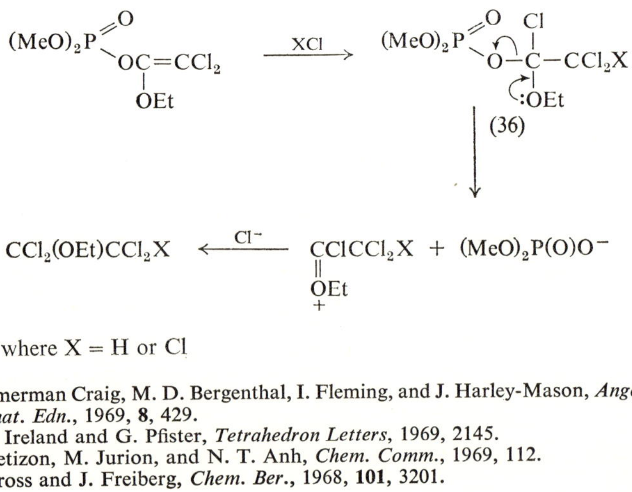

where X = H or Cl

[64] J. Cymerman Craig, M. D. Bergenthal, I. Fleming, and J. Harley-Mason, *Angew. Chem. Internat. Edn.*, 1969, **8**, 429.
[65] R. E. Ireland and G. Pfister, *Tetrahedron Letters*, 1969, 2145.
[66] M. Fetizon, M. Jurion, and N. T. Anh, *Chem. Comm.*, 1969, 112.
[67] H. Gross and J. Freiberg, *Chem. Ber.*, 1968, **101**, 3201.

N-methylphthalimide are among the products which can be isolated.[68] Involatile polymers are also formed which are presumably derivatives of polyphosphoric acids.

The photo-oxidation and photodephosphorylation of 2-methylnaphtha-1,4-hydroquinone bisphosphate (37) by riboflavin and oxygen have been investigated.[69] Singlet oxygen generated by triplet energy transfer from

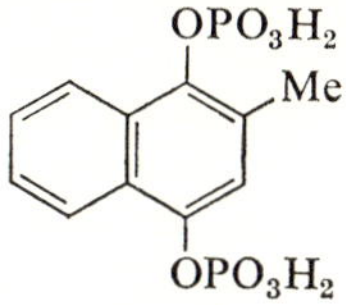

(37)

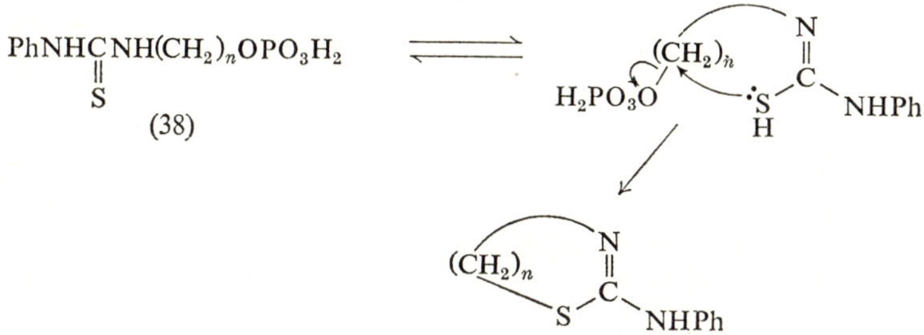

(38)

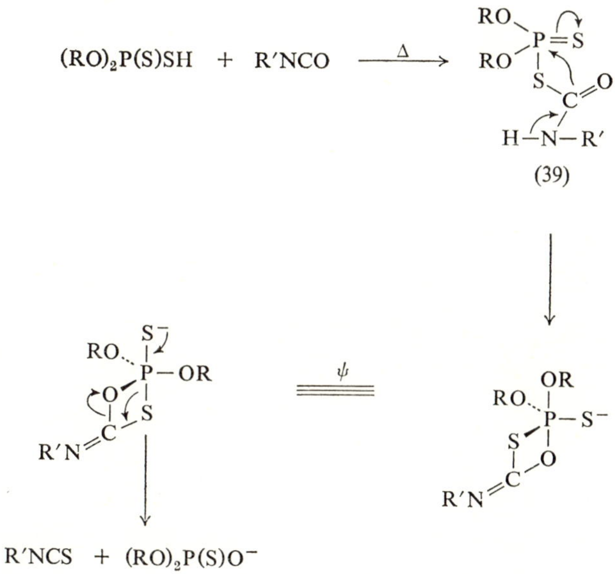

(39)

[68] G. Peiffer, A. Guillemonat, J. C. Traynard, and E. Gaydou, *Bull. Soc. chim. France*, 1969, 1304.

[69] Pill-Soon Song and T. A. Moore, *J. Amer. Chem. Soc.*, 1968, **90**, 6507.

riboflavin is the photochemically reactive species which oxidises and dephosphorylates (37). When the reaction is carried out in acetic acid as solvent, acetyl phosphate is formed.

A series of ω-(phenylthiocarbamylamino)alkyl phosphates (38) has been prepared by treating ω-aminoalkyl phosphates with phenyl isothiocyanate.[70] In the presence of acid, (38, n = 2 or 3) cyclises with elimination of phosphate to produce thiazolines or thiazines. As n increases the stability of (38) increases, and (38, n = 4) does not cyclise in acid.

OO-Dialkyl phosphorodithioic acids will add to isocyanates to give (39) which when heated to *ca.* 200° decompose to isothiocyanates.[71a] This reaction can be explained by the cyclisation of (39) to a pentaco-ordinate intermediate followed by pseudorotation and decomposition. The reaction of *OO*-dialkyl phosphorodithioic acids with cyanogen bromide, when *OO*-dialkyl phosphoroisothiocyanatidates are formed, probably proceeds by a similar mechanism.[71b]

(*ii*) *P—N Systems.* *N*-(2-Haloethyl)phosphoramidates[72] and thiophosphonamides[73] can be prepared by treating the appropriate phosphorochloridate with either a 2-haloethylamine or aziridine. *N*-(2-Bromoethyl)phosphoramidates undergo the expected displacement and dehydrohalogenation reactions,[72] while *N*-(ω-haloethyl)thiophosphonamides[73] and thiophosphoramidates[74] cyclise intramolecularly to 2-oxo-1,3,2-thiaza-phospholanes and -phosphorinanes.

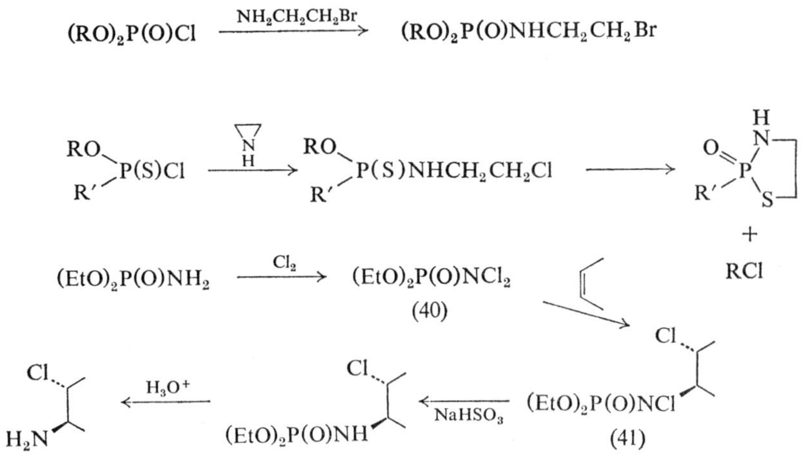

[70] E. Cherbuliez, B. Baehler, O. Espejo, E. Frankenfeld, and J. Rabinowitz, *Helv. Chim. Acta*, 1968, **49**, 2608.

[71] [(a)] G. F. Ottmann and H. Hooks jun., U.S.P. 3,409,656 (*Chem. Abs.*, 1969, **70**, 37,437). [(b)] R. R. Engel, *Canad. J. Chem.*, 1969, **47**, 1258.

[72] P. Savignac and P. Chabrier, *Compt. rend.*, 1968, **268**, *C*, 861.

[73] P. Chabrier and P. Savignac, *Compt. rend.*, 1968, **267**, *C*, 1166.

[74] P. Savignac, T. N. Thanh, and P. Chabrier, *Compt. rend.*, 1968, **266**, *C*, 1791; P. Savignac, T. N. Thanh, and P. Chabrier, *Compt. rend.*, 1968, **267**, *C*, 183.

The chlorination of *OO*-diethyl phosphoramidate with gaseous chlorine is a strongly exothermic reaction giving rise to *OO*-diethyl *NN*-dichlorophosphoramidate (40) which adds on to unsaturated hydrocarbons to give *N*-chloro-*N*-(chloroalkyl)phosphoramidates (41).[75a] Reduction of (41) followed by treatment with aqueous acid liberates 2-chloroalkylamines.

The reaction between isoquinoline, potassium cyanide, and a phosphorochloridate or phosphorothiochloridate leads to the formation of phosphorus-containing analogues of Reissert compounds.[75b] Acid hydrolysis

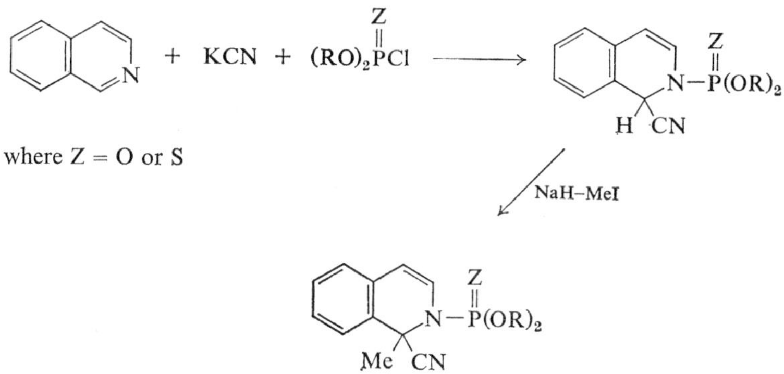

of the Reissert compounds affords isoquinoline, while they can be alkylated in the presence of base.

(*iii*) *P—S Systems.* The inhibition of acetyl cholinesterase by organophosphorus compounds (*e.g.* 42, $n = 2$) occurs as a result of phosphorylation of the active site of the enzyme. *O*-Dealkylation of the inhibitor reduces its effectiveness as a phosphorylating agent and hence reduces its inhibitory power.[76] Analogues of (42) have been prepared and their properties compared;[77] their effectiveness as inhibiting agents increases as n increases reaching a maximum at $n = 6$.[78] The findings are interpreted as indicating the presence of a hydrophobic site of limited size near the active site of the enzyme.

Trialkyl thiophosphates can be prepared in high yields from the reaction between a dialkyl phosphite and a salt of an *OO*-dialkyl thiophosphoric acid,[79] when the latter functions as an ambident nucleophile and dealkylates the phosphite. Direct alkylation of salts of *OO*-dialkyl thiophosphoric acid has also been investigated.[80]

75 (a) A. Zwierzak and A. Koziara, *Angew. Chem. Internat. Edn.*, 1968, **7**, 292. (b) D. M. Spatz and F. D. Popp, *J. Heterocyclic Chem.*, 1968, **5**, 497.

76 A. H. Aharoni and R. D. O'Brien, *Biochemistry*, 1968, **7**, 1538.

77 P. Bracha and R. D. O'Brien, *Biochemistry*, 1968, **7**, 1545.

78 P. Bracha and R. D. O'Brien, *Biochemistry*, 1968, **7**, 1555.

79 Y. A. Mandel'baum, P. G. Zaks, N. N. Mel'nikov, and V. V. Ivanov, *Khim. Org. Soedin. Fosfora, Akad. Nauk S.S.S.R., Otd. obshchei Tekh. Khim.*, 1967, 262 (*Chem. Abs.*, 1968, **69**, 2457).

80 N. T. Thanh, P. Chabrier, N. H. Phuong, and M. J. Ferrere, *Compt. rend.*, 1969, **268**, *C*, 1714.

$$(EtO)_2P(O)S(CH_2)_nNEt_2$$

(42)

$$(RO)_2P(O)H + (R'O)_2P(S)O^- \ NH_4^+ \longrightarrow (R'O)_2P(O)SR$$

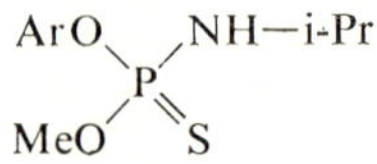

(44)

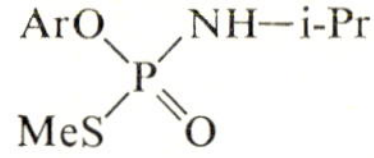

(45)

where Ar = 2,4-$Cl_2C_6H_3$—

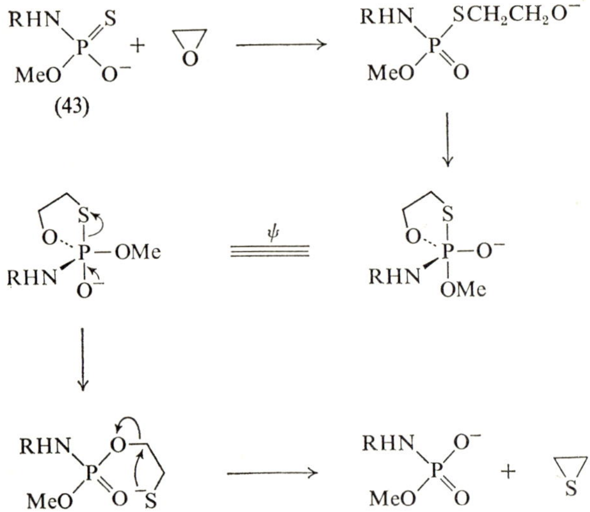

(43)

Aqueous solutions of sodium methyl *N*-cyclohexylphosphoramidothioate (43) on treatment with an excess of ethylene oxide liberate ethylene sulphide.[81] The migration of the phosphoryl residue from sulphur to oxygen can be explained in terms of the formation of a pentaco-ordinate intermediate followed by pseudorotation and expulsion of ethylene sulphide.

Stereochemical aspects of phosphorus chemistry are the subject of a recent review.[82] The optical isomers of *O*-2,4-dichlorophenyl *O*-methyl isopropylphosphoramidothioate (44) and *O*-2,4-dichlorophenyl *S*-methylisopropylphosphoramidothioate (45) have been synthesised.[83] Racemic (44) was demethylated with methylamine and the acid liberated was resolved as its quinine salt. Methylation of the optically active quinine salts with diazomethane gave (44) and (45). The value of the optical

[81] N. K. Hamer, *Chem. Comm.*, 1968, 1399.
[82] M. J. Gallagher and I. D. Jenkins, *Topics Stereochem.*, 1968, **3**, 1.
[83] J. N. Seiber and H. Tolkmith, *Tetrahedron*, 1969, **25**, 381.

rotation of the two compounds was strongly solvent dependent presumably due to the possibility of hydrogen bond formation between the solvent and the amido group. Racemic methyl 1-naphthyl phosphorothionic acid has been resolved into its optical antipodes by means of its ephedrine salt.[84]

Addition of the sulphenyl chloride (46) to an olefin gives an *OO*-diethyl-*S*-(*trans*-2-chloroalkyl) phosphorothioate (47) which on acetolysis with dry acetic acid and sodium acetate affords the diacetate of a *trans*-2-hydroxyalkanethiol (48).[85] Acetolysis with wet acetic acid and sodium acetate gives the *trans*-2-acetoxyalkanethiol (49) exclusively,[86] while the

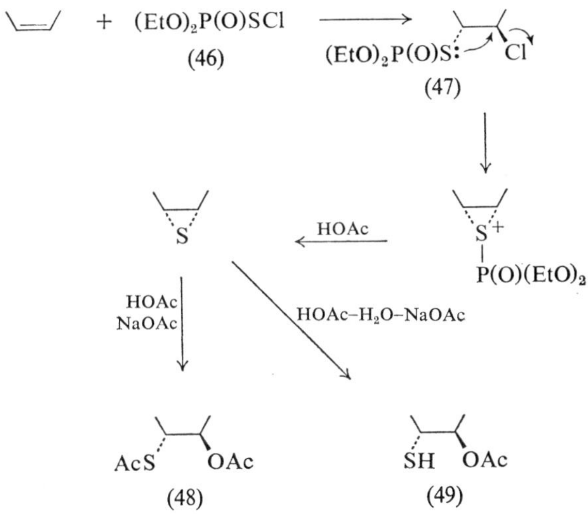

episulphide is produced with dry acetic acid in the absence of sodium acetate. The retention of configuration during acetolysis suggests the formation of an episulphonium ion which is dephosphorylated by acetic acid to the episulphide. The latter can then undergo attack by acetate ion or water to give the observed products.

OO-Dialkyl *S*-chlorothiolothionophosphates (50) on treatment with alcohols afford the corresponding esters.[87] With sodium alcoholates a more complex reaction takes place and *OOO*-trialkyl thiophosphates, salts of *OO*-dialkylthiophosphoric acid and bis-(*OO*-dialkyl thiophosphoryl) disulphides are among the products of the reaction. Grignard reagents displace halide ion from (50) and *OO*-dialkyl *S*-alkyl and *S*-aryl dithiophosphates can be prepared in this way.[88] *SS*-2-Acetaminoethyl *OO*-dialkyl thioperoxymonophosphorothioates (51) are of interest as compounds

[84] C. Donninger and D. H. Hutson, *Tetrahedron Letters*, 1968, 4871.
[85] B. Bochwic and A. Frankowski, *Tetrahedron*, 1968, **24**, 6653.
[86] B. Bochwic, A. Frankowski, and A. Kus, *Bull. Acad. polon. Sci., Ser. Sci. chim.*, 1968, **16**, 469 (*Chem. Abs.*, 1969, **70**, 76,958).
[87] L. Almasi and A. Hantz, *Monatsh.*, 1968, **99**, 1045.
[88] L. Almasi and A. Hantz, *Rev. Roumaine Chim.*, 1968, **13**, 653 (*Chem. Abs.*, 1969, **70**, 11,256).

offering protection against ionising radiation.[89] Only symmetrical disulphides can be isolated from the reaction between (50) and *S*-2-acetaminoethanethiol; however, on treating *OO*-dialkyl phosphorodithioic acid with 2-acetaminoethyl 2-acetaminoethane thiolsulphate (52), (51) was obtained in moderate yield.

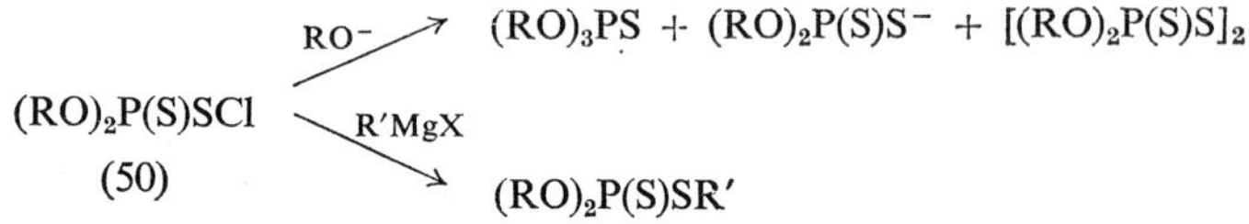

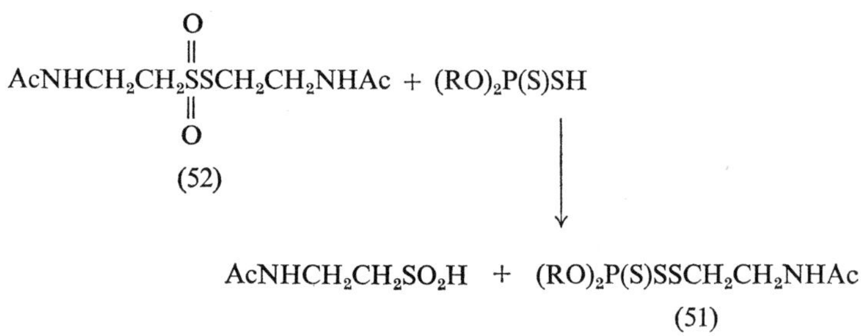

Changes in P–S force constants due to hybridisation, inductive effects, ionic charge, and π bonding have been discussed in a recent review,[90] and it has been shown that the P—S bond, unlike the P—O bond, never possesses full double-bond character. Internuclear double resonance has been used to investigate the reaction between thiophosphoryl halides and Grignard reagents.[91]

The mass spectrometric investigation of phosphorothionates,[92] phosphine oxides,[93] trimethylsilyl derivatives of nucleotides,[94] and tris(trimethylsilyl) phosphates[95] has been reported recently. One of the decomposition pathways of arylphosphine oxides and phosphorothionates requires the formation of bridged phosphofluorenyl ions;[92, 93] an alternative decomposition of phosphorothionates involves the rearrangement of P(:S)O to P(:O)S systems.[92] The mass spectrometry of volatile derivatives of phosphoric acid is being investigated as a potential method for identifying trace constituents in biological systems.

(*iv*) *Miscellaneous Reactions.* From a study of the kinetic parameters of the reaction between *OO*-diethyl phosphorodithioic acid and α,β-unsaturated

[89] D. G. Stoffey, *J. Org. Chem.*, 1968, **33**, 1651.
[90] J. Goubeau, *Angew. Chem. Internat. Edn.*, 1968, **8**, 328.
[91] R. Kosfeld, G. Hägele, and W. Kuchen, *Angew. Chem. Internat. Edn.*, 1968, **7**, 814.
[92] R. G. Cooks and A. F. Gerrard, *J. Chem. Soc.* (*B*), 1968, 1327.
[93] D. H. Williams, R. S. Ward, and R. G. Cooks, *J. Amer. Chem. Soc.*, 1968, **90**, 966.
[94] J. A. McCloskey, A. M. Lawson, K. Tsuboyama, P. M. Krueger, and R. N. Stillwell, *J. Amer. Chem. Soc.*, 1968, **90**, 4182.
[95] M. Zinbo and W. R. Sherman, *Tetrahedron Letters*, 1969, 2811.

ketones, esters, and nitriles, it has been proposed [96] that a 1,4-addition takes place with a six-centre transition state.

Phosphoryl compounds have been alkylated with triethyloxonium fluoroborate to water-sensitive quasiphosphonium compounds (53),[97] the stability of which increases as *n* increases. Unstable dialkyl t-butylperoxy phosphates (54) can be prepared by treating potassium hydroxide solutions of t-butylhydroperoxide with dialkyl phosphorochloridates.[98]

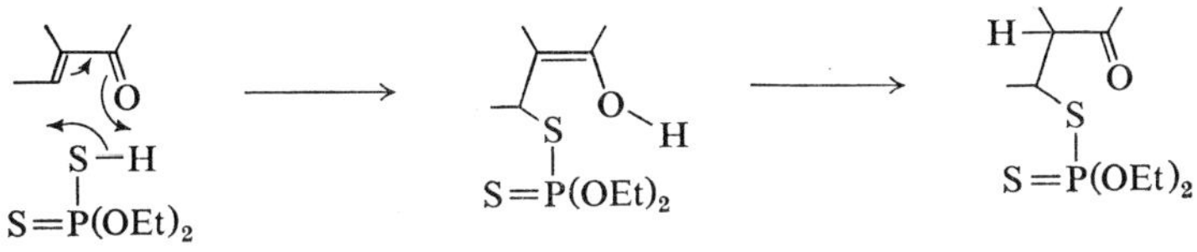

$Et_n\overset{+}{P}(OEt)_{4-n}BF_4^-$ (53)

$(RO)_2P(O)OO{-}t\text{-}Bu$ (54)

^{32}P-Labelled ferric phosphate will exchange phosphorus with trialkyl phosphates at elevated temperatures;[99] the rate of exchange is dependent on chain length with trimethyl phosphate exchanging the fastest. The mechanism of the exchange reaction is uncertain but it is postulated that pyrophosphates are involved.

D. Cyclic Derivatives of Phosphoric Acid.—The ^1H n.m.r. spectra of the two isomers of 5,5-dimethyl-2-oxo-2-piperidino-1,3,2-dioxaphosphorinane (55, X = H, Y = $C_5H_{10}N$) are essentially unchanged over a wide range of temperatures, suggesting that in this instance the ring is conformationally immobile.[100] The ^1H n.m.r. spectrum of (55, X = Cl, Y = CH_2Ph) [101] also shows little change with temperature apart from line broadening.[102] On the other hand, pronounced changes in the ^1H n.m.r. spectrum of (55, X = Br, Y = Me) occur with variation in temperature, the chemical shifts of groups attached to C(5), and to phosphorus moving to lower fields as the temperature is lowered.[102] The size of groups attached to phosphorus appears to have an important bearing on the ease of ring inversion as analogous phosphites (55, X = H, Y = lone pair) possess conformational mobility.[103]

[96] A. N. Pudovik and R. A. Cherkasov, *Zhur. obshchei Khim.*, 1968, **38**, 2532 (*Chem. Abs.*, 1969, **70**, 57,079).
[97] G. Sosnovsky and E. H. Zaret, *J. Org. Chem.*, 1969, **34**, 968.
[98] L. V. Nesterov and R. I. Mutalapova, *Tetrahedron Letters*, 1968, 51.
[99] L. Termens and K. H. Schweer, *Radiochim. Acta*, 1968, **9**, 125.
[100] R. S. Edmundson and E. W. Mitchell, *J. Chem. Soc.* (*C*), 1968, 3033.
[101] R. S. Edmundson and E. W. Mitchell, *J. Chem. Soc.* (*C*), 1968, 2091.
[102] R. S. Edmundson, *Tetrahedron Letters*, 1969, 1905.
[103] [(a)] J. H. Hargis, and W. G. Bentrude, *Tetrahedron Letters*, 1968, 5365. [(b)] D. W. White, G. K. McEwen and W. G. Bentrude, *Tetrahedron Letters*, 1968, 5369. [(c)] C. Bodkin and P. Simpson, *Chem. Comm.*, 1969, 829.

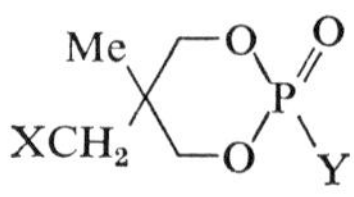

(55)

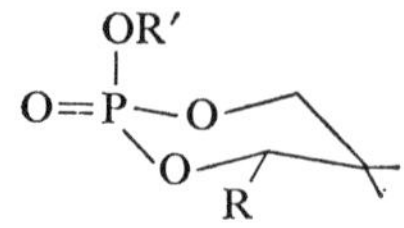

(56)

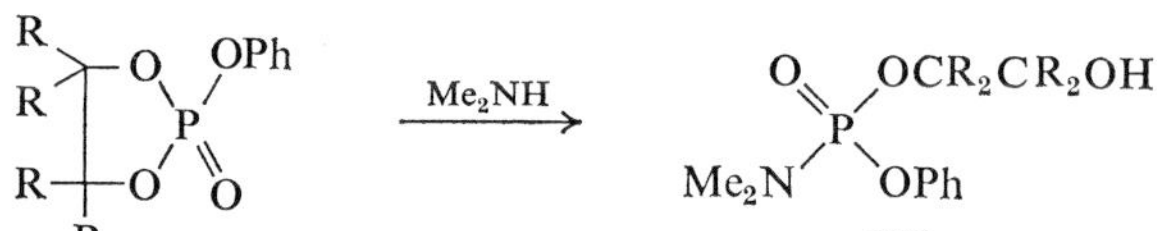

(57)

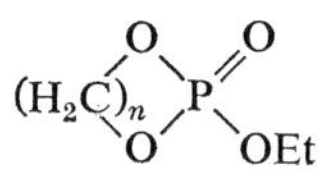

(58)

2-Oxo-2-phenoxy-1,3,2-dioxaphosphorinanes bearing substituents in the ring (*e.g.* 56, R = Ph, R′ = i-Pr), which have recently been prepared and resolved,[104] are assumed to have the chair conformation. Preliminary data on the acid hydrolysis of (56, R = Et, R′ = H) have appeared.[105] Dimethylamine will attack 2-oxo-2-phenoxy-1,3,2-dioxaphospholanes to give the phosphoramidate (57); no evidence has been found for the formation of a phosphoramidate containing an intact five-membered ring.[106] The effect of catalysts on the polymerisation of ethyl esters of cyclic phosphates (58) has been studied,[107] and the most effective catalyst when $n = 3$ is lithium aluminium hydride.

Sugar-like phosphates have been obtained by the action of anhydrous acid chlorides on the phospholene (59) prepared from glyoxal (60, R = H) and trimethyl phosphite.[108] Bromination of (59, R = Me or Ph) gives rise to α-bromo-α-ketol phosphates which can react further with excess trimethyl phosphite to give enediol bisphosphates.[109] The general reactions of pentaoxyphosphoranes are discussed in detail in Chapter 2.

[104] J. P. Majoral, A. Munoz, and J. Navech, *Compt. rend.*, 1968, **266**, *C*, 235.
[105] J. P. Majoral, J. Devillers, and J. Navech, *Compt. rend.*, 1969, **268**, *C*, 1077.
[106] M. Revel and J. Navech, *Compt. rend.*, 1969, **268**, *C*, 121.
[107] J. P. Majoral, F. Mathis, A. Munoz, J. P. Vives, and J. Navech, *Bull. Soc. chim. France*, 1968, 4455.
[108] F. Ramirez, S. L. Glaser, A. J. Bigler, and J. F. Pilot, *J. Amer. Chem. Soc.*, 1969, **91**, 496.
[109] F. Ramirez, K. Tasaka, N. B. Desai, and C. P. Smith, *J. Org. Chem.*, 1968, **33**, 25.

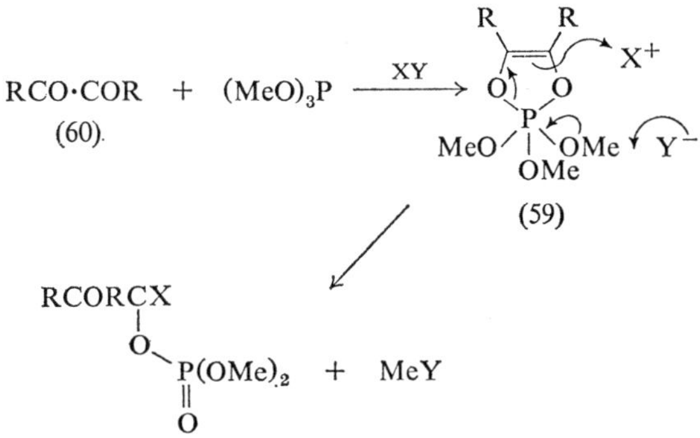

where X = H, RCO or Br
Y = Cl or Br

2 Phosphonic Acid and its Derivatives

A. Synthetic Methods.—Esters of acylphosphonic[110] and acylphosphinic[111] acids can be prepared by a variant of the Arbusov reaction when an acyl halide is treated with an alkyl phosphite or phosphonite. Oximes of acylphosphonic acids can be reduced to α-aminophosphonic acids (61).[110, 112] An alternative synthesis of (61) consists of the addition of dialkyl phosphites to Schiff bases.[113]

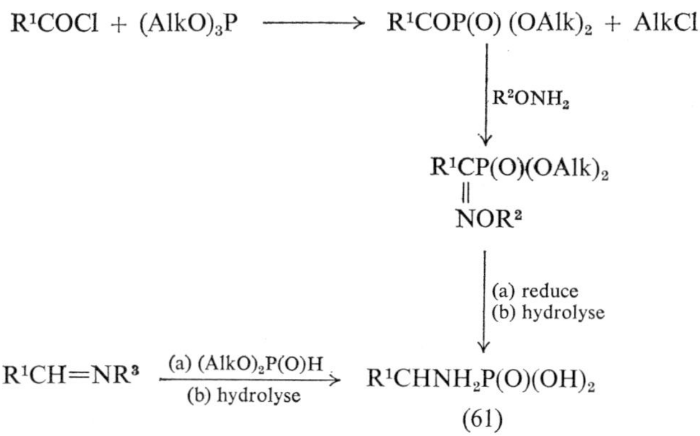

[110] K. D. Berlin, R. T. Claunch, and E. T. Gaudy, *J. Org. Chem.*, 1968, **33**, 3090.

[111] A. I. Razumov and M. B. Gazizov, *Zhur. obshchei Khim.*, 1967, **37**, 2738 (*Chem. Abs.*, 1968, **69**, 19,263).

[112] K. D. Berlin, N. K. Roy, R. T. Claunch, and D. Bude, *J. Amer. Chem. Soc.*, 1968, **90**, 4494.

[113] [(a)] H. Hoffmann and H. Förster, *Monatsh.*, 1968, **99**, 380. [(b)] N. S. Kozlov, V. D. Pak, and E. S. Elin, *Vesti Akad. Navuk. Belarus. S.S.R., Ser. khim. Navuk*, 1968, 113 (*Chem. Abs.*, 1969, **70**, 20,158).

Phosphite esters do not form pentaoxyphosphoranes with phenyl cyclobutenedione (62) as these would be derivatives of cyclobutadiene.[114] Quantitative yields of 1-alkoxy-3-dialkyl-phosphono-4-oxo-2-phenylcyclobutenes (63) are obtained, and the reactions are too rapid for the formation of phosphonium intermediates to be detected.

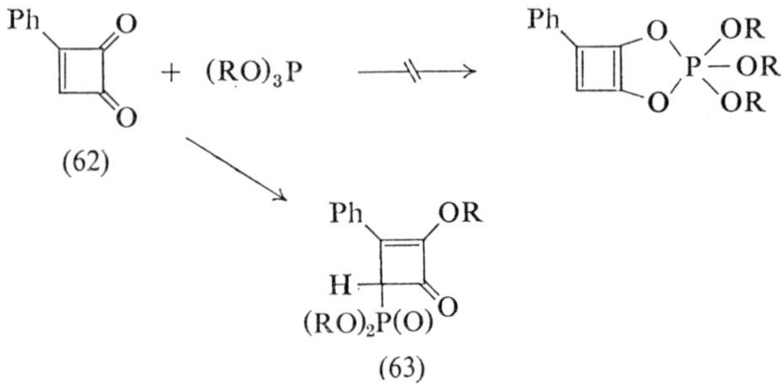

Phosphorus trichloride reacts with cholest4-en-3-one in the presence of benzoic acid to give a stable crystalline phostonyl chloride (64) in addition to the major product 3-chlorocholesta-3,5-diene.[115] This reaction has been cited as an example of attack by electrophilic phosphorus on carbonyl oxygen followed by attack by chloride ion at carbonyl carbon and then cyclisation. This mechanism appears to be unlikely as there is little proven precedent for electrophilic attack by phosphorus on carbonyl oxygen. Furthermore, addition of chloride ion to the carbonyl group prior to the cyclisation step would hinder attack by electron-rich phosphorus on the 4,5-double bond. An alternative mechanism would be addition of phosphorus trichloride to the 4,5-double bond followed by proton transfer from the benzoic acid to give the cation (65). Addition of chloride ion to the carbonyl group followed by cyclisation and partial hydrolysis would generate (64).

C-Phosphorylated formaldehyde acetals (66) are prepared by treating esters of orthoacids with tervalent phosphorus chlorides; a slow esterification of the phosphorus chloride also takes place.[116a] Similar products are obtained when dialkyl phosphites are used in place of the chloride.[117] Diethylmethylketen ketal reacts with monoalkyl esters of alkyl phosphonic and alkylphosphonothionic acids in an inert solvent to give ethyl propionate and dialkyl alkylphosphonates or *O*-alkyl *S*-alkylphosphonothiolate.[116b] It is proposed that addition of the phosphonate to the keten occurs followed by a six-centre rearrangement.

Phenothiaphosphinic acids (67) can be prepared from *p*-chlorinated diphenyl thioethers and phosphorus trichloride in the presence of aluminium

[114] R. C. De Selms, *Tetrahedron Letters*, 1968, 5545.
[115] J. A. Ross and M. D. Martz, *J. Org. Chem.*, 1969, **34**, 399.
[116] [(a)] W. Dietsche, *Annalen*, 1968, **712**, 21. [(b)] C. D. Hall, *J. Chem. Soc.* (*B*), 1968, 708.
[117] V. V. Moskva, A. Maikova, and A. I. Razumov, *Zhur. obshchei Khim.*, 1967, **37**, 1623 (*Chem. Abs.*, 1968, **68**, 39,732).

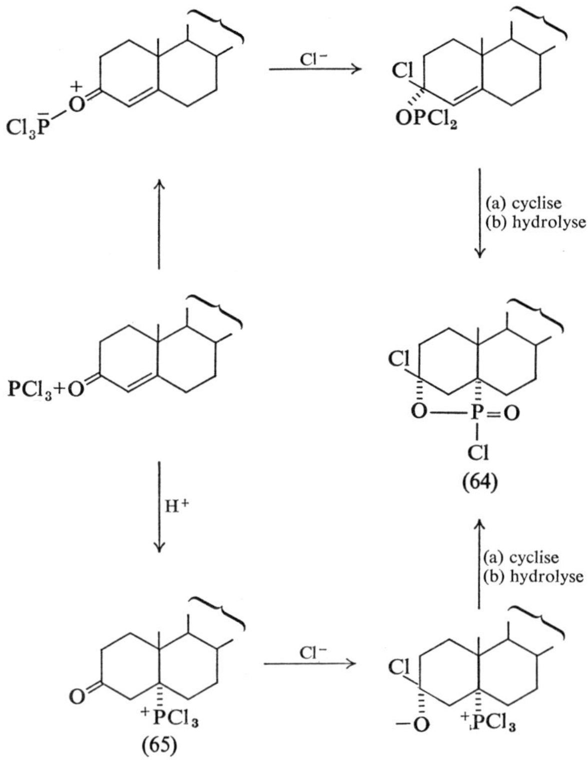

trichloride.[118a] When bromine is present in either the *ortho*- or *para*-positions, *p*-arylthiophenylphosphonic acids (68) are obtained.[118b] The formation of a *p*-phosphonic acid from either the *o*- or *p*-bromo-derivative is unusual and suggests the participation of a common symmetrical intermediate (*e.g.* 69).

Acridinium salts will add on the anion of diethyl phosphite to afford esters of 9,10-dihydroacridinephosphonic acids (70),[119] which are converted to the corresponding 9-arylidene derivatives by the action of aromatic aldehydes.[120] In the presence of Grignard reagents alkyl esters of (70) will rearrange, presumably *via* the anion.

The deoxygenation of aromatic nitro-compounds by triethyl phosphite leads to a complex mixture of products [121] which include triethyl phosphate, triethyl *N*-arylphosphorimidates, 3*H*-azepines,[122] and arylphosphonic

[118] [(a)] I. Granoth, A. Kalir, and Z. Pelah, *Israel J. Chem.*, 1968, **6**, 651. [(b)] I. Granoth, A. Kalir, Z. Pelah, and E. D. Bergmann, *Chem. Comm.*, 1969, 260.

[119] D. Redmore, *J. Org. Chem.*, 1969, **34**, 1420.

[120] W. S. Wadsworth jun. and W. D. Emmons, *J. Amer. Chem. Soc.*, 1961, **83**, 1733.

[121] R. J. Sundberg, B. P. Das, and R. H. Smith jun., *J. Amer. Chem. Soc.*, 1969, **91**, 658.

[122] J. I. G. Cadogan, R. K. Mackie, and M. J. Todd, *Chem. Comm.*, 1968, 736.

$$(RO)_2PCl + HC(OEt)_3 \longrightarrow (RO)_2P(O)CH(OEt)_2$$

(66)

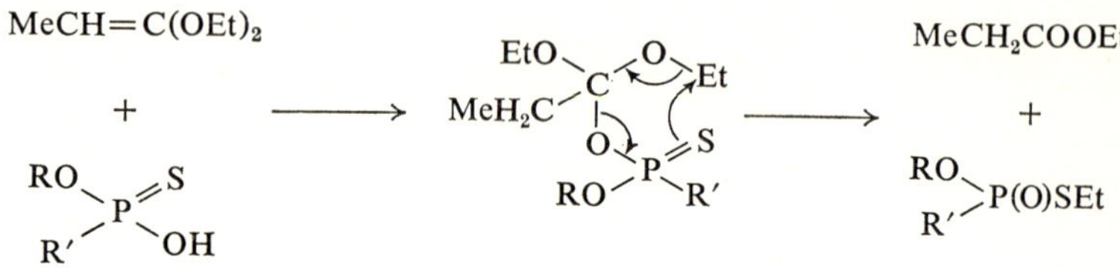

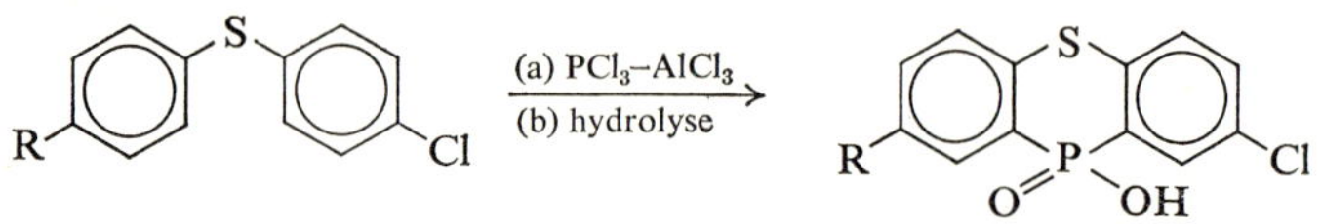

(67)

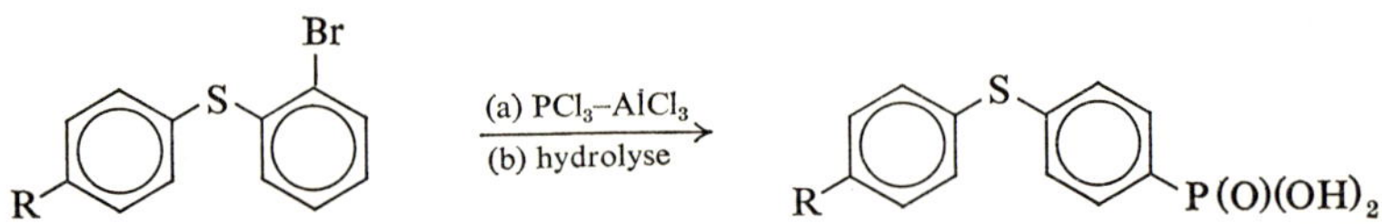

or *p*-isomer

(68)

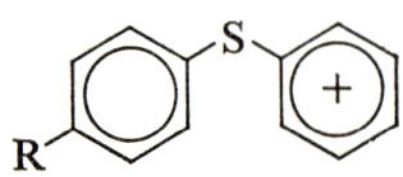

(69)

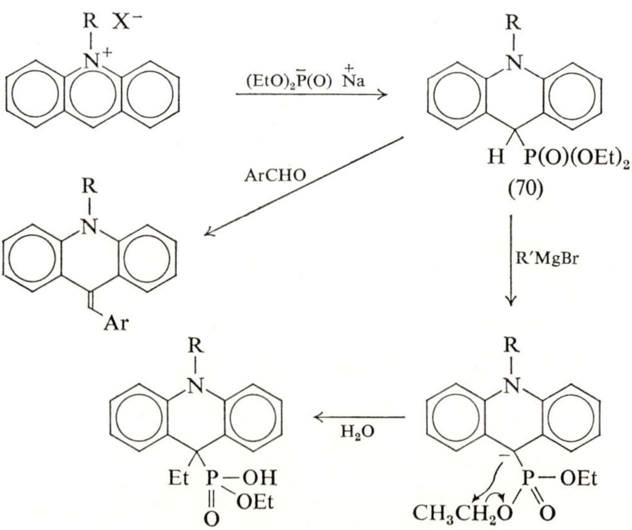

(70)

acids.[123] The mechanism of deoxygenation reactions is discussed in Chapter 10.

B. Hydrolysis of Phosphonate Esters.—The hydrolysis of diethyl 2-carboxyphenylphosphonate (71, R^1 = Et, R^2 = H) proceeds 10^7—10^8 times faster than the 4-isomer. On the other hand, the rate of hydrolysis of (71, R^1 = Et, R^2 = Me) is 10^5 times slower than the parent carboxylic acid.[124] The very rapid hydrolysis of (71, R^1 = Et, R^2 = H) suggests that neighbouring group participation is an important feature of this reaction. Studies with ^{18}O-labelled compounds and other kinetic information indicate that a pentaco-ordinate intermediate (72) is formed during the hydrolysis.

Diesters of 1-cyanovinylphosphonic acids (73) [125a] hydrolyse rapidly in aqueous sodium carbonate solution.[125b] Two reaction pathways are possible: either (73) is hydrolysed to a carbonyl compound and a diester of cyanomethylphosphonic acid, or attack by hydroxide on phosphorus takes place with the liberation of a phospho-diester and an acrylonitrile

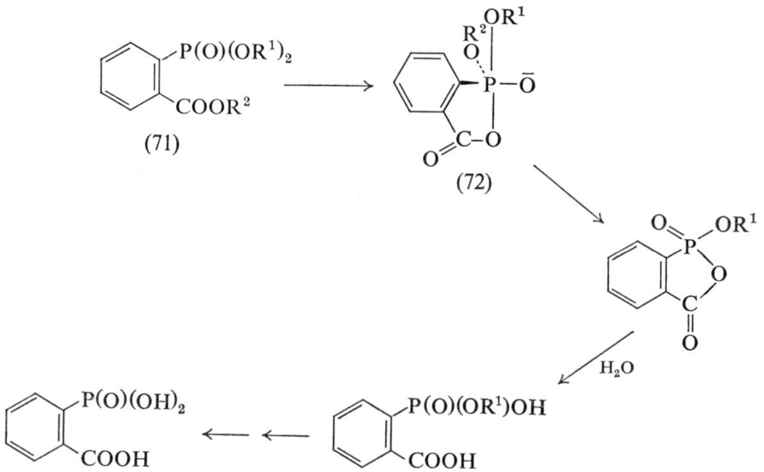

(71) (72)

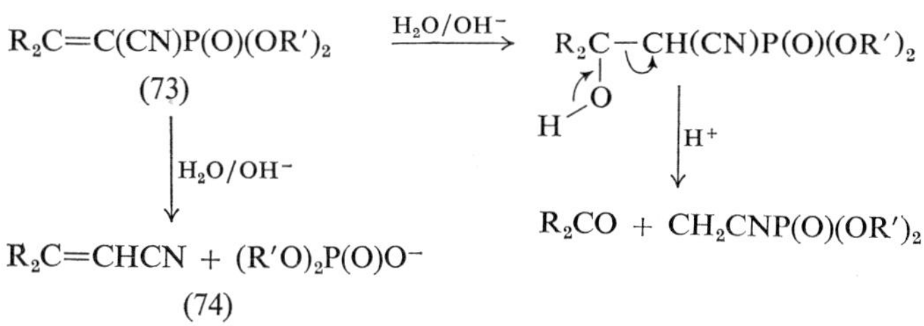

(73) (74)

[123] J. I. G. Cadogan, D. J. Sears, and D. M. Smith, *Chem. Comm.*, 1968, 1107.

[124] G. M. Blackburn and M. J. Brown, *J. Amer. Chem. Soc.*, 1969, **91**, 525.

[125] (a) D. Danion and R. Carrié, *Tetrahedron Letters*, 1968, 4537. (b) D. Danion and R. Carrié, *Compt. rend.*, 1968, **267**, *C*, 735.

(74). Alkaline hydrolysis of monoesters of (73) occurs exclusively by the second route, possibly by elimination of a metaphosphate derivative.

Data on the hydrolysis of diesters of phosphonic acids,[126] tetraethyl methylenediphosphonate,[127] and *S*-alkyl alkyl- and aryl-thio-phosphonic acids [128] have been published recently.

Cholinesterases which have been poisoned by phosphonic acids containing secondary alkoxy groups undergo relatively rapid dealkylation to the free phosphonic acid, and it has been suggested that a protonated enzyme is involved in this process.[129] The acid hydrolysis of alkyl esters of methylphosphonic acid, which has been measured [130] in aqueous dioxan, occurs with C—O bond fission. A qualitative agreement has been observed between the rate of ageing of phosphonylated cholinesterases and the rate of acid hydrolysis of the corresponding diester.

C. Reactions of Derivatives of Phosphonic Acid.—The synthesis and properties of aminoalkylphosphonic acids have been reviewed with particular regard to their complexing properties.[131, 132] Organosilicon compounds which contain phosphorus have also been reviewed recently.[133]

(*i*) *Dephosphorylation Reactions.* Anions of activated methylenephosphonates will add on to Schiff bases to afford 2-arylaminoethylphosphonates (75).[134] On heating in the presence of catalytic amounts of base (75, R = Ph), which is a P—XYZ system,[2] cleaves to liberate *trans*-stilbene and the corresponding *N*-arylphosphoramidate diester (76).[135] This decomposition has been represented as an intramolecular process which can only be true if the groups attached to the C=N of the Schiff base are in a *cis*-relationship to one another. No mention is made of the geometry of the Schiff base by the authors, and this aspect of the reaction is still unresolved.

Both chloromethylphosphonic dichloride (77) and bis(chloromethyl)-phosphinic chloride when heated with phosphorus pentachloride give rise to carbon tetrachloride and a phosphorus(III) compound.[136] Attack by $PCl_4^{(+)}$ on phosphoryl oxygen could occur followed by rapid chlorination of the chloromethyl groups. Fragmentation to carbon tetrachloride, phosphorus oxychloride, and phosphorus trichloride could then take place.

[126] H. Christol, M. Levy, and C. Marty, *J. Organometallic Chem.*, 1968, **12**, 459.
[127] V. E. Bel'skii and L. A. Kudryavtsevov, *Izvest. Akad. Nauk S.S.S.R., Ser. khim.*, 1968, 2160 (*Chem. Abs.*, 1969, **70**, 10,734).
[128] A. A. Neimysheva, V. I. Savchuk, and I. L. Knunyants, *Zhur. obshchei Khim.*, 1967, **37**, 1822 (*Chem. Abs.*, 1968, **68**, 58,979).
[129] H. P. Benschop and J. H. Keijer, *Biochim. Biophys. Acta*, 1966, **128**, 586.
[130] J. I. G. Cadogan, D. Eastlick, F. Hampson, and R. K. Mackie, *J. Chem. Soc.* (*B*), 1969, 144.
[131] M. I. Kabachnik, T. Y. Medved', N. M. Dyatlova, O. G. Arkhipova, and M. V. Rudomino, *Russ. Chem. Rev.*, 1968, **37**, 503.
[132] D. Forest and G. Thomas, *Bull. Soc. chim. France*, 1968, 3441.
[133] E. A. Chernyshev and E. F. Bugerenko, *Organometallic Chem. Rev.*, 1968, **3**, 469.
[134] M. Kirilov and J. Petrova, *Monatsh.*, 1968, **99**, 148.
[135] M. Kirilov and J. Petrova, *Chem. Ber.*, 1968, **101**, 3467.
[136] A. W. Frank, *Canad. J. Chem.*, 1968, **46**, 3573.

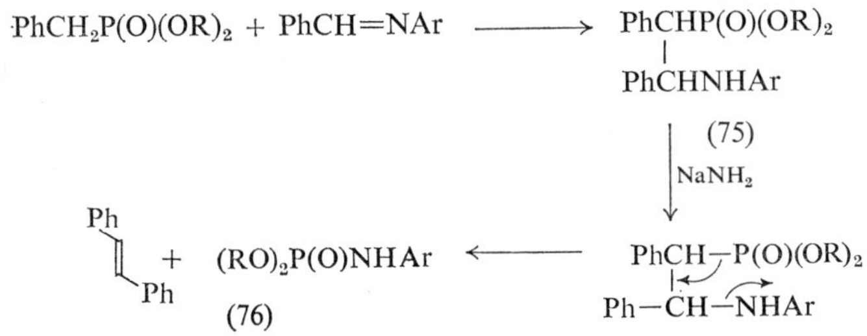

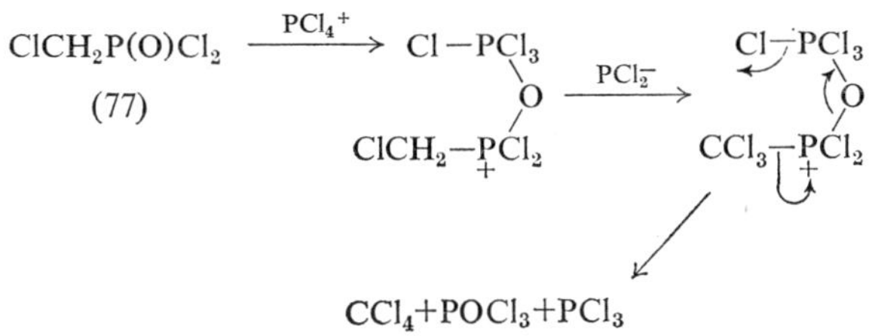

(*ii*) *α,β-Unsaturated Phosphonic Acids.* Unsaturated phosphonic acids may be prepared by the condensation of a carbonyl compound with a derivative of methylenephosphonic acid,[137] by elimination reactions,[138] by the reaction of a carbene with α-diazophosphonic acids,[139] or by an Arbusov reaction between a trialkyl phosphite and an acetylene halide.[140, 141] The phosphoryl group, like the carbonyl group, is electron withdrawing and hence α,β-unsaturated phosphonates can undergo the Michael [142] and the Diels–Alder reactions.[140, 141]

Esters of allenylphosphonic acid (79) can be isolated from the reaction between acetylene alcohols (78) and tervalent phosphorus halides.[143] Acetylene phosphates are formed initially in this reaction and then isomerise slowly to (79). Conditions for the selective hydrogenation of allenyl-

137 [(a)] A. N. Pudovik, G. E. Yastrebova, and V. I. Nikitina, *Zhur. obshchei Khim.*, 1968, **38**, 300 (*Chem. Abs.*, 1968, **69**, 27,480. [(b)] A. N. Pudovik, G. E. Yastrebova, and Y. Y. Samitov, *Zhur. obshchei Khim.*, 1968, **38**, 292 (*Chem. Abs.*, 1968, **69**, 106,815). [(c)] A. N. Pudovik, G. E. Yastrebova, V. I. Nikitina, and E. K. Mukhametzyanova, *Khim. Org. Soedin. Fosfora, Akad. Nauk S.S.S.R., Otd. obshchei Tekh. Khim.*, 1967, 18 (*Chem. Abs.*, 1968, **69**, 59,331).

138 V. S. Abramov, A. L. Shalman, and Z. V. Molodykh, *Zhur. obshchei Khim.*, 1968, **38**, 541 (*Chem. Abs.*, 1968, **69**, 27,487).

139 D. Seyferth, J. D. H. Paetsch, and R. S. Marmor, *J. Organometallic Chem.*, 1969, **16**, 185.

140 G. Peiffer, A. Guillemonat, J. C. Traynard, and M. Faure, *Compt. rend.*, 1969, **286**, *C*, 358.

141 D. Seyferth and J. D. H. Paetsch, *J. Org. Chem.*, 1969, **34**, 1483.

142 A. N. Pudovik and N. G. Khusainova, *Khim. Org. Soedin. Fosfora, Akad. Nauk S.S.S.R., Otd. obshchei Tekh. Khim.*, 1967, 53 (*Chem. Abs.*, 1968, **69**, 36,223).

143 [(a)] V. N. Pastushkhov, E. S. Vdovina, Y. A. Kondrat'ev, S. Z. Ivin, and V. V. Tarasov, *Zhur. obshchei Khim.*, 1968, **38**, 1408 (*Chem. Abs.*, 1968, **69**, 87,101). [(b)] V. N. Pastushkhov, Y. A. Arbisma, Y. A. Kondrat'ev, S. Z. Ivin, and A. S. Vasil'eva, *Zhur. obshchei Khim.*, 1968, **38**, 1405 (*Chem. Abs.*, 1968, **69**, 87,102).

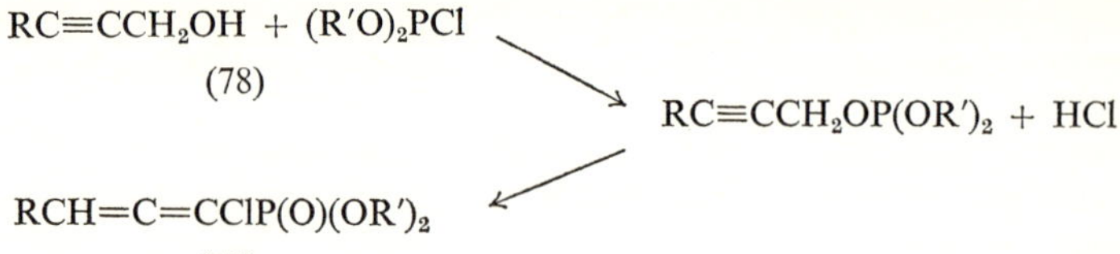

(78) (79)

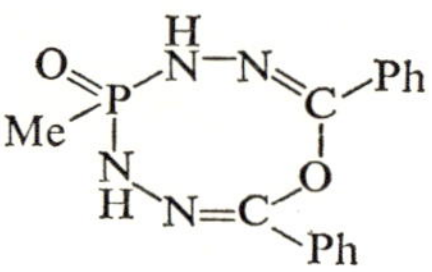

(80)

phosphonates to vinylphosphonates[144] and the epoxidation of vinylphosphonates[145] have been reported recently.

(*iii*) *P—N and P—S Systems.* The preparation of phosphonamidates by ester exchange[146] or halide ion displacement[147] has been described, and it is

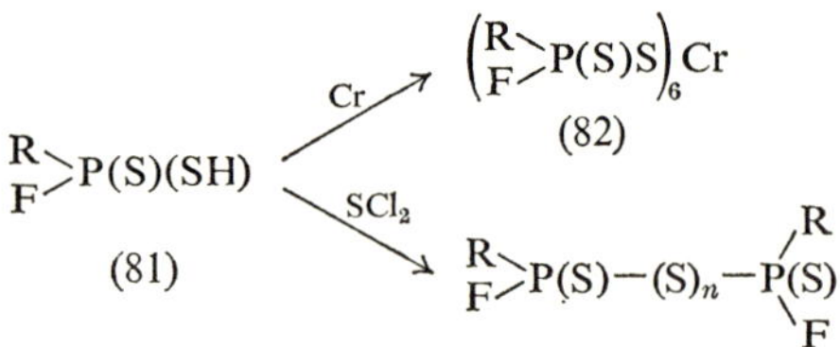

(81) (82) (83)

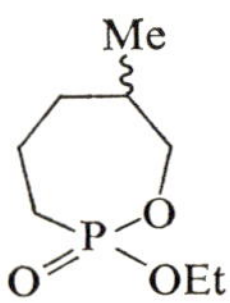

(84)

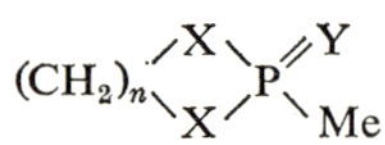

(85)

where X=O, S:Y=O, S, Se

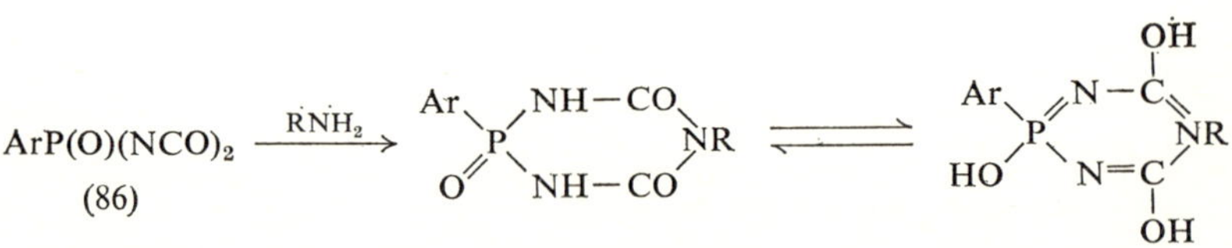

(86) (87)

[144] A. A. Petrov, B. I. Ionin, and V. M. Ignat'ev, *Tetrahedron Letters*, 1968, 15.

[145] K. Hunger, *Chem. Ber.*, 1968, **101**, 3530.

[146] P. M. Zavlin and T. F. Kondyurina, *Zhur. obshchei Khim.*, 1967, **37**, 2787 (*Chem. Abs.*, 1968, **69**, 2981).

[147] [a] N. N. Mel'nikov, N. V. Lebedeva, N. K. Daragan, and A. F. Grapov, *Zhur. obshchei Khim.*, 1967, **37**, 2760 (*Chem. Abs.*, 1968, **69**, 19,259). [b] N. N. Mel'nikov, L. V. Razvodovskaya, and A. F. Grapov, *Zhur. obshchei Khim.*, 1968, **38**, 2648 (*Chem. Abs.*, 1969, **70**, 78,083).

claimed that benzoyl hydrazine will react with methylphosphonodichloridate to give heterocyclic products (80) as well as the expected methylphosphonohydrazidate.[148]

Fluorinating agents will convert alkyl phosphonodithioic acid anhydrides into alkylphosphonodithiofluoridates (81)[149, 150] which will react with transition metals to form volatile complexes (82).[151] Sulphur halides act upon (81) to afford bis(alkylphosphonothiofluoride)sulphides (83).[152]

(*iv*) *Cyclic Derivatives of Phosphonic Acid.* The occurrence of geometric isomerism in a seven-membered ring phosphonate (84) has been reported.[153] Crude (84), which can be prepared by treating 1,5-dibromo-2-methylpentane with triethyl phosphite, has a ^{1}H n.m.r. spectrum containing two doublets for the 6-methyl group. The relative areas of these doublets correspond to the relative amounts of the *cis*- and *trans*-isomers which can be obtained by g.l.c.

Alkylphosphonodichloridates together with their sulphur and selenium analogues give cyclic products (85) with diols and dithiols.[154, 155] Arylphosphonodi-isocyanatidates (86) react with amines to give 1,3,5,2-triazaphosphorinanes (87) which are enolised to a considerable extent.[156]

(*v*) *Miscellaneous Reactions.* During the rearrangement of $\alpha\beta$-epoxyvinylphosphonates, the phosphonyl group migrates more readily than hydrogen or an alkyl group, but less readily than a phenyl group.[157, 158] The reactions of dialkyl aroylphosphonates with peracids,[159] thiamine (88),[160] and other thiazolium salts[161] all proceed with migration of a phosphonyl group.

The photo-reduction of β-ketophosphonates leads to β-hydroxyphosphonates,[162] and if the intermediate acetal radical is stabilised a pinacol is produced.

Tetraethyl dimethylaminomethylenediphosphonate (89) can be methylated to a strong acid which ionises to form a stable ylide (90).[163a] (89) has

148 N. N. Mel'nikov, O. B. Mikhailova, and A. F. Grapov, *Zhur. obshchei Khim.*, 1968, **38**, 2099 (*Chem. Abs.*, 1969, **70**, 47,539).
149 (*a*) H. W. Roesky, *Chem. Ber.*, 1968, **101**, 3679. (*b*) H. W. Roesky, *Z. Naturforsch. B*, 1969, **24**, 5.
150 V. N. Aleksandrov and V. I. Emel'yanov, *Zhur. obshchei Khim.*, 1967, **37**, 2714 (*Chem. Abs.*, 1968, **69**, 2982).
151 H. W. Roesky, *Angew. Chem. Internat. Edn.*, 1968, **7**, 815.
152 H. W. Roesky, *Inorg. Nuclear Chem. Letters*, 1969, **5**, 453.
153 K. Bergesen, *Acta Chem. Scand.*, 1968, **22**, 1366.
154 M. Wieber and H. U. Werther, *Monatsh.*, 1968, **99**, 1153.
155 A. N. Shapovalova and S. M. Shner, *Zhur. obshchei Khim.*, 1968, **38**, 317 (*Chem. Abs.*, 1968, **69**, 27,482).
156 G. Tomaschewski, C. Berseck, and G. Hilgetag, *Chem. Ber.*, 1968, **101**, 2073.
157 R. H. Churi and C. E. Griffin, *J. Amer. Chem. Soc.*, 1966, **88**, 1824.
158 M. Sprecher and D. Kost, *Tetrahedron Letters*, 1969, 703.
159 M. Sprecher and E. Nativ, *Tetrahedron Letters*, 1968, 4405.
160 A. Takamizawa, Y. Mori, H. Sato, and S. Tanaka, *Chem. and Pharm. Bull.* (*Japan*), 1968, **16**, 1773.
161 A. Takamizawa, Y. Hamashima, and H. Sato, *J. Org. Chem.*, 1968, **33**, 4038.
162 H. Tomioka, Y. Izawa, and Y. Ogata, *Tetrahedron*, 1969, **24**, 1501.
163 (*a*) H. Gross and B. Costisella, *Angew. Chem. Internat. Edn.*, 1968, **7**, 463. (*b*) H. Gross and B. Costisella, *ibid.*, 391.

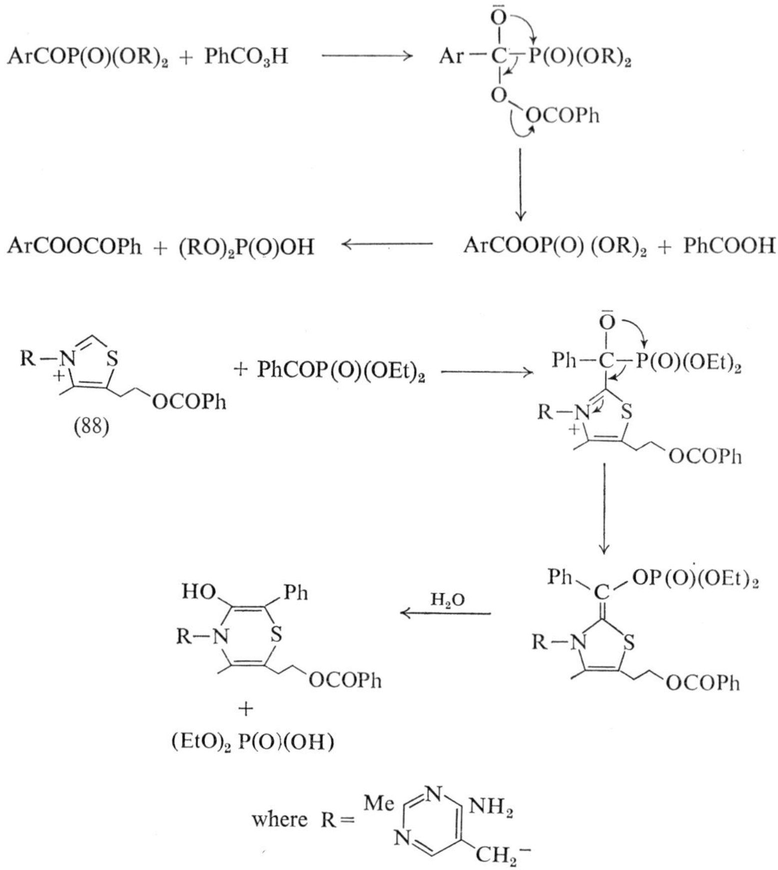

been used to convert aldehydes to carboxylic acids containing one more carbon atom.[163b] Methylenediphosphonic acid is an analogue of malonic acid and metalated derivatives of the former have been prepared.[164] 1-Hydroxyalkylidenediphosphonic acids (91)[165a] on treatment with acetic anhydride produce diacetates (92) which on treatment with base give bicyclic dimers (93).[165b]

The products (94) of the reaction between diethyl phosphonate and Grignard reagents contain tervalent phosphorus.[166a, b] Hydrolysis or alkylation of (94) leads to secondary or tertiary phosphine oxides. Grignard derivatives of dialkyl phosphites react with α-allenyl ketones to give

164 O. T. Quimby, J. D. Curry, D. A. Nicholson, J. B. Prentice, and C. H. Roy, *J. Organometallic Chem.*, 1968, **13**, 199.

165 (a) F. Kašpárek, *Monatsh.*, 1968, **99**, 2016. (b) O. T. Quimby and J. B. Prentice, U.S.P. 3,400,151 (*Chem. Abs.*, 1969, **70**, 78,147).

166 (a) H. R. Hays, *J. Org. Chem.*, 1968, **33**, 3690. (b) H. R. Hays, *ibid.*, 4201.

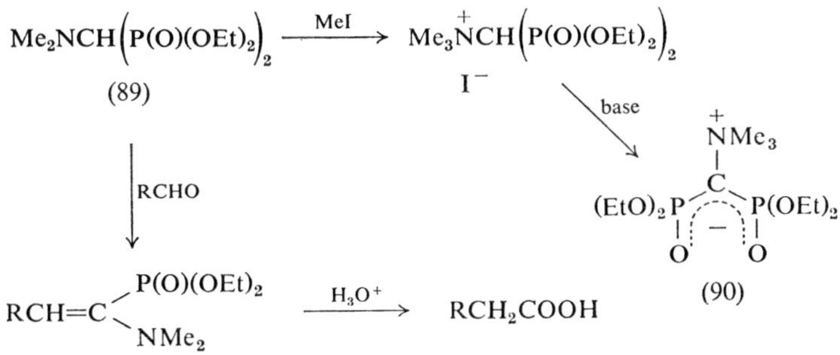

(89) (90)

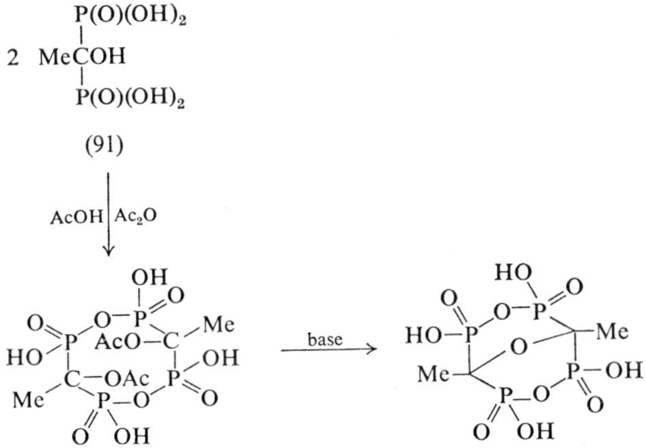

(91) (92) (93)

α,β-unsaturated γ-ketophosphonates.[167] This reaction can be represented as a 1,4-addition followed by isomerisation of the intermediate.

Spectroscopic evidence indicates the formation in tetrahydrofuran of a 1 : 1 complex between the phosphoryl oxygen atom of ethyl diphenylphosphinate and phenylmagnesium bromide.[168a] The reorganisation of this complex to a pentaco-ordinate intermediate obeys first-order kinetics and pseudorotation of the intermediate followed by expulsion of ethoxide leads to triphenylphosphine oxide.

This mechanism implies that a racemic product would be expected from the reaction between a Grignard reagent and an optically active phosphinate; however, inversion of configuration has been observed both in this reaction,[168b] and in the reaction between an optically active phosphinate and an amine.[168c] Moreover, treatment of dextrorotatory

[167] G. Peiffer, A. Guillemonat, and G. Buono, *Bull. Soc. chim. France*, 1969, 946.

[168] [(a)] H. R. Hays, *J. Amer. Chem. Soc.*, 1969, **91**, 2736. [(b)] O. Korpiun, R. A. Lewis, J. Chickos, and K. Mislow, *J. Amer. Chem. Soc.*, 1968, **90**, 4842. [(c)] A. Nudelman and D. J. Cram, *J. Amer. Chem. Soc.*, 1968, **90**, 3869. [(d)] H. P. Benschop, G. R. Van den Berg, and H. L. Boter, *Rec. Trav. chim.*, 1968, **87**, 387.

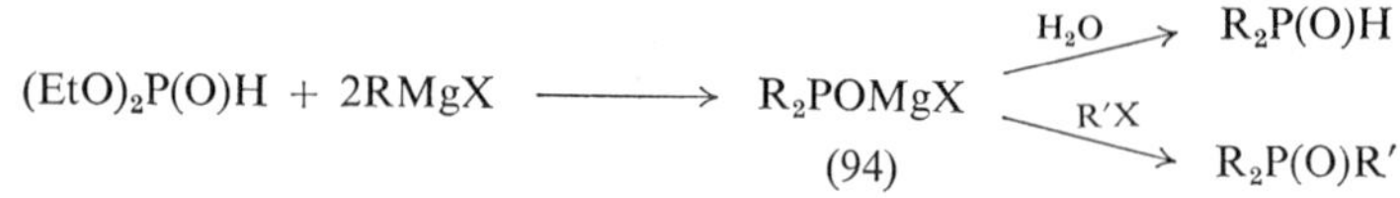

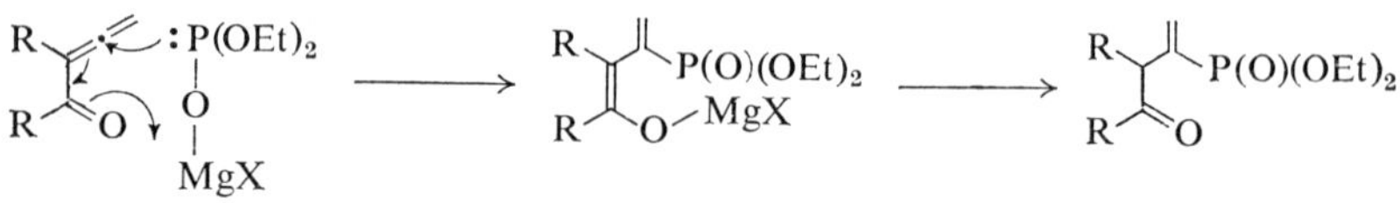

O-isopropyl *S*-methyl methylphosphonate (95) with phenylmagnesium bromide affords *O*-isopropyl methylphenylphosphinate (96) with inversion of configuration.[168d] (96) in turn can be converted into a phosphine oxide (97) by further treatment with a Grignard reagent. These reactions have been used to deduce that the most active form of the cholinesterase Sarin is *R*-(−)-isopropyl methylphosphonofluoridate (98).

Dimethyl phosphonate when heated with phenyl isocyanate gives *inter alia* carbon dioxide and diphenyl carbodi-imide.[169] This is an example of

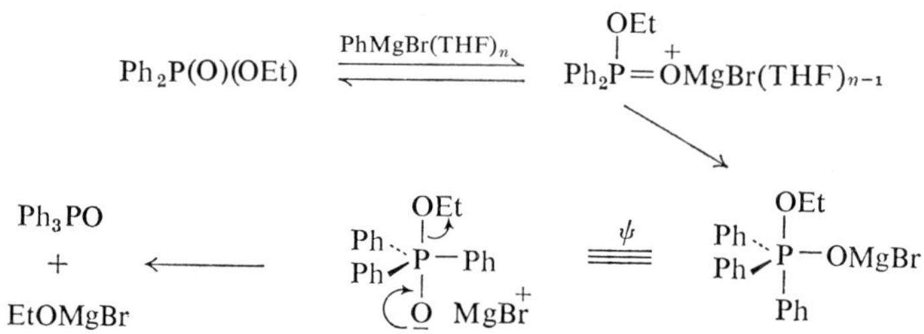

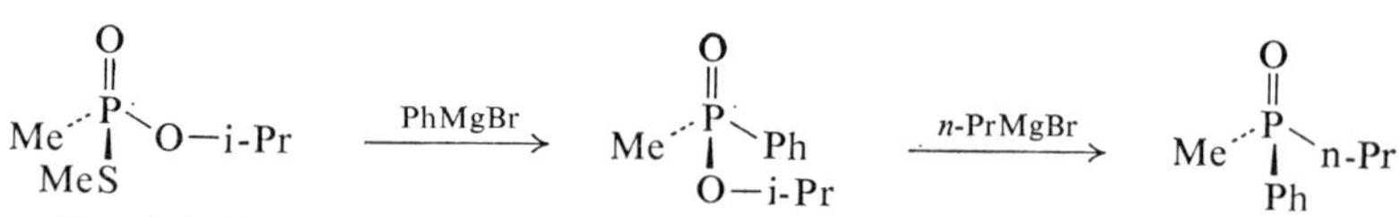

(S)−(+) (95) (R)−(−) (96) (S)−(−) (97)

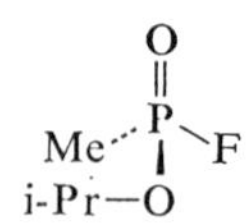

(R)(−) (98)

[169] F. K. Samigulin, I. M. Kafengauz, E. L. Gefter, and A. P. Kafengauz, *Zhur. obshchei Khim.*, 1968, **38**, 1766 (*Chem. Abs.*, 1969, **70**, 4237).

the well-characterised reaction between phosphine oxides and phenyl isocyanate.[170]

The equilibrium transalkylation between diesters of phosphonic acid and alkyl halides or tosylates has been studied as a possible method for the synthesis of phosphonate esters.[171] Alkyl methylphosphonochloridates will esterify phosgene oxime (99) to produce (100). Similar compounds can be obtained by the action of trialkyl phosphites on halogenonitroalkanes.[172] (100) is a P—XYZ system and hydrolyses rapidly in alkali.

Acyl nitriles of pyruvic and benzoyl formic acids will add on dialkyl phosphonates to give (101) which isomerise on heating to (102).[173] Presumably the isomerisation involves the formation of a pentaco-ordinate intermediate followed by pseudorotation and fission of the P—C bond. Similar rearrangements have been invoked for the Perkow reaction.[174]

(99) (100)

ArCOP(O)(OR)$_2$
(103)

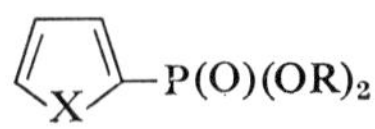

(104)

where X = NH, O or S

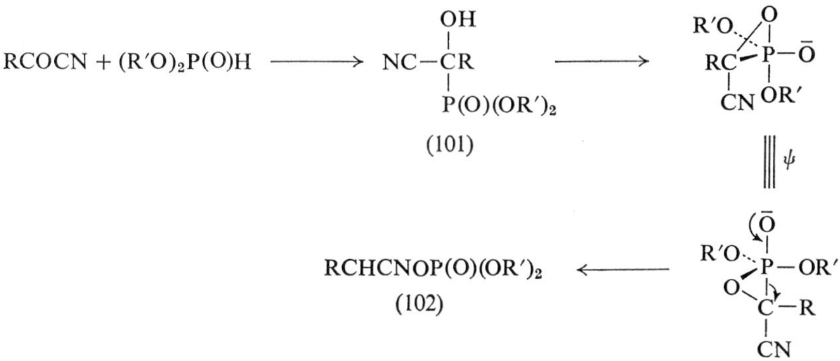

(101)

(102)

170 J. J. Monagle, T. W. Campbell, and H. F. McShane jun., *J. Amer. Chem. Soc.*, 1962, **84**, 4288.
171 H. M. Bell, *J. Org. Chem.*, 1969, **34**, 681.
172 R. N. Sterlin, B. N. Evplov, and V. M. Izmailov, *Zhur. Vses. Khim. Obshchest.*, 1968, **13**, 118 (*Chem. Abs.*, 1968, **69**, 19,254).
173 A. N. Pudovik, Y. Y. Samitov, I. V. Gur'yanova, and L. V. Banderova, *Khim. Org. Soedin. Fosfora, Akad. Nauk S.S.S.R., Otd. obshchei Tekh. Khim.*, 1967, 45 (*Chem. Abs.*, 1968, **69**, 67,489).
174 P. A. Chopard, V. M. Clark, R. F. Hudson, and A. J. Kirby, *Tetrahedron*, 1965, **21**, 1961.

Aroylphosphonates (103) are usually yellow although their carbon analogues (*e.g.* benzoyl formates) are colourless. An examination of the u.v. spectra of (103) indicates that both the $\pi \rightarrow \pi^*$ and $n \rightarrow \pi^*$ transitions occur at longer wavelengths than the carbon analogues, an effect which may be due to involvement of d orbitals in the phosphonyl bond.[175] The ^{1}H n.m.r. spectra of a number of π excessive heteroarylphosphonates (104) show that d_π–p_π bonding occurs between the phosphonyl group and the heteroaromatic ring. This effect is more important with (104, X = O or S) than with (104, X = NH).[176] Substituent effects on geminal P–C–H coupling constants in ^{1}H n.m.r. spectra have been measured for a number of organophosphorus compounds.[177]

3 Phosphinic Acid and its Derivatives

A. Synthetic Methods.—Secondary phosphines (105, R = C_6H_{11} or Ph) will react with hexafluoroacetone to generate esters of phosphinic acids.[178] It is suggested that this reaction takes place with attack by phosphorus on carbonyl carbon followed by rearrangement and aerial oxidation.

Bis(hydroxymethyl)phosphinic acid (106) can be chlorinated in high yield to produce bis(chloromethyl)phosphinyl chloride (107). Subsequent

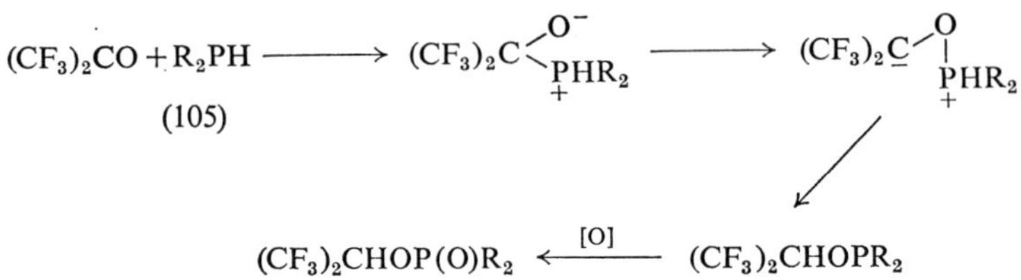

(105)

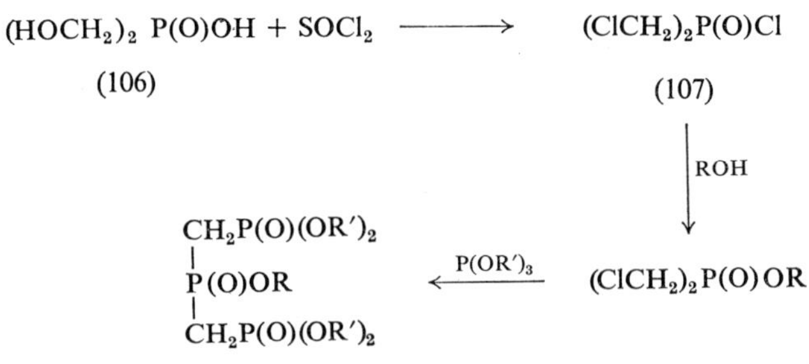

(106) (107) (108)

[175] K. Terauchi and H. Sakurai, *Bull. Chem. Soc. Japan*, 1969, **42**, 821.
[176] R. H. Kemp, W. A. Thomas, M. Gordon, and C. E. Griffin, *J. Chem. Soc.* (*B*), 1969, 527.
[177] M. J. Gallagher, *Austral. J. Chem.*, 1968, **21**, 1197.
[178] R. F. Stockel, *Chem. Comm.*, 1968, 1594.

treatment of (107) with phosphite esters leads to bis(dialkoxy methylphosphonyl)phosphinates (108).[179a, b] Bis(dialkoxy methylphosphonyl)-phosphine oxides[179c] and tris(dialkoxy methylphosphonyl)phosphinic acids[179d] can be prepared in an analogous manner.

B. Hydrolysis of Derivatives of Phosphinic Acid.—1-Ethoxyphosphole 1-oxide[180] dimerises readily to 1,8-diethoxy-3a,4,7,7a-tetrahydro-4,7-phosphinidenephosphindole 1,8-dioxide (109) whose structure has been confirmed by *X*-ray crystallography.[181] (109) can be hydrogenated to (110)[182] and both compounds undergo rapid hydrolysis of one of the two ester groups under either acidic or basic conditions.[183] ^{31}P-nuclear magnetic

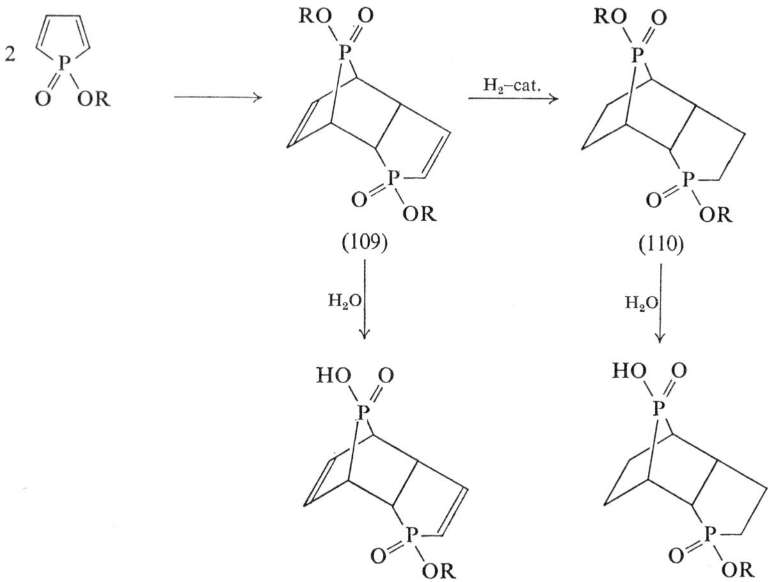

resonance studies indicate that the rapid hydrolysis occurs at the phosphorus atom in the bridge position, *i.e.* the position of maximum strain. While the pH–rate profile above pH 2 corresponds to that usually obtained during the hydrolysis of phosphate esters, the rate does not continue to rise in proportion to the acid strength below this pH value but reaches a maximum in 0·2M acid. During the hydrolyses of (109) and (110), water must enter and alcohol must leave from an apical position in the pentaco-ordinate intermediate. To achieve this, pseudorotation must occur and it is

179 (a) L. Maier, *Angew. Chem. Internat. Edn.*, 1968, **7**, 384. (b) L. Maier, *Helv. Chim. Acta*, 1969, **52**, 827. (c) L. Maier, *ibid.*, 845. (d) L. Maier, *ibid.*, 858.

180 D. A. Usher and F. H. Westheimer, *J. Amer. Chem. Soc.*, 1964, **86**, 4732.

181 Y. Y. H. Chiu and W. N. Lipscomb, *J. Amer. Chem. Soc.*, 1969, **91**, 4150.

182 (a) R. Kluger, F. Kerst, D. G. Lee, and F. H. Westheimer, *J. Amer. Chem. Soc.*, 1967, **89**, 3918. (b) R. Kluger, F. Kerst, D. G. Lee, E. A. Dennis, and F. H. Westheimer, *ibid.*, 3919.

183 R. Kluger and F. H. Westheimer, *J. Amer. Chem. Soc.*, 1969, **91**, 4143.

suggested[183] that in solutions of pH < 1 the pseudorotation process itself is rate limiting which would account for the deviation of the pH–rate profiles. Although the alkaline hydrolysis of the phosphinate ester (111, R = Et) might be expected to be very rapid due to relief of steric strain, the rate is approximately the same as that for the alkaline hydrolysis of triethyl phosphate.[184] Since the alkaline hydrolysis of (112, R = Et) is very rapid, relief of steric strain during the hydrolysis of (111, R = Et) must be counterbalanced by a retardation due to steric hindrance by the neighbouring methyl groups. A similar effect is observed in the hydrolysis of t-butyl phosphinates (113, R^1 = t-Bu, R^2 = H). One t-butyl group attached to phosphorus produces little steric hindrance; however, substitution of phosphorus by two t-butyl groups (*e.g.* 113, R^1 = R^2 = t-Bu) causes a sharp drop in the rate of hydrolysis.[185] For (113, R^1 = R^2 = t-Bu) one t-butyl group must occupy an equatorial position in the pentaco-ordinate intermediate (114) while the other t-butyl group occupies an apical position. In the transition state leading to this intermediate, steric hindrance by the 'equatorial' t-butyl group inhibits attack by hydroxide ion on phosphorus.

The effect of steric hindrance on the alkaline hydrolysis of acyclic and cyclic phosphinates has been determined.[186] In the series (115), (116), and (117), additional steric strain due to the introduction of a double bond into the five-membered ring in (116) causes an increase in the rate of alkaline hydrolysis. However, when the double bond is in the α,β-position relative to phosphorus (*e.g.* 117), the electron density on phosphorus is reduced and the rate of hydrolysis is lower than for (115). 1,4-Addition of dichlorophenylphosphine to diacetone alcohol leads to 3-hydroxy-2-oxo-2-phenyl-3,5,5-trimethyl-1,2-oxaphospholane (118).[187] Cleavage of the five membered ring occurs on alkaline hydrolysis and the rate is ten times slower than that for the alkaline hydrolysis of 2-ethoxy-1,2-oxaphospholane (119)[188] presumably due to steric hindrance.

No exchange of oxygen occurs during the hydrolysis of esters of diphenylphosphinic acid[189] and there is little electronic interaction between the phosphoryl group and aromatic rings. On the other hand, the hydrolytic behaviour of substituted aryl esters of diphenylphosphinic acid is very similar to that of aryl benzoates.[190] Kinetic parameters for the alkaline hydrolysis of phenyl bis(α-halogenoalkyl)phosphinates[191] and the acidic hydrolysis of diarylphosphinamides[192] have been published. It is of

[184] K. Bergesen, *Acta Chem. Scand.*, 1967, **21**, 1587.
[185] W. Hawes and S. Trippett, *Chem. Comm.*, 1968, 577.
[186] P. Haake, R. D. Cook, W. Schwarz, and D. R. McCoy, *Tetrahedron Letters*, 1968, 5251.
[187] K. Bergesen, *Acta Chem. Scand.*, 1969, **23**, 696.
[188] G. Aksnes and K. Bergesen, *Acta Chem. Scand.*, 1966, **20**, 2508.
[189] P. Haake, C. E. Diebert, and R. S. Marmor, *Tetrahedron Letters*, 1968, 5247.
[190] P. Haake, D. R. McCoy, W. Okamura, S. R. Alpha, S. Y. Wong, D. A. Tyssee, J. P. McNeal, and R. D. Cook, *Tetrahedron Letters*, 1968, 5243.
[191] V. E. Bel'skii, M. V. Efremova, and A. R. Panteleeva, *Izvest. Akad. Nauk S.S.S.R., Ser. khim.*, 1968, 2278 (*Chem. Abs.*, 1969, **70**, 28,109).
[192] G. Tomaschewski and G. Kuehn, *J. prakt. Chem.*, 1968, **38**, 222.

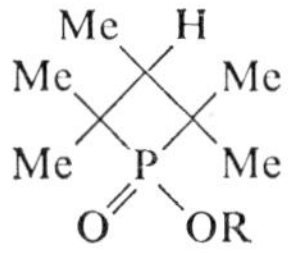
(111)

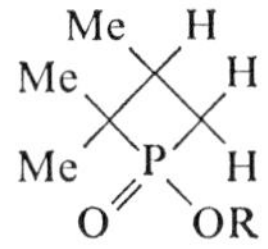
(112)

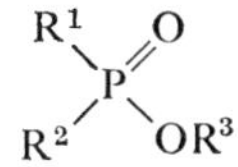
(113)

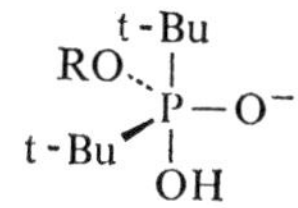
(114)

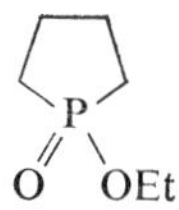
(115)

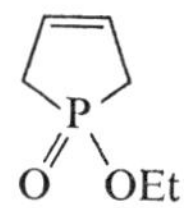
(116)

(117)

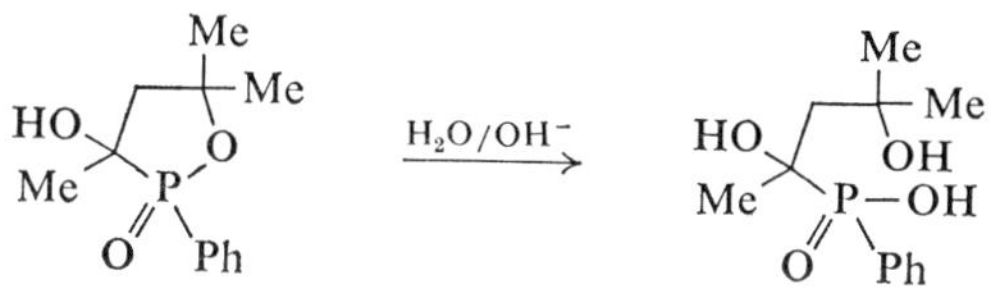
(118)

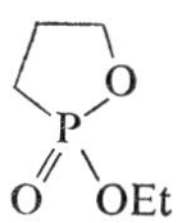
(119)

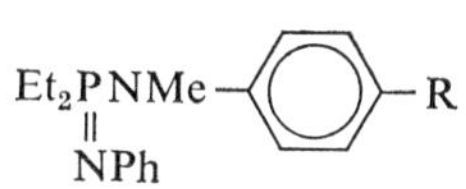
(120)

interest that the heat liberated on hydrolysis of a pyrophosphinate bond is considerably greater than that liberated on hydrolysis of pyrophosphates or triphosphates.[193] The basicities and rates of methylation of a series of diethylphosphinimidic amides (120) have been correlated with Hammett σ and ρ constants.[194]

C. P—S Systems.—*S*-4-Morpholinodithiophosphinates (120), which are obtained by treating thiophosphinous acids with 4-morpholinosulphenyl chloride (121),[195] will convert thiols into compounds containing S—S

[193] G. M. Kosolapoff and H. G. Kirksey, *Doklady Akad. Nauk S.S.S.R.*, 1967, 176 (*Chem. Abs.*, 1968, **69**, 18,452).
[194] V. A. Gilyarov, A. M. Maksudov, B. A. Korolev, B. I. Stepanov, and M. I. Kabachnik, *Izvest. Akad. Nauk S.S.S.R., Ser. khim.*, 1968, 2019 (*Chem. Abs.*, 1969, **70**, 10,930).
[195] L. Almasi and L. Paskucz, *Chem. Ber.*, 1969, **102**, 1489.

bridges. When the thiophosphinous acid is present in excess, dithiophosphinic anhydrides are formed in high yield.

The addition of elemental sulphur to alkyl methylphosphinites occurs by a radical mechanism as the reaction is inhibited by hydroquinone and other radical scavengers.[196] An unusual reaction occurs between diphenylchlorophosphine and salts of arylsulphinic acids,[197] the products being diphenylphosphinyl chloride (122) and *S*-aryl diphenylphosphinothiolates. The first step in this reaction may be oxygen transfer from the sulphinic acid to diphenylchlorophosphine as pentavalent phosphorus compounds do not deoxygenate sulphinic acids. Disilver cyanamide when heated with (122) or its thio-analogue gives symmetrically substituted products (123), the i.r. spectra of which contain bands indicative of a —N=C=N— system rather than a >N—C≡N system.[198] Hydrolysis of the carbodiimides (123) generates *NN*-diphosphinylureas.

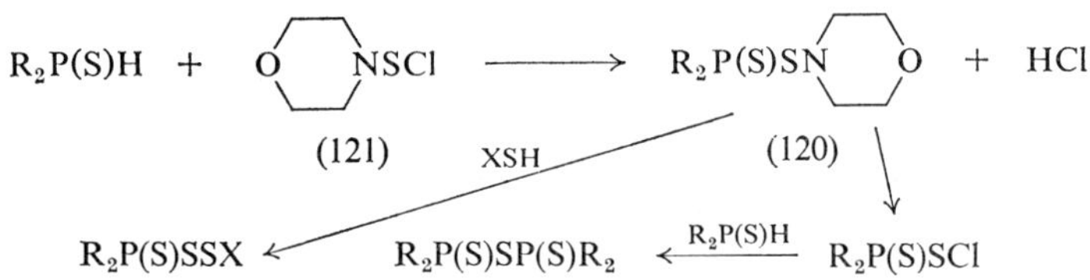

$$2Ph_2PCl + ArSO_2Na \longrightarrow Ph_2P(O)SAr + Ph_2P(O)Cl$$

(122)

$$2Ph_2P(X)Cl + Ag_2CN_2 \longrightarrow Ph_2P(X)N{=}C{=}NP(X)Ph_2$$

where X = O or S (123)

$$Ph_2P(O)SePh$$

(124)

D. Miscellaneous Reactions.—A number of methods for the synthesis of *Se*-phenyl diphenylselenophosphinates (124) have been described[199] and the properties of metal complexes of thiophosphinic together with selenophosphinic acids have been reviewed.[200]

When the benzyl ester of diphenylphosphinylformate (125) is treated with a solution of iodide ion in acetone, a high yield of the acetone adduct (126) is obtained.[201] This is an intermolecular S_EP-4 reaction[202] in which both the electrophilic and electrofugal centres are carbon atoms. Iodine is also an

196 W. A. Mosher and R. R. Irino, *J. Amer. Chem. Soc.*, 1969, **91**, 756.
197 H. Schindlbauer and W. Prikoszovich, *Monatsh.*, 1968, **99**, 1792.
198 A. Weisz and K. Utvary, *Monatsh.*, 1968, **99**, 2498.
199 N. Petragnani, V. G. Toscano, and M. Moura Campos, *Chem. Ber.*, 1968, **101**, 3070.
200 W. Kuchen and H. Hertel, *Angew. Chem. Internat. Edn.*, 1969, **8**, 89.
201 S. G. Warren and M. R. Williams, *Chem. Comm.*, 1969, 180.
202 S. G. Warren, *Angew. Chem. Internat. Edn.*, 1968, **7**, 606.

effective electrophile which will convert (125) in the presence of water or methanol to diphenylphosphinic acid or its methyl ester.

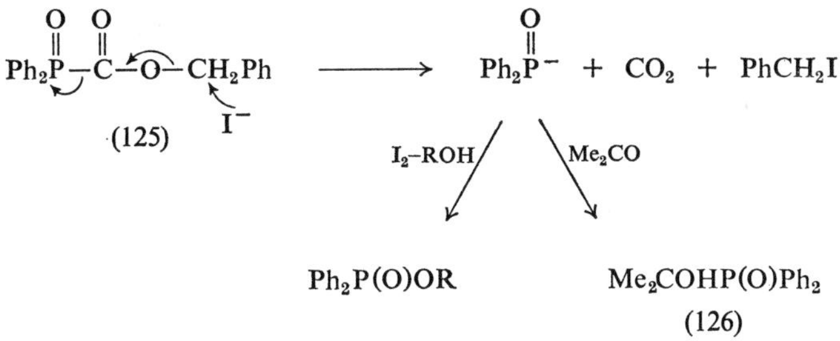

(125) (126)

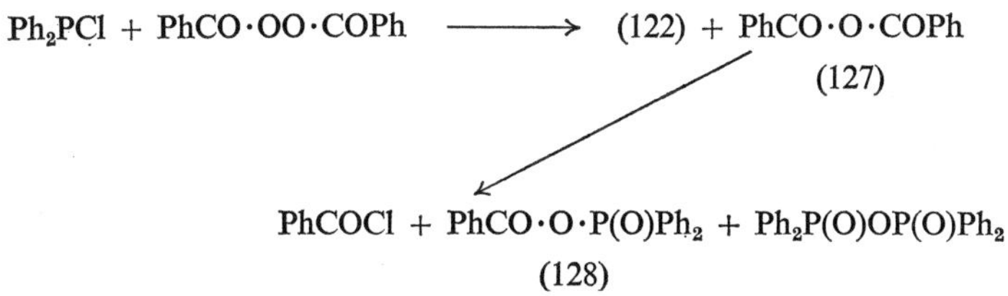

(127) (128)

A mixture of diphenylphosphinic anhydride, benzoyl chloride, and benzoic anhydride (127) is obtained when a benzene solution of equimolar quantities of diphenylphosphinous chloride and benzoyl peroxide are heated under reflux.[203] The first step in this reaction appears to be oxygen transfer from the peroxide to give diphenylphosphinyl chloride (122). Interaction of (122) and (127) affords the observed products together with the mixed anhydride (128) which can also be detected in this reaction.

The ^{1}H n.m.r. spectra of menthyl n-alkylphenylphosphinates can be used to ascertain the configuration at phosphorus.[204] The configuration (129) has been assigned to the diastereoisomer in which the resonance due to the protons of one of the methyl groups in the isopropyl residue is shifted upfield due to anisotropic shielding by the phenyl ring. The structures of the unsaturated hydrocarbons obtained on pyrolysis of diphenylphosphinates of menthol, neomenthol, and neocarvomenthol indicate that the elimination reaction proceeds by an E_1 mechanism.[205]

The mass spectrometric fragmentation pattern of dimethylphosphinic acid differs considerably from the fragmentation patterns obtained from other dialkyl-[206] and diaryl-phosphinic acids.[207] The base peak for dimethylphosphinic acid corresponds to (130, R = Me) and other peaks corre-

[203] G. Sosnovsky and D. J. Rawlinson, *J. Org. Chem.*, 1968, **33**, 2325.
[204] R. A. Lewis, O. Korpiun, and K. Mislow, *J. Amer. Chem. Soc.*, 1968, **90**, 4847.
[205] P. Mastalerz and Z. E. Golubski, *Roczniki Chem.*, 1967, **41**, 1527 (*Chem. Abs.*, 1968, **68**, 39,717).
[206] P. Haake and P. S. Ossip, *Tetrahedron*, 1968, **24**, 565.
[207] P. Haake, M. J. Frearson, and C. E. Diebert, *J. Org. Chem.*, 1969, **34**, 788.

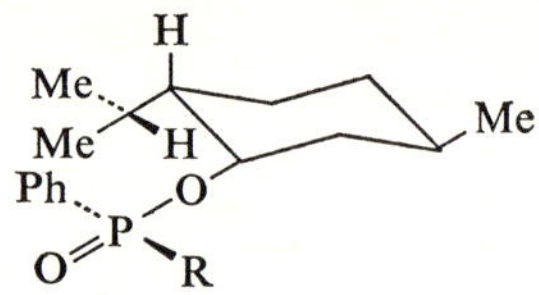

(129)

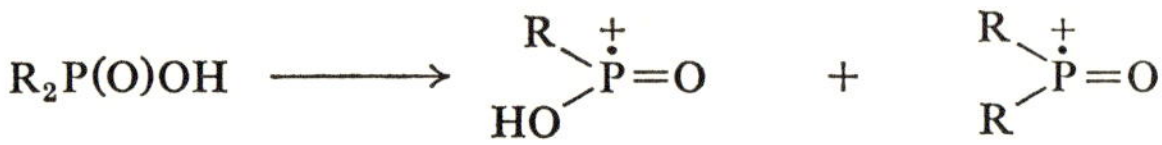

(130) (131)

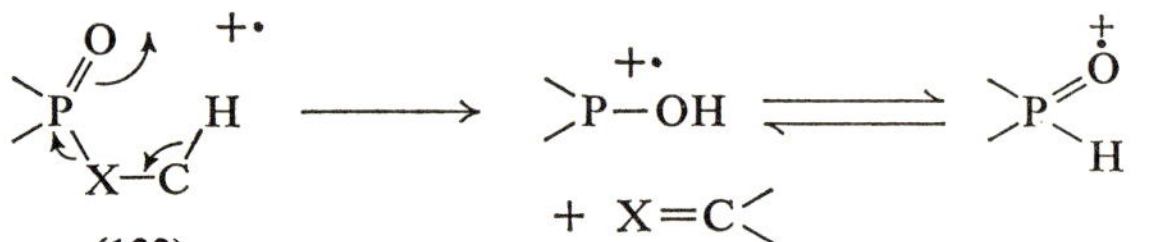

(132)

sponding to (131, R = Me) and PO^+ are observed.[206] For other dialkylphosphinic acids elimination of alkenes is observed and the base peak for diethylphosphinic acid is HPO_2H^+. Diarylphosphinic acids cyclise in the mass spectrometer to biphenyl-2,2′-phosphorus ions which then fragment.[207] A general scheme for the fragmentation of pentavalent phosphorus compounds (132) has been put forward.[206]

The ^{35}Cl frequencies in nuclear quadrupole spectra have been used to correlate the reactivities of dialkylphosphinyl chlorides with the nature of the substituents on phosphorus.[208] It is claimed that evidence has been obtained for hyperconjugation between C—H bonds and 3*d* orbitals of phosphorus.

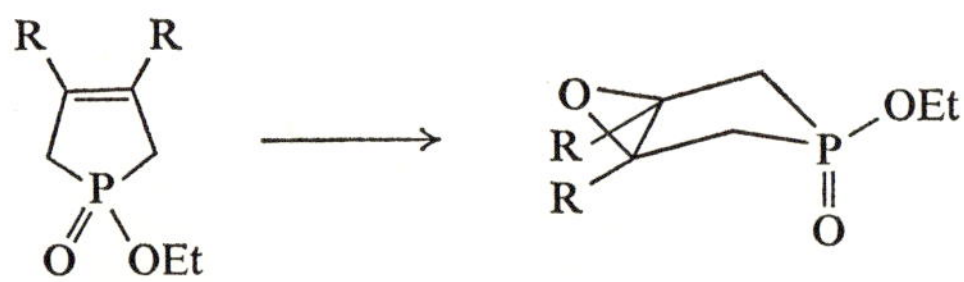

(133) (134)

From a study of the dipole moments of phospholenes (133) and their oxides (134), it has been shown[209] that the epoxidation of the phospholene ring takes place from the side opposite the phosphoryl group. This suggests that electronic factors rather than steric hindrance control the geometry of the product.

[208] A. A. Neimysheva, V. A. Pal'm, G. K. Semin, N. A. Loshadkin, and I. L. Knunyants, *Zhur. obshchei Khim.*, 1967, **37**, 2255 (*Chem. Abs.*, 1968, **69**, 2386).

[209] B. A. Arbusov, A. P. Anastas'eva, A. N. Vereshchagin, A. O. Vizel, and A. P. Rakov, *Izvest. Akad. Nauk S.S.S.R., Ser. khim.*, 1968, 1729 (*Chem. Abs.*, 1969, **70**, 29,008).

7

Phosphates and Phosphonates of Biochemical Interest

BY D. W. HUTCHINSON

1 Introduction

The widespread occurrence of phosphate and polyphosphate esters (*e.g.* nucleic acids and coenzymes*) in living organisms has always been a major factor in promoting their study. However, since this is one of the more prolific areas of biological chemistry, it is impossible to review all recent developments in the space of a single chapter. Many excellent reviews already exist[1] of the biochemical properties and reactions of phosphoric acids and their derivatives, and hence this review is selective in its choice of topics. Only reactions which involve the phosphorus atom of naturally occurring phosphates, polyphosphates, and phosphonates are considered and, for example, important developments in the elucidation of the secondary and tertiary structure of some types of t-RNA have been omitted as has the enzymology of phosphate and polyphosphate esters. Apart from recording new synthetic methods and structural determinations, an attempt has been made to describe some of the reactions mentioned in detail in order to rationalise some of the stages involved.

2 Nucleotides and their Phosphonate Analogues

Two excellent reviews[2,3] of oligonucleotide synthesis, both originally given as Plenary lectures at I.U.P.A.C. Symposia, have appeared recently. The first is an account of the techniques employed by Khorana and his group in their efforts to synthesise biologically specific nucleic acids. In this review, Khorana favours the use of dicyclohexyl carbodi-imide or aromatic sulphonyl chlorides for the synthesis of internucleotide bonds. The second review, by Cramer, mentions the use of picryl chloride as well as the two main condensing agents for internucleotide bond formation.

[1] Chapters on biological chemistry in *Annual Reports of the Chemical Society, Ann. Rev. Biochem., Progr. Nucleic Acid Res., Adv. Enzymol.*
[2] H. G. Khorana, *Pure Appl. Chem.*, 1968, **17**, 313.
[3] F. Cramer, *Pure Appl. Chem.*, 1969, **18**, 197.

* Standard abbreviations for biochemical compounds are used without definition in this chapter. A detailed key to the abbreviations may be found in the Instructions to Authors of the *Journal of Biological Chemistry*.

Both authors comment on the use of insoluble polymer supports for oligonucleotide synthesis.

A. Oligonucleotide Synthesis on Polymer Supports.—A highly cross-linked rigid polystyrene–divinyl benzene copolymer has been prepared [4] in a bead form which does not swell appreciably under normal reaction conditions and has been used as a support for the synthesis of moderate yields of oligonucleotides. With an aromatic sulphonyl chloride as condensing agent polythymidylic acid containing six monomeric units has been synthesised. Pentathymidine tetraphosphate (TpTpTpTpT) [5] and a variety of di(deoxyribonucleoside) monophosphates and trinucleoside diphosphates [6] have also been synthesised with the aid of insoluble polymer supports.

It has been suggested [7] that polyphosphate anhydrides, principally derivatives of cyclic trimetaphosphate (1), are the active phosphorylating

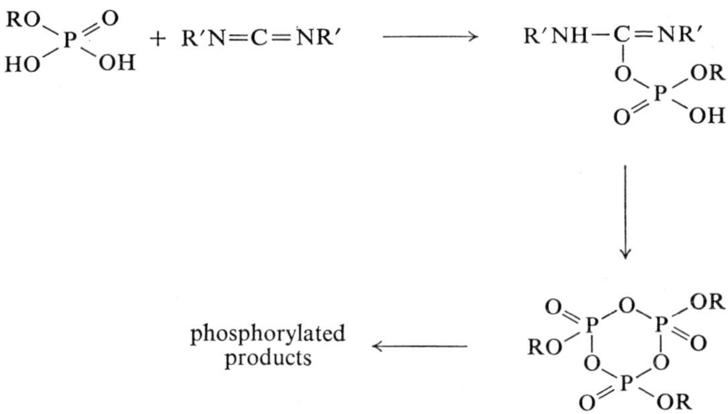

(1)

agents in phosphorylation reactions with carbodi-imides. Evidence has now been presented [8] from reactions carried out on polymer supports, that the co-operative interaction of at least three residues is needed before carbodi-imide-promoted phosphorylations can occur, supporting the earlier suggestion.[7] A similar interaction of three or more phosphate residues also appears to be necessary in reactions promoted by aromatic sulphonyl chlorides. The rate of reaction between (1) and an alcohol appears to be slow and the overall reaction proceeds at a much faster rate if (1) undergoes prior attack by a sulphonyl chloride.

[4] F. Cramer and H. Köster, *Angew. Chem. Internat. Edn.*, 1968, **7**, 473.
[5] L. R. Melby and D. R. Strobach, *J. Org. Chem.*, 1969, **34**, 421.
[6] L. R. Melby and D. R. Strobach, *J. Org. Chem.*, 1969, **34**, 427.
[7] G. Weimann and H. G. Khorana, *J. Amer. Chem. Soc.*, 1962, **84**, 4329.
[8] G. M. Blackburn, M. J. Brown, M. R. Harris, and D. Shire, *J. Chem. Soc.* (*C*), 1969, 676.

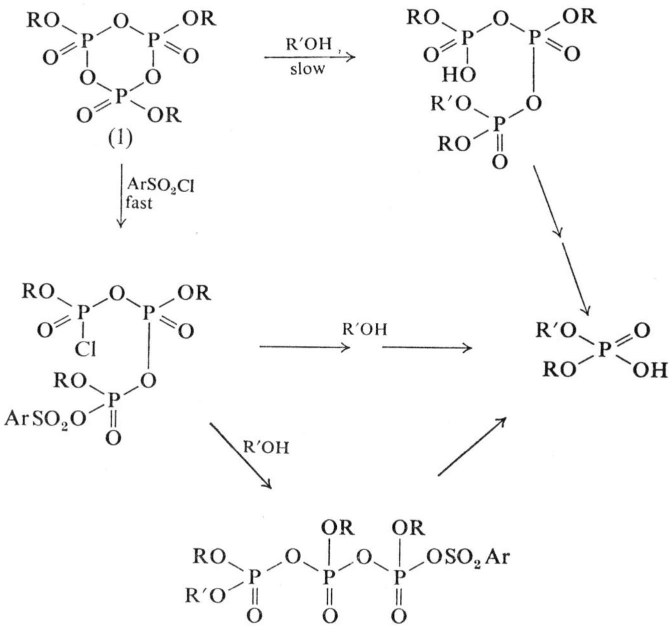

B. Mononucleotides.—The following methods have been reported for the synthesis of nucleotides:

(*i*) 2-cyanoethyl phosphate and dicyclohexyl carbodi-imide; (2a) [9] and oligomers,[10] (2b),[11] [3, R = 9*H*-purine-6(1*H*-thione)],[12]

(*ii*) phosphorus oxychloride; (2c),[13] (2d),[13] (2e),[14] (2f),[14] (2g),[15]

(*iii*) diphenyl phosphorochloridate; (2h),[16]

(*iv*) di(2,2,2-trichloroethyl) phosphorochloridate (4); (2i),[17]

(*v*) P^1-diphenyl P^2-(4-morpholino) pyrophosphorochloridate (5); (2j),[18]

(*vi*) tri-imidazolylphosphine oxide (6);[19] (2k) [20] and oligomers.[20–22]

9 M. P. Mertes and J. Smrt, *Coll. Czech. Chem. Comm.*, 1968, **33**, 3304.
10 M. P. Mertes, A. Hóly, and J. Smrt, *Coll. Czech. Chem. Comm.*, 1968, **33**, 3313.
11 A. R. Hanze, *Biochemistry*, 1968, **7**, 932.
12 J. P. Bell, *Canad. J. Chem.*, 1969, **47**, 1095.
13 K. Kusashio and M. Yoshikawa, *Bull. Chem. Soc. Japan*, 1968, **41**, 142.
14 A. Yamakazi, I. Kumashiro, and T. Takenishi, *Chem. and Pharm. Bull.* (*Japan*), 1968, **16**, 338.
15 A. Yamazaki, I. Kumashiro, and T. Takenishi, *Chem. and Pharm. Bull.* (*Japan*), 1968, **16**, 1561.
16 M. M. Vigdorchik, M. N. Preobrazhenskaya, and N. N. Suvorov, *Tetrahedron Letters*, 1968, 4645.
17 K. H. Scheit, *Biochim. Biophys. Acta*, 1968, **157**, 632.
18 M. Ikehara and M. Murao, *Chem. and Pharm. Bull.* (*Japan*), 1968, **16**, 1330.
19 F. Cramer, H. Schaller, and H. A. Staab, *Chem. Ber.*, 1961, **94**, 1612.
20 K. H. Scheit, *Chem. Ber.*, 1968, **101**, 1141.
21 K. H. Scheit, *Biochim. Biophys. Acta*, 1968, **166**, 285.
22 K. H. Scheit, *Biochim. Biophys. Acta*, 1969, **182**, 10.

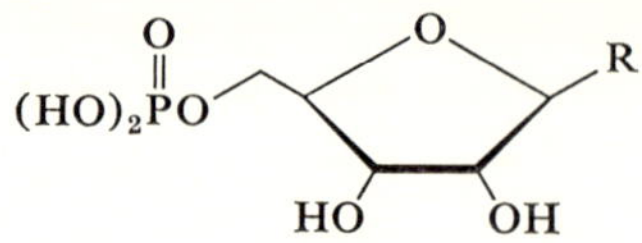

(2)

R = (a) 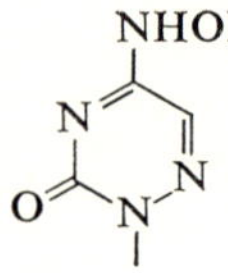

(b) 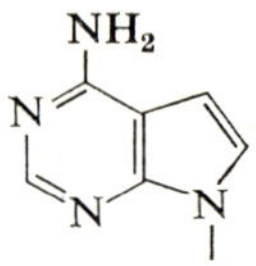

(c) hypoxanthine

(d) guanine

(e) 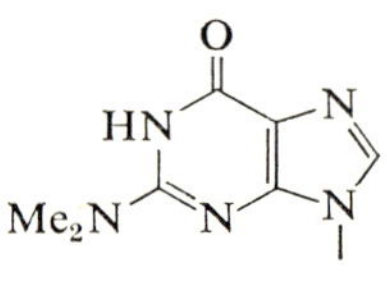

(f) 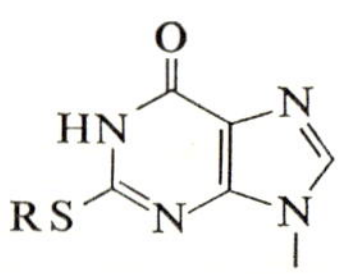

(g) 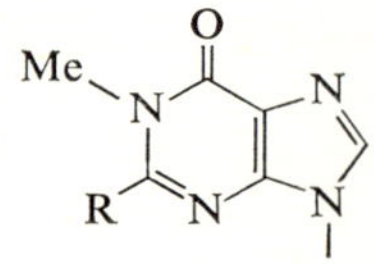

(h) 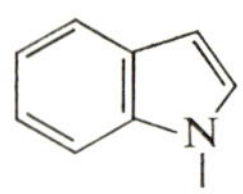

(i) 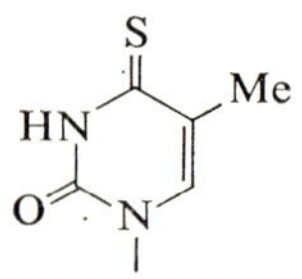

(j) 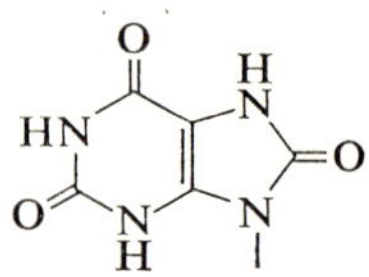

(k) 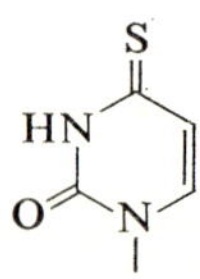

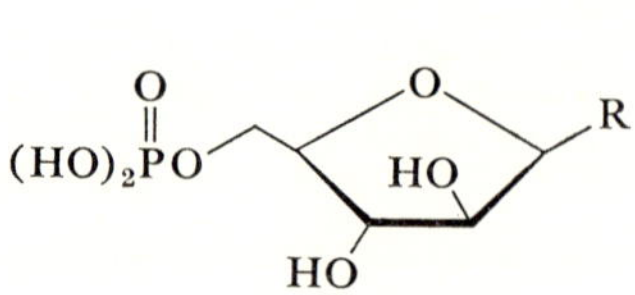

(3)

$$(CCl_3CH_2O)_2\overset{O}{\overset{\|}{P}}Cl$$

(4)

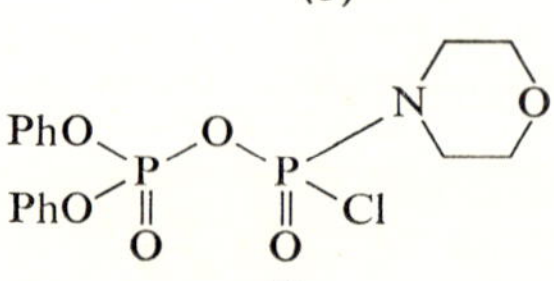

(5)

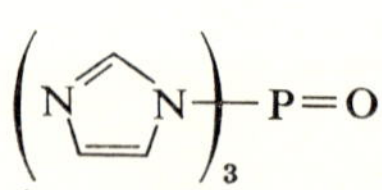

(6)

Selective phosphorylation of nucleosides at the 5′-position has been reported with (4) and the preparation of di(2,2,2-trichloroethyl) thymidine 5′-phosphate has been achieved in 75% yield from the unprotected nucleoside.[23] Phosphorylation reactions with (5) [24] are carried out in anhydrous media followed by mild acid treatment to remove the protecting groups. Moderate yields of purine nucleotides have been obtained in this manner.

arabino-Cytidine 2′,5′-cyclic phosphate (8), prepared by cyclisation of N^4-benzoyl-*arabino*-cytidine 5′-phosphate (7) [25] with carbodi-imide followed by removal of the benzoyl group, is unreactive towards acid, base, and the nucleases present in crude snake venom.[26] In the ^{1}H n.m.r. spectrum of (8), phosphate anisotropy has only a small effect on the 2′- and 5′-protons but strongly deshields the 3′-proton. A paramagnetic shift of 60 Hz is observed for (8) compared with the 3′-proton of *arabino*-cytidine.

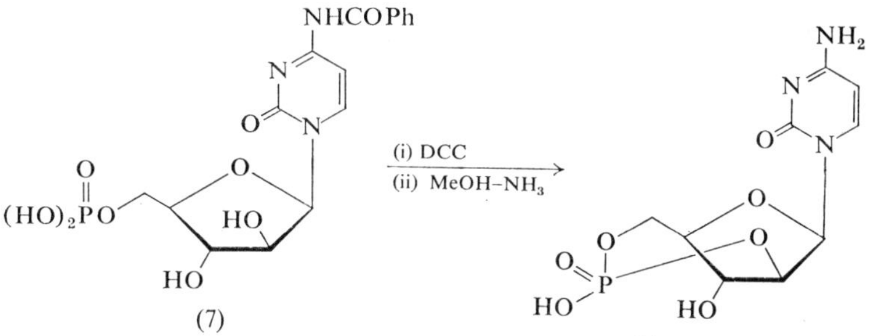

(7)

(8)

A direct synthesis of 3′,5′-cyclic mononucleotides has been achieved [27] by the action of a solution of phosphorus oxychloride in trimethyl phosphate on the parent nucleosides. The byproducts which are usually formed during reactions with phosphorus oxychloride in pyridine solution are greatly reduced in absence of the tertiary base.[28]

C. Nucleoside Polyphosphates.—A general synthesis of γ-^{32}P-labelled nucleoside 5′-triphosphates consists of the treatment of the appropriate P^1-(nucleoside-5′) P^2-(4-morpholino) pyrophosphate (9) with ^{32}P-labelled orthophosphate.[29] Enzymic assay of the products showed that at least 99·8% of the label was in the γ-position.

The 5′-triphosphates of the nucleoside antibiotics tubericidin (10, R = H),[30] toyocamycin (10, R = CN),[30] sangivamycin (10, R = $CONH_2$),[30, 31]

[23] A. Franke, K. H. Scheit, and F. Eckstein, *Chem. Ber.*, 1968, **101**, 2998.
[24] M. Ikehara and E. Ohtsuka, *Chem. and Pharm. Bull.* (*Japan*), 1963, **11**, 961.
[25] W. J. Wechter, *J. Medicin. Chem.*, 1967, **10**, 762.
[26] W. J. Wechter, *J. Org. Chem.*, 1969, **34**, 244.
[27] M. Heubert-Habart and L. Goodman, *Chem. Comm.*, 1969, 740.
[28] M. Yoshikawa, T. Kata, and T. Takenishi, *Tetrahedron Letters*, 1967, 5065.
[29] A. Adam and J. G. Moffatt, *Biochemistry*, 1968, **7**, 875.
[30] S. C. Uretski, G. Acs, E. Reich, M. Mori, and L. Altwerger, *J. Biol. Chem.*, 1968, **243**, 306.
[31] R. J. Suhadolnik, T. Uematsu, H. Uematsu, and R. G. Wilson, *J. Biol. Chem.*, 1968, **243**, 2761.

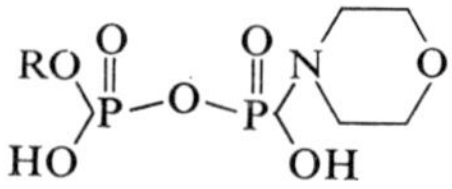

(9)

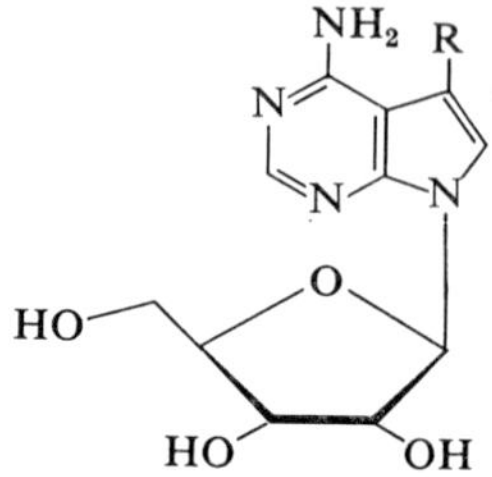

(10)

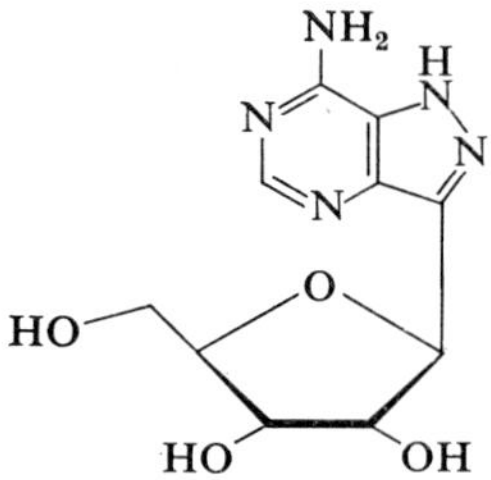

(11)

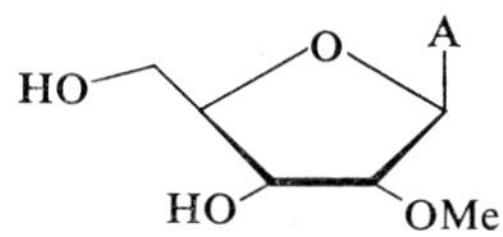

(12)

and formycin (11) [32] have been synthesised and all show some biological activity. Both (10, R = $CONH_2$) [31] and (11) [32] could replace ATP in RNA polymerase systems using GTP, CTP, UTP, and a DNA primer. The coding properties of several tubericidin-containing polyribonucleotides were indistinguishable from those of the corresponding adenosine-containing polymers.[30]

Polyribonucleotides have been synthesised from the 5′-pyrophosphates of 8-substituted purine ribonucleotides [33] and 2′-*O*-methyladenosine (12) [34] with the aid of polynucleotide phosphorylase. The 5′-pyrophosphate of (12) was prepared from the corresponding monophosphate using rabbit muscle myokinase.

The free 5′-hydroxyl of the RNA chain of the RNA bacteriophages and T_4 m-RNA bears a triphosphate group, and consequently the terminal nucleosides are liberated as 2′(3′)-monophosphate 5′-triphosphates (13) [35] on hydrolysis. These have been prepared from nucleoside 2′,3′-cyclic phosphate 5′-phosphoromorpholidates (14) and inorganic pyrophosphate, the method being analogous to that used for the synthesis of CoA.[36]

[32] M. Ikehara, K. Murao, F. Harada, and S. Nishimura, *Biochim. Biophys. Acta*, 1968, **155**, 82.
[33] M. Ikehara, I. Tazawa, and T. Fukui, *Biochemistry*, 1969, **8**, 736.
[34] F. Rottman and K. Heinlen, *Biochemistry*, 1968, **7**, 2634.
[35] E. Messens and M. van Montagu, *F.E.B.S. Letters*, 1968, **1**, 326.
[36] J. G. Moffatt and H. G. Khorana, *J. Amer. Chem. Soc.*, 1961, **83**, 663.

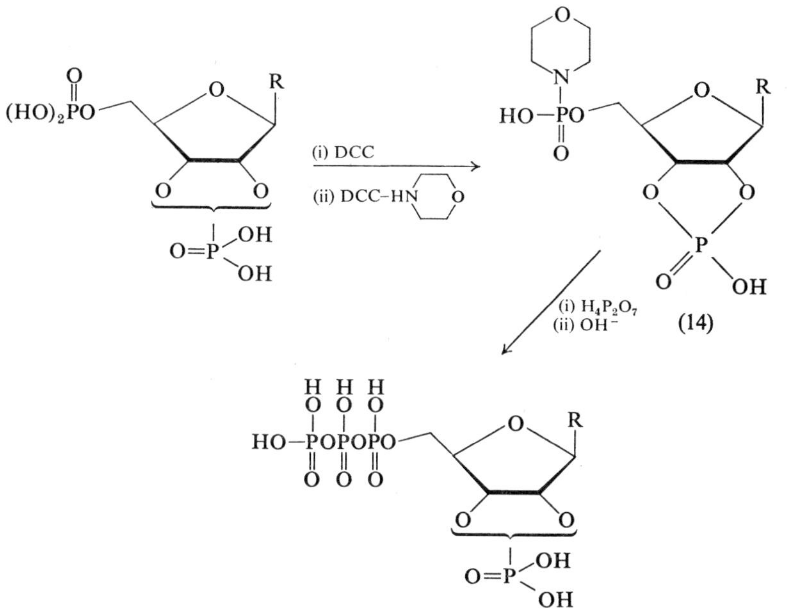

(14)

(13)

N-Benzyloxycarbonylglycine ethyl orthoester (15) will react with ADP or ATP in the presence of acid to form 2′(3′)-*ON*-benzyloxycarbonylglycyladenosine 5′-tri- and 5′-pyro-phosphates. Catalytic hydrogenolytic removal of the protecting groups gives the glycyl derivatives (16) which may enable 'wrong' amino-acids to be introduced on to specific t-RNAs.[37] Compound (16) may be also obtained by the action of pyro- or orthophosphate on the phosphoromorpholidate of the cyclic orthoester (17).

D. Oligonucleotides.—Most of the syntheses of oligoribo- and oligodeoxyribo-nucleotides which have been reported recently have made use of either dicyclohexyl carbodi-imide[38–43] or aromatic sulphonyl chlorides[44–49] as

[37] J. Žemlička and S. Chládek, *Coll. Czech. Chem. Comm.*, 1968, **33**, 3293.
[38] B. E. Griffin and C. B. Reese, *Tetrahedron*, 1968, **24**, 2537.
[39] A. Hóly, J. Smrt, and F. Šorm, *Coll. Czech. Chem. Comm.*, 1968, **33**, 3809.
[40] F. Kathawala and F. Cramer, *Annalen*, 1968, **712**, 195.
[41] F. Cramer and G. Schneider, *Annalen*, 1968, **717**, 193.
[42] E. Ohtsuka, K. Murao, M. Ubasawa, and M. Ikehara, *J. Amer. Chem. Soc.*, 1969, **91**, 1537.
[43] A. Franke, F. Eckstein, K. H. Scheit, and F. Cramer, *Chem. Ber.*, 1968, **101**, 944.
[44] [a] S. A. Narang, J. M. Michniewicz, and S. K. Dheer, *J. Amer. Chem. Soc.*, 1968, **90**, 2702. [b] *J. Amer. Chem. Soc.*, 1969, **91**, 936.
[45] A. Kumar and H. G. Khorana, *J. Amer. Chem. Soc.*, 1969, **91**, 2743.
[46] J. Hachmann and H. G. Khorana, *J. Amer. Chem. Soc.*, 1969, **91**, 2749.
[47] R. L. Letsinger and K. K. Ogilvie, *J. Amer. Chem. Soc.*, 1969, **91**, 3350.
[48] R. L. Letsinger, K. K. Ogilvie, and P. S. Miller, *J. Amer. Chem. Soc.*, 1969, **91**, 3360.
[49] H. P. M. Fromageot, C. B. Reese, and J. E. Sulston, *Tetrahedron*, 1968, **24**, 3533.

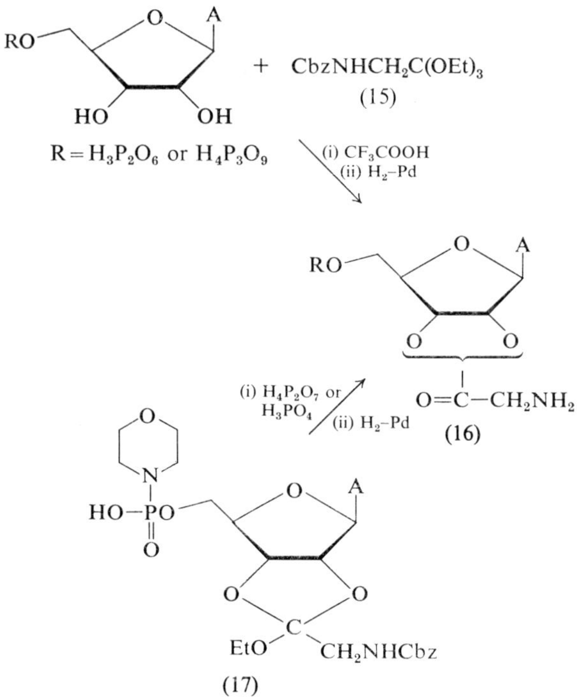

condensing agents, although phenyl phosphorodichloridate[50] and diphenyl phosphorochloridate[51] have also been used. Dinucleoside phosphates have been obtained from the reaction between nucleotide salts and either 5′-chloro-5′-deoxynucleosides[52] or purine cyclonucleosides.[53] When uridine 2′(3′)-phosphate is heated to 160° in the absence of a solvent, dinucleoside diphosphates and trinucleoside triphosphates are formed with both 2′ → 5′ and 3′ → 5′ bonds.[54, 55] This has been suggested as a method for the preparation of uridine oligonucleotides possessing a high specific molar radioactivity.

GpU is selectively alkylated on N(7) of the guanine moiety after treatment with an excess of dimethyl sulphate.[56] The zwitterionic species (18) is readily separable from GpU by ion-exchange chromatography. 7-MeGpU is completely resistant to RNase T_1 under conditions which led to the

[50] C. B. Reese and R. Saffhill, *Chem. Comm.*, 1968, 767.
[51] H. Nishimura, T. Sekiya, and T. Ukita, *Biochim. Biophys. Acta*, 1969, **174**, 653.
[52] P. C. Srivastava, K. L. Nagpal, and M. M. Dhar, *Experientia*, 1969, **25**, 356.
[53] K. L. Nagpal and M. M. Dhar, *Tetrahedron Letters*, 1968, 47.
[54] J. Morávek, J. Kopecký, and J. Škoda, *Coll. Czech. Chem. Comm.*, 1968, **33**, 4120.
[55] J. Morávek, J. Kopecký, and J. Škoda, *Coll. Czech. Chem. Comm.*, 1968, **33**, 4407.
[56] C. B. Reese and J. E. Sulston, *Biochim. Biophys. Acta*, 1967, **149**, 293.

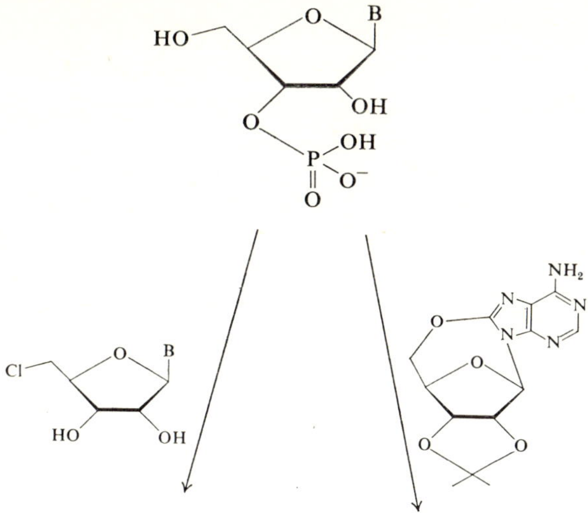

Dinucleoside phosphates

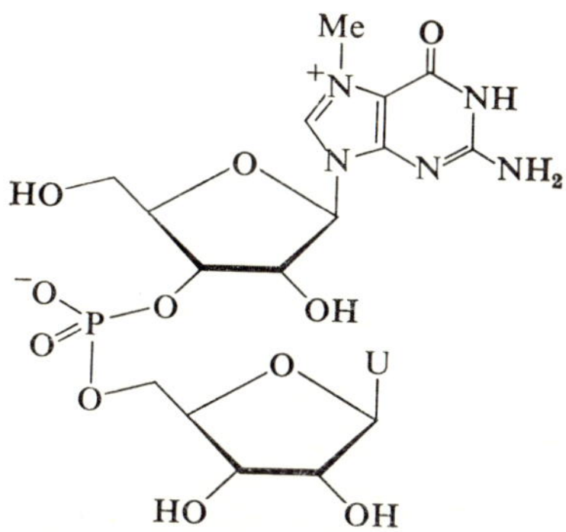

(18)

total breakdown of GpU, an observation which could be of use in the sequence determination of RNAs.

E. Nucleoside Thiophosphates.—Polyribonucleotides with a thiophosphate backbone have been used in an elegant series of investigations to study the mechanism of action of pancreatic RNase.[57–60] Tri-imidazolyl 1-phosphine sulphide (19) will convert a suitably protected nucleoside into its 5′-phosphorodi-imidazolothionate (20). Removal of the protecting groups

[57] F. Eckstein and H. Gindl, *Biochim. Biophys. Acta*, 1967, **149**, 35.
[58] F. Eckstein and H. Gindl, *Chem. Ber.*, 1968, **101**, 1670.
[59] F. Eckstein and H. Gindl, *F.E.B.S. Letters*, 1968, **2**, 262.
[60] H. Matzura and F. Eckstein, *European J. Biochem.*, 1968, **3**, 448.

followed by partial hydrolysis yields the 5′-phosphoromonoimidazolothioate (21) which can readily be converted into the nucleoside 5′-triphosphorothioate (NTPS) by the action of inorganic pyrophosphate. Using a DNA-dependent RNA polymerase, UTPS and ATPS as substrates and a poly-d(AT) template, poly (AS·US) was obtained,[59, 60] which nearest-neighbour analysis showed to be a faithful copy of the template. Poly (US) was formed from UDPS with the aid of polynucleotide phosphorylase from *E. coli* when the substrate was present in low concentration, as high concentrations of substrate result in complete inhibition of the polymerisation reaction. Both polymers underwent slow hydrolysis by pancreatic RNase, or snake venom or spleen phosphodiesterases.

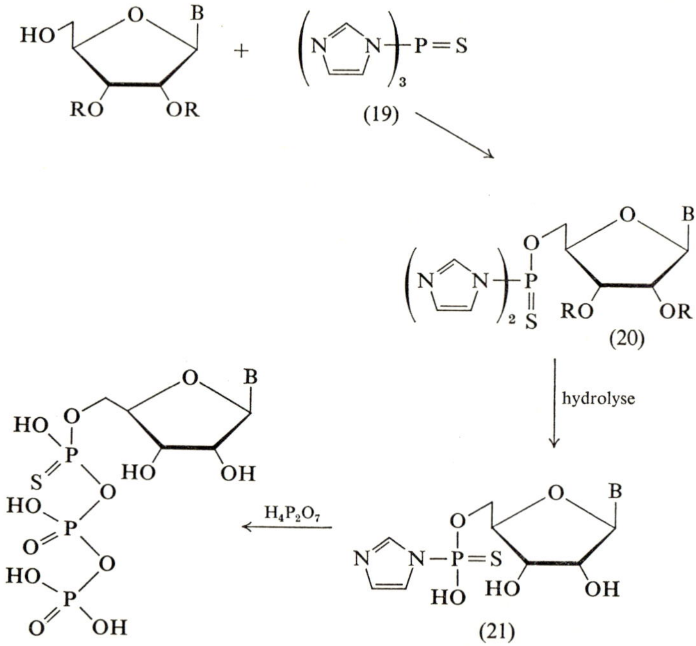

Uridine 2′,3′-*O*-cyclophosphorothioate [58] contains an asymmetric phosphorus atom, and the two diastereoisomers can be separated by crystallisation of their triethylammonium salts.[61] The apparent first-order rate constants for both acid and alkaline hydrolysis are indistinguishable for the two isomers. There is a large amount of exchange of sulphur for oxygen in the acid-catalysed hydrolysis, suggesting the participation in this reaction of quinquecovalent intermediates and concomitant pseudorotation.[62] On the other hand, no loss of sulphur occurs from either isomer during hydrolysis of the cyclic phosphorothioate with pancreatic RNase, which

[61] F. Eckstein, *F.E.B.S. Letters*, 1968, **2**, 85.
[62] D. B. Boyd, *J. Amer. Chem. Soc.*, 1969, **91**, 1200.

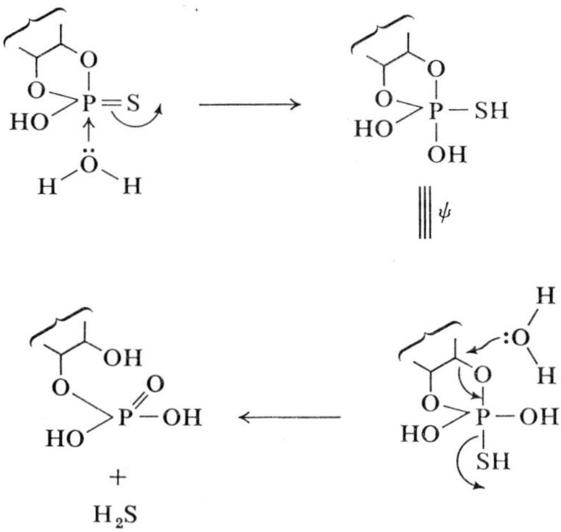

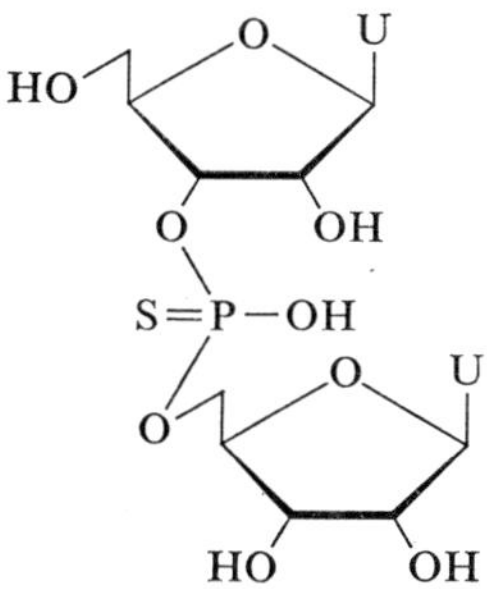

(22)

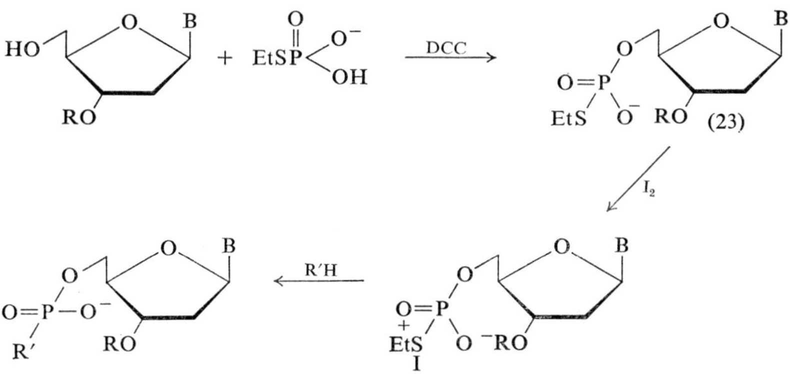

where R′ = OH, H_2PO_4, $H_3P_2O_7$, 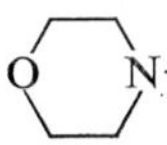—, CCl_3CH_2O—, 4-$NO_2C_6H_4O$—

indicates[60] that there is no pseudorotation in a quinquecovalent intermediate during enzymic hydrolysis.

Ribonuclease T_1 is capable of converting a mixture of diuridine-3′ → 5′ phosphorothioate (22)[63] and guanosine 2′,3′-cyclic phosphate into guanylyl-diuridinephosphorothioate (GpUpSU)[64] which has the weak ability to stimulate the binding of [^{14}C]valyl-t-RNA to ribosomes.

Treatment of adenosine with thiophosphoryl chloride in solution in triethyl phosphate[65] gives adenosine 5′-phosphorothioate which can undergo reaction with a variety of enzymes, albeit very slowly; for example it is phosphorylated by ATP and myokinase 0·3% as fast as AMP.

Nucleoside 5′-*S*-ethyl phosphorothiolates (23) can be prepared from the carbodi-imide-promoted condensation between pyridinium *S*-ethyl phosphorothiolate and the protected nucleoside.[66] The *S*-ethyl group can readily be removed by oxidation with iodine when it can be displaced by a variety of suitable nucleophiles. This appears to be a promising method for the synthesis of nucleotide derivatives under comparatively mild conditions.

F. Phosphonic Analogues of Nucleotides.—Oxidation of nucleosides by a solution of dicyclohexyl carbodi-imide in dimethyl sulphoxide is a simple method for the preparation of nucleoside 5′-aldehydes.[67] 2′,3′-*O*-Isopropylidene uridine 5′-aldehyde prepared in this manner has been condensed with diphenyl triphenylphosphoranylidenemethylphosphonate (24) to give the *trans*-vinyl phosphonate (25).[68]

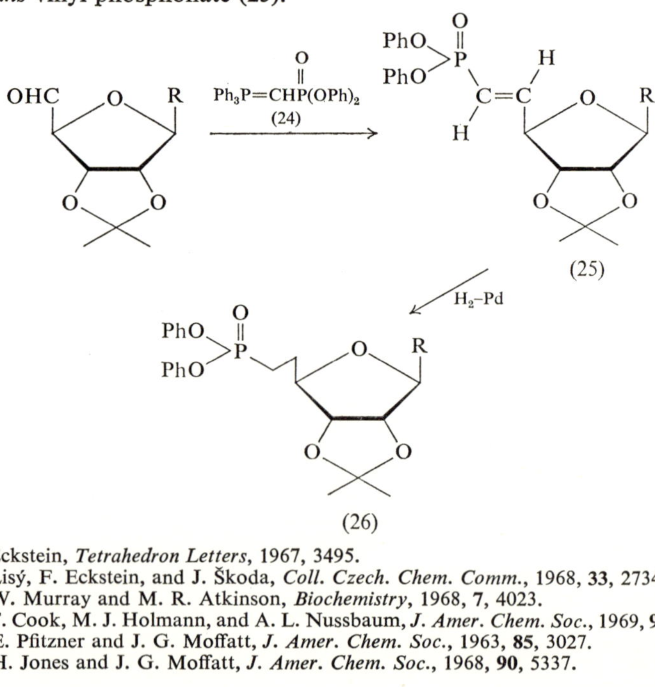

(25)

(26)

[63] F. Eckstein, *Tetrahedron Letters*, 1967, 3495.
[64] V. Lisý, F. Eckstein, and J. Škoda, *Coll. Czech. Chem. Comm.*, 1968, **33**, 2734.
[65] A. W. Murray and M. R. Atkinson, *Biochemistry*, 1968, **7**, 4023.
[66] A. F. Cook, M. J. Holmann, and A. L. Nussbaum, *J. Amer. Chem. Soc.*, 1969, **91**, 1522.
[67] K. E. Pfitzner and J. G. Moffatt, *J. Amer. Chem. Soc.*, 1963, **85**, 3027.
[68] G. H. Jones and J. G. Moffatt, *J. Amer. Chem. Soc.*, 1968, **90**, 5337.

Catalytic hydrogenation of (25) gave the diphenyl ester of 6′-deoxyhomouridine 6′-phosphonic acid (26). Alkaline hydrolysis removed one phenyl group and, following acid-catalysed removal of the isopropylidene group, the second could be removed by snake venom phosphodiesterase. Alternatively, the ester groups could be removed by treatment with sodium benzoxide in dimethyl sulphoxide. A similar series of reactions has been carried out in the adenosine series. These phosphonates have been converted into analogues of nucleoside coenzymes and may provide some insight into their mode of action.

Fusion of 1,2,3-tri-*O*-acetyl-5,6-dideoxy-6-di-*O*-ethylphosphono-D-*ribo*-hexofuranose (27, R = OAc) with 2,6-dichloropurine gave an anomeric mixture of the phosphonic nucleotide analogue (28) containing slightly more α- than β-anomer.[69] A similar result was obtained with 6-chloropurine except that the α : β ratio was even higher. In general, fusion of glycosyl halides bearing an acetoxy-group at C(2) with the salt of a heterocyclic base gives a predominantly *trans*-product (*i.e.* β in the case of ribose derivatives.)[70]

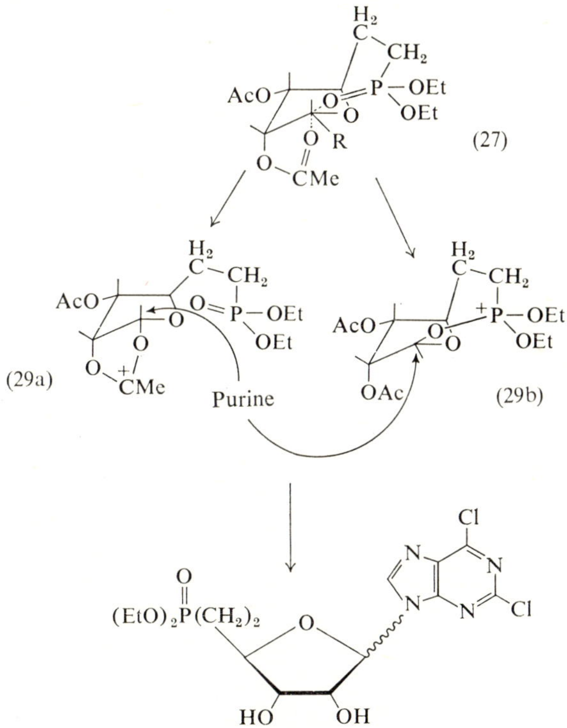

(27) (29a) (29b) (28)

[69] J. A. Montgomery and K. Hewson, *Chem. Comm.*, 1969, 15.
[70] B. R. Baker, Ciba Foundation Symposium, 'Chemistry and Biology of Purines', 1957, p. 120.

With (27, R = Cl) the α-anomer predominates and it has been suggested that there may be an interaction between the oxygen of the ethyl phosphonate group to give the intermediate (29b) as well as (29a).

G. Nucleotide Structure.—A compilation of o.r.d. data of anomeric nucleosides and nucleotides has been published;[71] in general, the presence of a 5′-phosphate group has no effect on the sign of the Cotton effect but the numerical value of the rotation is smaller for the nucleotides than for the nucleosides.

The interaction and conformation of nucleoside 2′-, 3′-, and 5′-monophosphates[72] and dinucleoside mono- and di-phosphates[73] have been studied by ^{1}H n.m.r. at 100 MHz at varying concentrations and pH. The 5′-phosphate group specifically deshields the 8-H proton of purine nucleotides and the 6-H proton of pyrimidine nucleotides. Comparison of the ^{1}H n.m.r. data for ribose and deoxyribose 5′-nucleotides suggests the presence of intramolecular hydrogen bonding between the 2′-hydroxyl and the N(3) of a purine or the 2-keto-group of a pyrimidine. From the ^{1}H n.m.r. data on the dinucleoside derivatives, a general conformational model has been proposed in which all the nucleosidyl units have the *anti*-conformation with respect to the sugar–base torsion angles and the turn of the (3′ → 5′) screw axis is right handed. The general features of this model are similar to those of a single strand in the DNA helix.

H. Sequence Studies.—A detailed review[74] has been published on the techniques available for the sequence determination of both t-RNA and 5*S*-RNA. A method for the determination of base sequence in oligodeoxyribonucleotides has been described[75] which consists of specific oxidation of a free 3′-hydroxy group with acetic anhydride–dimethyl sulphoxide.[76] The 3′-keto 5′-nucleotide (30) can then eliminate the terminal base and the next lower oligomer bearing a 3′-phosphate group. Treatment of this oligomer with phosphatases uncovers the 3′-hydroxy-group and the cycle can be repeated. Depurination does not seem to be excessive but a drawback is the 'decay of signal strength' caused by incomplete reaction at the oxidation step, particularly with derivatives of guanosine.

Alkaline degradation of apurinic and apyrimidinic acid gives, in addition to oligonucleotides and inorganic phosphate, appreciable amounts of 2-oxocyclopent-1-enyl phosphate (31).[77] This probably arises by an elimination reaction followed by opening of the sugar ring and condensation of the 1-aldehyde group with the 5-methylene group.

[71] T. Nishimura, B. Shimizu, and I. Iwai, *Biochim. Biophys. Acta*, 1968, **157**, 221.
[72] M. P. Schweizer, A. D. Broom, P. O. P. Ts'o, and D. P. Hollis, *J. Amer. Chem. Soc.*, 1968, **90**, 1042.
[73] P. O. P. Ts'o, N. S. Kondo, M. P. Schweizer, and D. P. Hollis, *Biochemistry*, 1969, **8**, 997.
[74] D. Dutting, *Fortschr. Chem. org. Naturstoffe*, 1968, **26**, 356.
[75] T. Gabriel, W. Y. Chen, and A. L. Nussbaum, *J. Amer. Chem. Soc.*, 1968, **90**, 6833.
[76] J. D. Albright and L. Goldman, *J. Amer. Chem. Soc.*, 1965, **87**, 4214.
[77] A. S. Jones, A. M. Mian, and R. T. Walker, *J. Chem. Soc.* (*C*), 1968, 2042.

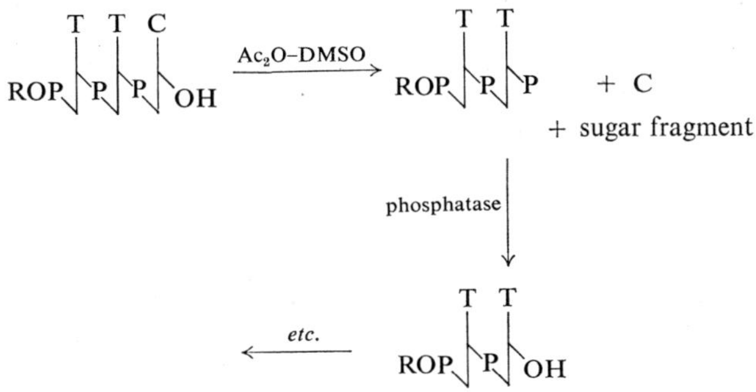

T T C
ROP P P OH
Ac_2O–DMSO
T T
ROP P P
+ C
+ sugar fragment
phosphatase
T T
ROP P OH
etc.

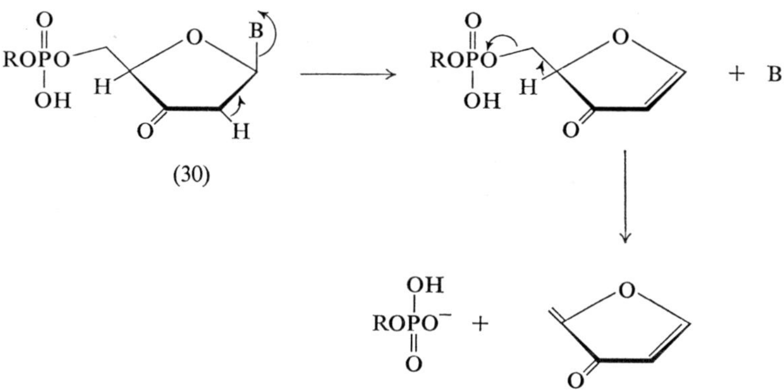

(30)

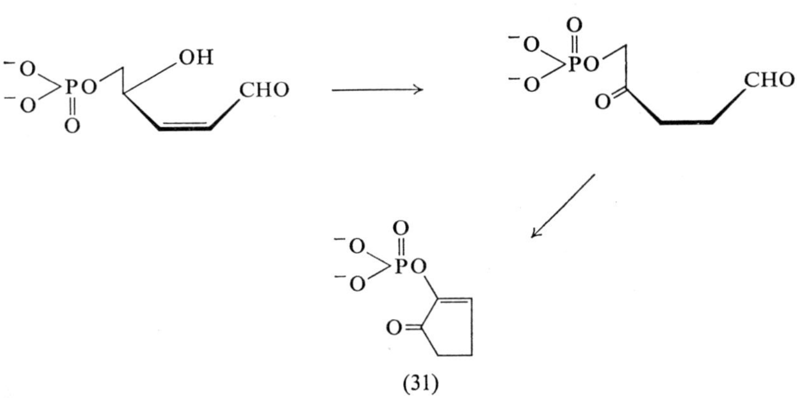

(31)

Partial dephosphorylation of salts of ribonucleoside 2′(3′)-phosphates by dimethylformamide acetals (32) has been observed,[78] provided that the cation does not contain an easily removable proton (*e.g.* tetra-alkylammonium ion). If the cation contains such a proton the 2′,3′-cyclic phosphate is obtained. The phosphoryl group is eliminated as orthophosphate and presumably an intermediate such as (33) is formed which can either eliminate metaphosphate, or undergo attack by the adjacent hydroxy-group of the nucleotide to yield the cyclic phosphate. The former process is probably followed in the presence of a proton source.

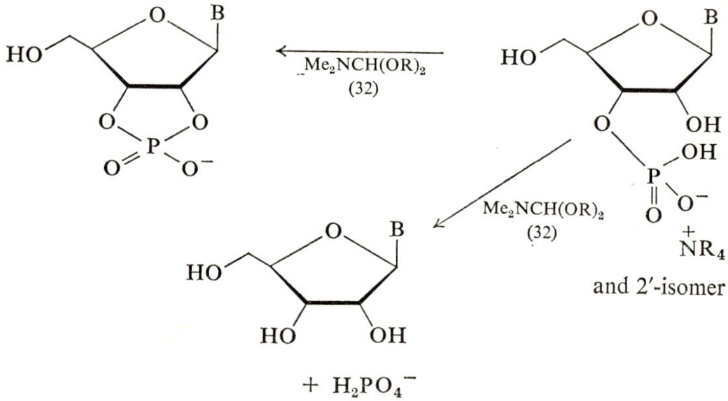

and 2′-isomer

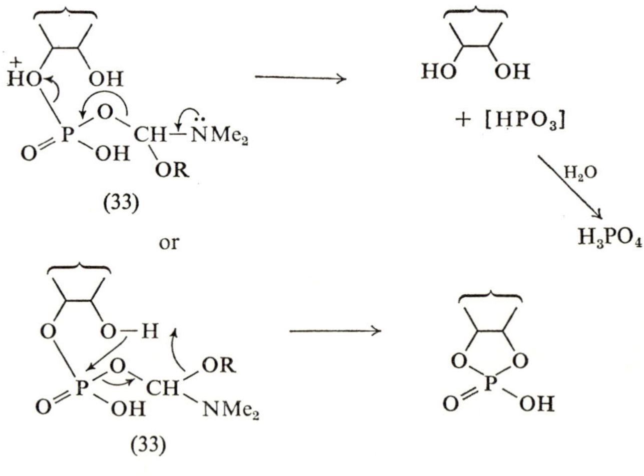

Dephosphorylation of nucleoside 2′(3′)-phosphates can also be achieved by heating the nucleotides with benzene boronic acid in dimethylformamide

[78] J. Žemlička and S. Chládek, *Tetrahedron Letters*, 1969, 715.

at >120°,[79] the benzene boronate being obtained in high yield. In the absence of boronic acid the cyclic phosphates are formed.

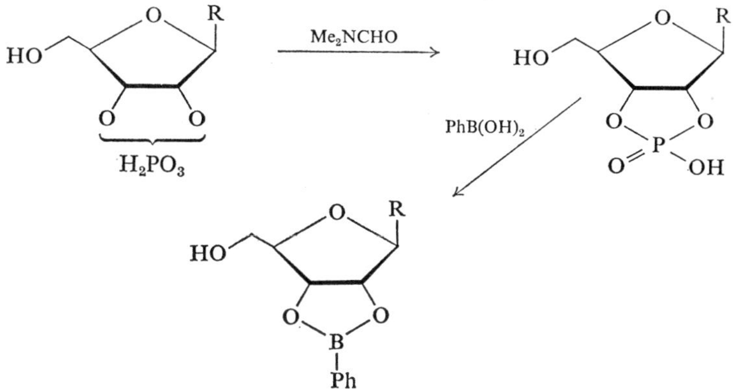

I. Analytical Techniques.—Chromatographic methods have been described for the determination of adenosine 3′,5′-cyclic phosphate[80] and the nucleotide composition of RNA.[81]

A procedure has been developed[82] for the enzymatic hydrolysis of calf thymus DNA absorbed on DEAE-cellulose. Large oligomers have been obtained by stepwise elution and this technique may be of use in sequence studies on DNA. Another aid for the fractionation of mixtures of oligonucleotides is chromatography on cellulose containing covalently bound nucleotides.[83] In this manner, celluloses which contain nucleotides bound to the cellulose by ester linkages and which will specifically bind one nucleotide or polynucleotide can be prepared.

The mass spectrometry of trimethylsilyl derivatives of nucleotides could be a powerful tool for the determination of minor bases in t-RNA and other polynucleotides, and preliminary reports on this technique have now appeared.[84]

3 Coenzymes and Cofactors

A. Phosphoenolpyruvate.—Phosphoenolpyruvate (PEP) is classified as an 'energy-rich' compound with a free energy of hydrolysis, $\Delta G^{\circ\prime} = -13$ kcal mole^{-1} compared with the value $\Delta G^{\circ\prime} = -3$ kcal mole^{-1} for alkyl phosphates.[85] However, the ratio of hydrolytic rate constants of PEP and

[79] A. M. Yurkevich, I. I. Kolodkina, L. S. Varshavskaya, V. I. Borodulina-Shvetz, I. P. Rudakova, and N. A. Preobazhenski, *Tetrahedron*, 1969, **25**, 477.
[80] P. K. Dighe, D. N. Pahnja, and D. H. Shah, *J. Chromatog.*, 1969, **40**, 449.
[81] J. L. East, *Analyt. Biochem.*, 1968, **24**, 409.
[82] G. W. Rushizky, I. H. Stavenski, and A. E. Greco, *Biochemistry*, 1968, **7**, 406.
[83] P. T. Gilham, *Biochemistry*, 1968, **7**, 2809.
[84] J. A. McCloskey, A. M. Lawson, K. Tsuboyama, P. M. Krueger, and R. N. Stillwell, *J. Amer. Chem. Soc.*, 1968, **90**, 4182.
[85] H. R. Mahler and E. H. Cordes, 'Biological Chemistry', Harper and Row, New York, 1966, p. 201.

methyl phosphate is very much smaller than would be expected from the relative activation energies of the two compounds. The kinetics and mechanisms of the hydrolysis of PEP and its ethyl carboxylate ester have now been examined.[86] In the pH range 1–7 the monoanions of both compounds are hydrolysed by the expulsion of the non-selective, highly reactive, monomeric metaphosphate in agreement with the hydrolysis of other monoanions.[87, 88] The rate of hydrolysis of PEP and its ethyl ester are, however, greater than predicted and it appears that the carboxyl group has little or no kinetic effect.

These observations can best be accommodated by proposing a six-membered cyclic reaction (34) for the hydrolysis. A similar cyclic mechanism has been put forward[89] for the hydrolysis of the monoanion of acetyl phosphate.

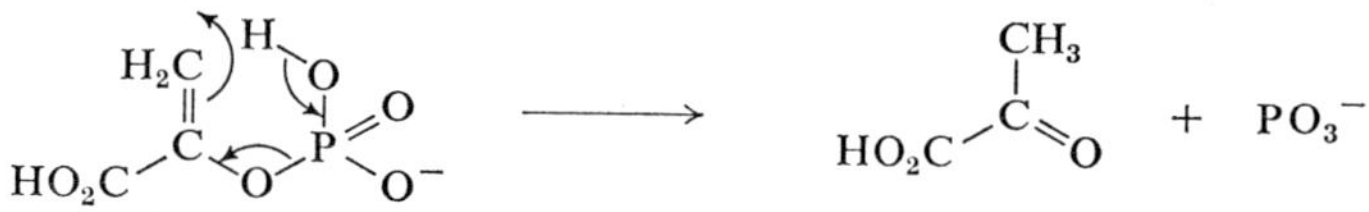

(34)

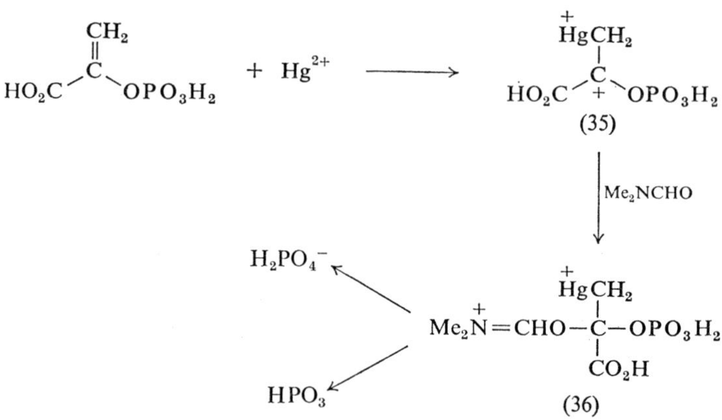

The hydrolysis of PEP occurs at an appreciable rate only at elevated temperatures; however, the addition of mercury(II) ions increases the hydrolysis rate dramatically ($>10^6$) and other metal ions have a slight catalytic effect.[90] These results are consistent with the addition of a mercuric cation to the carbon–carbon double bond of PEP to generate the carbonium ion (35). This can decompose with either C—O bond fission (leading to

[86] S. J. Benkovic and K. J. Schray, *Biochemistry*, 1968, **7**, 4090.
[87] J. D. Chanley and E. Feaguson, *J. Amer. Chem. Soc.*, 1963, **85**, 1181.
[88] A. J. Kirby and A. G. Varvoglis, *J. Amer. Chem. Soc.*, 1967, **89**, 415.
[89] G. Di Sabato and W. P. Jencks, *J. Amer. Chem. Soc.*, 1961, **83**, 4393.
[90] S. J. Benkovic and K. J. Schray, *Biochemistry*, 1968, **7**, 4097.

orthophosphate) or P—O bond fission (leading to monomeric metaphosphate). A similar species may be an intermediate in the oxidation of hydroquinone phosphates.[91]

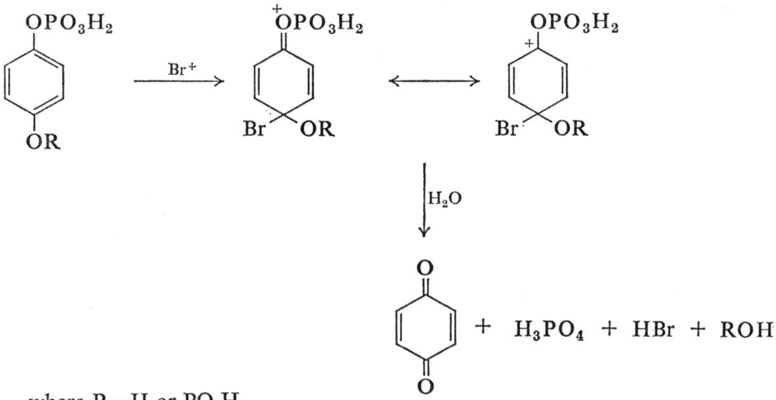

where R = H or PO_3H_2

For the mercury(II) ion-catalysed hydrolysis of PEP in methanol–water, the relative amount of P—O bond cleavage, as measured by the product ratio of methyl phosphate to orthophosphate, increases with pH. A higher relative percentage of methyl phosphate is produced after addition of dimethylformamide to the reaction. Presumably (35) is stabilised by dimethylformamide to give (36), and a similar participation by dimethylformamide has been observed in the oxidation of hydroquinone phosphates.[92]

B. Coenzyme A.—The sulphydryl group, essential for the metabolic role of Coenzyme A (37), is not essential for the latter stages of CoA biosynthesis as D-oxypantetheine 4′-phosphate[93] and D-desulphopantetheine 4′-phosphate (40)[94] can be converted enzymically into oxy-CoA (38) and desulpho-CoA (39) respectively. Compound (40) has been prepared by the phosphorylation of desulphopantetheine with dibenzyl phosphorochloridate followed by catalytic hydrogenation; (40) was then converted into 3′-dephosphodesulpho-CoA by coupling with adenosine 5′-phosphoromorpholidate. By use of an enzyme preparation from beef liver, (40) was converted into 3′-dephospho-(39) which was in turn phosphorylated enzymically to give (39).

The chemical synthesis of the 4-phosphate of pantothenyl alcohol (41) by the 2-cyanoethylphosphate–carbodi-imide method has been reported.[95]

[91] V. M. Clark and D. W. Hutchinson, *Progr. Org. Chem.*, 1968, **7**, 75.
[92] J. S. Cohen and A. Lapidot, *J. Chem. Soc.* (*C*), 1967, 1210.
[93] C. J. Stewart and W. J. Ball jun., *Biochemistry*, 1966, **5**, 3883.
[94] C. J. Stewart, J. O. Thomas, W. J. Ball jun., and A. R. Aguirre, *J. Amer. Chem. Soc.*, 1968, **90**, 5000.
[95] Y. Hosokawa, M. Tomikawa, O. Nagase, and M. Shimizu, *Chem. and Pharm. Bull.* (*Japan*), 1969, **17**, 202.

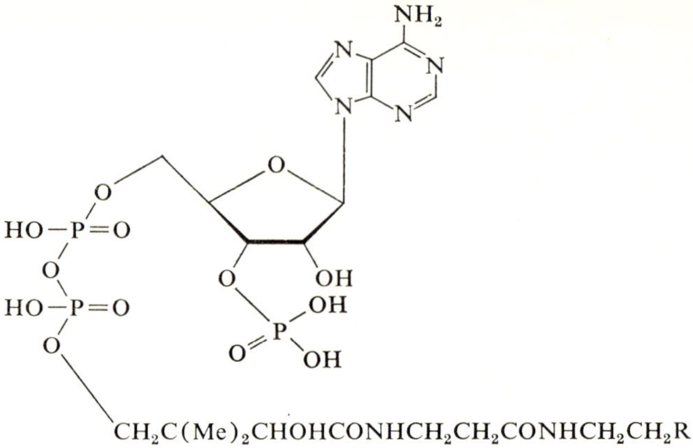

R = SH CoA (37)
R = OH oxy-CoA (38)
R = H desulpho-CoA (39)

$$(HO)_2P(=O)OCH_2C(Me)_2CHOHCONHCH_2CH_2CONHCH_2CH_3$$

(40)

$$(40) + ATP \rightleftharpoons \text{dephospho } (39) + PP_i$$
$$\text{dephospho } (39) + ATP \rightleftharpoons (39) + ADP$$

$$HOCH_2C(Me)_2CHOHCONHCH_2CH_2CH_2OH$$

(41)

Analogues of CoA modified in the cysteamine moiety have been prepared [96] by the method of Moffatt and Khorana [36] for the formation of the pyrophosphate bond. The 2′,3′-cyclic phosphate was cleaved specifically with RNase T_2 to give the 3′-isomers.[97]

C. Nicotinamide Nucleotides.—The participation in mitochondrial oxidative phosphorylation of NAD^+ derivatives which are oxygenated in the pyridine ring has been proposed.[98] The synthesis of 3-carbamoyl-6-pyridone adenine dinucleotide (42) has now been reported [99] but it is not deoxygenated by rat liver mitochondria. 3-Carbamoyl-6-pyridone mononucleotide was prepared by phosphorylating the 2′,3′-*O*-isopropylidene nucleoside with 2-cyanoethyl phosphate and dicyclohexyl carbodi-imide. The configuration

[96] M. Shimizu, O. Nagase, Y. Hosokawa, and H. Tagawa, *Tetrahedron*, 1968, **24**, 5241.
[97] A. M. Michelson, *Biochim. Biophys. Acta*, 1964, **93**, 71.
[98] D. E. Griffiths, *Fed. Proc.*, 1963, **22**, 1064.
[99] D. Orth and T. Wieland, *Chem. Ber.*, 1969, **102**, 196.

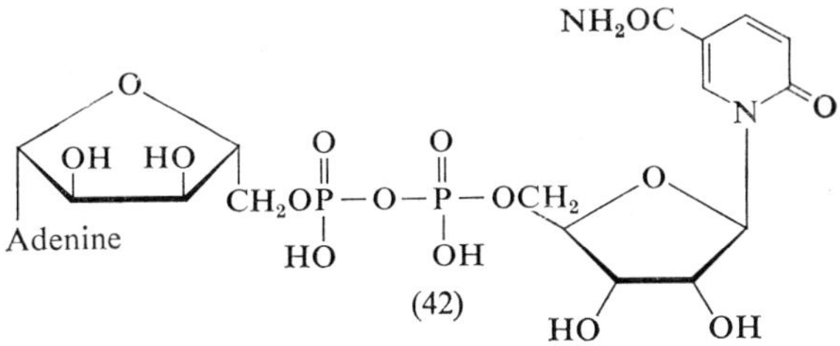

(42)

of the nucleoside and resulting nucleotides was assumed to be β from the '*trans*-rule' [70] and also from ^{1}H n.m.r. data. The 2- and 4-pyridone forms of NAD have also been shown not to be reduced in mitochondria.[100]

The pK_a value for NAD^+ has been determined [101] as $3{\cdot}88 \pm 0{\cdot}02$ at 25° over a range of ionic strengths.

D. Vitamin B_{12}.—A precursor of vitamin B_{12} in *Propionibacterium Shermanii* has been shown [102, 103] to be the 5′-phosphate [104] by *X*-ray diffraction analysis, and is converted presumably by a phosphatase into the vitamin. Mild heating of the bacteria appears to inactivate the phosphatase and enables the 5′-phosphate to be obtained in an increased yield.

E. Nucleoside Diphosphate Sugars.—Previous investigations have suggested that D-glucuronic acid is an intermediate in the biosynthesis of the branched chain sugar D-apiose (3-*C*-hydroxymethyl-D-*erythro*-furanose) (43).

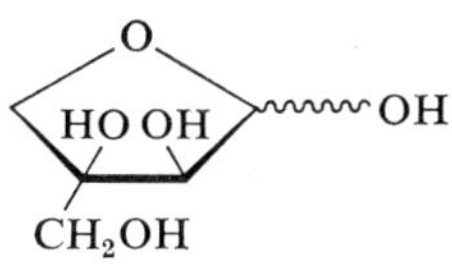

(43)

UDP-apiose has now been isolated [105] from parsley and its formation from UDP-D-glucuronic acid in cell-free extracts has been demonstrated.[106] Thus the loss of C(6) from the glucuronic acid and concomitant rearrangement followed by loss of C(3) occurs when the sugar is nucleotide-bound.

Among the nucleoside diphosphate sugars which have been isolated recently are UDP-D-rhamnose [107] and ADP-D-mannitol.[108] The biochemical function of the latter is obscure as no polymers containing mannitol are

100 D. E. Griffiths, Abstracts of VIIth Congress of Biochemistry, Tokyo, 1967.
101 C. E. Moore jun. and A. L. Underwood, *Analyt. Biochem.*, 1969, **29**, 149.
102 H. C. Friedmann, *J. Biol. Chem.*, 1968, **243**, 2065.
103 C. L. Coulter, S. W. Hawkinson, and H. C. Friedmann, *Biochim. Biophys. Acta*, 1969, **177**, 293.
104 F. Wagner, *Biochem. J.*, 1962, **336**, 99.
105 H. Sandermann jun. and H. Grisebach, *Biochim. Biophys. Acta*, 1968, **156**, 435.
106 H. Sandermann jun., G. T. Tisue, and G. Grisebach, *Biochim. Biophys. Acta*, 1968, **165**, 550.
107 G. A. Barber, *Biochim. Biophys. Acta*, 1968, **165**, 68.
108 B. Scher and V. Ginsburg, *J. Biol. Chem.*, 1968, **243**, 2385.

known in nature. In contrast the analogous compound CDP-ribitol is an intermediate in the synthesis of polyribitol phosphate.[109]

An unresolved mixture of the α- and β-anomers of UDP-*N*-acetylmuramic acid has been prepared[110a] by coupling uridine 5′-phosphoromorpholidate with an anomeric mixture of *N*-acetylmuramic acid 1-phosphates. One-half of the mixture can accept [^{14}C]-alanine to give UDP-*N*-acetylmuramyl-L-alanine. The latter has been identified[110b] as a precursor of a mucopeptide in the cell walls of *Bacillus cereus*.

4 Aminophosphonates

A. Occurrence and Biosynthesis.—2-Aminoethylphosphonic acid (AEP) (44) and its analogues (45, 46) occur in marine organisms,[111] protozoa,[112] and

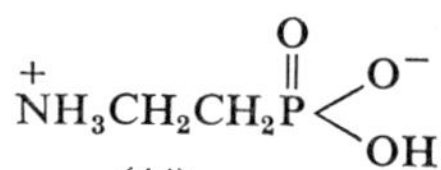

(44)

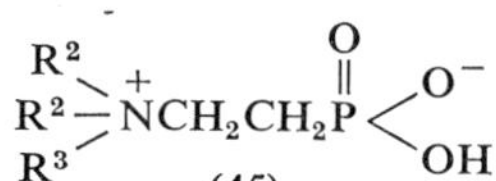

(45)

where R^1, R^2, R^3 = H or Me

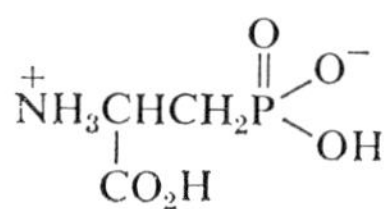

(46)

in certain phospholipid extracts of animal tissues.[113] The biosynthesis of these naturally occurring phosphonates is of interest as the carbon–phosphorus bond must be formed *in vivo* from a derivative of phosphoric acid. All chemical methods so far developed for the synthesis of phosphonic acids require tervalent phosphorus precursors (*e.g.* the Arbusov reaction[114]). However, no such tervalent phosphorus compounds have been found in nature up to the present. Any scheme for the biosynthesis of aminophosphonates must therefore require a phosphate or polyphosphate ester as a precursor for the C—P bond and this scheme must include a step when attack by a carbanion (or its equivalent) occurs at a phosphoryl centre.

The occurrence of 2-aminoethylphosphonic acids in lipid fractions led to the suggestion[115] that phosphatidyl ethanolamine might be the precursor of AEP in *Tetrahymena pyriformis* and that phosphatidyl serine could be the precursor of 1-amino-2-phosphonopropionic acid (46).

However, more recent work[116–118] indicates that although ethanolamine and serine are incorporated rapidly into phospholipids, they are poor

[109] N. Ishimoto and J. L. Strominger, *J. Biol. Chem.*, 1966, **241**, 639.
[110] (a) H. Heymann, R. Turdin, B. K. Lee, and S. S. Barkulis, *Biochemistry*, 1968, **7**, 393. (b) T. Nakatani, Y. Araki, and E. Ito, *Biochim. Biophys. Acta*, 1968, **156**, 210.
[111] J. S. Kittredge, E. Roberts, and D. G. Simonsen, *Biochemistry*, 1962, **1**, 624.
[112] J. S. Kittredge and E. Roberts, *Science*, 1969, **164**, 37.
[113] (a) M. Horiguchi and M. Kandatsu, *Nature*, 1959, **184**, 901. (b) M. Horiguchi and M. Kandatsu, *Bull. Agric. Chem. Soc.* (*Japan*), 1960, **24**, 565.
[114] A. J. Kirby and S. G. Warren, 'The Organic Chemistry of Phosphorus', Elsevier, Amsterdam, 1967, p. 37.
[115] W. Segal, *Nature*, 1965, **208**, 1284.
[116] G. A. Thompson jun., *Biochemistry*, 1967, **6**, 2015.
[117] C. R. Liang and H. Rosenberg, *Biochim. Biophys. Acta*, 1968, **156**, 437.
[118] M. Horiguchi, J. S. Kittredge, and E. Roberts, *Biochim. Biophys. Acta*, 1968, **165**, 164.

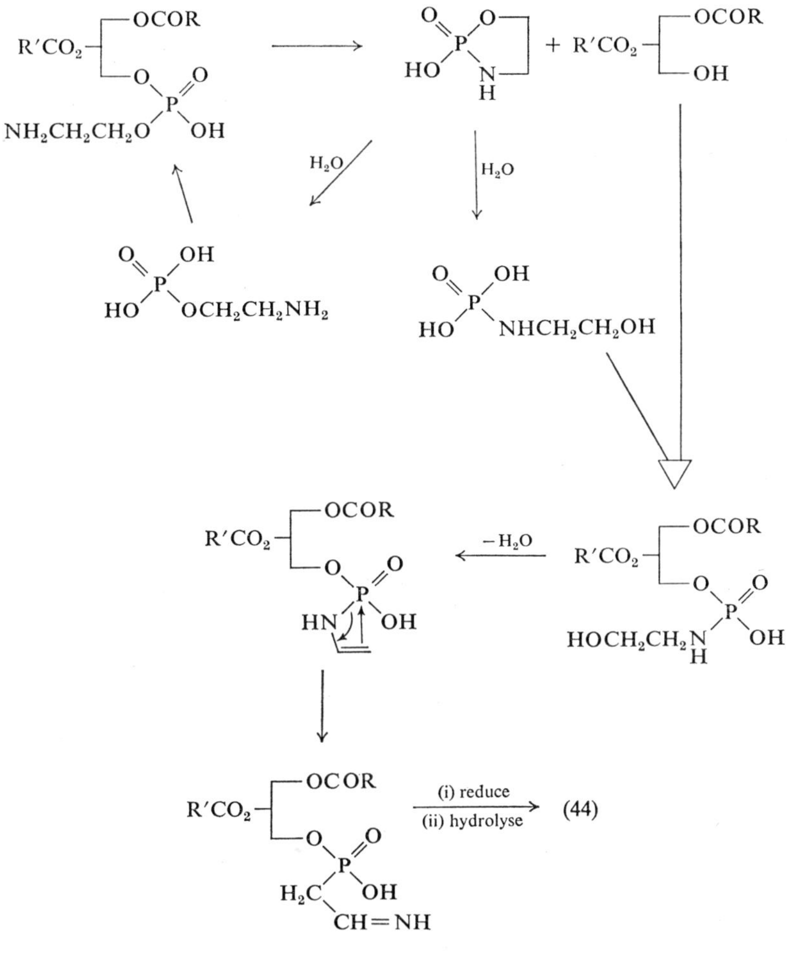

precursors of phosphonolipids. Furthermore, experiments[119] using various glucoses specifically labelled with ^{14}C indicate that phosphoenolpyruvate and/or oxaloacetate are the precursors of AEP. This has subsequently been confirmed and the following biosynthetic scheme put forward.[117, 118, 120] As a result of the rearrangement in the initial step in this scheme C(1) of AEP is derived from C(3) of the phosphoenol pyruvate and the C(2) of AEP is derived from C(2) of the latter; C(1) of the phosphoenol pyruvate is eliminated entirely in a final decarboxylation. This mechanism has been found experimentally using labelled precursors.[119]

[119] A. Trebst and F. Geike, *Z. Naturforsch.*, 1967, **22b**, 989.
[120] W. A. Warren, *Biochim. Biophys. Acta*, 1968, **156**, 340.

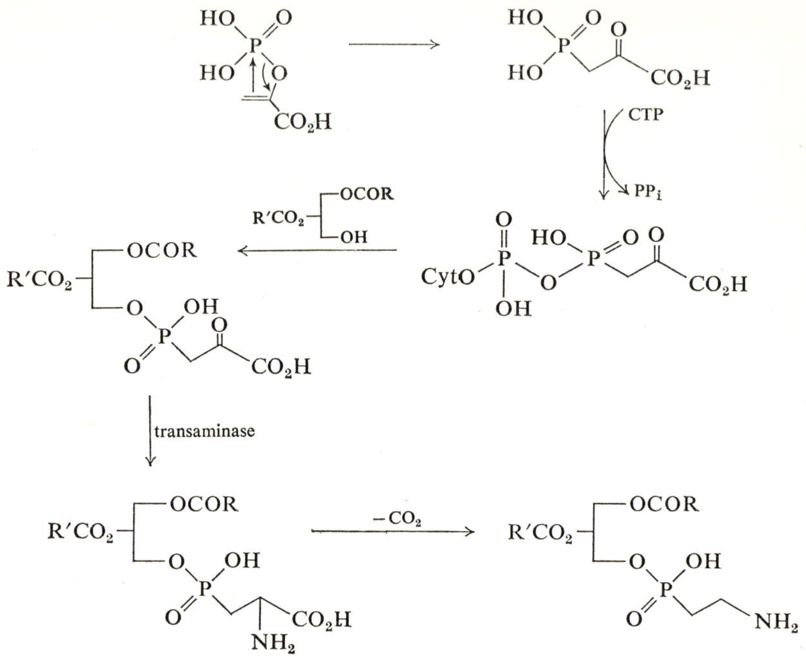

Extracts of *Tetrahymena*, *E. coli*, and *Anthopleura elegantissima* are all capable of transaminating AEP with α-ketoglutarate [121] and the mechanism of formation of the phosphonolipids *via* the cytidine 5′-pyrophosphate has analogies in the biosynthesis of phospholipids. No reports of a rearrangement of an enol phosphate to a ketophosphonate have been recorded, however, although the photochemical and thermal rearrangement of enol-benzoates to β-diketones is well known.[122]

B. Synthesis of Lipid Analogues.—The discovery that amino-phosphonic acids can be constituents of lipids has initiated the synthesis of analogues of lecithins, cephalins, and ceramide.[123–127] Phosphonolipids which contain an ether instead of a fatty acid group [116] have also been synthesised.[128, 129]

121 E. Roberts, D. G. Simonsen, M. Horiguchi, and J. S. Kittredge, *Science*, 1968, **159**, 886.
122 C. L. McIntosh, *Canad. J. Chem.*, 1967, **45**, 2267.
123 E. Baer and R. Robinson, *Canad. J. Biochem.*, 1968, **46**, 1273.
124 E. Baer and H. Basu, *Canad. J. Biochem.*, 1968, **46**, 351.
125 E. Baer and H. Basu, *Canad. J. Biochem.*, 1968, **46**, 1279.
126 E. Baer and B. C. Pal, *Canad. J. Physiol. Pharmacol.*, 1968, **46**, 525.
127 E. Baer and G. R. Sarma, *Canad. J. Biochem.*, 1969, **47**, 224.
128 G. K. Chacko and D. J. Hanahan, *Biochim. Biophys. Acta*, 1969, **176**, 190.
129 J. Berecoechean, M. Fauve, and J. Anatol, *Bull. Soc. Chim. Biol.*, 1968, **50**, 1561.

C. Catabolism of Phosphonic Acids.—The isolation [130] of a strain of *Bacillus cereus* which can utilise AEP as its sole source of phosphorus indicates that living organisms are capable of catabolising phosphonic acid derivatives despite the high stability of the C—P bond under normal hydrolytic conditions. The C—P bond undergoes ready fission, however, when AEP is treated with ninhydrin, when acetaldehyde and inorganic phosphate are the main products.[131] An intermediate in this reaction is presumed to be 2-phosphono-acetaldehyde (47) which is a P—XYZ system [91] and as

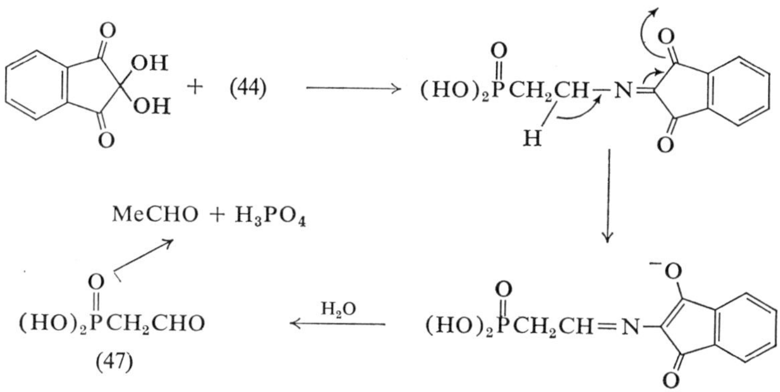

such will undergo cleavage of the C—P bond without difficulty. Compound (47) has been identified [132] as an intermediate in the degradation of AEP by *B. cereus* and has recently been synthesised [133] from 2-acetoxy-2-chloroethylphosphonyl dichloride (48). This intermediate is stable in 6N-hydrochloric acid and at high pH, but hydrolyses at pH 5. Similar behaviour has been observed [134] with 2-hydroxyethylphosphonic acid which can be obtained by the diazotisation of AEP in aqueous solution.

$$CH_3CO_2CH{=}CH_2 \xrightarrow{PCl_5-SO_2} \underset{(48)}{CH_3CO_2CHClCH_2\overset{O}{\overset{\|}{P}}Cl_2} \xrightarrow{H_2O} (47)$$

5 Oxidative Phosphorylation

Mitochondrial oxidative phosphorylation has continued to arouse considerable interest and the main, and rival theories of oxidative phosphorylation—the chemical ($\sim$X) and the chemiosmotic—have recently been compared.[135] Much effort is being expended on the study of uncouplers and inhibitors in an effort to separate the various steps in the enzyme

130 H. Rosenberg and J. M. La Nauze, *Biochim. Biophys. Acta*, 1967, **141**, 79.
131 A. J. De Koning, *Biochim. Biophys. Acta*, 1966, **130**, 521.
132 J. M. La Nauze and H. Rosenberg, *Biochim. Biophys. Acta*, 1968, **165**, 438.
133 A. F. Isbell, L. F. Englert, and H. Rosenberg, *J. Org. Chem.*, 1969, **34**, 755.
134 D. W. Hutchinson and B. D. Place, unpublished observations.
135 D. W. Deamer, *J. Chem. Educ.*, 1969, **46**, 198.

complex; however, various model chemical reactions have also been investigated as possible begetters of 'high-energy' intermediates.

A. Hydroquinone Esters.—Acyl transfer following the oxidation of carboxylic esters of hydroquinones has been demonstrated,[136] the products include anhydrides and esters. Oxidation at room temperature of a pyridine solution of durohydroquinone monoacetate (49), orthophosphate, and ADP by bromine or a high potential quinone gives rise to a low yield of ATP[137] presumably through a derivative of acetyl phosphate.

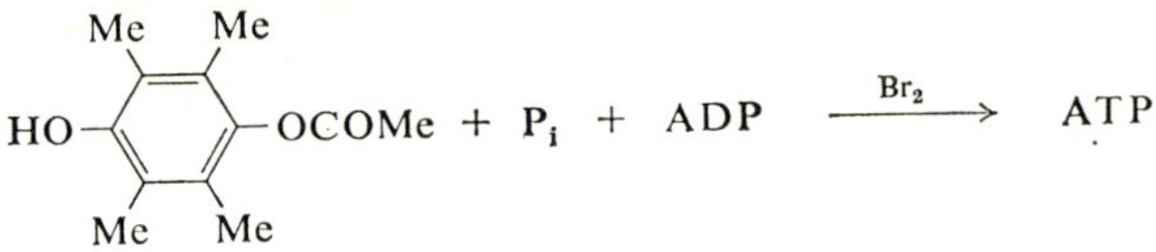

(49)

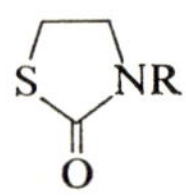

(50)

$HSCH_2CO_2H$

(51)

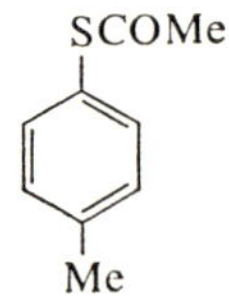

(52)

where R = H, Ph, Me, Ac

Related syntheses of ATP have been achieved by the oxidation of thiazolid-2-ones (50)[138] and mercaptoacetic acid (51)[139] with bromine, although the mechanisms of these two phosphoryl transfer reactions have yet to be elucidated; it is most probable that sulphenyl bromides are involved as intermediates. *S*-Acetyl-*p*-thiocresol (52) is another compound capable of generating a sulphenyl bromide with bromine, and ATP has been synthesised[140] from (52), ADP, and orthophosphate following the addition of bromine.

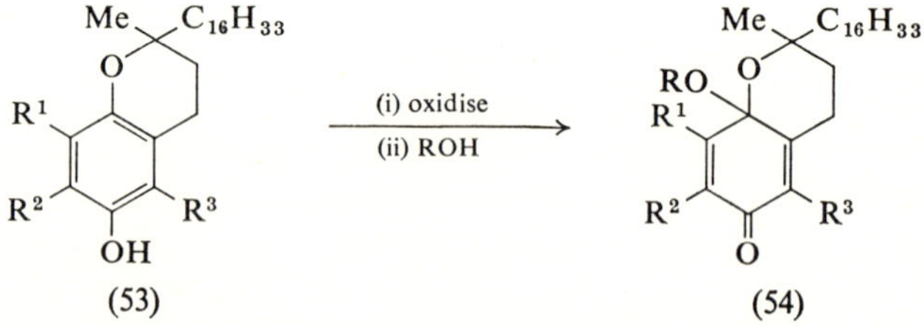

(53) (54)

[136] V. M. Clark, M. R. Eraut, and D. W. Hutchinson, *J. Chem. Soc.* (*C*), 1969, 79.
[137] T. Wieland and H. Aquila, *Angew. Chem. Internat. Edn.*, 1968, **7**, 213.
[138] T. Wieland and H. Aquila, *Chem. Ber.*, 1968, **101**, 3031.
[139] T. Wieland and E. Bäuerlein, *Angew. Chem. Internat. Edn.*, 1968, **7**, 893.
[140] E. Bäuerlein and T. Wieland, *Chem. Ber.*, 1969, **102**, 1299.

In ethanolic solution, tocopherols (53) can be converted into (54, R = Et) by the action of iron(III) chloride.[141] A similar intermediate (54, R = PO_3H_2) has been postulated as being formed when a solution of (53) and orthophosphate in pyridine is oxidised with bromine. Protonation of the oxygen atom in the chroman ring of (54, R = PO_3H_2) generates a phosphorylating agent of the P—XYZ type[91] which explains the formation of ATP and ADP when nucleoside phosphates are added to this reaction.

The enzymic oxidation by horse radish peroxidase of durohydroquinone monophosphate to duroquinone and orthophosphate in ^{18}O-enriched water occurs with negligible C—O bond fission.[142] This implies (see section 2) that the reaction proceeds by nucleophilic attack on phosphorus before the P—O bond is broken, and hence the transfer of phosphate from hydroquinone phosphates *in vivo* could be a more efficient process than model chemical studies indicate.[143] It appears that the relative amounts of P—O and C—O bond fission during the oxidative dephosphorylation of hydroquinone phosphates depend on the oxidising agent employed, and oxidation by periodate in aqueous solution proceeds largely with P—O fission.[144]

B. Metal Complexes.—The final phosphorylation step in mitochondrial oxidative phosphorylation is associated with cytochrome-*a*. A model reaction has been devised[145] in which ATP is produced from orthophosphate and AMP or ADP during the oxidation by molecular oxygen of a ferrohaemochrome in the presence of imidazole. In this case 1-phosphoimidazole

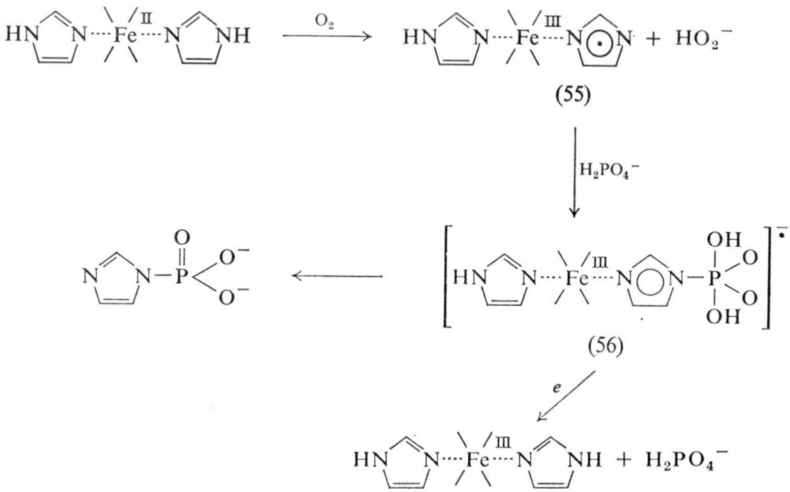

141 C. Martius and H. Eilingsfeld, *Annalen*, 1957, **607,** 159.
142 J. Wodak, *J. Amer. Chem. Soc.*, 1968, **90**, 2991.
143 [a] A. Lapidot and D. Samuel, *Biochim. Biophys. Acta*, 1962, **65**, 164. [b] A. Lapidot and D. Samuel, *J. Amer. Chem. Soc.*, 1964, **86**, 1886.
144 C. A. Bunton and J. Hellyer, *Tetrahedron Letters*, 1969, 187.
145 T. A. Cooper, W. S. Brinigar, and J. H. Wang, *J. Biol. Chem.*, 1968, **243**, 5854.

is the phosphorylating agent, and similar intermediates are formed from 2-methyl- and 2,4-dimethylimidazole but not from 1,2-dimethylimidazole. It is suggested that the initial oxidation of di-imidazoleferrohaemochrome is a two-electron process generating the radical (55) which reacts with orthophosphate to give (56). Reduction of the latter (by an unspecified mechanism) produces 1-phosphoimidazole. The experimental evidence outlined above supports the proposal[146] that phosphohistidine[147] (which could arise from such a sequence of reactions *in vivo*) is involved in oxidative phosphorylation.

6 Sugar Phosphates

A. Pentoses.—When 5,6-*O*-isopropylidene-L-ascorbic acid (57) is treated with phosphorus oxychloride in wet pyridine, a mixture of products is formed from which the 2- and 3-phosphates of L-ascorbic acid (previously synthesised as their monophenyl esters[148]) together with L-ascorbic acid 3-pyrophosphate and bis(L-ascorbic acid)-3,3′-phosphate (58) can be isolated.[149, 150] Compound (58) is acid labile and is readily hydrolysed by hydrochloric acid to an equimolar mixture of L-ascorbic acid and its 3-phosphate. When L-ascorbic acid 2-phosphate is heated in dilute hydrochloric acid at 100°, phosphoryl migration to the 3-position is observed, and it is tempting to speculate that the highly strained 2,3-cyclic phosphate (59) may be an intermediate in this reaction.

When D-ribose 5-phosphate (60) is heated in an aqueous buffer at pH 5·5 in the presence of an organic base (*e.g.* pyridine) 4-hydroxy-5-methyl-2,3-dihydrofuran-3-one (61) is formed.[151] The methyl group of (61) does not

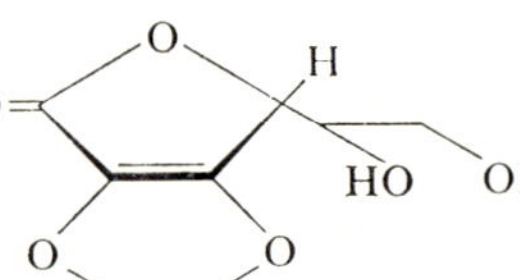

(57)

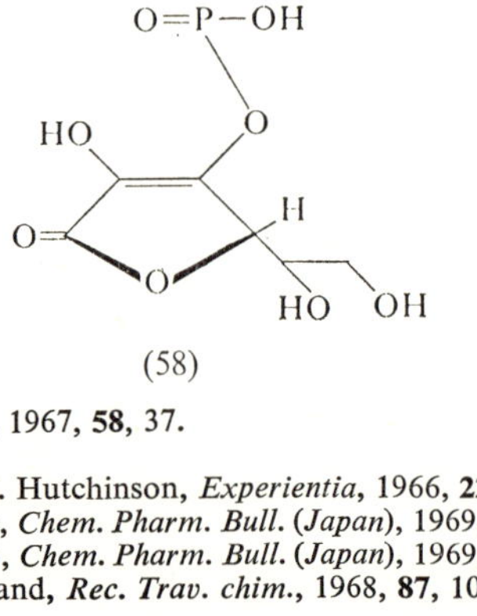

(58)

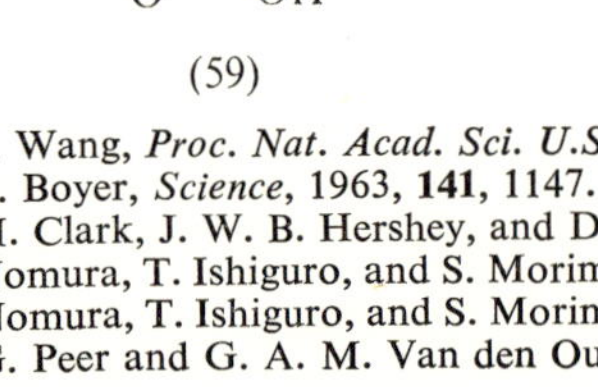

(59)

[146] J. H. Wang, *Proc. Nat. Acad. Sci. U.S.A.*, 1967, **58**, 37.
[147] P. D. Boyer, *Science*, 1963, **141**, 1147.
[148] V. M. Clark, J. W. B. Hershey, and D. W. Hutchinson, *Experientia*, 1966, **22**, 425.
[149] H. Nomura, T. Ishiguro, and S. Morimoto, *Chem. Pharm. Bull. (Japan)*, 1969, **17**, 381.
[150] H. Nomura, T. Ishiguro, and S. Morimoto, *Chem. Pharm. Bull. (Japan)*, 1969, **17**, 387.
[151] H. G. Peer and G. A. M. Van den Ouweland, *Rec. Trav. chim.*, 1968, **87**, 1017.

originate from C(1) of the ribose 5-phosphate, and a base-catalysed elimination of phosphate followed by loss of water has been put forward as a possible mechanism for this reaction.

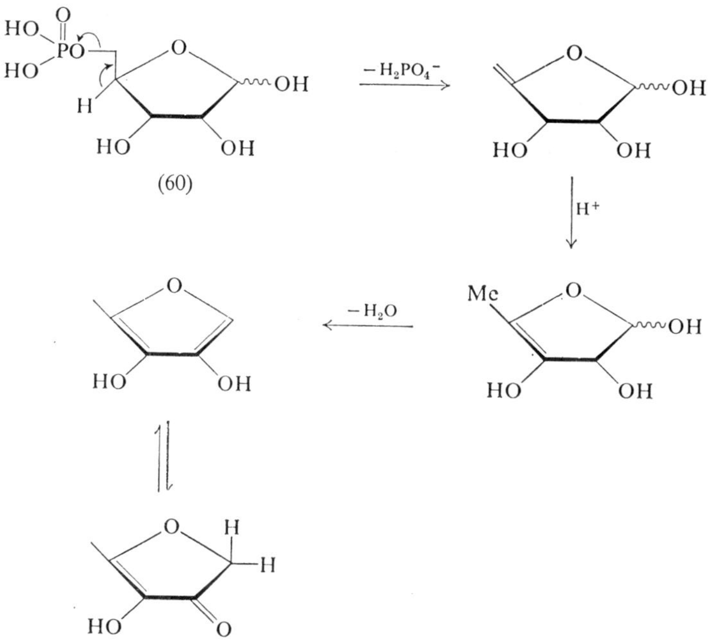

(60)

(61)

The first aminopentose to be isolated from natural sources was 3-amino-3-deoxy-D-ribose (62) which is a constituent of the antibiotic puromycin.[152] An investigation of the biosynthesis of this amino-sugar in a puromycin-producing strain of *Streptomyces alboniger* revealed the presence not of (62) but of 2-amino-2-deoxy-D-ribose 5-phosphate (63) and 2-amino-2-deoxy-D-lyxose 5-phosphate (64).[153] It is not thought that (63) and (64) have any function in puromycin biosynthesis as it has been shown that adenosine is converted into 3′-amino-3′-deoxyadenosine in *Helminthosporium* without fission of the adenine–ribose bond.[154]

Among the products of the enzymic oxidative decarboxylation of D-glucuronic acid 6-phosphate are (60) and D-ribulose 5-phosphate, and the existence of an isomerase which interconverts (60) and D-ribulose 5-

[152] C. W. Waller, P. W. Fryth, B. L. Hutchings, and J. H. Williams, *J. Amer. Chem. Soc.*, 1953, **75**, 2025.

[153] P. F. Rebello, B. M. Pogell, and P. P. Mukherjee, *Biochim. Biophys. Acta*, 1969, **177**, 468.

[154] R. J. Suhadolnik, 'Antibiotics', Vol. 2, eds. D. Gottlieb and P. D. Shaw, Springer-Verlag, Berlin, 1967, p. 448.

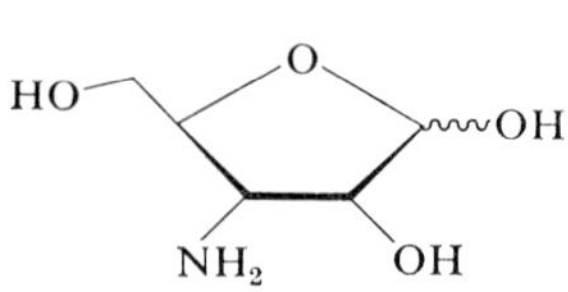

(62)

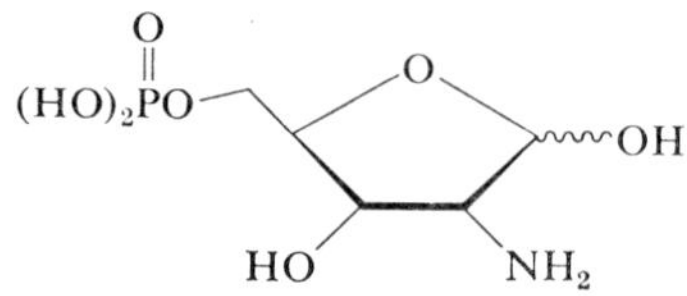

(63)

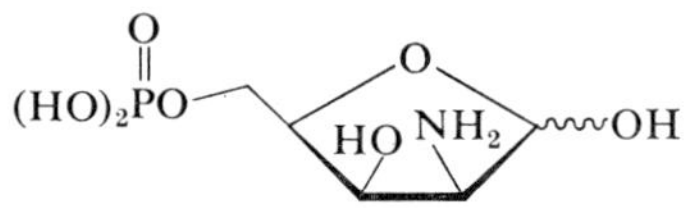

(64)

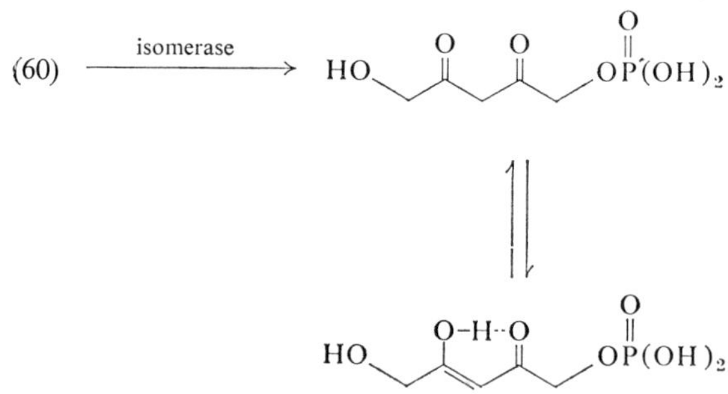

(65)

phosphate has been claimed.[155] Recent studies [156] on the isomerase show that (60) is converted into a compound which has a strong absorption in the u.v. at 280 nm. This cannot be D-ribulose 5-phosphate and it is probably the enol form (65) of the β-diketone phosphate.

B. Hexoses and Other Sugars.—In a new variant [157] of the preparation of sugar phosphates by nucleophilic displacement, primary tosyl groups are replaced by diphenyl phosphate ion. The yields after removal of the protecting groups are, however, low.

The formation of D-xylo-hexos-5-ulose 6-phosphate (66) has been inferred [158] following the removal of protecting groups from (67); the final product in this reaction is glucitol 6-phosphate. Compound (66) is of interest as a putative intermediate in the biosynthesis of inositol.

C. Phosphonates.—Analogues of sugars with phosphorus as the ring heteroatom can be prepared [159] from sugar 5-deoxy 5-phosphonate esters (68) which can be readily obtained by means of the Arbusov reaction.

155 B. L. Horecker, P. Z. Smyrniotis, and J. E. Seegmiller, *J. Biol. Chem.*, 1951, **193**, 383.
156 F. C. Knowles and N. G. Pon, *J. Amer. Chem. Soc.*, 1968, **90**, 6536.
157 A. K. Chatterjee and D. L. MacDonald, *J. Org. Chem.*, 1968, **33**, 1584.
158 D. E. Kiely and H. G. Fletcher jun., *J. Org. Chem.*, 1968, **33**, 3723.
159 R. L. Whistler and C. C. Wang, *J. Org. Chem.*, 1968, **33**, 4455.

Reduction of (68) yields the 5-phosphino-compound (69) which is isomerised by acid to (70), oxidation of which produces the phosphine oxide (71) and the phosphinic acid (72).

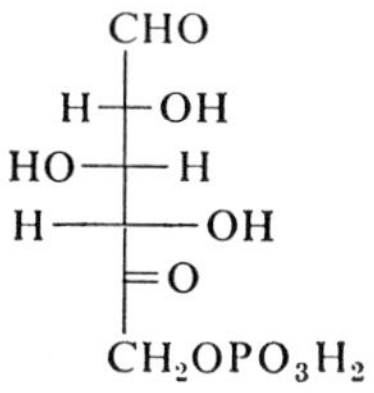

(66)

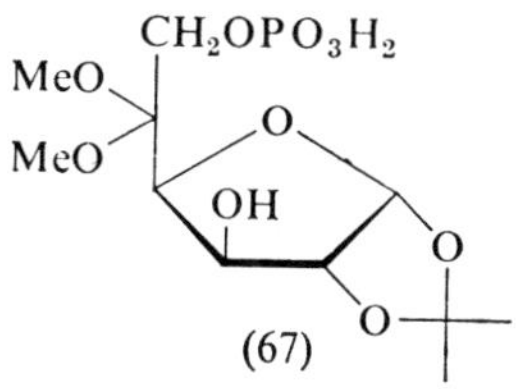

(67)

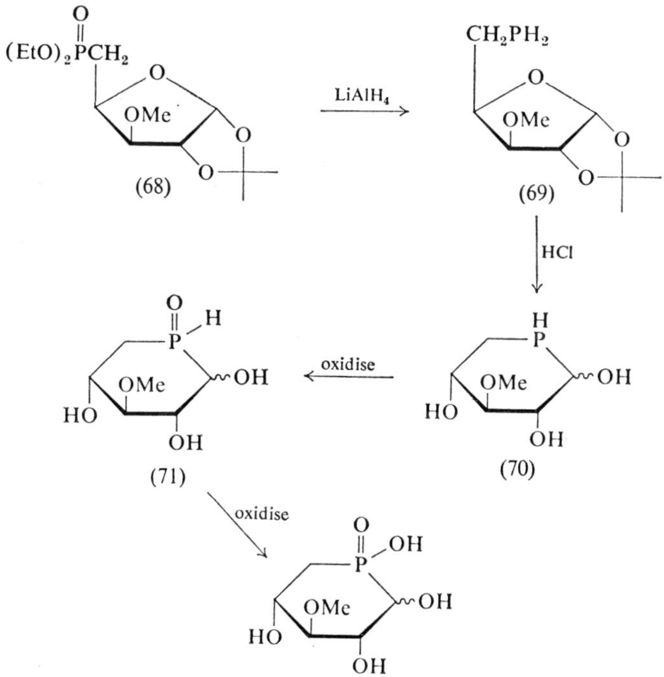

(68) (69) (70) (71) (72)

7 Inositol Phosphates

The structure of phytic acid from cereal grains, which has been the subject of some controversy, has been confirmed as *myo*-inositol hexaphosphate (73) rather than the cyclic pyrophosphate (74) by ^{31}P n.m.r.[160] and by chemical means.[161] The single axial phosphate can be distinguished in the

[160] L. F. Johnson and M. E. Tate, *Canad. J. Chem.*, 1969, **47**, 63.
[161] S. J. Angyal and A. F. Russell, *Austral. J. Chem.*, 1969, **22**, 383.

^{31}P n.m.r. spectrum as a well-resolved doublet with the five equatorial phosphates appearing as a sextet at higher field.[160] Treatment of (73) with dicyclohexyl carbodi-imide yielded (74) with a very different ^{31}P n.m.r. spectrum. Methylation of phytic acid with diazomethane gives rise to a dodecamethyl ester which is only possible for (73).[161] The ^{31}P n.m.r. spectrum of the major component of chicken blood 'phytic acid' showed that it was the 1,3,4,5,6-pentaphosphate (75).[160] A synthesis of *myo*-inositol

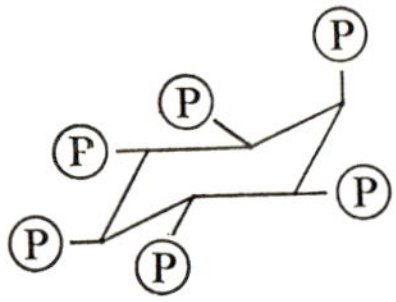

(73)

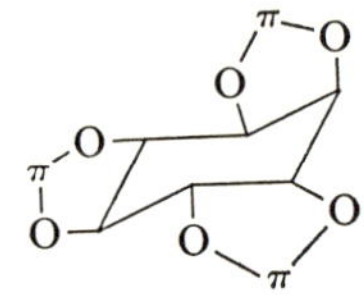

(74)

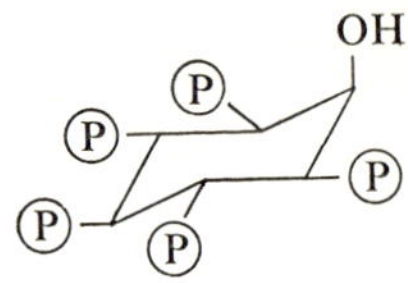

(75)

where (P) = OPO_3H_2

π = 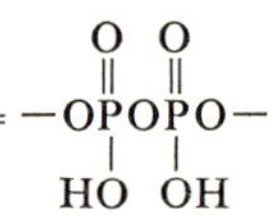

pentaphosphates [but not (75)] consists of phosphorylation of the mono-benzyl ethers with an excess of the triethylammonium salt of *N*-benzoyl-phosphoramidic acid followed by hydrogenolysis.[162] Other methods of phosphorylation were found to be inferior. Phytic acid is hydrolysed during the germination of seeds to provide inorganic phosphate for the developing embryo. The enthalpy change on hydrolysis of the phosphate esters of *myo*-inositol has been measured,[163] and there is little significant difference between the enthalpies for inositol hexa-, penta-, tetra-, and tri-phosphates. This confirms structure (73) for phytic acid as a variation in enthalpies between the homologues would be expected for structure (74).

8 Terpene Phosphates

The spontaneous decomposition of geranyl (77) and neryl (76) diphenyl phosphates[164] in an inert solvent gives rise to a complex mixture of acyclic and cyclic terpene hydrocarbons.[165] The formation of cyclic hydrocarbons, (78a) and (78b), from (77) is unusual as allylic carbonium ions preserve their configuration.[166] It is suggested[165] that participation by the P=O bond occurs and that (77) rearranges to linaloyl phosphate (79) which cyclises with ease.[167] Terpene hydrocarbons are also found when the two

[162] S. J. Angyal and A. F. Russell, *Austral. J. Chem.*, 1969, **22**, 391.
[163] J. K. Raison and W. J. Evans, *Biochim. Biophys. Acta*, 1968, **170**, 448.
[164] J. A. Miller and H. C. S. Wood, *J. Chem. Soc.* (*C*), 1968, 1837.
[165] R. C. Haley, J. A. Miller, and H. C. S. Wood, *J. Chem. Soc.* (*C*), 1969, 264.
[166] W. G. Young, S. H. Sharman, and S. Winstein, *J. Amer. Chem. Soc.*, 1960, **82**, 1376.
[167] W. Rittersdorf and F. Cramer, *Tetrahedron*, 1968, **24**, 43.

diphenyl phosphate esters are reduced with sodium borohyride[165] whereas aryl Grignard reagents give alkyl benzenes. When allylic diphenyl phosphates are solvolysed in phenol a carbonium ion is formed[168] and coumarans and chromans are produced in high yield.[164] Thus the reaction of 2,3,5-trimethylhydroquinone and phytyl diphenyl phosphate gives α-tocopherol.

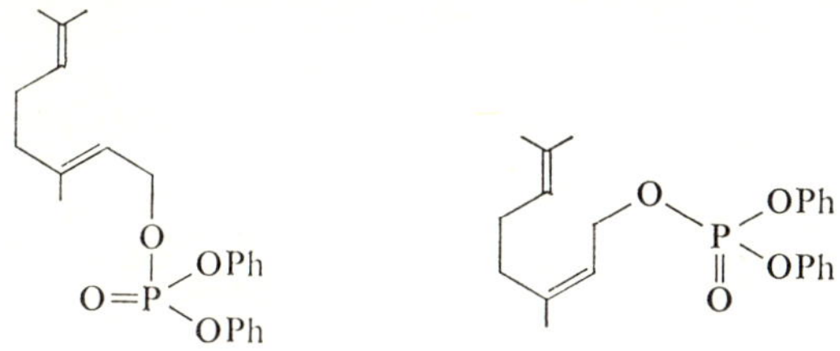

(77) (76)

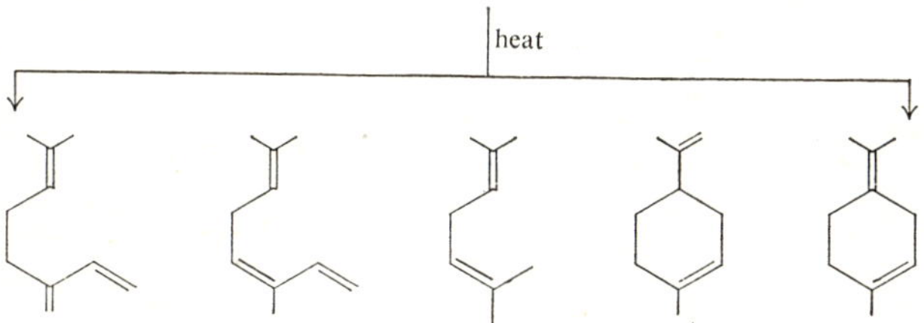

(78a) (78b)

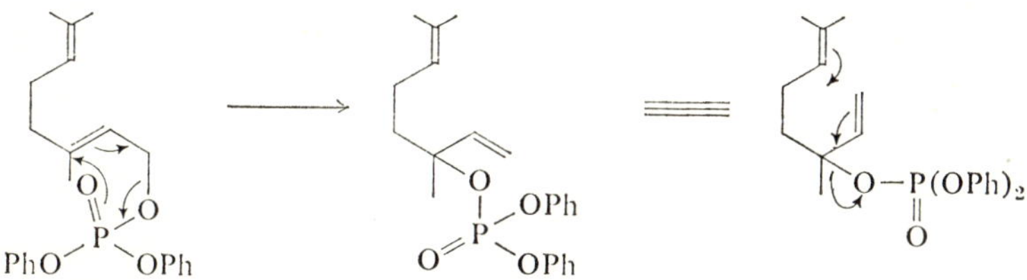

(77) (79)

An intermediate in the biosynthesis of squalene[169] was believed to be compound (80);[170] however, this has been synthesised[171] and shown to be inactive. A reinvestigation[172] of the intermediate has now shown it to be structure (81) which meets the stereochemical requirements of squalene biosynthesis.[169]

Trans-β-carotene has been prepared by the use of retinyl phosphonate (82) as a 'Wittig' reagent.[173]

[168] J. A. Miller, *J. Chem. Soc. (B)*, 1968, 1427.
[169] J. W. Cornforth, *Angew. Chem. Internat. Edn.*, 1968, **7**, 903.
[170] H. C. Rilling, *J. Biol. Chem.*, 1966, **241**, 3233.
[171] E. J. Corey and P. R. Ortiz de Montellano, *Tetrahedron Letters*, 1968, 5113.
[172] H. C. Rilling and W. W. Epstein, *J. Amer. Chem. Soc.*, 1969, **91**, 1041.
[173] J. D. Surmatis and R. Thommen, *J. Org. Chem.*, 1969, **34**, 559.

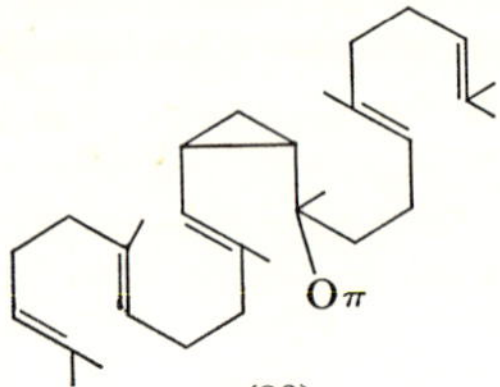

(80)

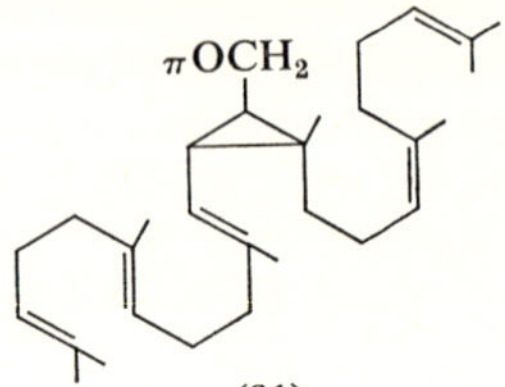

(81)

where π = 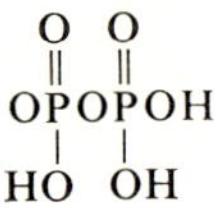

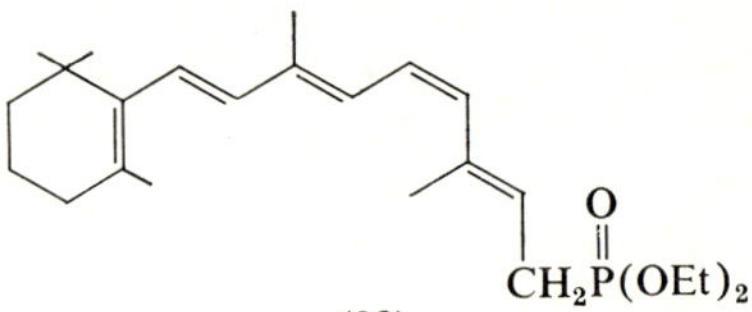

(82)

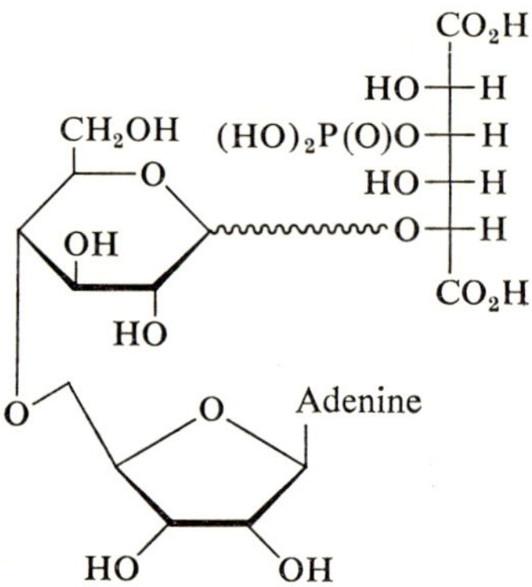

(83)

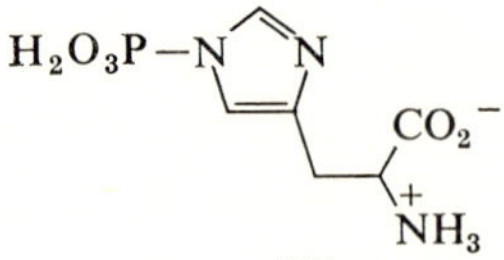

(85)

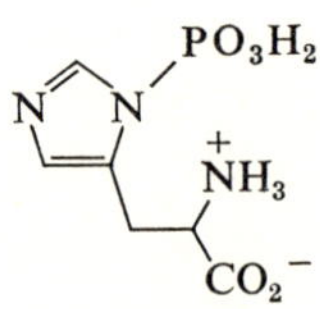

(84)

(86)

9 Other Phosphorus Compounds of Biological Interest

The centrifuged culture media from various strains of *Bacillus thuringiensis* have insecticidal properties and initial investigation [174] revealed that the exotoxin contained adenine, one phosphate group, and allomucic acid. Further investigations [175, 176] have shown the structure to be (83) and have also shown that the exotoxin inhibits a DNA-dependent RNA polymerase from *E. coli*.[177]

Phosphohistidines have been found in a variety of natural sources; for example 1-phosphohistidine (84) is the main phosphorylated compound at the active site of bovine liver nucleoside diphosphate kinase.[178] The syntheses of (84), 3-phosphohistidine (85), 1,3-diphosphohistidine, and α-1,3-triphosphohistidine have been reported and their chemical properties investigated.[179] For both mono- and di-phosphoryl derivatives of histidine the phosphoryl group on N(1) was hydrolysed faster than on N(3), possibly due to participation by the α-amino-group. In aqueous solution, (84) is rapidly converted to (85), histidine, and inorganic phosphate.

Alkaline phosphatase from *E. coli* will catalyse the phosphorylation of a number of alcohol substrates by cysteamine *S*-phosphate (86).[180] From the kinetics of the reaction, it is postulated that phosphoryl transfer occurs through a phosphorylated enzyme complex, and hence the types of acceptor compounds should be independent of the donor.[181]

174 R. P. M. Bond, *Chem. Comm.*, 1969, 338.
175 K. Šebesta, K. Horská, and J. Vanková, *Coll. Czech. Chem. Comm.*, 1969, **34**, 891.
176 J. Farkaš, K. Šebesta, K. Horská, Z. Samek, L. Dolejš, and F. Šorm, *Coll. Czech. Chem. Comm.*, 1969, **34**, 1118.
177 K. Šebesta and K. Horská, *Biochim. Biophys. Acta.*, 1968, **169**, 281.
178 O. Wålinder, *Acta Chem. Scand.*, 1969, **23**, 339.
179 D. E. Hultquist, *Biochim. Biophys. Acta*, 1968, **153**, 329.
180 H. Neumann, *J. Biol. Chem.*, 1968, **243**, 4671.
181 H. Neumann, *European J. Biochem.*, 1969, **8**, 164.

8

Ylides and Related Compounds

BY S. TRIPPETT

1 Methylenephosphoranes

A. Preparation.—Ethylene oxide has been used [1] in the generation of ylides from phosphonium salts. The position of the equilibrium depends on the stability of the ylide, which can be trapped in olefin synthesis by carbonyl compounds. Thus the propylphosphonium salt with benzaldehyde and ethylene oxide at 150° gave 1-phenylbut-1-ene (74%; *cis* : *trans* : : 3 : 2).

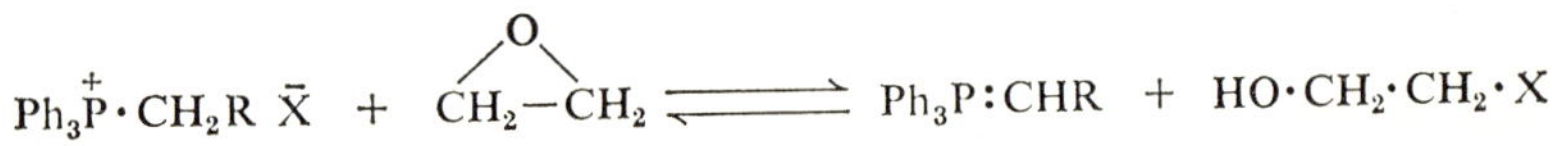

Ylides have been generated by electrolysis of phosphonium salts using carbon electrodes with a fine glass filter between the anodic and cathodic chambers and carbonyl compounds as solvent.[2] High yields of olefins, based on salt, were formed.

The preparation of stable ylides from dichlorotriphenylphosphorane and reactive methylene compounds in the presence of triethylamine [3] has been extended [4] to various heterocyclic ylides. Oxidation of the salts (1) with lead tetra-acetate gave the stable ylides (2).[5]

$$\underset{(1)}{Ph_3\overset{+}{P}\cdot CH_2X\ \overset{-}{Y}} \xrightarrow{Pb(OAc)_4} \underset{(2)}{Ph_3P{:}CX\cdot Y}$$

X	CO_2Et	$SO_2\cdot C_6H_4\cdot Me$-*p*	CO·Ph
Y	SeCN	Br	SeCN
%	80	25	67

Details have appeared of the synthesis of 2-*H*-chromens [6] (3) and of 2,5-dihydrofurans [7] (4) by the addition of alkoxides to triphenylvinylphosphonium bromide followed by intramolecular olefin syntheses of the resulting ylides.

[1] J. Buddrus, *Angew. Chem. Internat. Edn.*, 1968, **7**, 536.
[2] T. Shono and M. Mitani, *J. Amer. Chem. Soc.*, 1968, **90**, 2728.
[3] L. Horner and H. Oediger, *Chem. Ber.*, 1958, **91**, 437.
[4] J. J. Pappas and E. Gancher, *J. Heterocyclic Chem.*, 1969, **6**, 265.
[5] E. Zbiral and H. Hengstberger, *Monatsh.*, 1968, **99**, 429.
[6] E. E. Schweizer, J. G. Liehr, and D. J. Monaco, *J. Org. Chem.*, 1968, **33**, 2416.
[7] E. E. Schweizer and J. G. Liehr, *J. Org. Chem.*, 1968, **33**, 583.

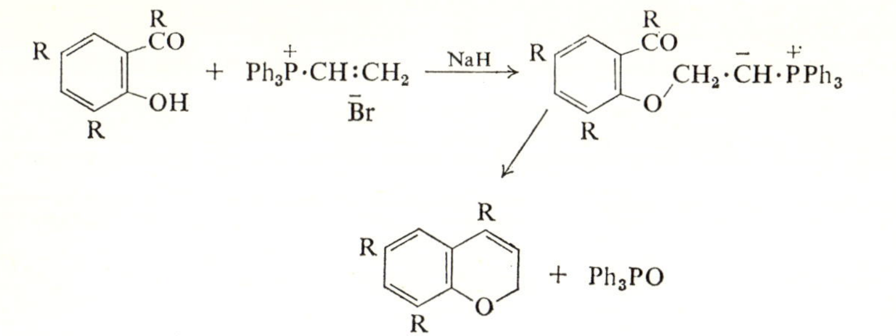

(3)

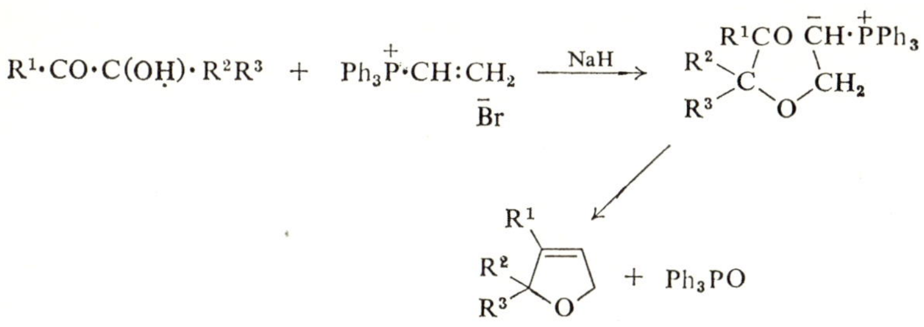

(4)

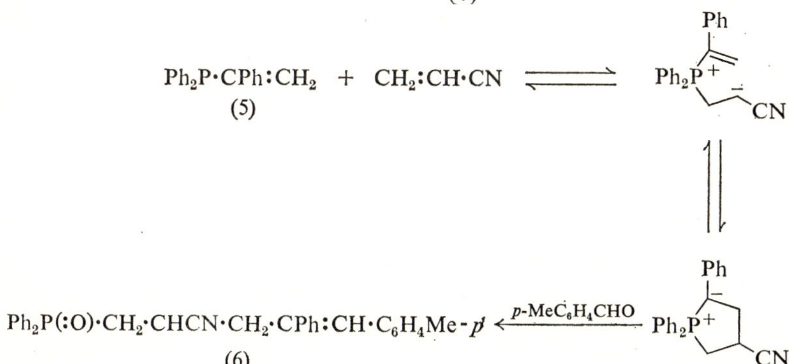

(5) (6) (7)

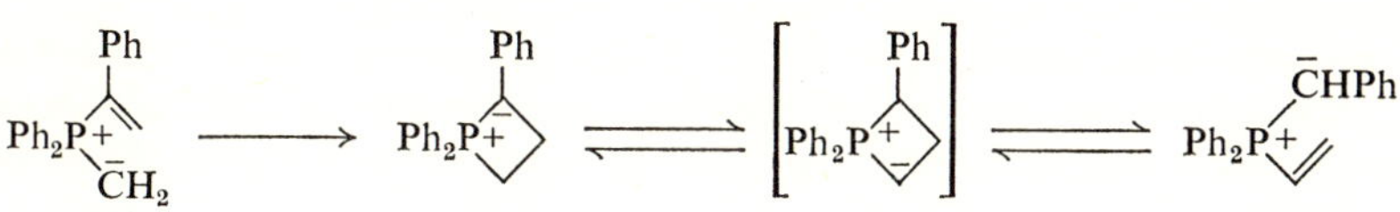

(10) (8) (9)

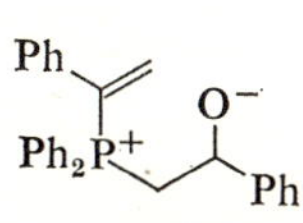

(11)

The formation of cyclic ylides by the intramolecular addition of nucleophiles to vinylphosphonium salts has also been observed.[8] Thus diphenyl-1-phenylvinylphosphine (5) with acrylonitrile and *p*-tolualdehyde gave the *trans*-oxide (6) formed *via* the cyclic ylide (7). Evidence was also presented for the formation of the ylides (8) and (9) from the methylenephosphorane (10), but no similar cyclisation occurred with the betaine (11) from (5) and styrene oxide.

B. Reactions.—(*i*) *Inorganic Reagents.* The reactive ylides (12) with nitric oxide[9] gave complex mixtures of nitrile, aldehyde, and olefin, the relative proportions varying widely with the nature of the substituent R. The suggested mechanism is given below.

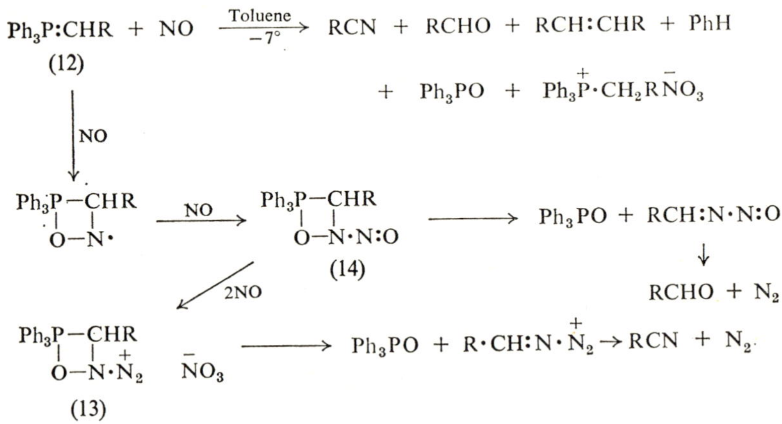

Formation of the diazonium nitrate (13) from (14) is analogous to the conversion of nitrosobenzene into benzenediazonium nitrate.[10] An olefin results from the reaction of aldehyde with unchanged ylide.

Nitriles have been obtained by the action of nitrosyl chloride on ylides in the presence of a base presumably *via* the betaines (15).[11] The phenacyclidenephosphorane (12; R = CO·Ph) with nitrosyl chloride gave the oximinosalt (16) pyrolysis of which gave benzoyl cyanide.[12] These reactions are analogous to those using nitrite esters[13] and additional examples of these reactions have appeared.[14, 61]

Oxidation of the stable ylides (17) with aqueous potassium permanganate gave α-diketones in 18—75% yield.[15] The remarkable olefins (18) were

[8] M. P. Savage and S. Trippett, *J. Chem. Soc.* (*C*), 1968, 591.
[9] K.-Y. Akiba, M. Imanari, and N. Inamoto, *Chem. Comm.*, 1969, 166.
[10] J. M. Tedder and G. Theaker, *Tetrahedron*, 1959, **5**, 288.
[11] A. M. Van Leusen, A. J. W. Iedema, and J. Strating, *Chem. Comm.*, 1968, 440.
[12] K. Akiba, C. Eguchi, and N. Inamoto, *Bull. Chem. Soc. Japan*, 1967, **40**, 2983.
[13] E. Zbiral and L. Fenz, *Monatsh.*, 1965, **96**, 11.
[14] E. Zbiral and L. Berner-Fenz, *Tetrahedron*, 1968, **24**, 1363.
[15] E. Zbiral and M. Rasberger, *Tetrahedron*, 1968, **24**, 2419.

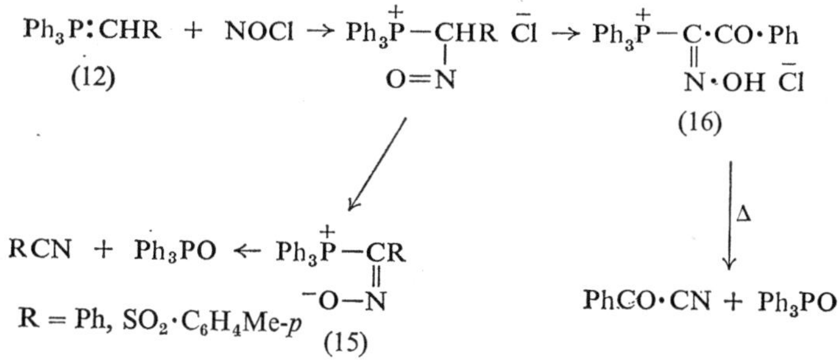

$$Ph_3P{:}CR^1{\cdot}CO{\cdot}R^2 \xrightarrow{KMnO_4} Ph_3PO + R^1{\cdot}CO{\cdot}CO{\cdot}R^2$$

(17)

obtained[16] in 50—80% yields from thiocyanogen and the β-ketoalkylidene ylides (19), transfer of CN from one sulphur to the other being postulated in the cyclic intermediate (20). Phosphorus and oxygen must be *trans* in (20) or phosphine oxide elimination would be expected to occur. Proton transfer took place in the intermediate (21; $R^1 = R^2 = R^3 = H$) to give the stable ylide (22) while enolisation of the intermediate (23; $R^1 = Pr$, $R^2 = R^3 = H$) followed by elimination of phosphine oxide gave the allene (24).

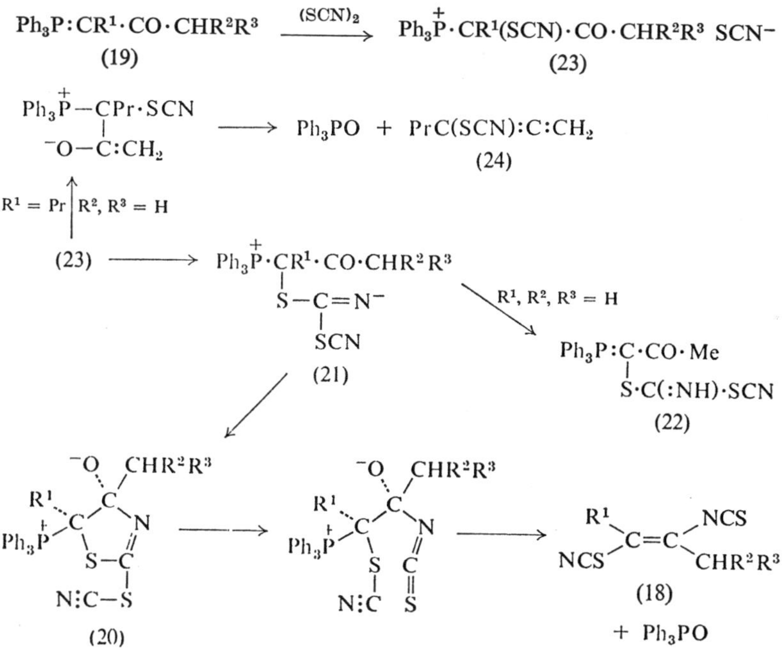

[16] E. Zbiral and H. Hengstberger, *Annalen*, 1969, **721**, 121.

(*ii*) *Halides*. Full details and many more examples have appeared of the intramolecular alkylation of ylides with alkyl halides and tosylates to give cycloalkylphosphonium salts.[17, 18] The ylide may be generated in the conventional way from a phosphonium salt,[17] *e.g.* in the synthesis of (25), or produced as an intermediate in the reaction of a dihalide (26) with two molecules of methylenetriphenylphosphorane.[18] The resulting salts (27) have been isolated as such or, where difficult to separate from the accompanying methyltriphenylphosphonium salt, used directly in olefin synthesis. In this way the toluene sulphonate (28) gave 51% of the olefin (29).

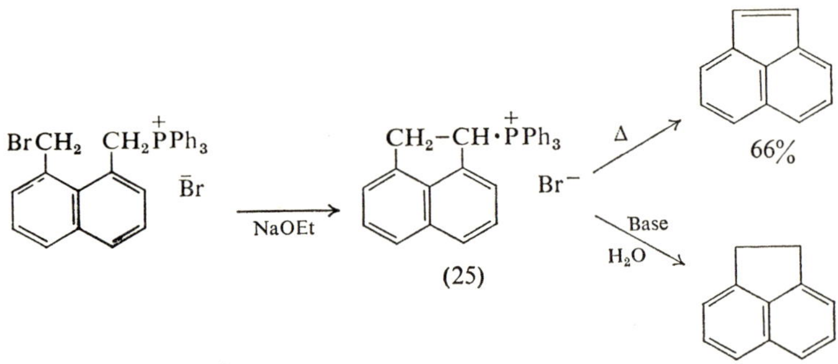

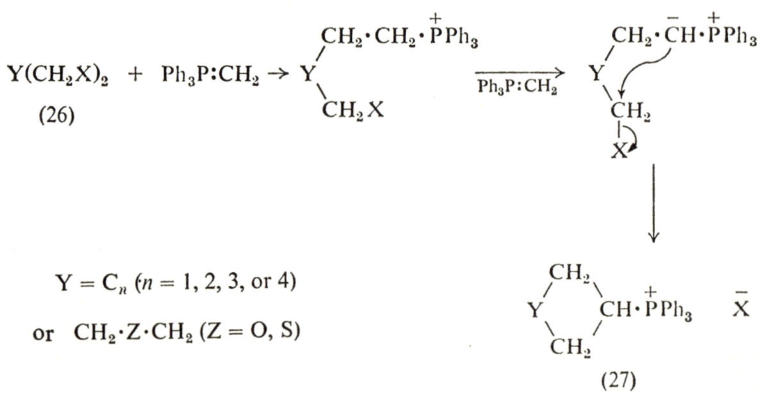

$Y = C_n$ (n = 1, 2, 3, or 4)

or $CH_2{\cdot}Z{\cdot}CH_2$ (Z = O, S)

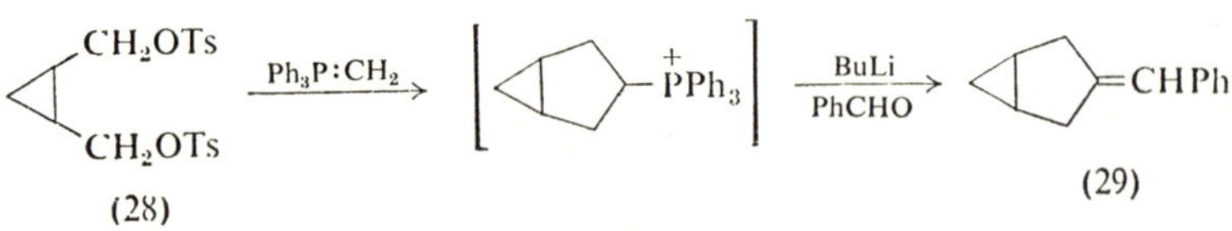

[17] H. J. Bestmann, R. Härtz, and H. Haberlein, *Annalen*, 1968, **718**, 33.
[18] H. J. Bestmann and E. Kranz, *Chem. Ber.*, 1969, **102**, 1802.

Ring closure also occurred in the reactions of bis(arylidenephosphoranes) with suitable bis(bromomethyl)-aromatic systems to give benzocycloalkene bisphosphonium salts, *e.g.* (30).[19] Alkaline hydrolysis of these was frequently accompanied by transannular attack of ylide on phosphonium salt with the expulsion of triphenylphosphine. Thus hydrolysis of the salt (30) gave 44% of the hydrocarbon (31).

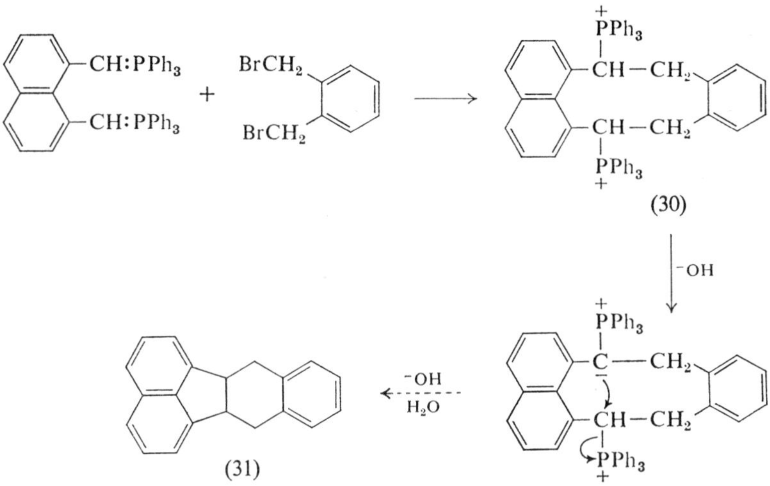

The reactions of the series of chloroketones $PhCO{\cdot}(CH_2)_n{\cdot}Cl$ with methylenetriphenylphosphorane have been investigated.[20] Phenacyl chloride gave a complex mixture of products similar to that obtained in various base-catalysed reactions involving phenacyl chloride and the first step is probably formation of the enolate anion (32). 4-Chlorobutyrophenone gave 5-chloro-2-phenyl-1-pentene (30%) and 1-cyclopropyl-1-phenylethylene (35%), produced *via* the corresponding ketone, while 5-chlorovalerophenone gave only the corresponding olefin (60%). For the reactions of phenacyl chloride with stable ylides see the review by Maercker.[21]

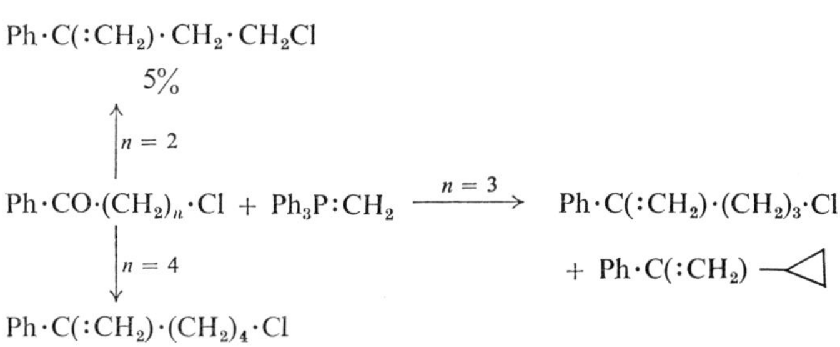

19 H. J. Bestmann and D. Rupert, *Angew. Chem. Internat. Edn.*, 1968, **7**, 637.
20 D. J. Pasto, K. Garves, and J. P. Sevenair, *J. Org. Chem.*, 1968, **33**, 2975.
21 A. Maercker, *Org. Reactions*, 1965, **14**, 270.

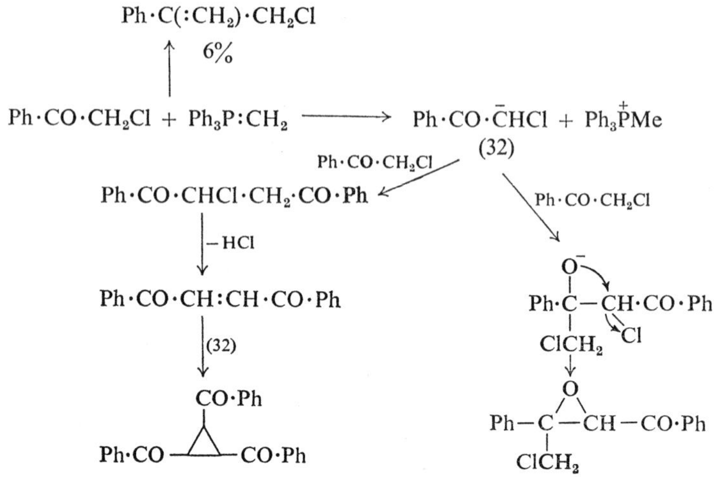

In the reactions of ylides with acyl chlorides partial kinetic resolutions have been observed [22] both at the phosphorus of an alkylidenephosphorane having a chiral phosphorus and in the alcohol of a carboalkoxymethylene-triphenylphosphorane. Both kinetic resolution and partial asymmetric synthesis occurred in the reactions of the ylide (33) with optically active acid chlorides to give allenic carboxylic esters. The optical yield of the acid (34) was 48·5%. From considerations of the probable steric course of the reactions it was possible to assign absolute configurations to many of the products.

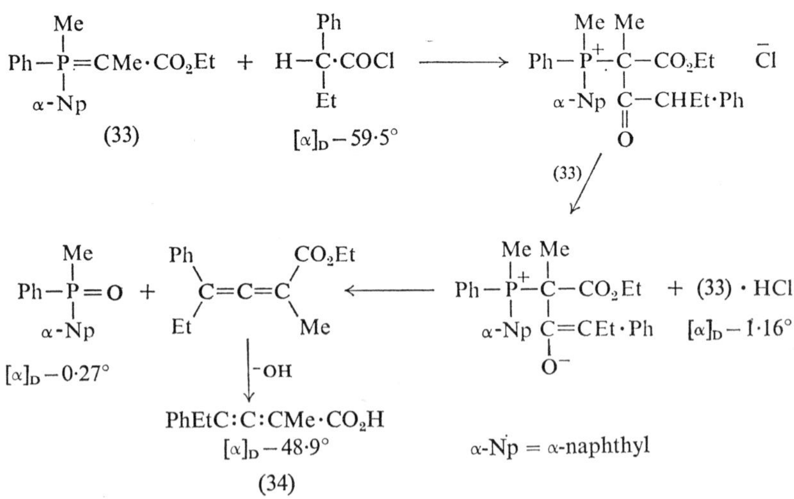

[22] H. J. Bestmann and I. Tomoskozi, *Tetrahedron*, 1968, **24**, 3299.

The acylation of allylidenetriphenylphosphorane at the γ-position with ethyl chloroformate has been extended to the use of other acyl halides.[14] The structure of the resulting salts depends on the nature of the substituents, (35) being favoured by an α-methyl substituent. With $R^2 = H$ nitrosation with ethyl nitrite gave the γ-oximino-salts (36) from which the nitrosophosphoranes (37) were obtained on treatment with base. Nitrosation of the salts (35; R^3 = OMe, R^1 = H, Me) was accompanied by migration of the ester methyl to the oxime oxygen to give the salts (38). Alkaline hydrolysis

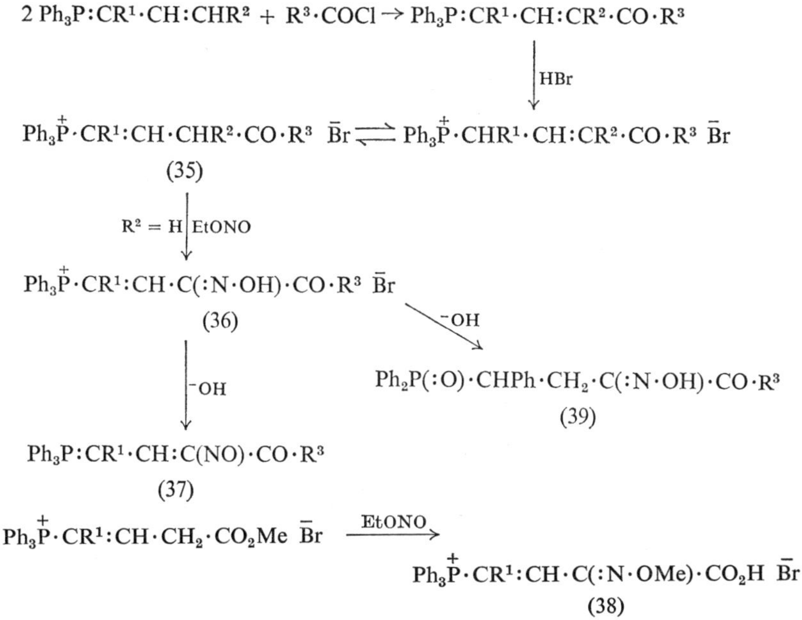

of the vinylphosphonium salts (36; R^1 = H, R^3 = Me, $CHMe_2$) was accompanied by migration of phenyl from phosphorus to the α-carbon as discussed to give the oxides (39). (See Chapter 1, Part II, section 2A.)

Sulphenyl chlorides and alkylidenephosphoranes gave the α-alkylthioalkylidenephosphoranes (40) formed by transylidation from the initial adducts.[23, 24] These were either isolated (R^1 = CO_2Et, CN, CO·Ph *etc.*) or (R^1 = H or Me) used directly in olefin syntheses to give the thiovinyl ethers (41). With the methylenephosphorane and benzenesulphenyl chloride the phosphorane (40) reacted with a second molecule of sulphenyl chloride to give the stable ylide (42).

α-Chloroketones have been obtained by hydrolysis with aqueous sodium carbonate in the cold of the salts resulting from the reaction of the stable

[23] H. Saikachi and S. Nakamura, *Yakugaku Zasshi*, 1968, **88**, 715.
[24] T. Mukaiyama, S. Fukuyama, and T. Kumamoto, *Tetrahedron Letters*, 1968, 3787.

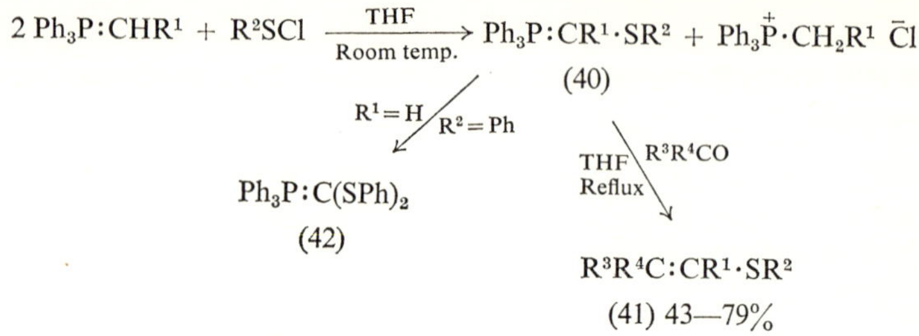

ylides (17) with iodosobenzene dichloride.[25] Only one chlorine of cyanuric chloride was utilised in reactions with stable ylides but the other chlorides in the resulting ylides (43) could then be replaced by amino residues.[26]

$$\underset{(17)}{Ph_3P{:}CR^1{\cdot}CO{\cdot}R^2} + PhICl_2 \longrightarrow Ph_3\overset{+}{P}{\cdot}CClR^1{\cdot}CO{\cdot}R^2\ \bar{C}l \xrightarrow{Na_2CO_3-H_2O} \underset{46\text{—}88\%}{R^1CHCl{\cdot}CO{\cdot}R^2}$$

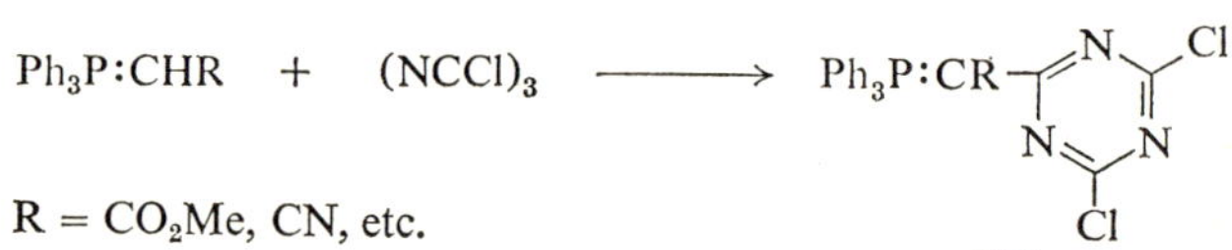

(43)

Chlorodimethylphosphine with the trimethylphosphoranes (44) gave the ylides (45)[27,28] and the stable ylide (46) was formed from methylene-triphenylphosphorane and diphenyl phosphorochloridate.[29] Vinylphosphonates were obtained from (46) and aldehydes at 110°.

$$2\ Me_3P{:}CHR + Me_2PCl \longrightarrow \underset{(45)}{Me_3P{:}CR{\cdot}PMe_2} + Me_3\overset{+}{P}{\cdot}CH_2R\ \bar{C}l$$
$$(44)\ R = H,\ SiMe_3$$

$$2Ph_3P{:}CH_2 + (PhO)_2P({:}O)Cl \longrightarrow \underset{(46)}{Ph_3P{:}CH{\cdot}P({:}O)(OPh)_2} + Ph_3\overset{+}{P}Me\ \bar{C}l$$

(*iii*) *Carbonyls.* The relative proportion of *trans*-olefin produced when the phosphoranes (47) reacted with aliphatic aldehydes decreased as *n* increased from 0 to 1 to 2.[30] Thus the $\gamma\delta$-unsaturated ylide (48) with ethyl 8-formyl-octanoate in benzene in the presence of lithium chloride gave *cis*-9,*cis*-12- and *trans*-9,*cis*-12-pentadecadienoates in the ratio of 1 : 2 which changed to

[25] E. Zbiral and M. Rasberger, *Tetrahedron*, 1969, **25**, 1871.
[26] J. J. Pappas and E. Gancher, *J. Heterocyclic Chem.*, 1968, **5**, 123.
[27] H. Schmidbauer and W. Tronich, *Chem. Ber.*, 1968, **101**, 3545.
[28] D. R. Mathiason and N. E. Miller, *Inorg. Chem.*, 1968, **7**, 709.
[29] G. H. Jones, E. K. Hamamura, and J. G. Moffatt, *Tetrahedron Letters*, 1968, 5731.
[30] L. D. Bergelson, A. A. Bezzubov, V. A. Vaver, and M. M. Shemyakin, *Izvest. Akad. Nauk S.S.S.R. Ser. khim.*, 1968, 2558.

1 : 3·6 in the presence of lithium iodide, while reaction of the δε-unsaturated ylide (49) with (50) in benzene in the presence of lithium iodide was not stereospecific.

$$Ph_3P{:}CH{\cdot}(CH_2)_n{\cdot}CH{:}CH{\cdot}R$$

(47)

$$Et{\cdot}\underset{cis}{CH{:}CH}{\cdot}CH_2{\cdot}CH{:}PPh_3 + OHC{\cdot}(CH_2)_7{\cdot}CO_2Et$$

(48)

$$\longrightarrow Et{\cdot}\underset{cis}{CH{:}CH}{\cdot}CH_2{\cdot}CH{:}CH{\cdot}(CH_2)_7{\cdot}CO_2Et$$

$$C_6H_{13}{\cdot}\underset{cis}{CH{:}CH}{\cdot}CH_2{\cdot}CH_2{\cdot}CH{:}PPh_3 + OHC{\cdot}(CH_2)_3{\cdot}CO_2Et$$

(49) (50)

$$\longrightarrow C_6H_{13}{\cdot}\underset{cis}{CH{:}CH}{\cdot}CH_2{\cdot}CH_2{\cdot}\underset{cis,\ trans}{CH{:}CH}{\cdot}(CH_2)_3{\cdot}CO_2Et$$

Competitive reactions involving the ylides $Ph_3P{:}CHR$ (R = Et, Ph) and pairs of *p*-substituted benzaldehydes in benzene in the absence of halide salts, have given data [31] which support the view that under these conditions the overall stereochemistry is determined by the relative rates of formation of the intermediate betaines, that is betaine formation is effectively irreversible.

The ease with which acraldehyde polymerises in the presence of strong base has been overcome [32] in olefin syntheses involving acraldehyde and reactive phosphoranes by passing the vapour of the aldehyde in nitrogen over a stirred and salt-free benzene solution of the phosphorane. In this way yields of dienes of *ca.* 50% were obtained.

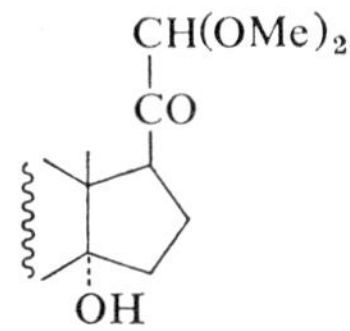

(51)

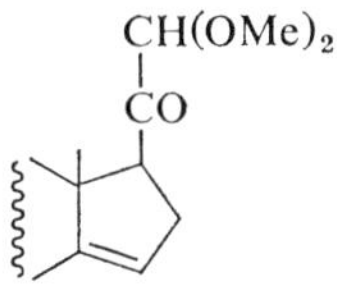

(52)

Several reports of the failures of attempted Wittig olefin syntheses have appeared. Steroidal keto-acetals of part structures (51) and (52) did not give the corresponding methylene-acetals when treated with methylene-triphenylphosphorane.[33] Furfuraldehyde with butylidenetriphenylphosphorane in dimethylformamide gave only 7% of the expected olefin, the

[31] L. D. Bergelson, L. I. Barsukov, and M. M. Shemyakin, *Zhur. obshchei Khim.*, 1968, **38**, 846.
[32] L. Crombie, P. Hemesley, and G. Pattenden, *J. Chem. Soc.* (*C*), 1969, 1016.
[33] F. Sondheimer, W. McCrae, and W. G. Salmond, *J. Amer. Chem. Soc.*, 1969, **91**, 1228.

major products being furoic acid and furfuryl alcohol.[34] In ether only the products of a Cannizzaro reaction were obtained. Finally, the β-acetoxy-ketone (53) with the phosphorane (54) in dimethylformamide gave only (55), the elimination product.[35]

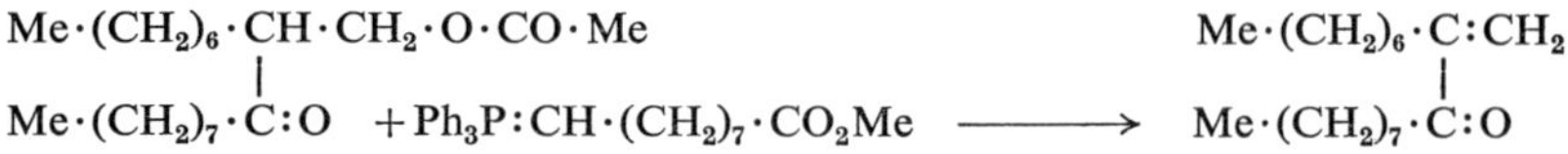

(53) (54) (55)

The product (56) isolated[36] in low yield from the reaction of carbon suboxide with the diphenylmethylenephosphorane (57) proved to be identical with that previously obtained[37] from diphenylketen and the phosphorane (58). The suggested route is *via* the four-membered phosphorane (59).

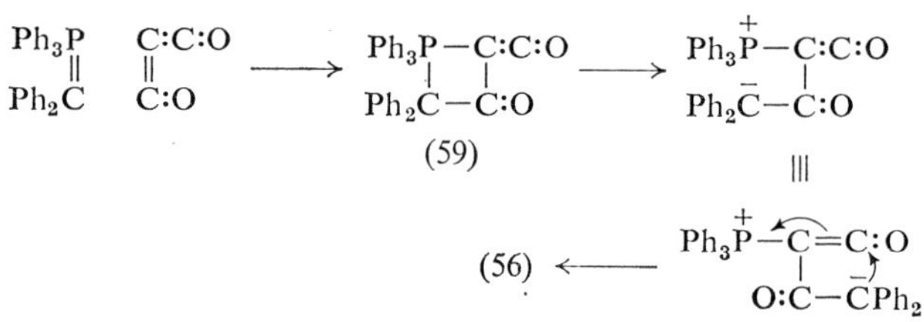

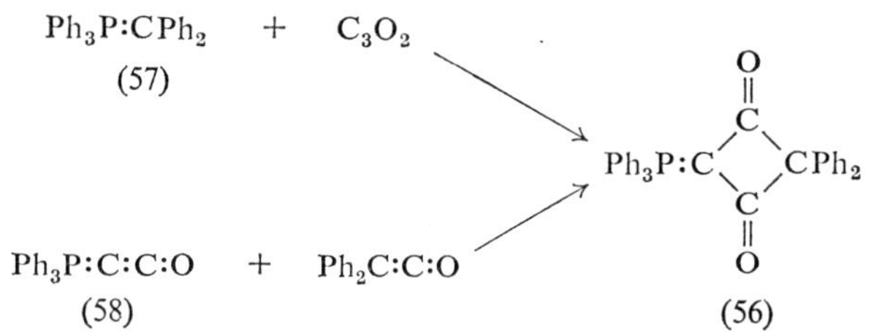

The synthesis of the allenecarboxylic ester (60) from the stable ester phosphorane (61) and keten[38] has been extended[39, 40] to other stable phosphoranes with diphenylketen or mesitylphenylketen. Unlike those formed from reactive phosphoranes,[41] the intermediates (62) could not be isolated. Cyclopentadienylidenetriphenylphosphorane with diphenylketen gave a compound to which structure (63) has been assigned.[39]

[34] A. K. Sen Gupta, *Tetrahedron Letters*, 1968, 5205.
[35] A. K. Sen Gupta and D. A. Mitchard, *Tetrahedron Letters*, 1968, 5207.
[36] H. F. Van Woerden, H. Cerfontain, and C. F. Van Valkenburg, *Rec. Trav. chim.*, 1969, **88**, 158.
[37] G. H. Birum and C. H. Matthews, *J. Amer. Chem. Soc.*, 1968, **90**, 3842.
[38] H. J. Bestmann and H. Hartung, *Chem. Ber.*, 1966, **99**, 1198.
[39] G. Aksnes and P. Frøyen, *Acta Chem. Scand.*, 1968, **22**, 2347.
[40] D. A. Phipps and G. A. Taylor, *Chem. and Ind.*, 1968, 1279.
[41] G. Wittig and A. Haag, *Chem. Ber.*, 1963, **96**, 1535.

$$Ph_3P{:}CMe{\cdot}CO_2Et + CH_2{:}C{:}O \longrightarrow CH_2{:}C{:}CMe{\cdot}CO_2Et$$

(61) (60) 71%

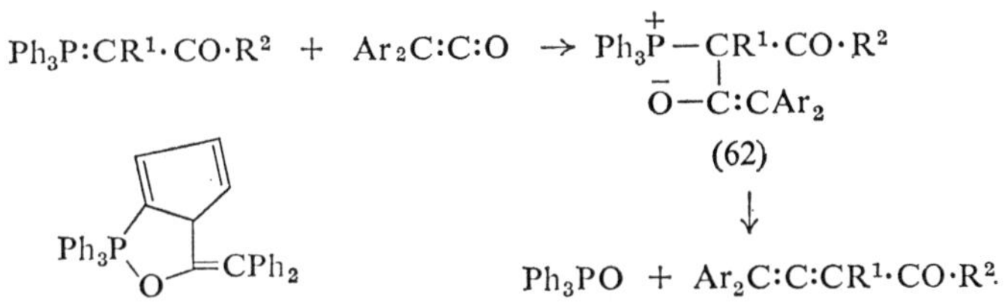

(63)

Phthalimide and the ester phosphorane (64) at 140° gave successively[42] the lactam (65; R = H) and the amine (66; R = H). As *N*-methylphthalimide gave only the lactam (65; R = Me) the second stage with phthalimide probably proceeds *via* the *N*-acylimine (67). Analogous reactions occurred with succinimide and *N*-methylsuccinimide. In both cases the products from the *N*-methylimides were mixtures of geometrical isomers.

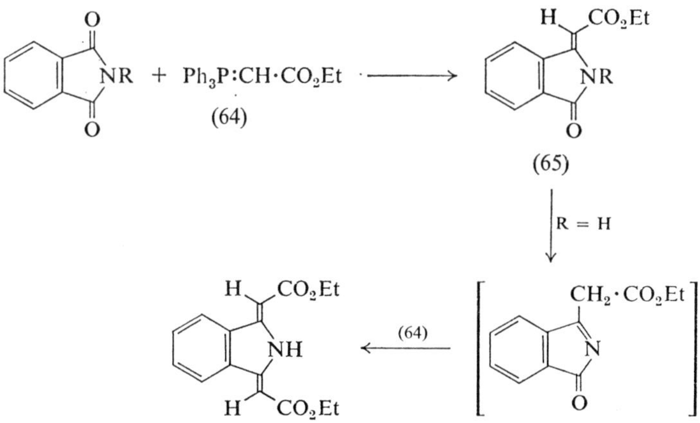

(66) (67)

The reactions of quinones with ylides are complicated by the susceptibility of the initially formed quinonemethides to Michael addition. Thus *p*-benzoquinones with the ester phosphorane (68)[43] gave the ylides (69) formed by addition of a second molecule of (68) to the quinonemethides (70). Similarly while the 1,4-naphthoquinones (71; R = H, NHPh) gave the expected quinonemethides (72), 2-methoxynaphthoquinone (71; R = OMe)

[42] W. Flitsch and H. Peters, *Tetrahedron Letters*, 1969, 1161.

[43] H. J. Bestmann and H. J. Lang, *Tetrahedron Letters*, 1969, 2101.

with the benzylidenephosphorane (73) or with (68) gave the products (74) and (75) formed with migration of the methyl groups.[43] In the latter case the intermediate (76) could be isolated.

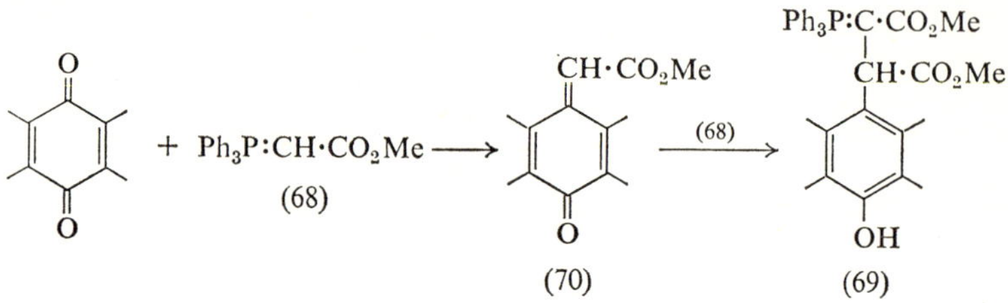

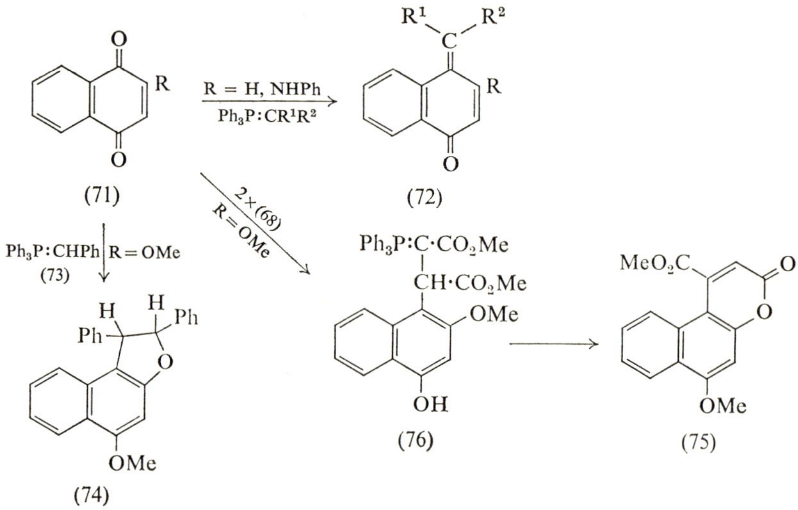

The dihydrofuran (74) was presumably formed *via* the corresponding stilbene in a way analogous to the formation of the dihydrofurans (77) from *o*-quinones and ylides.[43, 44] With the ester phosphorane (68) the intermediate olefins (78) cyclised to the coumarins (79). By the slow addition of the benzylidenephosphorane to an excess of phenanthraquinone the unstable *o*-quinonemethide (80; $R^1 = H$, $R^2 = Ph$) could be isolated and a number of analogous compounds were obtained using one equivalent of several stable phosphoranes.[44] These *o*-quinonemethides with further ylide gave the cyclobutenes (81).

The benzylideneylide (73) with tetrachloro-*o*-benzoquinone gave the benzodioxole (82)[43] previously obtained from the quinone and phenyldiazomethane.

[44] W. W. Sullivan, D. Ullman, and H. Shechter, *Tetrahedron Letters*, 1969, 457.

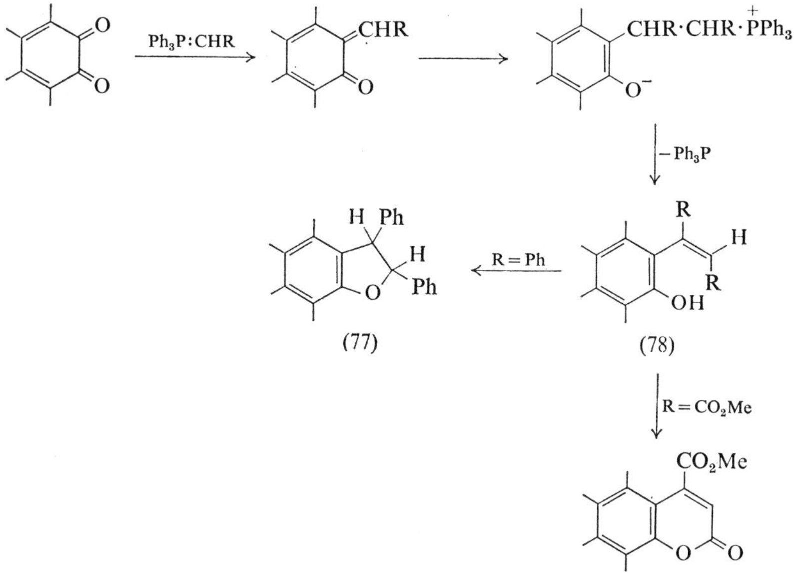

(77) (78) (79)

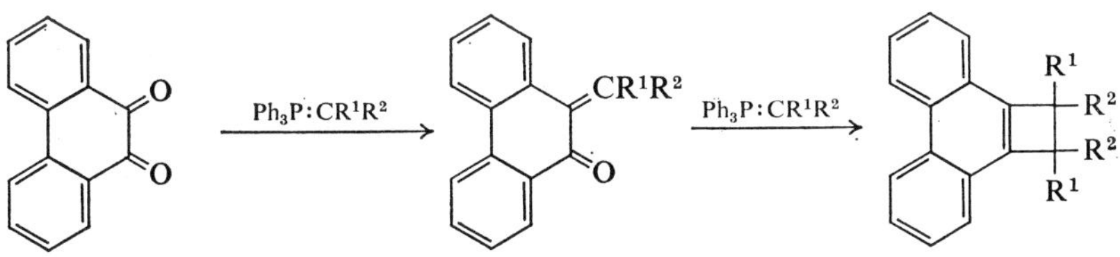

(80) (81)

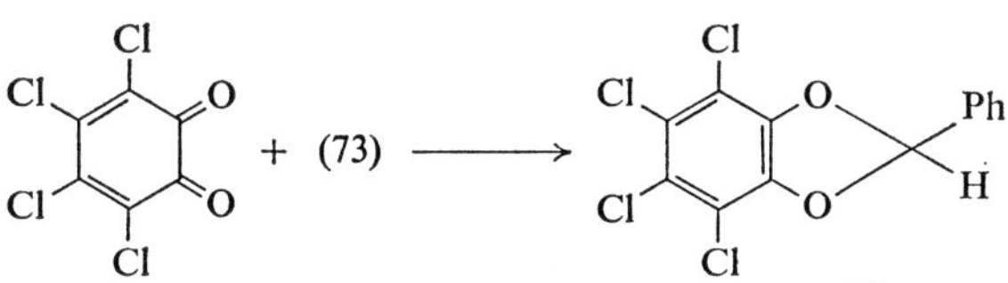

(82)

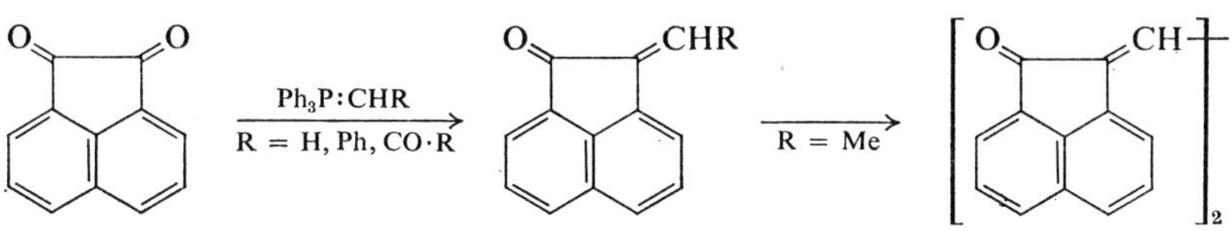

(83) (84)

Acenaphthenequinone with ylides [45] gave only the quinonemethides (83). The yield using methylenetriphenylphosphorane was only 5% while the ethylidenephosphorane gave a compound assigned the structure (84).

(*iv*) *Esters.* Esters of pivalic acid are resistant to attack by ylides and esterification with pivalic acid has been used to protect quinols in ylide formation and in reactions with ylides.[46] The groups were subsequently removed with lithium aluminium hydride.

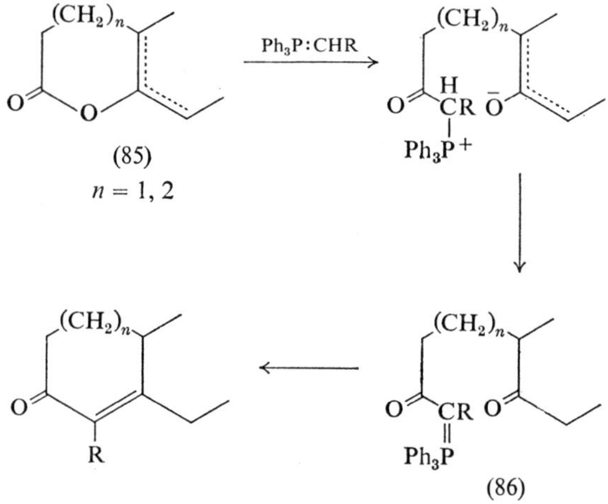

Cycloalkenones were obtained [47] from enol-lactones (85) on treatment with reactive ylides presumably *via* the ketones (86). With the enol-lactone (87) the ketone (88) was the major product. It was converted into testosterone acetate (89; total yield 50%) on treatment with aqueous potassium hydroxide in methanol.

Isopropylidenetriphenylphosphorane with *S*-ethyl thiobenzoate [48] gave the thiovinyl ether (90; R = Et). The phenyl ester also gave the thioether (91) formed by attack of the thiophenoxy anion on the α-carbon of the intermediate (92). The thiovinyl ethers were formed by nucleophilic addition of the thiol anions to the carbonyl of (92) followed by the elimination of phosphine oxide. Other nucleophiles can be used; the salt (92; X = I) with aqueous potassium cyanide in tetrahydrofuran at room temperature gave the nitrile (93).

(*v*) 1,3-*Dipoles.* Full details have appeared of the reactions of ylides with nitrile oxides.[49, 50] The 1 : 1-P^V-adducts (94) can be isolated when R^1, R^2

[45] O. Tsuge, M. Tashiro, and I. Shinkai, *Bull. Chem. Soc. Japan*, 1969, **42**, 181.

[46] W. E. Bondinell, S. J. DiMari, B. Frydman, K. Matsumoto, and H. Rapoport, *J. Org. Chem.*, 1968, **33**, 4351.

[47] C. A. Henrick, E. Böhme, J. A. Edwards, and J. H. Fried, *J. Amer. Chem. Soc.*, 1968, **90**, 5926.

[48] T. Kumamoto, K. Hosoi, and T. Mukaiyama, *Bull. Chem. Soc. Japan*, 1968, **41**, 2742.

[49] H. J. Bestmann and R. Kunstmann, *Chem. Ber.*, 1969, **102**, 1816.

[50] R. Huisgen and J. Wulff, *Chem. Ber.*, 1969, **102**, 1833.

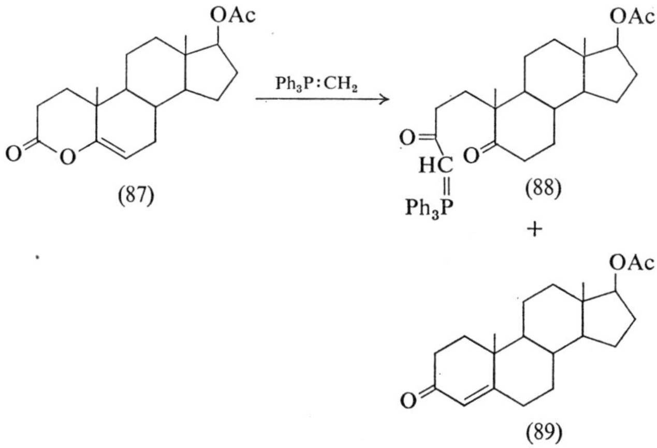
OAc
Ph3P:CH2
(87)
O=
HC
Ph3P
(88)
+
(89)

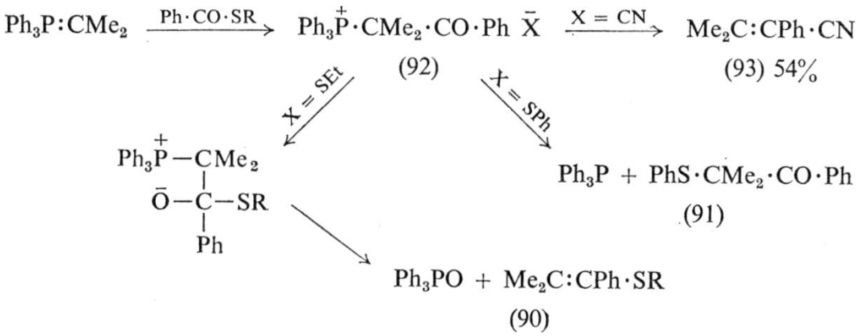
Ph3P:CMe2
Ph·CO·SR
Ph3P·CMe2·CO·Ph X
(92)
X = CN
Me2C:CPh·CN
(93) 54%
X = SEt
X = SPh
Ph3P—CMe2
O—C—SR
Ph
Ph3P + PhS·CMe2·CO·Ph
(91)
Ph3PO + Me2C:CPh·SR
(90)

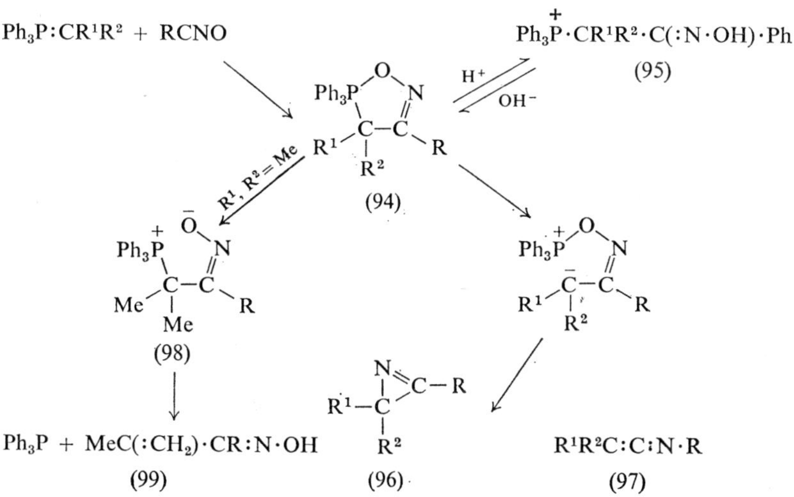
Ph3P:CR1R2 + RCNO
Ph3P·CR1R2·C(:N·OH)·Ph
(95)
H+
OH−
(94)
R1, R2 = Me
(98)
Ph3P + MeC(:CH2)·CR:N·OH
(99)
(96)
R1R2C:C:N·R
(97)

are hydrogen or alkyl groups and can also be obtained from the action of alkali on the oximinophosphonium salts (95).[51, 52] When R^1 is able to stabilise an adjacent carbanion the ring opens and the resulting betaines eliminate phosphine oxide to give azirines (96) and/or ketenimines (97) which may react with further ylide. Pyrolysis of the stable adducts occurs in the same way or *via* the betaines (98) to give the oximes (99). Of the many remarkable examples of these reactions two only can be mentioned. Cyclopropylidenetriphenylphosphorane and benzonitrile oxide gave the stable adduct (100) from which 2-phenyl-1-azaspiropent-1-ene (101) was

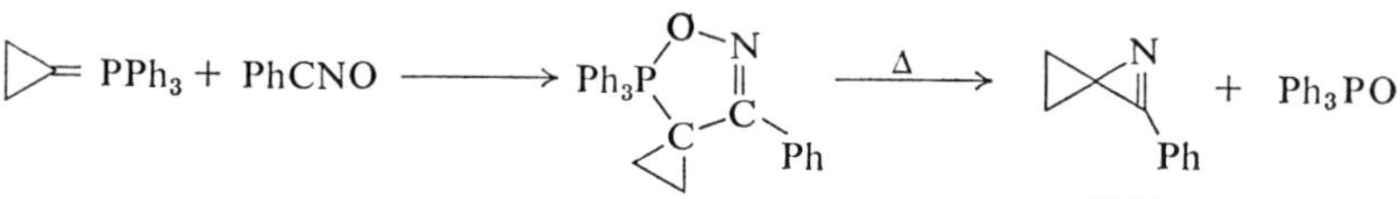

(100) 61% (101)

obtained on pyrolysis in 84% yield.[49] The same nitrile oxide with the ester phosphorane (68) gave the stable phosphorane (102) which existed as the imine in the solid but as an equilibrium of imine and enamine in solution.[49]

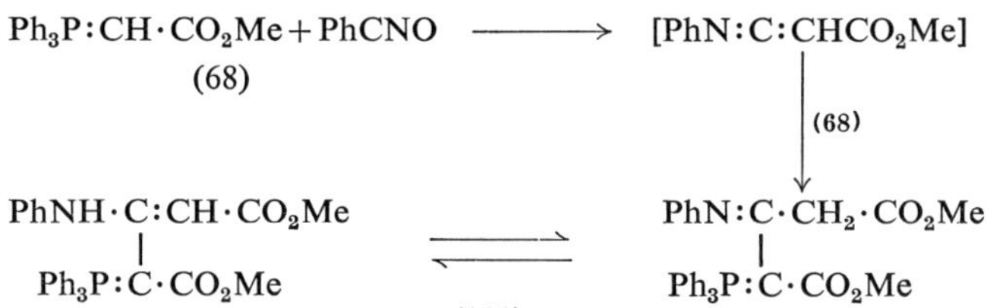

(102)

Details have also appeared of the formation of stable 1,2,5-P^V-oxazaphospholidines (103) from reactive ylides and nitrones.[53]

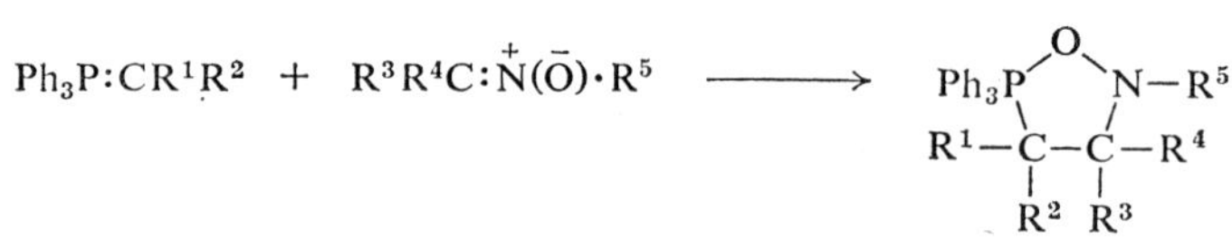

(103)

The ester phosphoranes (104; R = Me, Et) with nitrileimines[54] gave the stable ylides (105) formed by proton transfer in the intermediates (106). Pyrolysis of (105) gave the pyrazoles (107) in high yield presumably *via* (106).

[51] G. Gandiano, R. Mondelli, P. P. Ponti, C. Ticozzi, and A. Umani-Ronchi, *J. Org. Chem.*, 1968, **33**, 4431.
[52] M. Masaki, K. Fukui, and M. Ohta, *J. Org. Chem.*, 1967, **32**, 3564.
[53] R. Huisgen and J. Wulff, *Chem. Ber.*, 1969, **102**, 746.
[54] J. Wulff and R. Huisgen, *Chem. Ber.*, 1969, **102**, 1841.

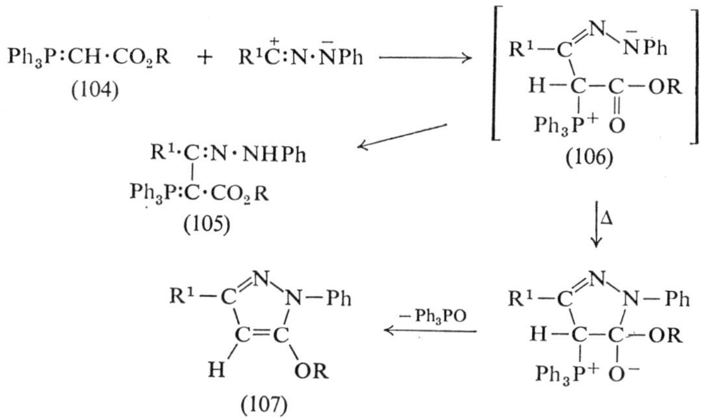

(104) (105) (106) (107)

The mesoionic oxazolone (108) behaved as an acylaminoketen when refluxed in benzene with the benzylidenephosphorane (73) and gave the acylaminoallene (109).[54]

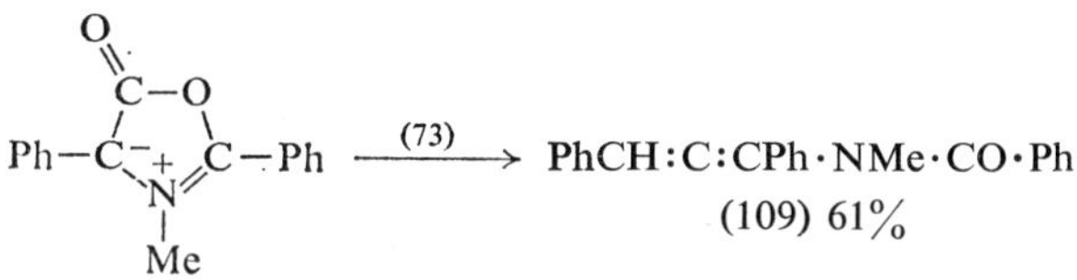

(108) (109) 61%

(*vi*) *Miscellaneous.* Mannich bases *C*-alkylate β-ketoylides.[55] Thus the ester phosphorane (64) with the amine (110) gave the phosphorane (111). An exceptional case was the reaction of the phenacylidenephosphorane (112) with the naphthalene (113) which gave the chromen (114) presumably by way of the ylide (115).

While the ester phosphoranes (116) are *C*-alkylated by alkyl halides they are *O*-alkylated by triethyloxonium borofluoride to give the *cis*- and *trans*-vinylphosphonium salts (117).[56]

Conjugated nitro-olefins with the ester phosphorane (68) gave the stable ylides (118).[57] On attempted reaction with aldehydes these dissociated to starting materials. Methylenetriphenylphosphorane with conjugated nitro-olefins gave small yields of nitrocyclopropanes (119).

The initial adducts (120) from the β-ketophosphoranes (17) and acyl azides normally cyclised to form (121) from which triazoles were obtained on elimination of phosphine oxide.[58, 59] When $R^2 = R^3 = OEt$, and

[55] M. von Strandtmann, M. P. Cohen, C. Puchalski, and J. Shavel jun., *J. Org. Chem.*, 1968, **33**, 4306.

[56] H. J. Bestmann, R. Saalfrank, and J. P. Snyder, *Angew. Chem. Internat. Edn.*, 1969, **8**, 216.

[57] J. Asunskis and H. Schechter, *J. Org. Chem.*, 1968, **33**, 1164.

[58] G. L'Abbé and H. J. Bestmann, *Tetrahedron Letters*, 1969, 63.

[59] G. L'Abbé, P. Ykman, and G. Smets, *Bull. Soc. chim. belges*, 1969, **78**, 147.

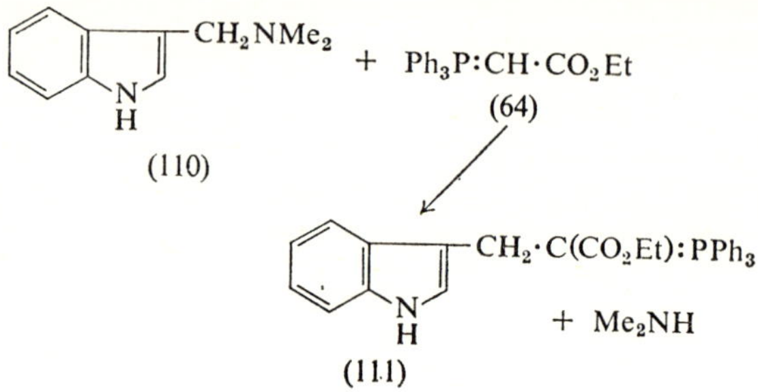

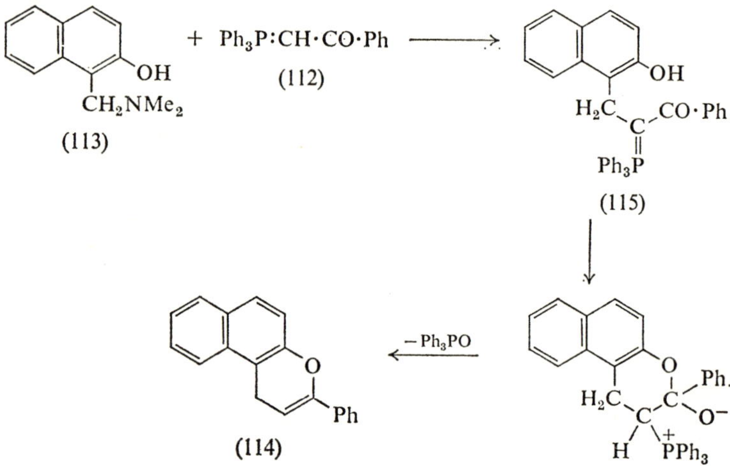

$$Ph_3P{:}CR{\cdot}CO{\cdot}OR^1 + Et_3\overset{+}{O}\ \overset{-}{B}F_4 \longrightarrow Ph_3\overset{+}{P}{\cdot}CR{:}C(OEt){\cdot}OR^1\ \overset{-}{B}F_4$$

(116) (117)

$$Ph_3P{:}CH{\cdot}CO_2Me + RCH{:}CR{\cdot}NO_2 \longrightarrow Ph_3P{:}C(CO_2Me){\cdot}CHR{\cdot}CHR{\cdot}NO_2$$

(68) (118)

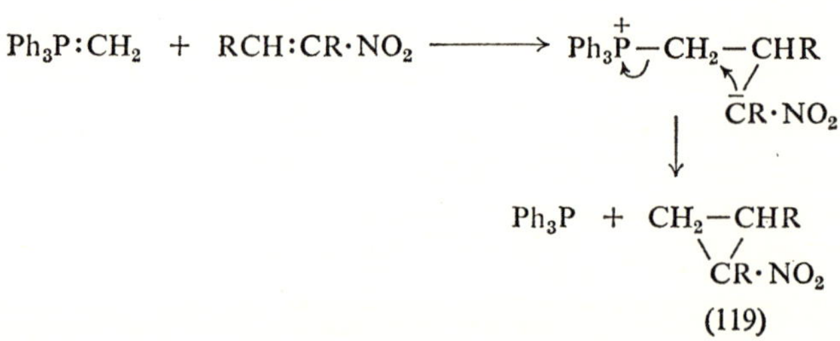

$R^1 = H$ cyclisation on phosphorus followed by fragmentation gave diazoacetic ester and the phosphineimine (122). The large negative entropies of activation of the reaction sequences leading to triazoles have been interpreted [59] in terms of a highly concerted addition to form the intermediates (121).

Diethyl azodicarboxylate with the ester phosphorane (68) gave triphenylphosphine, the phosphineimine (122), and the enamine (123) probably *via* the imine (124).[60]

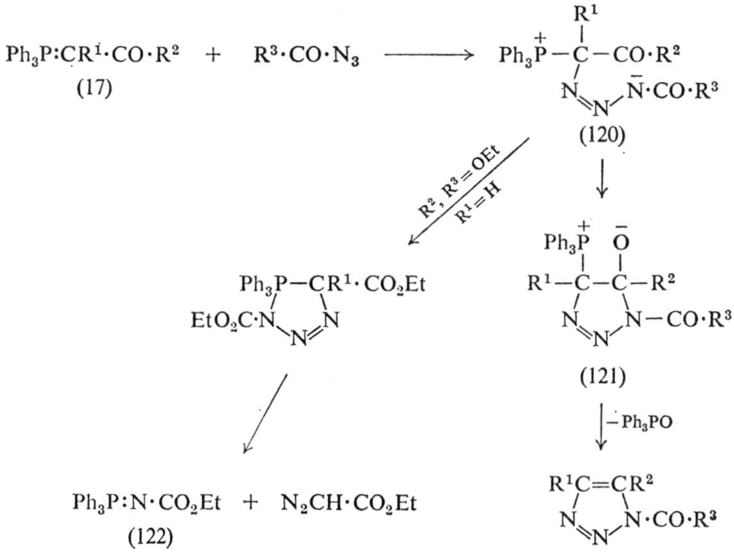

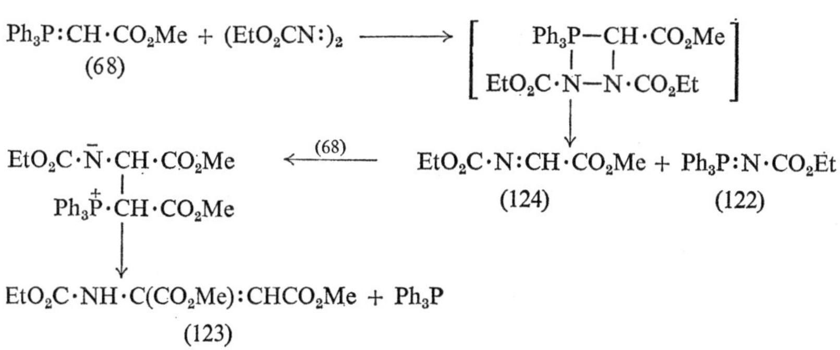

The nitrososulphonamide (125) with reactive ylides gave the corresponding nitriles in good yields.[61] Of the suggested mechanisms, that *via* the hydrazones (126) is supported by the formation of benzonitrile in 64% yield by the action of base on the hydrazone (126; R = Ph).

[60] H. J. Bestmann and R. Zimmermann, *Chem. Ber.*, 1968, **101**, 2185.
[61] A. Nürrenbach and H. Pommer, *Annalen*, 1969, **721**, 34.

$$\mathrm{RCH_2\cdot\overset{+}{P}Ph_3\ \bar{X} + ON\cdot NMe\cdot SO_2\cdot Ar \xrightarrow{\text{Base}} R\cdot CN}$$

(125)

$\mathrm{Ar = C_6H_4\cdot Me\text{-}}p$ 60—80%

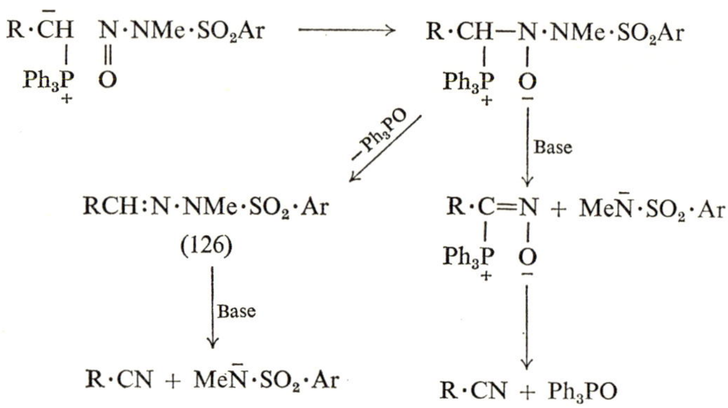

β-Carotene was obtained together with vitamin A aldehyde when the salt (127) was treated with *p*-nitrosodimethylaniline in the presence of triethylamine,[61] and was also produced from the salt (127; $X = HSO_4$) and the imine (128) in the presence of boron trifluoride,[62] this being an example of a general synthesis of olefins from imines and phosphoranes in the presence of electron acceptors.

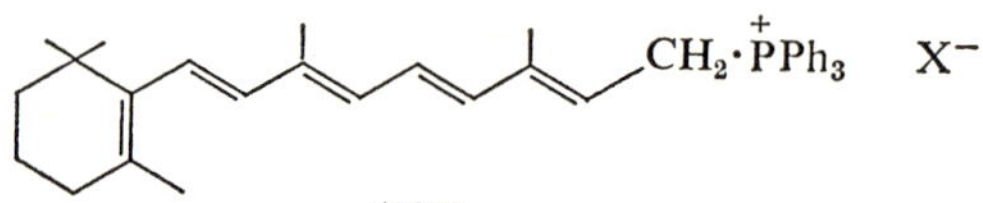

(127)

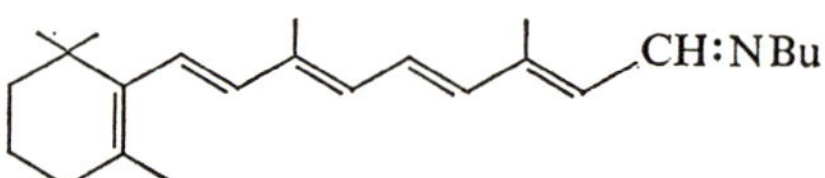

(128)

The oxophospholans (129) previously postulated as intermediates in the reactions of phosphoranes with epoxides have been isolated as stable compounds.[54, 63, 64]

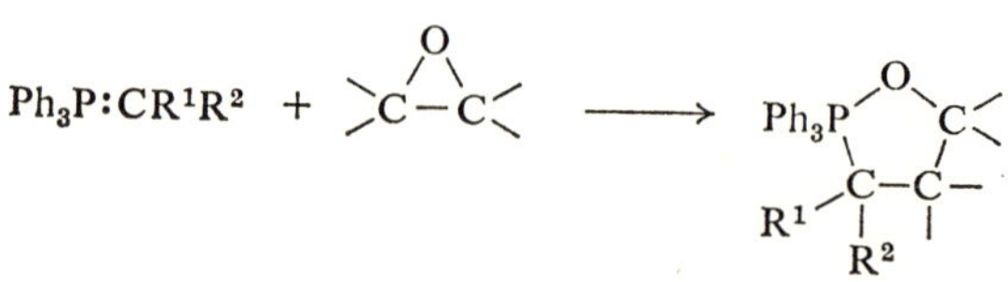

(129)

[62] Neth. Appl., 6,606,913.
[63] H. J. Bestmann, T. Denzel, and R. Kunstmann, *Tetrahedron Letters*, 1968, 2895.
[64] A. R. Hands and A. J. H. Mercer, *J. Chem. Soc.* (*C*), 1967, 1099.

Phenacylidenetriphenylphosphorane has been used as a ligand in transition metal complexes.[65] From i.r. evidence it is bonded to the metal through carbon.

For the ^{1}H n.m.r. spectra of stable ylides see Chapter 11, Section 1E.

2 Phosphoranes of Special Interest

The structure of methylenetriphenylphosphorane has been established by *X*-ray analysis.[66] The $P\cdot CH_2$ bond (1·661 Å) is much shorter than the phosphorus—carbon single bond and is close to the sum of the double bond radii of phosphorus and carbon. This, taken together with evidence from i.r. and ^{31}P n.m.r. studies, was interpreted in terms of overlap of the empty *d* orbital of phosphorus with the carbon *p* orbital, the overlap being close to carbon and giving the observed polarity to the bond.

Cyclopropylidenetriphenylphosphorane (130) reacted normally in olefin synthesis.[63, 67] With fluorenone the very unstable olefin (131) was trapped in the presence of fluorene or diethyl malonate as (132) and (133) respectively.[63] Diethyl fumarate and (130) gave diethyl bis(cyclopropyl)maleate (134) while with epoxides (130) formed particularly stable oxaphospholans, *e.g.* (135) from cyclohexene oxide.[63] For the reaction with benzonitrile oxide see this chapter, section 1B(v).

The ratio of the ethers (136) and (137) produced[68] when the salt (138) was treated with base appeared to depend on the solvent, strong proton donors favouring (137) which was not formed by way of (136). The allyl and cyclopropyl salts were also shown not to be intermediates and no satisfactory mechanism for the reactions could be proposed.

Cyclopropylmethylenetriphenylphosphorane reacted normally in olefin synthesis with no evidence of ring opening.[69, 70] However, while the salt (139) with benzaldehyde and sodium hydride in dimethylformamide gave the expected olefins (140), in ethanol in the presence of sodium ethoxide the major product was a mixture of the geometrical isomers of (141), showing that under these conditions equilibrium was established between the ylides (142) and (143).[71]

The phosphoranylideneketen (58) added to ketens to give[72] the stable phosphoranes (144) while with aldehydes and reactive ketones, *e.g.* hexafluoroacetone, the analogous phosphoranes (145) were obtained presumably *via* the methyleneketens (146). With carbon disulphide the phosphorane

[65] P. A. Arnup and M. C. Baird, *Inorg. Nuclear Chem. Letters*, 1969, **5**, 65.
[66] J. C. J. Bart, *J. Chem. Soc. (B)*, 1969, 350.
[67] E. E. Schweizer, C. J. Berninger, and J. G. Thompson, *J. Org. Chem.*, 1968, **33**, 336.
[68] E. E. Schweizer, C. J. Berninger, D. M. Crouse, R. A. Davis, and R. S. Logothetis, *J. Org. Chem.*, 1969, **34**, 207.
[69] A. Maercker, *Angew. Chem. Internat. Edn.*, 1967, **6**, 557.
[70] E. E. Schweizer, J. G. Thompson, and T. A. Ulrich, *J. Org. Chem.*, 1968, **33**, 3082.
[71] E. E. Schweizer, W. S. Creasy, K. K. Light, and E. T. Schaffer, *J. Amer. Chem. Soc.*, 1969, **34**, 212.
[72] G. H. Birum and C. N. Matthews, *J. Amer. Chem. Soc.*, 1968, **90**, 3842.

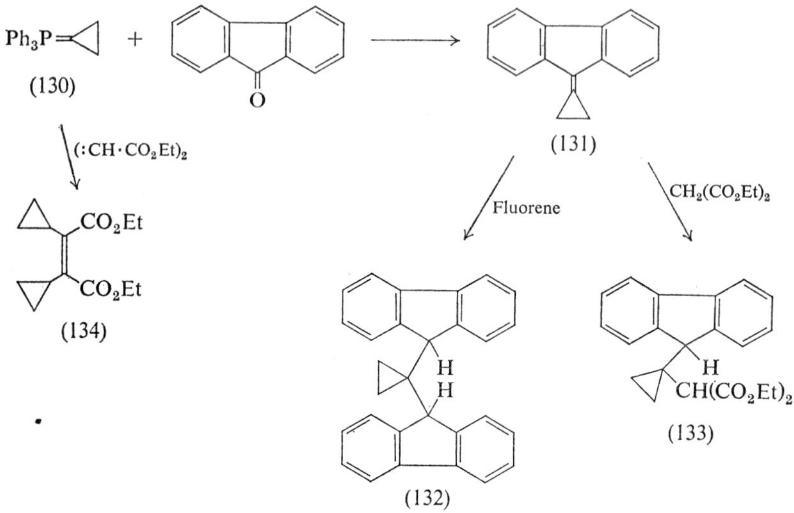

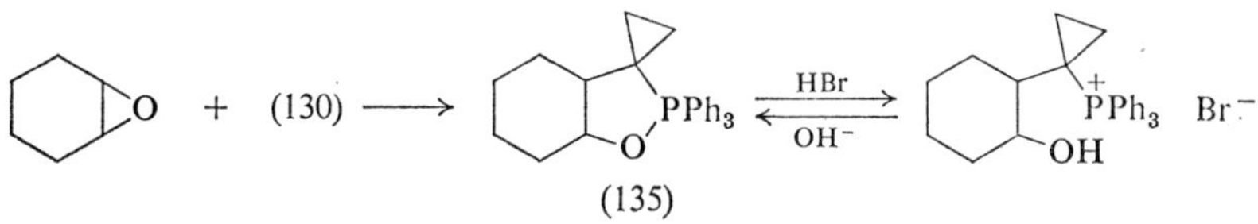

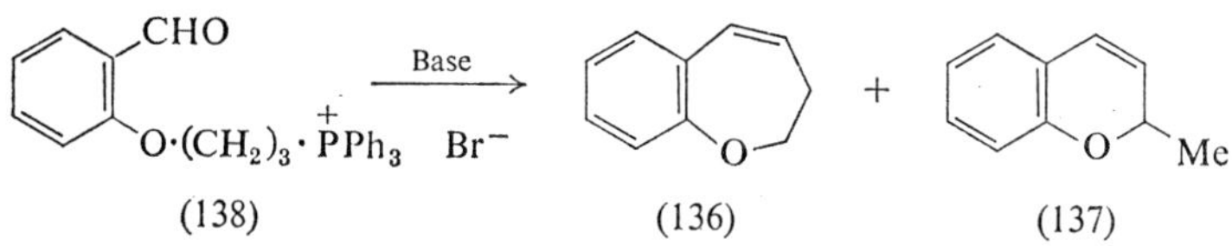

(58) gave the thioanalogue (147) and with aryl isocyanates the stable phosphoranes (148). Treatment of (58) with methyl iodide led to the dimer (149).

The carbodiphosphorane (150) with aryl isothiocyanates formed[73] the betaines (151) which on pyrolysis eliminated phosphine sulphide and gave the phosphoranylideneketenimines (152). These with methyl iodide formed salts (153) analogous to (149) above.

Details have appeared of the remarkable work of Schmidbauer and his collaborators on the synthesis of trialkylalkylidenephosphoranes as stable distillable liquids.[74] They are monomeric in benzene and very sensitive to oxygen and water. Trimethylmethylenephosphorane (154) reacted as expected with methyl iodide and with metal halides[27, 28] and was metallated

[73] G. H. Birum and C. N. Matthews, *Chem. and Ind.*, 1968, 653.
[74] H. Schmidbauer and W. Tronich, *Chem. Ber.*, 1968, **101**, 595.

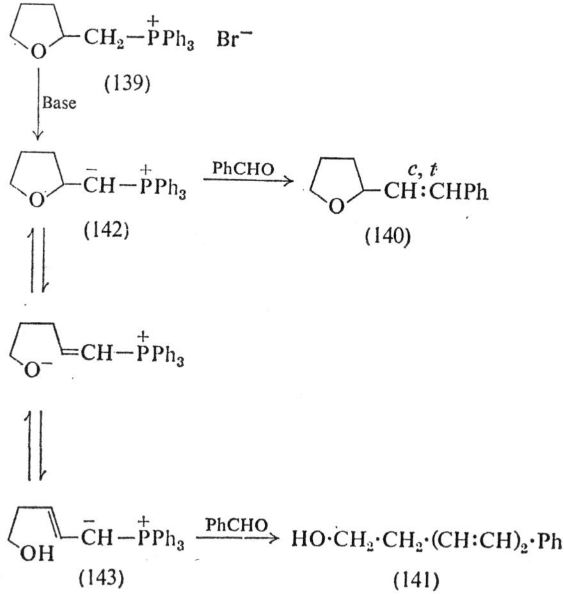

(139) (142) (140) (143) (141)

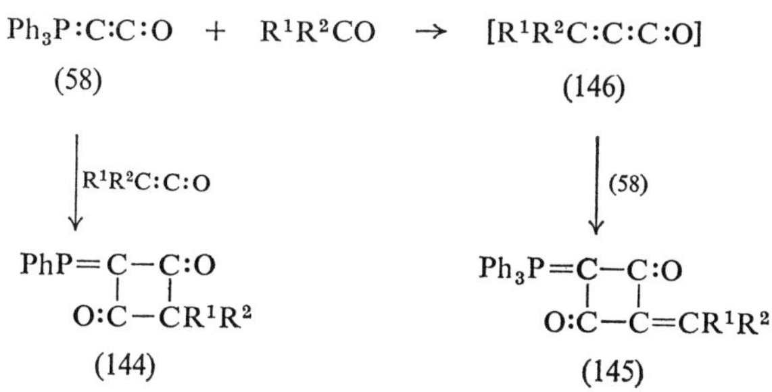

(58) (146) (144) (145)

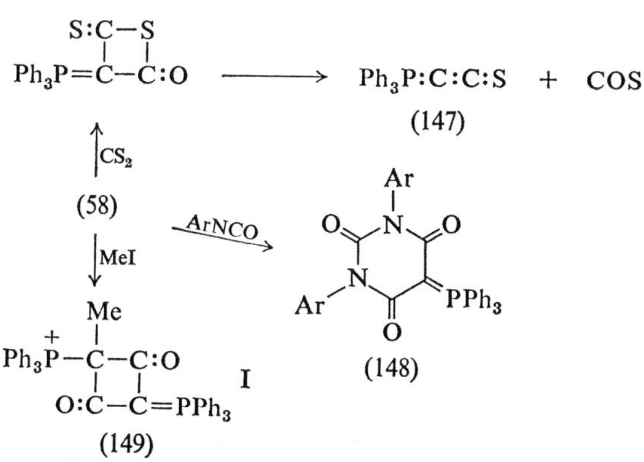

(147) (58) (148) (149)

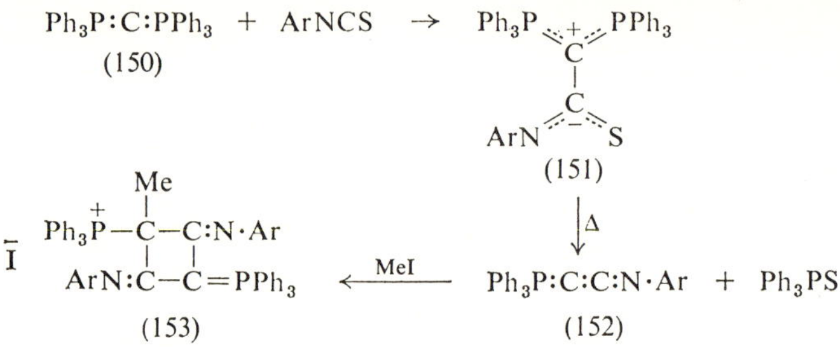

with butyl-lithium[75] to give the lithiomethylphosphorane (155). The various phosphoranes (156*a*; X = Ge, Sn, P, As, or Sb) were also obtained as shown.[27] The phosphorus and germanium compounds were stable at 120°, but the arsenic and antimony analogues disproportionated slowly at this temperature and the tin compound spontaneously at room temperature to give the phosphoranes (154) and (156*b*). Similar migrations of trimethylsilyl groups have been observed.

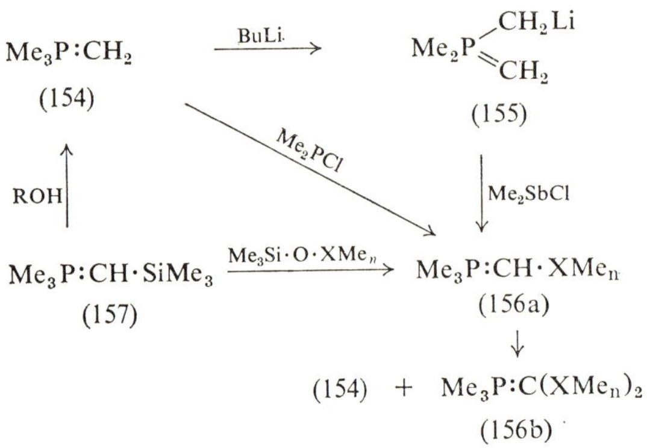

$$\mathrm{Me_3\overset{+}{P}{\cdot}CHMe{\cdot}SiMe_3\ \overset{-}{Cl}} \xrightarrow{\mathrm{BuLi}} \mathrm{EtMe_2P{:}CH{\cdot}SiMe_3}$$

(158) (159)

Thus the salt (158) with butyl-lithium gave the trimethylsilyl*methylene* ylide (159)[76] while the phosphorane (160) disproportionated[77] at 100° to give the bis(trimethylsilyl)methylenephosphorane (161; X = Si) *via* the phosphoranes (157) and (162). Such disproportionations occurred spontaneously when the lithiomethylphosphorane (163) was treated with trimethylgermanium and trimethyltin chlorides.

The stabilising influence of trimethylsilyl groups on ylides which these rearrangements imply is not due to a steric effect (*cf.* t-Bu) since it is also shown by the silyl group.[78] Thus the silylmethylphosphonium salt (164)

[75] H. Schmidbauer and W. Tronich, *Chem. Ber.*, 1968, **101**, 3556.
[76] H. Schmidbauer and W. Tronich, *Angew. Chem. Internat. Edn.*, 1968, **7**, 220.
[77] H. Schmidbauer and W. Malisch, *Chem. Ber.*, 1969, **102**, 83.
[78] H. Schmidbauer and W. Malisch, *Angew. Chem. Internat. Edn.*, 1969, **8**, 372.

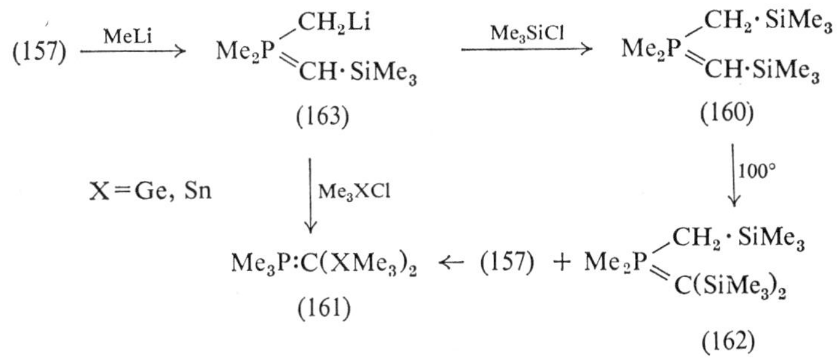

when treated with the ethylidenephosphorane (165) gave entirely the silylmethylenephosphorane (166) which disproportionated on heating to give the bis(silyl)methylenephosphorane (167). This stabilising effect is probably due to $(p\text{–}d)\pi$ bonding.

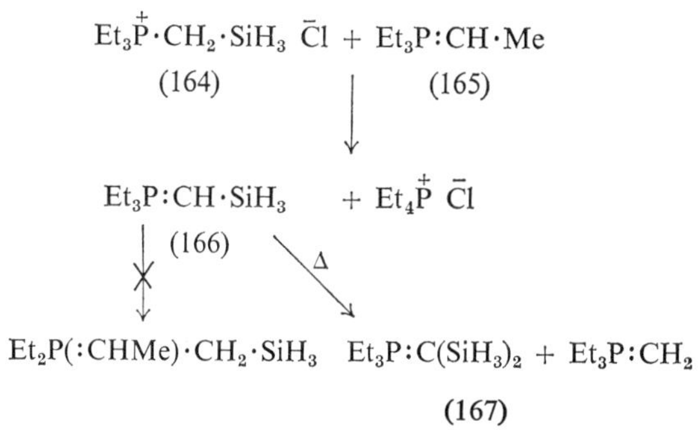

3 Selected Applications of the Wittig Olefin Synthesis

A. Macrocyclics.—The condensation of bisphosphoranes with dicarbonyl compounds has been applied to the synthesis of 9, 10, and 11-membered rings. Biphenyl-2,2′-dialdehyde (168) with the phosphorane from the salt (169) gave the 11-membered *cis,trans* isomer (170) [79] and with the phosphoranes from the salts (171; X = O, S) formed the non-aromatic 10-π electron systems (172; X = O, S).[80] Similar reaction with the furan bisphosphonium salt (173) gave the [10]-annulene (174) in two isomeric forms having the internal hydrogens eclipsed or staggered.[81]

Self-condensation of the phosphorane from the salt (175) gave the complex mixture [82] from which small amounts of triepoxy-[18]-annulene,

[79] R. H. Mitchell and F. Sondheimer, *Tetrahedron Letters*, 1968, 2873.
[80] A. P. Bindra, J. A. Elix, P. J. Garratt, and R. H. Mitchell, *J. Amer. Chem. Soc.*, 1968, **90**, 7372.
[81] A. P. Bindra, J. A. Elix, and M. V. Sargent, *Tetrahedron Letters*, 1968, 4335.
[82] J. A. Elix, *Chem. Comm.*, 1968, 343.

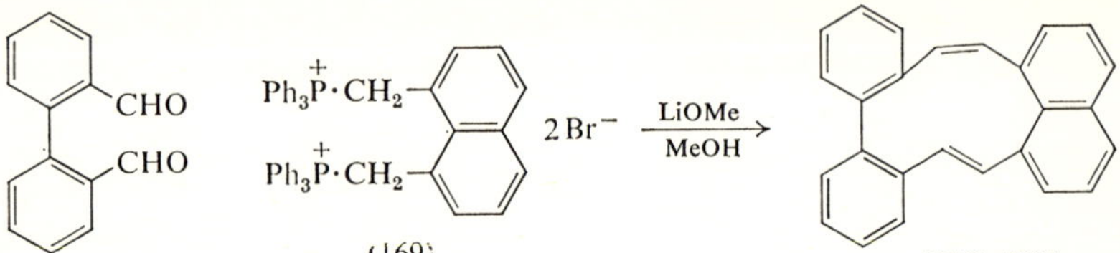

(168) (169) (170) 58%

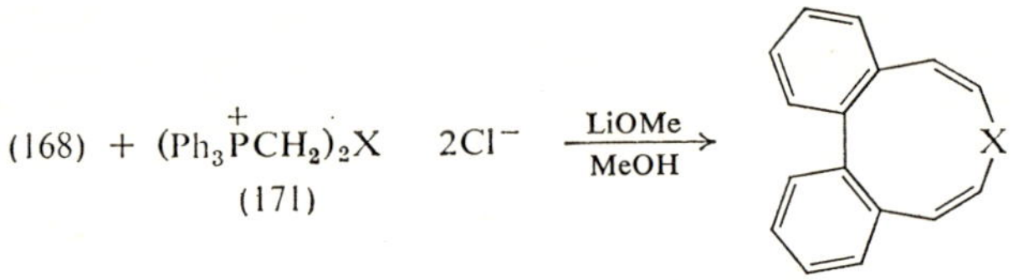

(172)

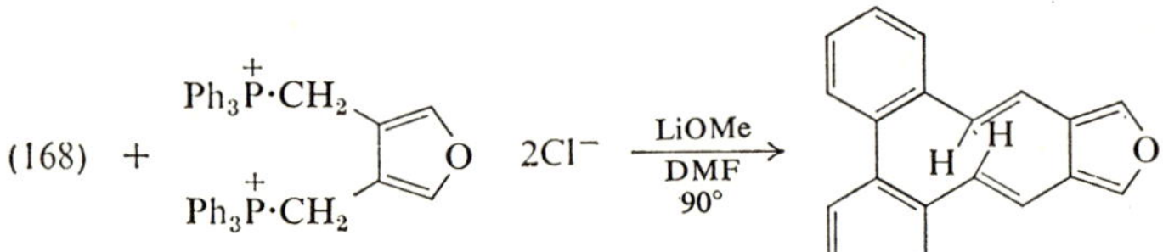

(173) (174) 3 + 10%

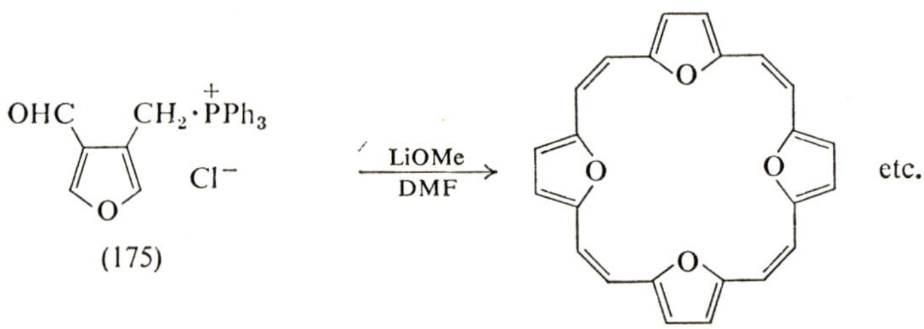

(176)

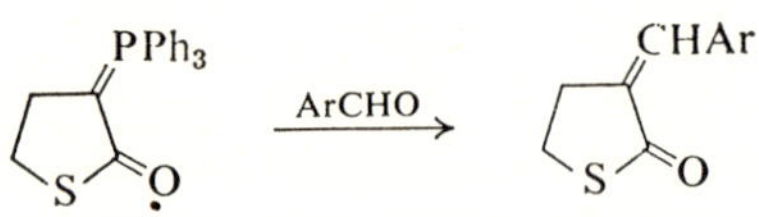

(177)

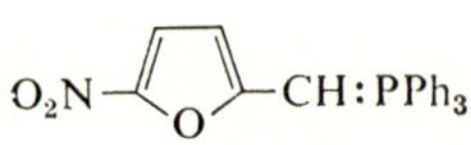

(178)

two isomers of tetraepoxy-[24]-annulene [neither of which may be the all-*cis* isomer (176)], and two isomers of pentaepoxy-[30]-annulene have been isolated.

B. Heterocyclics.—α-Arylidene-γ-thiobutyrolactones have been obtained[83] from the stable phosphorane (177).

Many 2-furylethylenes have been prepared[84, 85] either using the stable phosphorane (178) or from substituted furfuraldehydes and phosphoranes.

C. Natural Products.—*H. cecropia* juvenile hormone (179) has been synthesised by the sequence of reactions outlined[86] employing three successive Wittig olefin syntheses. The first of these gave olefin containing

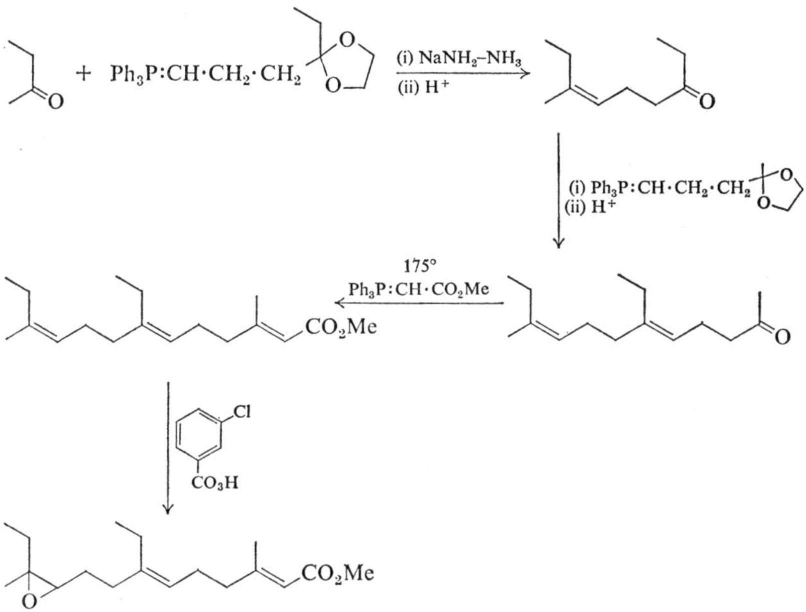

(179)

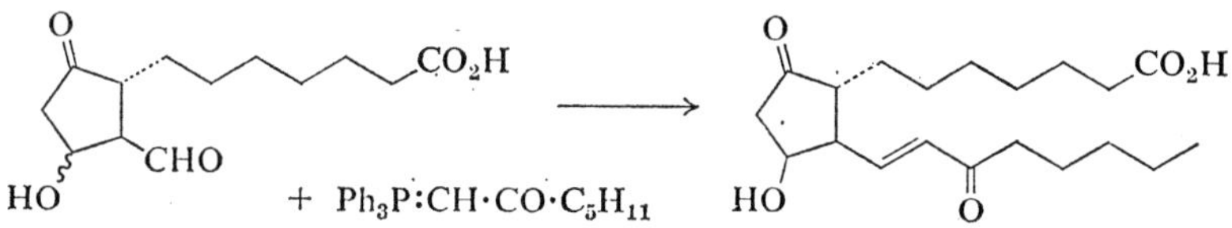

(181)

(180)

[83] H. Zimmer, F. Hauper, S. P. Kharidia, H. Pauling, R. G. Gailey, T. Pampalone, T. C. Purcell, and R. Walter, *Tetrahedron Letters*, 1968, 5435.
[84] S. Yoshina, A. Tanaka, and K. Yamamoto, *Yakugaku Zasshi*, 1968, **88**, 65.
[85] H. Saikachi and S. Nakamura, *Yakugaku Zasshi*, 1968, **88**, 110.
[86] H. Schulz and I. Sprung, *Angew. Chem. Internat. Edn.*, 1969, **8**, 271.

63% of the desired *cis*-isomer while the second and third were not stereoselective. However, in all cases the desired isomers were separated by fractional distillation.

15-Dehydroprostaglandin E_1 (180) was obtained[87] together with its 11-epi-isomer from the aldehydes (181) and hexanoylmethylenetriphenylphosphorane.

Conditions leading predominantly to *cis*-olefin ('salt-free' ylide or dimethylformamide in the presence of iodide) have been used in the syntheses of ^{14}C-labelled oleic[88] and erucic acids,[89] natural insecticides,[90] and *cis*-jasmone and related *cis*-rethrones.[91]

D. Carbohydrates.—Ylides have been used both in one-carbon chain extensions and in the synthesis of branched-chain carbohydrates. Acetals were usually used as protecting groups. Thus 2,3:4,5-di-*O*-isopropylidene-*aldehydo*-L-arabinose with the ylide (182; X = S) gave the thioether (183) readily converted into the acetal (184).[92] The ylide (182; X = O) has also been used in this and similar reactions.[93] The branched-chain sugar (185) was obtained from methylenetriphenylphosphorane and the corresponding ketone,[94] while the phosphorane (186) was used to prepare vinylphosphonates from protected nucleoside 5′-aldehydes.[95]

$$\underset{}{R{\cdot}CHO} + \underset{(182)}{Ph_3P{:}CH{\cdot}XMe} \longrightarrow \underset{(183)}{R{\cdot}CH{:}CH{\cdot}XMe} \dashrightarrow \underset{(184)}{R{\cdot}CH_2{\cdot}CH(OMe)_2}$$

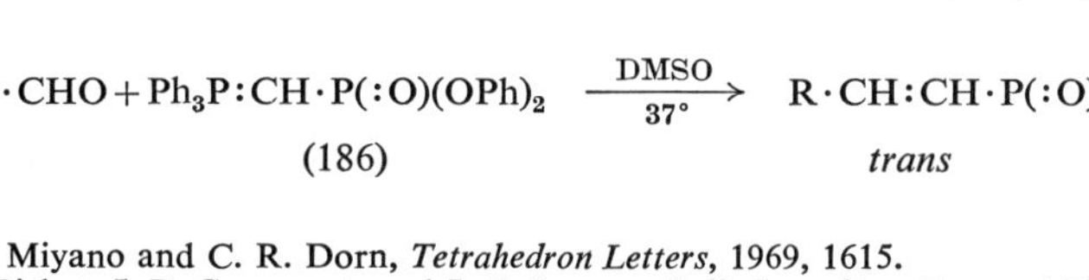

(185) 60%

$$R{\cdot}CHO + \underset{(186)}{Ph_3P{:}CH{\cdot}P({:}O)(OPh)_2} \xrightarrow[37^\circ]{DMSO} \underset{trans}{R{\cdot}CH{:}CH{\cdot}P({:}O)(OPh)_2}$$

87 M. Miyano and C. R. Dorn, *Tetrahedron Letters*, 1969, 1615.
88 L. Pichat, J. P. Guermont, and J. C. Levron, *Bull. Soc. chim. France*, 1969, 1198.
89 L. Pichat, J. C. Levron, and J. P. Guermont, *Bull. Soc. chim. France*, 1969, 1200.
90 P. E. Sonnet, *J. Org. Chem.*, 1969, **34**, 1147.
91 L. Crombie, P. Hemesley, and G. Pattenden, *J. Chem. Soc. (C)*, 1969, 1024.
92 J. M. J. Tronchet, S. Jaccard-Thorndahl, and Br. Baehler, *Helv. Chim. Acta*, 1969, **52**, 817.
93 Yu. A. Zhdanov and V. G. Alekseeva, *Zhur. obshchei Khim.*, 1968, **38**, 2594; *Carbohydrate Res.*, 1969, **10**, 184.
94 D. G. Lance and W. A. Szarek, *Carbohydrate Res.*, 1969, **10**, 306.
95 G. H. Jones and J. G. Moffatt, *J. Amer. Chem. Soc.*, 1968, **90**, 5337.

4 Synthetic Applications of Phosphonate Carbanions

A study has been made[96] of the variation in the proportion of *cis* (187) and *trans* (188) esters produced from the phosphonates (189) and aliphatic aldehydes as the groups R^1 and R^2 varied. As previously established, when $R^1 = H$ only *trans*-ester was formed but with $R^1 = Me$ the amount of *cis*-ester increased with the size of R^2 then decreased when $R^2 = Bu^t$. The

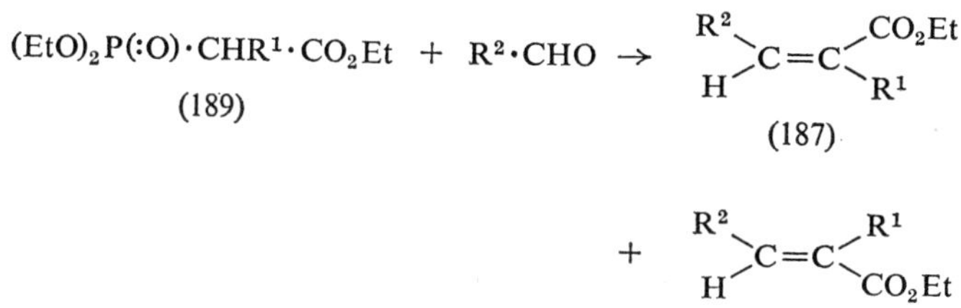

(189) (187) (188)

effect was more pronounced as R^1 increased in size and with $R^1 = R^2 =$ i-Pr only *cis*-ester was formed. The same phosphonates (189) with α-keto-esters gave predominantly the substituted maleic esters.[97]

The variation in the proportions of isomers produced from diethyl cyanomethylphosphonate and aryl alkyl ketones as the alkyl group varies has also been studied.[98, 99] Among other substituted phosphonate esters used in olefin synthesis were (190),[100] (191),[101] and (192).[102] The enamines resulting from the last were readily hydrolysed to acylphosphonates and further to carboxylic acids which were obtained in overall yields of 44—68% from the starting aldehydes.

$$\underset{(190)}{(EtO)_2P(:O)\cdot CH_2\cdot SO_2Et} + R\cdot CHO \xrightarrow{NaH} \underset{70\text{—}96\%}{R\cdot CH:CH\cdot SO_2\cdot Et}$$

$$\underset{(191)}{(EtO)_2P(:O)\cdot CH(CO_2Et)\cdot (CH_2)_n\cdot CO_2Et} + R\cdot CHO \xrightarrow[THF]{NaH} \underset{53\text{—}86\%}{R\cdot CH:C(CO_2Et)\cdot (CH_2)_nCO_2Et}$$

$$\underset{(192)}{[(EtO)_2P(:O)]_2CH\cdot NMe_2} + R\cdot CHO \xrightarrow[DIOXAN]{NaH} R\cdot CH:C(NMe)_2\cdot P(:O)(OEt)_2$$

$$R\cdot CH:C(NMe)_2\cdot P(:O)(OEt)_2 \xrightarrow{H^+} R\cdot CH_2\cdot CO\cdot P(:O)(OEt)_2 \xrightarrow{H^+} R\cdot CH_2\cdot CO_2H$$

[96] T. H. Kinstle and B. Y. Mandanas, *Chem. Comm.*, 1968, 1699. See also K. Sasaki, *Bull. Chem. Soc. Japan*, 1968, **41**, 1252.
[97] R. K. Huff, C. E. Moppett, and J. K. Sutherland, *J. Chem. Soc.* (*C*), 1968, 2725.
[98] G. Jones and R. F. Maisey, *Chem. Comm.*, 1968, 543.
[99] D. Danion and R. Carrie, *Tetrahedron Letters*, 1968, 4537.
[100] I. C. Popoff, J. L. Dever, and G. R. Leader, *J. Org. Chem.*, 1969, **34**, 1128.
[101] K. Sato, M. Hirayama, T. Inoue, and S. Kikuchi, *Bull. Chem. Soc. Japan*, 1969, **42**, 250.
[102] H. Gross and B. Costisella, *Angew. Chem. Internat. Edn.*, 1968, **7**, 391.

Diethyl cyclohexyliminovinylphosphonate (193) has been used in 'formylolefination' of carbonyl compound.[103] The anion reacted readily with aldehydes and ketones at room temperature and acid hydrolysis of the resulting aldimines in a two-layer system gave $\alpha\beta$-unsaturated aldehydes in overall 63—86% yield.

Details have appeared of the use of phosphonamides in olefin synthesis.[104] The diastereoisomeric β-hydroxyphosphonamides (194) can be separated by fractional crystallisation and undergo *cis*-elimination on heating, *e.g.* in refluxing benzene, to give isomerically pure olefins. Alternatively the β-hydroxyphosphonamides (194; R^4 = H) can be obtained in largely one diastereoisomeric form by reduction of the β-ketophosphonamides (195).

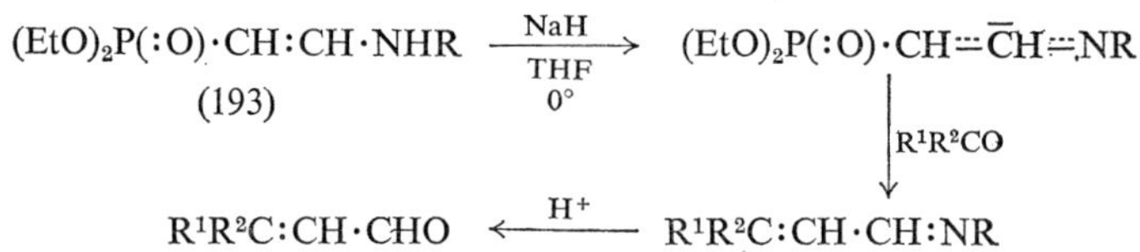

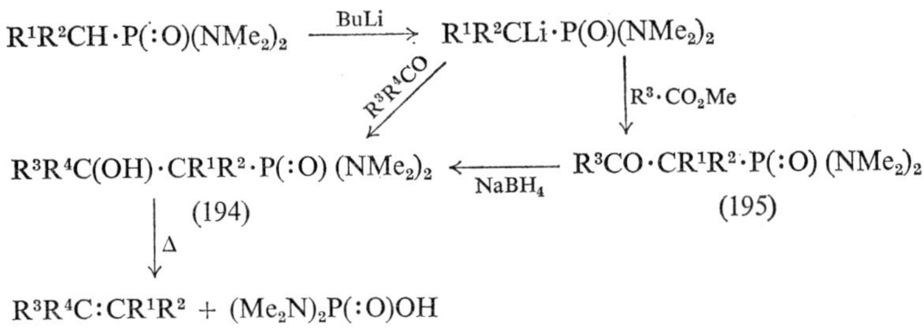

Phosphonate carbanions have been used in carbohydrate chemistry for the synthesis of branched-chain sugars [105, 106] and for chain extension to give the phosphonates (196).[107] Among natural products synthesised with the aid of phosphonate carbanions were *H. cecropia* juvenile hormone [108] and β-carotene.[109]

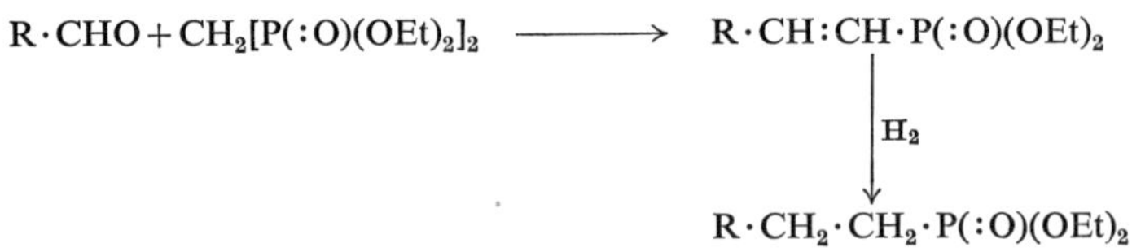

(196)

[103] W. Nagata and Y. Hayase, *J. Chem. Soc.* (*C*), 1969, 460.
[104] E. J. Corey and G. T. Kwiatkowski, *J. Amer. Chem. Soc.*, 1968, **90**, 6816.
[105] A. Rosenthal and P. Catsoulacos, *Canad. J. Chem.*, 1968, **46**, 2868.
[106] A. Rosenthal and D. A. Baker, *Tetrahedron Letters*, 1969, 397.
[107] J. A. Montgomery and K. Hewson, *Chem. Comm.*, 1969, 15.
[108] K. H. Dahm, H. Roeller, and B. M. Trost, *Life Sciences*, 1968, **7**, 129.
[109] J. D. Surmatis and R. Thommen, *J. Org. Chem.*, 1969, **34**, 559.

Aliphatic tertiary nitroso compounds with the phosphonates (197) gave aldimines[110] while amidines were obtained from the α-aminophosphonates (198) and aromatic nitroso compounds.[111] The benzylphosphonates (199; R = Et, i-Pr) with the Schiff's bases (200) and sodamide in refluxing ether gave stilbene; the intermediates (201) could be isolated under milder conditions.[112]

$$(EtO)_2P(:O)\cdot CH_2R' + R^2\cdot NO \xrightarrow{NaH} R^2\cdot N:CHR' \quad 38\text{—}80\%$$

(197) $R' = CO_2Et$ or $CMe:CH\cdot CO_2Me$

$$\underset{(198)}{(PhO)_2P(:O)\cdot CHR^1\cdot NHR^2} + R^3\cdot NO \xrightarrow{KOH} \underset{45\text{—}97\%}{R^3N:CR^1\cdot NHR^2}$$

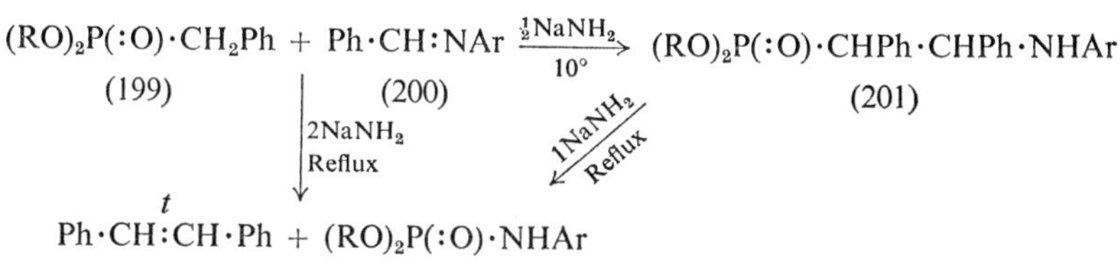

5 Ylide Aspects of Iminophosphoranes

A. Preparation.—Isotopic labelling with ^{15}N has been used[113] to show that the β and γ nitrogens are eliminated when the adduct (202) of triphenylphosphine and toluene-*p*-sulphonyl azide is pyrolysed to give the iminophosphorane (203).

$$\underset{(202)}{Ph_3P:\overset{\gamma}{N}\cdot\overset{\beta}{N}:\overset{\alpha}{N}\cdot SO_2\cdot C_6H_4\cdot Me\text{-}p} \xrightarrow{\Delta} \underset{(203)}{Ph_3P:\overset{\alpha}{N}\cdot SO_2\cdot C_6H_4\cdot Me\text{-}p}$$

The potassium salt of aminodiphenylphosphine with methyl iodide in liquid ammonia gave the iminophosphorane (204).[114] Similar phosphoranes were obtained[115] when hydrazine hydrochloride was treated with dibromo- or di-iodo-triphenylphosphorane, while the dichlorophosphorane gave the bisphosphonium salt (205). This, and the corresponding trialkylphosphonium salts, with base formed the bis(iminophosphoranes) (206).[116]

[110] J. A. Maassen, T. A. J. W. Wajer, and T. J. de Boer, *Rec. Trav. chim.*, 1969, **88**, 5.
[111] H. Zimmer, P. J. Bercz, and G. E. Heuer, *Tetrahedron Letters*, 1968, 171.
[112] M. Kirilov and J. Petrova, *Chem. Ber.*, 1968, **101**, 3467.
[113] H. Bock and M. Schnöller, *Chem. Ber.*, 1969, **102**, 38.
[114] O. Schmitz-Dumont and H. Klieber, *Z. Naturforsch.*, 1968, **23b**, 1604.
[115] R. Appel and G. Siegemund, *Z. anorg. Chem.*, 1968, **361**, 203.
[116] R. Appel, B. Blaser, and G. Siegemund, *Z. anorg. Chem.*, 1968, **363**, 176.

$$Ph_2PNHK \xrightarrow{2MeI} [Ph_2PMe{\cdot}N{:}PMePh_2]^+ \; I^-$$
(204)

$$Ph_3PX_2 + N_2H_4{\cdot}2HCl \xrightarrow{X=Br,I} [Ph_3P{\cdot}N{:}PPh_3]^+ \; \bar{X}$$

$$Ph_3PX_2 + N_2H_4{\cdot}2HCl \xrightarrow{X=Cl} Ph_3\overset{+}{P}{\cdot}NH{\cdot}NH{\cdot}\overset{+}{P}Ph_3 \; 2\bar{C}l$$
(205)

$$(205) \xrightarrow{NH_3-KNH_2} R_3P{:}N{\cdot}N{:}PR_3$$
(206)

While trimethylgermanium azide with trimethylphosphine gave the expected iminophosphorane the corresponding tin azide with a range of phosphines formed only the azides (207), presumably *via* disproportionation of the tin azide.[117]

$$Me_3SnN_3 + R_3P \longrightarrow Me_2Sn(N_3){\cdot}N{:}PR_3 + [Me_4Sn]$$
(207)

Some interesting iminochlorophosphoranes have been obtained as thermally stable distillable liquids by the action of carbon tetrachloride on amino-t-butyl[118] and aminodi-t-butylphosphines.[119] The formation of the iminophosphoranes (208) from the *N*-lithiophosphines (209) and trimethylmetal chlorides did not involve *N*-metallation followed by rearrangement as X and Y did not become equivalent during the reactions.

$$R_2PCl \xrightarrow{NH_3} R_2PNH_2 \xrightarrow[Me_3XCl]{BuLi} R_2PNH{\cdot}XMe_3 \xrightarrow{BuLi} R_2P{\cdot}NLi{\cdot}XMe_3$$
(209)

$$R_2PNH_2 \xrightarrow{CCl_4} R_2PCl{:}NH$$

$$R_2PNH{\cdot}XMe_3 \xrightarrow{CCl_4} R_2PCl{:}N{\cdot}XMe_3$$

$$(209) \xrightarrow{Me_3YCl} R_2P(YMe_3){:}N{\cdot}XMe_3$$
(208)

R = t-Bu; X, Y = Si, Ge, Sn, or Pb

B. Reactions.—Iminotrimethylphosphorane (210) has been obtained as shown as a distillable crystalline compound.[120] The tetrameric adducts (211) with dialkylzincs and dialkylcadmiums had cubic structures. Pyrolysis of the 1,1-adducts with Group III metal trialkyls gave the iminophosphoranes

[117] H. Schmidbauer and W. Wolfsberger, *Chem. Ber.*, 1968, **101**, 1664.
[118] O. J. Scherer and P. Klusman, *Angew. Chem. Internat. Edn.*, 1968, **7**, 541.
[119] O. J. Scherer and G. Schieder, *Chem. Ber.*, 1968, **101**, 4184.
[120] H. Schmidbauer and G. Jonas, *Chem. Ber.*, 1968, **101**, 1271.

(212). Similar adducts have been obtained with other iminophosphoranes.[121] The cyclic compounds (213) were obtained[122] from the aminoimine (214) and the cyclic salts (215) from the bis(iminophosphorane) (216).[123]

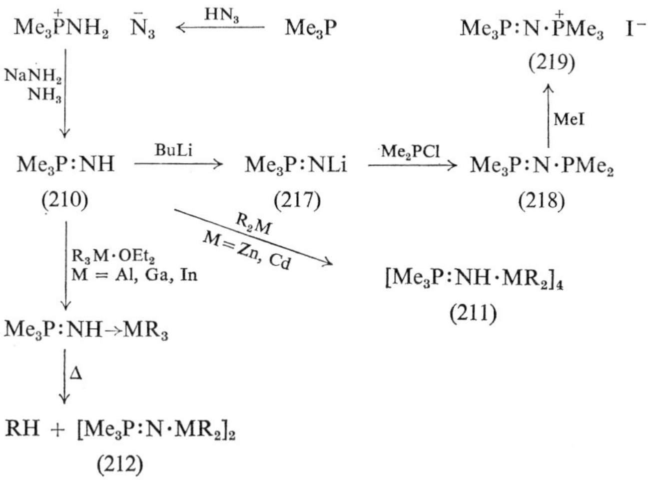

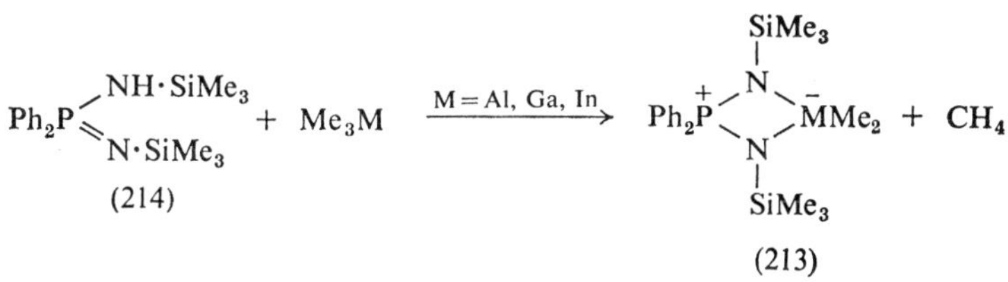

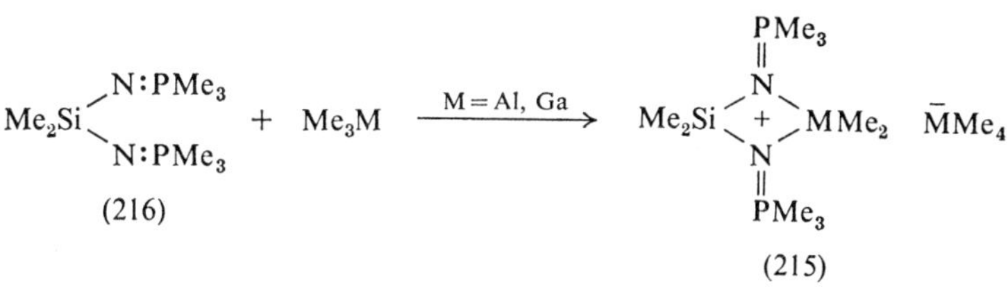

The *N*-lithioiminophosphorane (217) with chlorodimethylphosphine[120] gave the phosphorane (218) which with methyl iodide formed (219) analogous to (204) above.

The substituted ureas (220) were obtained from iminotriphenylphosphorane and aryl isocyanates[124] or isothiocyanates.[125] The thiourea (220;

[121] W. Wolfsberger and H. Schmidbauer, *J. Organometallic Chem.*, 1969, **17**, 41.
[122] H. Schmidbauer, K. Schwirten, and H.-H. Pickel, *Chem. Ber.*, 1969, **102**, 564.
[123] H. Schmidbauer, W. Wolfsberger, and K. Schwirten, *Chem. Ber.*, 1969, **102**, 556.
[124] I. N. Zhmurova and A. P. Martynyuk, *Zhur. obshchei. Khim.*, 1968, **38**, 876.
[125] I. N. Zhmurova, A. P. Martynyuk, and G. I. Derkach, *Zhur. obshchei. Khim.*, 1968, **38**, 163.

R = Ph) was also obtained by the dichlorophosphorane route. Arylsulphinyliminophosphoranes (221) were formed from the same iminophosphorane and sulphinyl chlorides.[126, 127] Oxidation gave the corresponding sulphonyl compounds.

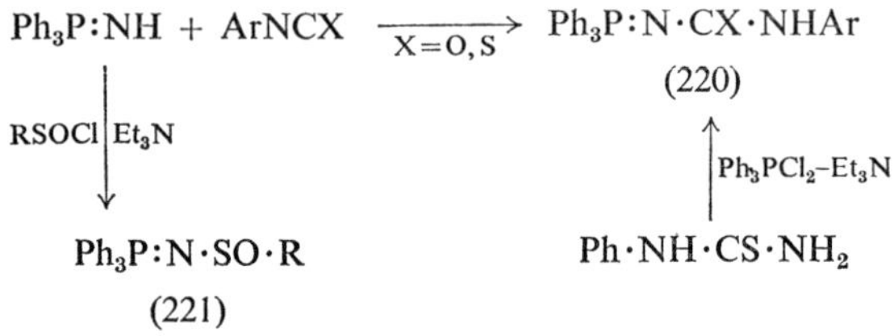

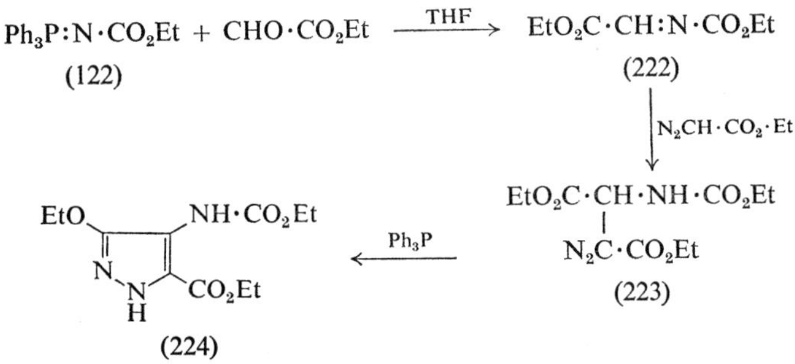

The highly reactive imine [222; see this chapter, section 1, B(vi)] has been generated *in situ* from the phosphorane (122) and ethyl glyoxylate and trapped in the presence of nucleophiles.[128] Diazoacetic ester gave the diazoester (223) from which the pyrazole (224) was obtained on refluxing in toluene with triphenylphosphine. The same phosphorane (122) has also been used[129] to introduce the nitrile group into nucleophilic heterocyclic nuclei, being particularly useful where these are acid sensitive. Thus indole with (122) in the presence of the boron trifluoride–ether complex gave the 3-nitrile (225). These reactions may involve the electrophilic adduct (226).

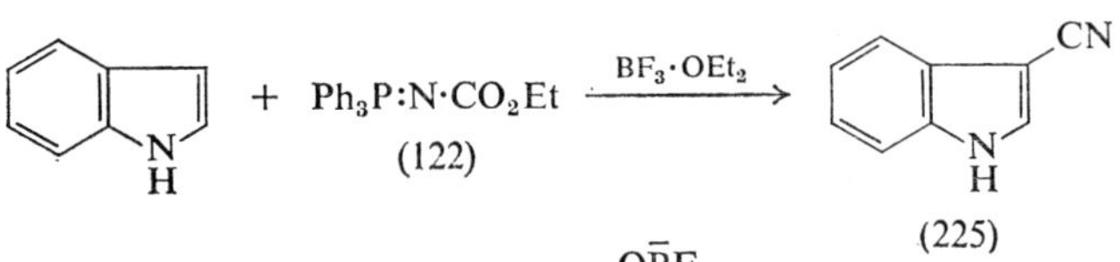

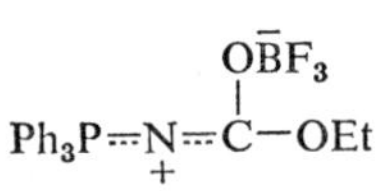

(226)

[126] A. Senning and P. Kelly, *Naturwiss.*, 1968, **55**, 543.
[127] I. N. Zhmurova and A. P. Martynyuk, *Zhur. obshchei. Khim.*, 1967, **37**, 2706.
[128] H. Plieninger and D. von der Brück, *Tetrahedron Letters*, 1968, 4371.
[129] D. von der Brück, A. Tapia, R. Riechel, and H. Plieninger, *Angew. Chem. Internat. Edn.*, 1968, **7**, 377.

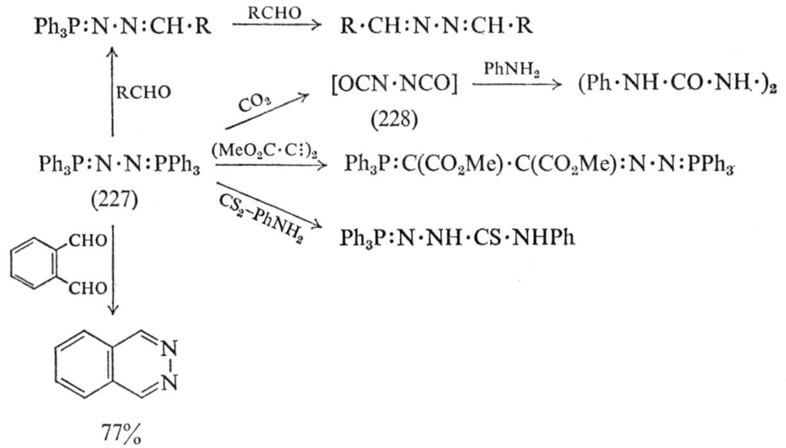

The ylide reactions of the bis(iminophosphorane) (227) have been explored [130] and are as shown. Outstanding are the reactions with carbon dioxide, in which the product (228) was trapped with aniline, and that with dimethyl acetylene dicarboxylate which presumably involves a four-membered intermediate in a reaction analogous to those of this ester with stable methylenephosphoranes.

The lithium salt of benzoylhydrazonotriphenylphosphorane (229) gave high yields of the 1,3,4-oxadiazoles (230) when treated with aromatic acid chlorides.[131] Symmetrical 2,5-diaryl-1,3,4-oxadiazoles can be obtained in one step by treating the salt (231) with 2 moles of aromatic acid chloride in the presence of an excess of triethylamine.

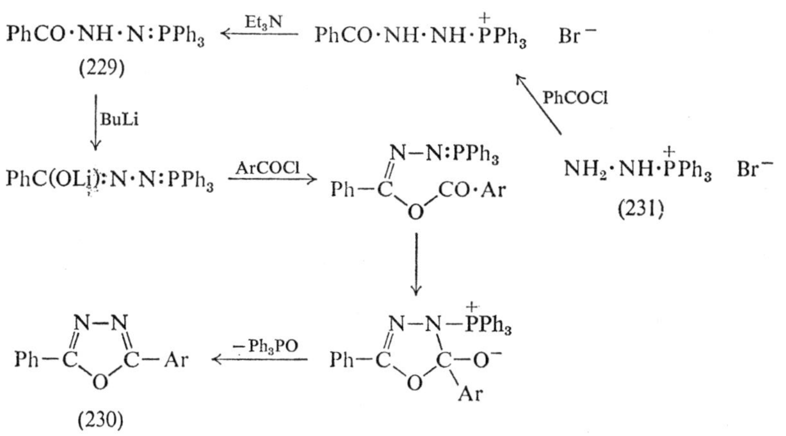

[130] R. Appel and G. Siegemund, *Z. anorg. Chem.*, 1968, **363**, 183.
[131] C. C. Walker and H. Shechter, *J. Amer. Chem. Soc.*, 1968, **90**, 5626.

The phenyliminophosphoranes (232) coupled with tetracyanoethylene[132] at the *para* position to give the iminophosphoranes (233).

Diphenylcarbodi-imide was the major product from benzonitrile oxide and the iminophosphorane (234) but some 8% of the 1,2,4-oxadiazole (235) was also obtained.[133] The suggested route to (235), involving an initial equilibrium between (234) and the carbodi-imide and the four-membered compound (236), was supported by the reaction of *p*-nitrobenzonitrile oxide with the *p*-methoxyiminophosphorane (237) which gave the symmetrical carbodi-imide (238) and the *p*-nitroiminophosphorane.

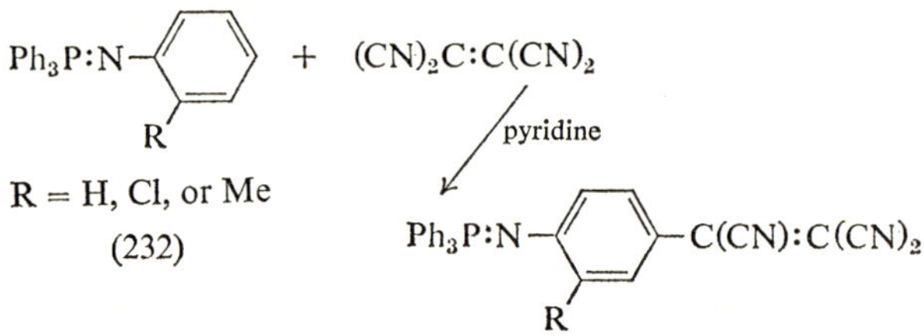

R = H, Cl, or Me

(232)

(233) 16—51%

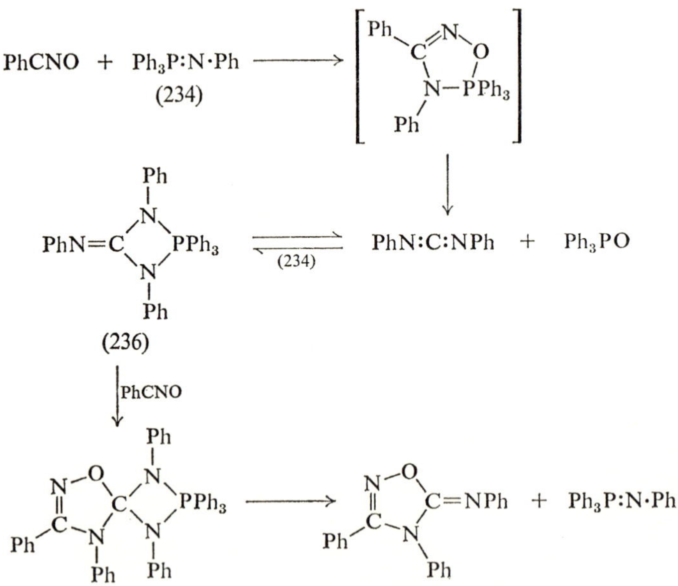

(235)

[132] I. N. Zhmurova, R. I. Yurchenko, and A. V. Kirsanov, *Zhur. obshchei. Khim.*, 1968, **38**, 2078.

[133] R. Huisgen and J. Wulff, *Chem. Ber.*, 1969, **102**, 1848.

$$p\text{-}NO_2\cdot C_6H_4\cdot CNO + 2\ p\text{-}MeO\cdot C_6H_4\cdot N{:}PPh_3$$
(237)

$$\longrightarrow p\text{-}MeO\cdot C_6H_4\cdot N{:}C{:}N\cdot C_6H_4\cdot OMe\text{-}p$$
(238) 60%

$$+ p\text{-}NO_2\cdot C_6H_4\cdot N{:}PPh_3 + Ph_3PO$$
63% 74%

The nitrone (239) and phenyliminotriethylphosphorane gave the imine (240) in a reaction involving migration of benzoyl in the decomposition of the intermediate.[133] The nitrileimine (241) with (234) formed the betaine (242) while proton migration in the corresponding compound from (243) and the nitrileimine (244) gave the iminophosphorane (245). The same nitrileimine with (237) formed the stable free radical (246) as shown.

$$PhCO\cdot CH{:}\overset{+}{N}(\bar{O})\cdot Ph + Et_3P{:}N\cdot Ph \longrightarrow$$
(239)

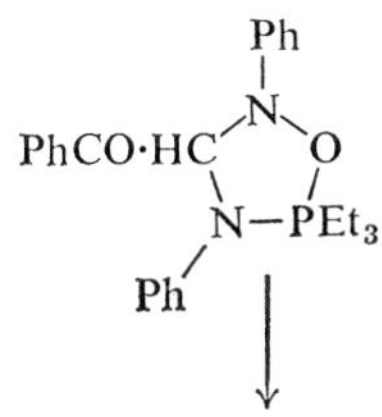

$$PhCO\cdot NPh\cdot CH{:}N\cdot Ph + Et_3PO$$
(240)

$$p\text{-}MeO\cdot C_6H_4\cdot C{:}\overset{+}{N}\cdot\bar{N}\cdot C_6H_4\cdot NO_2\text{-}p + Ph_3P{:}N\cdot Ph$$
(241) (234)

$$\longrightarrow p\text{-}MeO\cdot C_6H_4\cdot C(N(Ph)\cdot\overset{+}{P}Ph_3){:}N\cdot\bar{N}\cdot C_6H_4\cdot NO_2\text{-}p$$
(242) 89%

$$Ph\overset{+}{C}{:}N\cdot\bar{N}\cdot Ph + Ph_3P{:}NH \longrightarrow PhC(N{:}PPh_3){:}N\cdot NH\cdot Ph$$
(244) (243) (245)

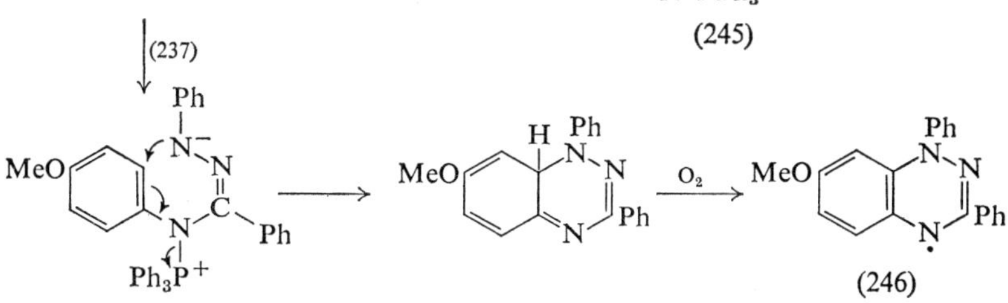

(246)

9
Phosphazenes

BY R. KEAT AND R. A. SHAW

1 Introduction

Interest, academic as well as industrial, in the phosphazenes has continued to grow. This is clearly reflected in the growth of publications in this field. In 1962 a comprehensive review[1] covering the literature since their discovery in 1834 quoted 397 references. Currently, *ca.* 100 papers and patents appear annually.

Perhaps the most significant feature of phosphazene chemistry in recent years has been the relatively rapid accumulation of structural data, obtained from n.m.r. spectroscopic and from *X*-ray crystallographic studies. This is particularly true of derivatives of cyclic oligomeric phosphazenes. However, no detailed structural data have, as yet, been reported on simple monomers of the type $R_3P{=}NR'$. The structural data available can often be rationalised in terms of bonding schemes which emphasise the rôle of the 3*d*-orbitals of the phosphorus atoms, and in terms of various physical properties. However, generalisations on the important subject of halogen-atom displacement at phosphorus are still difficult to make. Aspects of this latter subject have recently been reviewed,[2a, b] particularly with regard to the reactions and structures of derivatives of hexachlorocyclotriphosphazatriene (1).

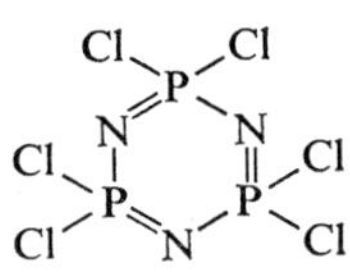

(1)

For the purpose of this report, the collective name phosphazene will be taken to include monomers and cyclic or linear polymers containing the $>P{=}N-$ unit. The naming of these compounds will be based on the nomenclature suggested by Shaw, Fitzsimmons, and Smith.[1] The literature is covered for the eighteen months up to June, 1969.

[1] R. A. Shaw, B. W. Fitzsimmons, and B. C. Smith, *Chem. Rev.*, 1962, **62**, 247.
[2] (a) R. A. Shaw, *Endeavour*, 1968, **27**, 74. (b) *Rec. Chem. Progr.*, 1967, **28**, 243.

2 Synthetic Routes to Phosphazenes

A. Acyclic Derivatives.—The Kirsanov reaction, whereby 2 moles of hydrogen halide are eliminated between halogenophosphoranes, R_nPX_{5-n} (X = halogen; R = other groups, *e.g.* Ph; n = 0—3), and organic or inorganic compounds containing one or more primary amino-groups, is still frequently used for the synthesis of new monomeric phosphazenes, *e.g.*;

$$P(:O)F_2NH_2 + PCl_5 \longrightarrow P(:O)F_2N{=}PCl_3 + 2HCl \quad \text{(refs. 3a, 4a)} \quad (1)$$

$$P(:O)F_2NH_2 + PCl_2Ph_3 \longrightarrow P(:O)F_2N{=}PPh_3 + 2HCl \quad \text{(ref. 4b)} \quad (2)$$

$$P(:S)XX'NH_2 + PCl_5 \longrightarrow P(:S)XX'N{=}PCl_3 + 2HCl \quad (X = X' = F)$$
$$(X = F, X' = Cl; X = X' = Cl)$$
$$\text{(ref. 3b)} \quad (3)$$

Solvolysis of $P(:O)F_2N{=}PCl_3$ by $H{\cdot}COOH$, $P(:O)F_2OH$, or FSO_2OH, occurs preferentially at the P=N linkage to give —NH— bridged derivatives of phosphorus or sulphur,[3a] *e.g.*:

$$P(:O)F_2N{=}PCl_3 + P(:O)F_2OH \longrightarrow P(:O)F_2NHP(:O)Cl_2 + P(:O)F_2Cl$$

Similarly, the trichlorophosphazenyl derivative (2) is obtained from $N_3C_3F_2{\cdot}NH_2$ and phosphorus pentachloride.[5] Solvolysis of (2) by formic acid gives $N_3C_3F_2NHP(:O)Cl_2$. Triphenylphosphazenyl derivatives, structurally related to the product of reaction (2), may be obtained from $Ph_3P{=}NH$.

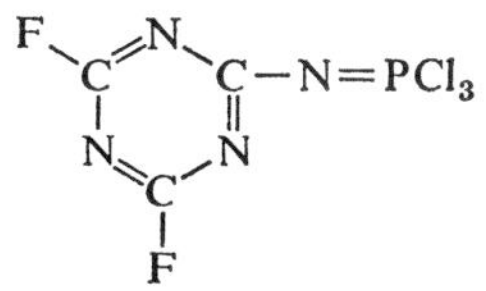

(2)

$$Ph_3P{=}NH + RSOCl \xrightarrow{NEt_3} Ph_3P{=}N{-}SOR + HCl$$
$$(R = CCl_3{-} \text{ or } p\text{-}CH_3C_6H_4{-}) \quad \text{(ref. 4b)}$$

$$Ph_3P{=}NH + RNCS \longrightarrow Ph_3P{=}NSCNHR$$
$$(R = \text{alkyl or aryl}) \quad \text{(ref. 4c)}$$

The derivatives, RNH_2 [R = FSO_2—,[6] $F_2P(:S)$—,[6,7] and ClFP(:S)—,[7]] react with the mixed halogenophosphorane, PCl_2F_3, eliminating hydrogen chloride to give $RN{=}PF_3$. It may be noted that $FSO_2N{=}PF_3$ is reported to be not accessible by fluorination of the chloride, $FSO_2N{=}PCl_3$.[6]

[3] [(a)] O. Glemser, H. W. Roesky, and P. R. Heinze, *Inorg. Nuclear Chem. Letters*, 1968, **4**, 179. [(b)] H. W. Roesky, *Chem. Ber.*, 1968, **101**, 3679.

[4] [(a)] S. Kongpricha and W. C. Preuse, U.S.P. 3,377,142/1968 (*Chem. Abs.*, 1968, **68**, 114,741). [(b)] A. Senning and P. Kelly, *Naturwiss*, 1968, **55**, 543; [(c)] I. N. Zhmurova, A. P. Martynyuk, and G. I. Derkach, *Zhur. obshchei. Khim.*, 1968, **38**, 163 (*Chem. Abs.*, 1968, **69**, 52,209).

[5] H. W. Roesky and H. H. Giere, *Inorg. Nuclear Chem. Letters*, 1968, **4**, 639.

[6] M. Lustig, *Inorg. Chem.*, 1969, **8**, 443.

[7] H. W. Roesky and L. F. Grimm, *Inorg. Nuclear Chem. Letters*, 1969, **5**, 13.

Phosphorus pentachloride reacts with a range of fluoroanilines to provide derivatives of the type 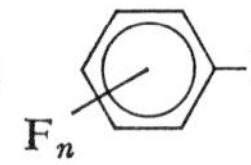 $-N{=}PCl_3$ (n = 2, 3, or 4).[8] No reaction was, however, observed with pentafluoroaniline. The products are all dimeric in benzene solution, but the 2,3,4,5- and 2,3,5,6-tetrafluoro-isomers are monomeric in carbon tetrachloride. The distinction between monomers and dimers is particularly marked in the observed ^{31}P chemical shifts. Shifts of +80 p.p.m. for dimers and of +30—40 p.p.m. for monomers (relative to 85% H_3PO_4 solution) were reported. As observed for other derivatives of this type, ammonolysis gives rise to ionic species, formulated as

$$[RNH{-}P(NH_2)_2{\cdots}N{\cdots}P(NH_2)_2{-}NHR]^+Cl^-\ (R=$$ 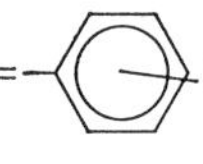 F_n).

The monomer–dimer equilibrium was also of interest in phosphazenes of the type, $MeCl_2P{=}NAr$ (Ar = aromatic group), obtained from tetrachloromethylphosphorane and substituted aniline hydrochlorides.[9a] When Ar = o-C_6H_4Cl, or o-C_6H_4Br, the products were monomeric in the liquid state, but dimeric in the solid state as shown by i.r. spectroscopy. Reactions of hydrazine dihydrochloride with phosphoranes, Ph_3PX_2, in the presence of aluminium chloride, are markedly dependent on the nature of the halogen atom, X:[9b]

$$Ph_3PCl_2 + N_2H_4,2HCl \xrightarrow[200^\circ]{AlCl_3} [Ph_3PNH{\cdot}NHPPh_3]^{++}2Cl^-$$

$$Ph_3PBr_2 + N_2H_4,2HCl \xrightarrow[200^\circ]{AlCl_3} [Ph_3P{-}N{=}PPh_3]^+Br^- + NH_4Br + Br_2 + HCl$$

$$Ph_3PI_2 + N_2H_4,2HCl \xrightarrow[200^\circ]{AlCl_3} [Ph_3P{-}N{=}PPh_3]^+I^- + NH_4^+I_3^- + I_2 + HCl$$

Other cationic species containing NH_2-groups within the cation have been used to synthesise P=N linkages. Thus, guanidine hydrochloride, $NH{=}C(NH_2)_2,HCl$, and Ph_nPCl_{5-n} (n = 0—2) give

$Ph_nCl_{3-n}P{=}N{-}C({=}NH){-}NH_2,HCl$ and $(Ph_nCl_{3-n}P{=}N)_2C{=}NH$, depending on the molar ratio of reactants.[10] Pyrolysis of these compounds gives rise to polymeric $(NPCl_2)_x$ (x = large number; n = 0) and/or oligomeric $N_3P_3Ph_6$ (n = 2) phosphazenes. In the same way, the carbonium ions of the salts $R^+C(NH_2)_2SbCl_6^-$ (R = Me, CCl_3, Ph, or NMe_2) and

8 K. Utvary and M. Bermann, *Monatsh.*, 1968, **99**, 2369.

9 (a) I. N. Zhmurova and I. Yu. Dolgushina, *Khim. Org. Soedineniya Fosfora, Akad. Nauk. S.S.S.R.*, 1967, 195 (*Chem. Abs.*, 1968, **69**, 76,781). (b) R. Appel and G. Siegemund, *Z. anorg. Chem.*, 1968, **361**, 203.

10 F. G. Sherif, *J. Inorg. Nuclear Chem.*, 1968, **30**, 1707.

$[C(NH_2)_3]^+SbCl_6^-$ gave in nitromethane solution with 2 or 3 moles of phosphorus pentachloride respectively the salts formulated as [11]

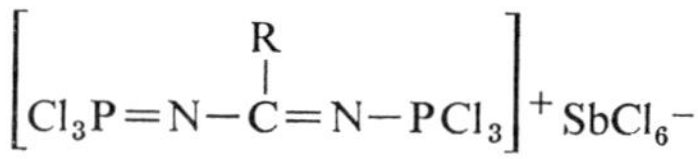

(3)

and

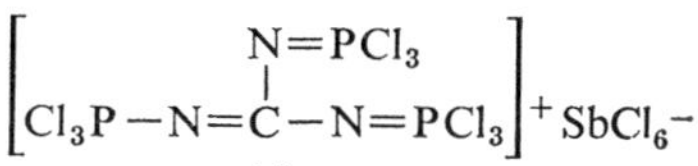

(4)

The structures (3) and (4) are consistent with the ^1H and ^{31}P n.m.r. data. In the case of (3) (R = Me), a four-bond spin–spin coupling, $J(P{-}N{-}C{-}C{-}H)$, of 2·55 Hz was taken as evidence of conjugation through the P—N—C—N—P skeleton. Evidence in favour of conjugation between the adjacent nitrogen and carbon atoms, and of slow rotation on the n.m.r. time-scale about this bond, as in (5) was obtained from the observation of different *cis* and *trans* $P{-}N{-}C{-}N{-}C{-}H$ spin–spin coupling constants of 2·2 and 3·2 Hz respectively.

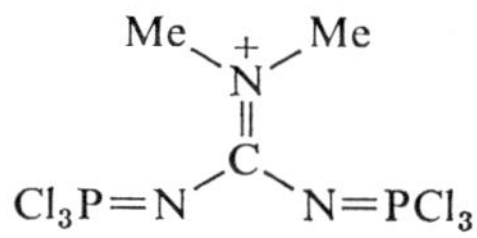

(5)

Further examples of phosphazene formation by the reaction of phosphorus(III) compounds with covalent organic and inorganic azides have been reported:

$$RN_3 + Ph_3P \longrightarrow RN{=}PPh_3 + N_2$$

(R = benzothiazolyl, benzimidazolyl, etc.) (ref. 12) (4)

$$ArN_3 + P(NC_2H_4)_3 \longrightarrow ArN{=}P(NC_2H_4)_3 + N_2$$

(Ar = aromatic group) (ref. 13) (5)

$$(RO)_2P(:O)N_3 + P(OR)_3 \longrightarrow (RO)_2P(:O)N{=}P(OR)_3 + N_2$$

(R = alkyl) (ref. 14) (6)

Products containing the grouping $=P(OR)_3$ (R = Alk) can isomerise on heating, *e.g.*,

$$(MeO)_2P(O)N{=}P(OMe)_3 \longrightarrow (MeO)_2P(:O)\cdot NMe\cdot P(:O)(OMe)_2$$

(ref. 14)

$$Me_3SiNH\cdot PBu^t_2 + Me_3SiN_3 \longrightarrow Me_3SiNH\cdot Bu^t_2P{=}NSiMe_3 + N_2$$

(ref. 15)

$$Bu^t_2PNH_2 + Me_3SiN_3 \longrightarrow Bu^t_2P(NH_2){=}NSiMe_3\ (A) + Bu^t_2P(NHSiMe_3){=}NSiMe_3\ (B) + Bu^t_2\overset{+}{P}(NH_2)_2\overset{-}{N_3}\ (C)$$

(ref. 15)

[11] A. Schmidpeter, K. Dull, and R. Böhm, *Z. anorg. Chem.*, 1968, **362**, 58.
[12] J. A. Van Allan and G. A. Reynolds, *J. Heterocyclic Chem.*, 1968, **5**, 471.
[13] G. I. Derkach and S. K. Mikhailik, *Khim. Org. Soedineniya Fosfora, Akad. Nauk S.S.S.R.*, 1967, 59 (*Chem. Abs.*, 1968, **69**, 2757).
[14] V. A. Shokol, N. K. Mikhailyuchenko, and G. I. Derkach, *Khim. Org. Soedineniya Fosfora, Akad. Nauk. S.S.S.R.*, 1967, 78 (*Chem. Abs.*, 1968, **69**, 10,065).
[15] O. J. Scherer and G. Schieder, *Chem. Ber.*, 1968, **101**, 4184.

Products (*B*) and (*C*) are believed to be formed by the reaction sequence:

$$(A) + Me_3SiN_3 \longrightarrow (B) + HN_3$$

$$Bu^t{}_2PNH_2 + HN_3 \longrightarrow Bu^t{}_2P(NH_2){=}NH \xrightarrow{HN_3} (C)$$

Notable work on the mechanism of the azide synthesis of phosphazenes has been carried out by Bock and co-workers.[16, 17] They established that the decomposition of intermediates such as (6) proceeds by elimination of nitrogen molecules containing γ- and β-nitrogen atoms. This was demonstrated by ^{15}N-labelling at the γ- and β-positions, and a study of the mass spectrum of the nitrogen eliminated in each case. These observations add weight to the suggestion that decomposition proceeds through a four-centred intermediate of the type (7).

$$Ph_3P{=}\underset{\gamma}{N}{-}\underset{\beta}{N}{=}\underset{\alpha}{N}{-}SO_2{\cdot}p\text{-tolyl}$$

(6)

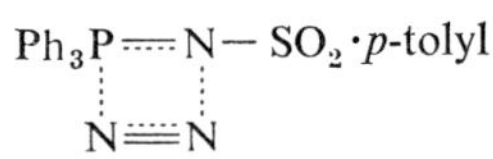

(7)

The i.r. spectra of these ^{15}N-labelled derivatives were also discussed in detail as well as the perdeuteriophenyl-derivatives [*e.g.* $(C_6{}^2H_5)_3P{=}N{-}N{=}N{-}SO_2$-*p*-tolyl]. These workers have also discussed the syntheses and i.r. spectra of ^{15}N- and ^2H-labelled derivatives of the type $(C_6X_5)_3P{=}N{-}N{=}CY_2$ ($X = {}^1H$ or ^2H, $N = {}^{14}N$ or ^{15}N, $Y = {}^1H$ or ^2H).[18]

Numerous other methods of monophosphazene synthesis have been reported. Compounds of the type, $[R_2P(:S)]_2NH$ (R = alkyl or phenyl) give on chlorination $[ClPR_2{=}N{\cdot}\overset{+}{P}R_2Cl]\overset{-}{Cl}$, and on subsequent ammonolysis, $[H_2N{\cdot}PR_2{=}N\overset{+}{P}R_2NH_2]\overset{-}{Cl}$.[19] Diazomethane methylates similar compounds on the sulphur atom:

$$Me_2P(:S)NHP(:S)R_2 + CH_2N_2 \longrightarrow MeS{\cdot}P(Me_2){=}N{\cdot}P(:S)R_2 \quad (R = Ph, Me)$$

and further reaction of the products with methyl iodide gives rise to further *S*-methylation at the terminal sulphur atom to give the cationic complex, $[MeS{\cdot}P(Me_2){=}N\overset{+}{P}(SMe)R_2]\overset{-}{I}$. The ^1H and ^{31}P n.m.r. spectra of these salts were discussed in detail and the ^1H spectra exhibited 'virtual coupling'. It was suggested that *J*(*P*—C—*H*) and *J*(P—N—P—C—*H*) are of opposite sign.[20] It has been shown that certain aminophosphines provide monophosphazenes by an exothermic reaction with carbon tetrachloride:[15]

[16] H. Bock and M. Schnöller, *Angew. Chem. Internat. Edn.*, 1968, **7**, 636.
[17] H. Bock and M. Schnöller, *Chem. Ber.*, 1969, **102,** 38.
[18] H. Bock, M. Schnöller, and H. tom Dieck, *Chem. Ber.*, 1969, **102**, 1363.
[19] A. Schmidpeter and J. Ebeling, *Chem. Ber.*, 1968, **101**, 815.
[20] A. Schmidpeter, H. Brecht, and J. Ebeling, *Chem. Ber.*, 1968, **101**, 390.

$$Me_3MNHPBu^t_2 + CCl_4 \longrightarrow Me_3MN{=}PClBu^t_2 + CHCl_3 \quad (M = Si, Ge, Sn)$$

or

$$(Me_3SiNH)_2PBu^t + CCl_4 \longrightarrow (Me_3SiNH)Bu^tP(Cl){=}NSiMe_3 + CHCl_3$$

This reaction probably proceeds *via* a nucleophilic attack by the phosphorus on the chlorine atom.

Triphenylphosphine and S_4N_3Cl undergo a complex reaction in the absence of solvent, to give the known salt, $[Ph_3P{=}N{-}PPh_3]^+Cl^-$. In benzene solution, under anhydrous conditions, a salt tentatively formulated as $[(Ph_3P{=}N)_3S]^{3+}3Cl^-$ is obtained. The Lewis acid character of this salt appears to be demonstrated by the fact that it forms an adduct, $[(Ph_3P{=}N)_3S]^{3+}$ $3Cl^-,2PPh_3$, with triphenylphosphine.[21] The compound $P(:S)(NCS)_3$ and chlorine give a red-brown solid in petrol at $-20°$, from which $Cl_3P{=}N{\cdot}CCl_3$ can be isolated.[22] This compound gives phosphorus pentachloride and cyanuric chloride on heating. The vibrational spectra of this, and other related derivatives, $Cl_3P{=}NR$ [$R = CCl_3$, $CCl_2{\cdot}CCl_3$, and $CCl(CCl_3)_2$], have been studied and the asymmetric and symmetric $P{=}N{-}C$ stretching modes identified. It was suggested that the difference in these frequencies indicates changes in the $P{=}\widehat{N}{-}C$ bond angle.[23]

B. Cyclic Derivatives.—Modifications of the original synthesis of cyclic oligomeric phosphazenes from ammonium chloride and phosphorus pentachloride continue to be reported. A patent [24] shows that the time for this reaction may be reduced by the introduction of an anhydrous metallic salt capable of complexing with ammonia. Salts claimed to be effective in a variety of solvents include $CoCl_2$, $AlCl_3$, $MnCl_2$, $CuCl_2$, $SnCl_4$, $MgCl_2$, $ZnCl_2$, $TiCl_4$, $CdCl_2$, and $CrCl_3$. Related to this is the report [25] that addition of metals to the PCl_5–NH_4Cl reaction mixture in inert solvents may increase or decrease the rate of hydrogen chloride evolution. The metals Zn, Co, Al, Cu, and Fe, increase the rate, whereas Ni, Mg, Ti, Mn, and Sn decrease the rate. Since the metals are thought to form their anhydrous chlorides in solution, the claim that Mg, Ti, Mn, and Sn slow the rate of hydrogen chloride evolution appears to contradict the claims of the patent.[24] Phosphoryl chloride is found to be a good catalyst for this reaction [26] and an 86% yield of cyclic derivatives was obtained in one case. Two Japanese patents indicate that the petrol-insoluble linear polymers from the PCl_5–NH_4Cl reaction may be converted to cyclic oligomers $(NPCl_2)_n$ on

[21] H. Prakash and H. H. Sisler, *Inorg. Chem.*, 1968, **7**, 2200.
[22] E. Fluck and F. L. Goldman, *Z. anorg. Chem.*, 1968, **356**, 307.
[23] D. P. Khomenko, G. G. Dyadyusha, and E. S. Koslov, *Spectroscopy Letters*, 1968, **1**, 245.
[24] N. L. Paddock and H. T. Searle, U.S.P. 3,407,047/1968 (*Chem. Abs.*, 1969, **70**, 39,397).
[25] E. Kobayashi, *Kogyo. Kagaku Zasshi*, 1967, **70**, 628 (*Chem. Abs.*, 1968, **68**, 56,122).
[26] J. Emsley and P. B. Udy, Chemical Society Annual Meeting, Nottingham, 1969, Abstract 7.16.

reaction with further NH_4Cl,[27] and that the yield of cyclic material from the PCl_5–NH_4Cl reaction may be improved by using finely divided NH_4Cl.[28] The reactivity of $Ph_3\overset{+}{P}ClPCl_6^{-}$, $C_5H_5N{\cdot}PCl_5$, $PCl_4^+SbCl_6^-$, and PCl_5 towards NH_4Cl to form linear salt-like phosphazenes has been compared.[29] The reactions in nitrobenzene solution were followed by ^{31}P n.m.r. spectroscopy and showed that the reactivity increases in the order $PCl_6^- < C_5H_5NPCl_5 < PCl_5 < PCl_4^+$.

The mass spectra of several oligomeric halogenocyclophosphazenes have been studied in detail. Homologues in the series $(NPCl_2)_n$ (n = 3—8),[30] (n = 3, 4),[31] and $(NPF_2)_n$ (n = 3—16)[32] break down to give a series of cyclic and acyclic ions. In the series of nongeminal derivatives, $N_3P_3Cl_xBr_{6-x}$ (x = 0—5),[33] the parent ion is of low intensity and the base peak arises from the loss of one bromine atom. Again, cyclic and acyclic ions were identified and the relatively large drop in ionisation potential on the introduction of one bromine atom into $N_3P_3Cl_6$ was interpreted in terms of increased base strengths at the ring nitrogen atoms. Appearance potential data gave ΔH_f° for $N_3P_3Cl_5Br$ as $-168{\cdot}6$ kcal/mole, compared with ΔH_f° for $N_3P_3Cl_6$ = $-175{\cdot}9$ kcal/mole.

Several examples of six-membered phosphazene ring systems have been obtained by the cyclisation of compounds containing a P—N—P fragment, *e.g.*:

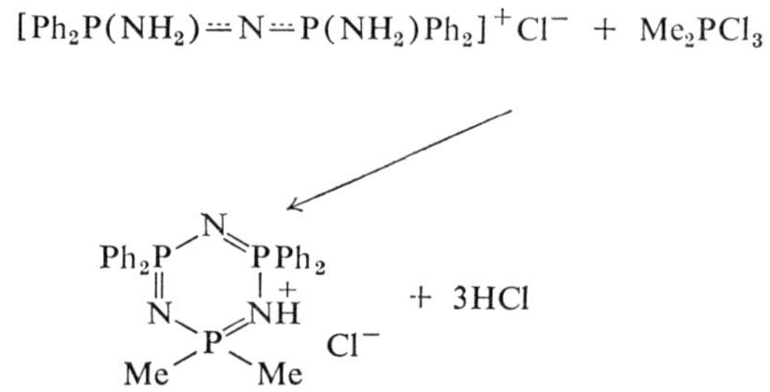

This result[34] differs from previous reports on analogous reactions with PCl_5, $PhPCl_4$, and Ph_2PCl_3, where the hydrochloride was not normally isolated. It is understandable, however, in terms of basicity substituent constants Me > Ph > Cl (*cf.* section 6). By use of related reactions it is also possible to introduce boron as a member of the ring system:[35a]

[27] M. Ura and K. Ogihara, Jap.P. 14,693/1967 (*Chem. Abs.*, 1968, **68**, 97,204).
[28] E. Kobayashi, Jap.P. 14,694/1967 (*Chem. Abs.*, 1968, **68**, 80,074).
[29] W. Lehr and M. Schwarz, *Z. anorg. Chem.*, 1968, **363**, 43.
[30] C. E. Brion and N. L. Paddock, *J. Chem. Soc.* (*A*), 1968, 388.
[31] C. D. Schmulbach, A. G. Cook, and V. R. Miller, *Inorg. Chem.*, 1968, **7**, 2463.
[32] C. E. Brion and N. L. Paddock, *J. Chem. Soc.* (*A*), 1968, 392.
[33] G. E. Coxon, T. F. Palmer, and D. B. Sowerby, *J. Chem. Soc.* (*A*), 1969, 358.
[34] M. Bermann and K. Utvary, *J. Inorg. Nuclear Chem.*, 1969, **31**, 271.
[35] [(a)] M. Becke-Goehring and H.-J. Müller, *Z. anorg. Chem.*, 1968, **362**, 51. [(b)] A. Schmidpeter and K. Stoll, *Angew. Chem. Internat. Edn.*, 1968, **7**, 549.

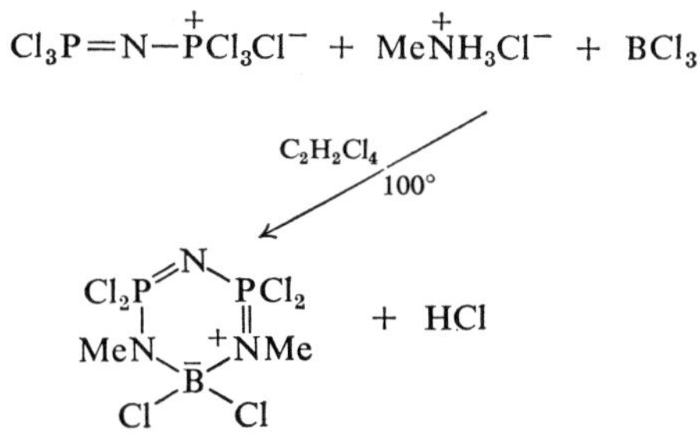

The same product was obtained with PCl_5 instead of $[Cl_3P{=}N{-}\overset{+}{P}Cl_3]Cl^-$ and appears to be covalent in character in spite of the formal charges on some of the ring atoms.

Cationic complexes of silicon, germanium, and tin containing the phosphazene linkage have been prepared[35b] by the reaction:

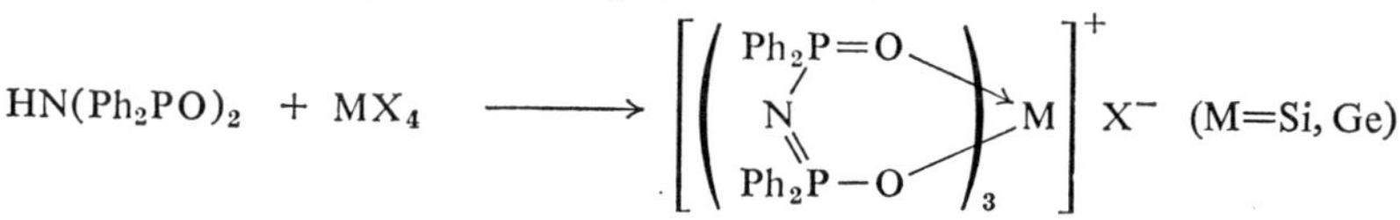

The reaction has not been observed where M = Sn (X = Cl, Br) and a two-ring derivative, 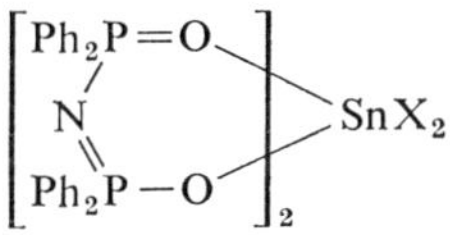 is formed. The third ring can only be introduced using $NaN(Ph_2PO)_2$ at elevated temperature. Salts with numerous anionic species replacing X have been prepared. When M = Sn, the cations are hydrolytically stable and can be used to precipitate large anions, *e.g.* MnO_4^-.

Novel syntheses of compounds containing the elements P, N, and C as ring atoms have been described.[36]

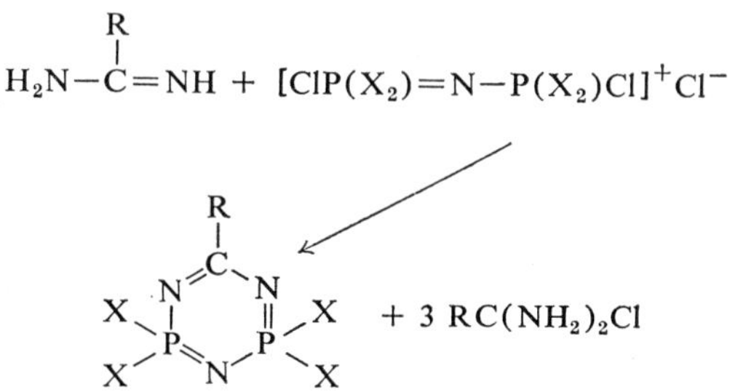

R =	$PhCH_2$	Ph	NH_2	NMe_2	$PhCH_2$	NMe_2
X =	Ph	Ph	Ph	Ph	Me	Me

[36] A. Schmidpeter and J. Ebeling, *Chem. Ber.*, 1968, **101**, 3883.

The corresponding hydrochlorides are also readily isolated from guanidine hydrochlorides:

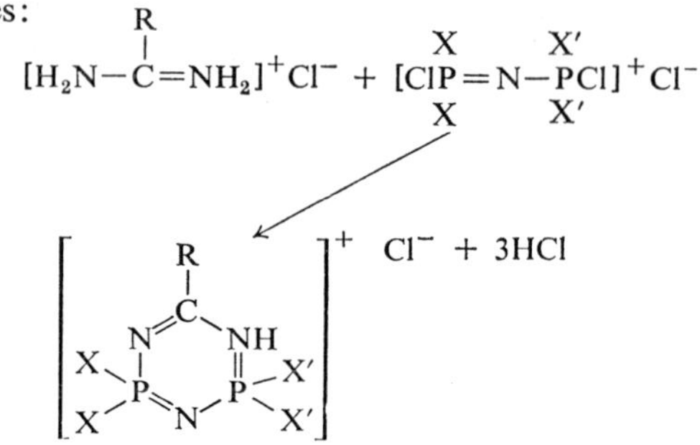

R =	Me	$PhCH_2$	Ph	Me	Me
X =	Ph	Ph	Ph	Me	Ph
X′ =	Ph	Ph	Ph	Me	Me

these can be dehydrochlorinated by treatment with an excess of ammonia. At 40°, there is a rapid exchange on the n.m.r. time scale of the proton

between the sites indicated, but at −60°, signals due to the different P—Me groups can be identified from the ^{1}H n.m.r. spectrum. A second route to the diphosphazatriazines lies in the elimination of aniline from linear aminophosphazenes and PhNHCH=NPh:

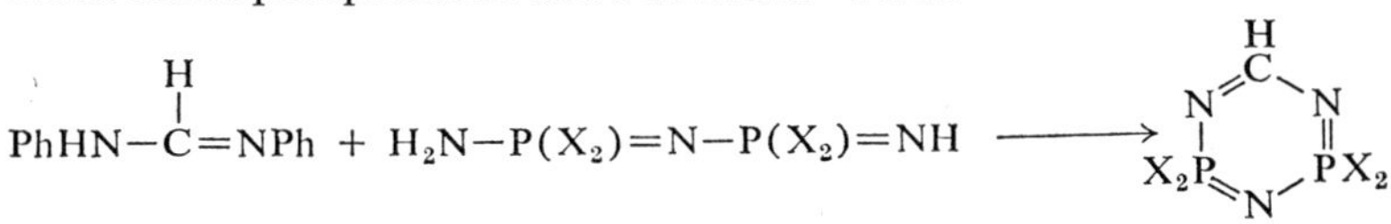

$+ 2PhNH_2 (X = Ph$ or $Me)$

These ring systems may also be prepared by the reactions:[37]

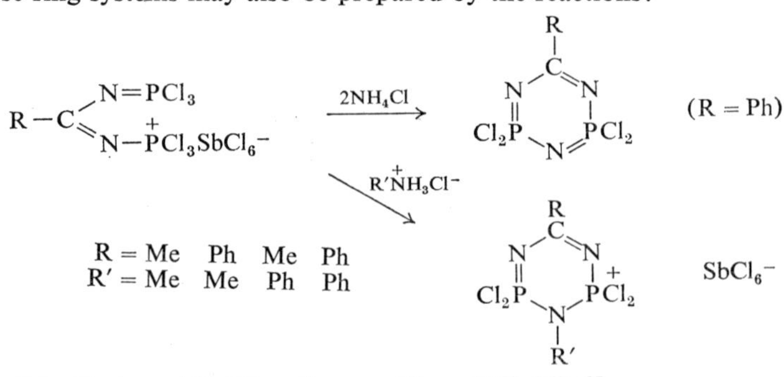

R = Me Ph Me Ph
R′ = Me Me Ph Ph

[37] A. Schmidpeter and R. Böhm, *Z. anorg. Chem.*, 1968, **362**, 65.

Where R = Me, R′ = Me and R = Me, R′ = Ph, relatively large *J(P—N—C—C—H)* coupling constants were observed from the ^{1}H n.m.r. spectra (R′ = Me, 4·3 Hz; R′ = Ph, 4·1 Hz) and cited as evidence of a conjugative interaction within the P—N—C system. There is strong evidence that this is the case in the compound (8) for its crystal structure has been studied [38] and it was shown that the ring system was planar.

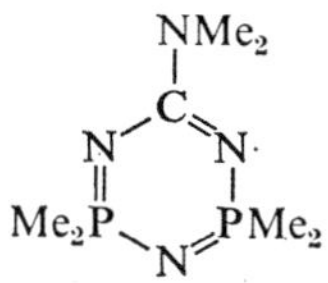

(8)

The ^{1}H n.m.r. spectra of species containing P, N, and S as component atoms of a six-membered ring have also been discussed.[39]

Compounds of the type, $NH_2PR_2{=}N{-}PR_2{=}NH$, can be cyclised by a variety of reagents, *e.g.* reactions (7), (8), and (9):

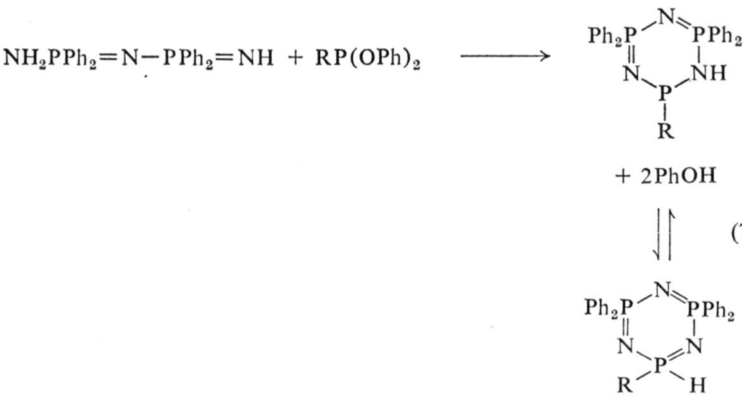

(7) Ref. 40*a*

Strong evidence for the predominance of the tautomer with a direct P—H bond was obtained from the large value of *J(P—H)* (500—700 Hz). This appears to be the first example of a cyclotriphosphazatriene with a P—H bond.

Related equilibria $P^{III} \leftrightharpoons P^{V}$, where the quinquevalent form contains a P—H bond, have been recently established for a number of mononuclear phosphorus compounds.[40b]

This derivative was also obtained from guanidine (R = H) and its dimethyl derivative (R = Me).[41]

[38] U. Klement and A. Schmidpeter, *Z. Naturforsch*, 1968, **23b**, 1610.

[39] L. Siekmann, H. O. Hoppen, and R. Appel, *Z. Naturforsch*, 1968, **23b**, 1156.

[40] [(a)] A. Schmidpeter and J. Ebeling, *Angew. Chem. Internat. Edn.*, 1968, **7**, 209. [(b)] *Cf.* R. Wolf, J. F. Brazier, D. Houalla, and M. Sanchez, Int., Symposium on Organo-phosphorus Chemistry, Paris, May, 1969.

[41] A. Schmidpeter and J. Ebeling, *Chem. Ber.*, 1968, **101**, 2602.

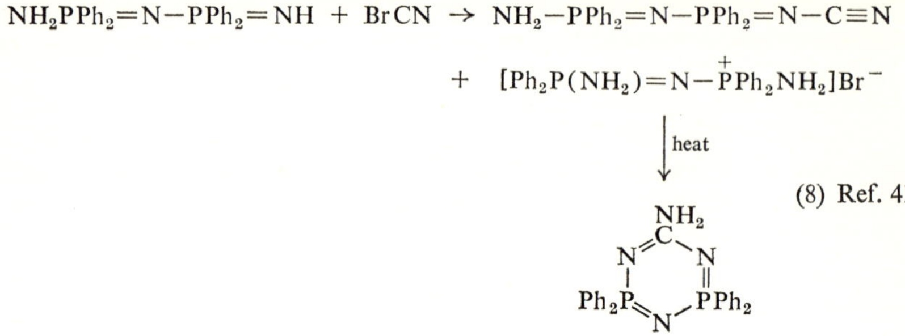

(8) Ref. 41

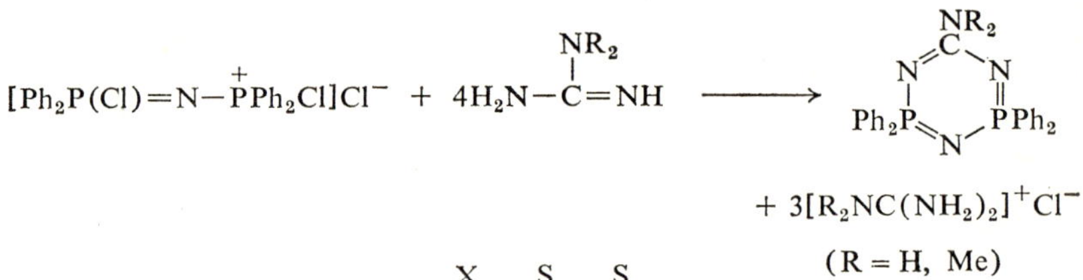

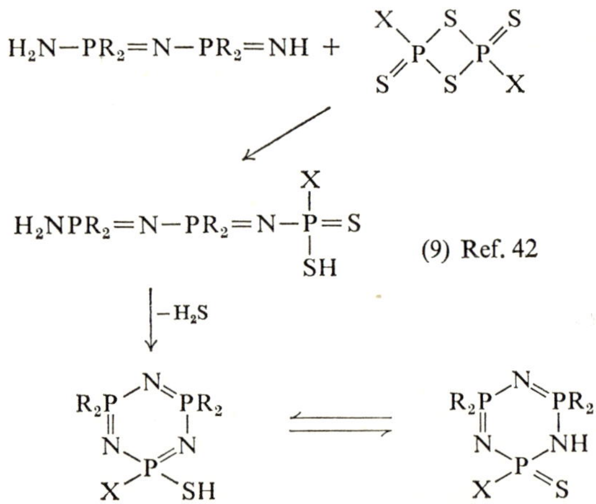

(9) Ref. 42

R = Ph, X = MeO·C_6H_4—
R = Ph, X = PhS—
R = Me_2N, X = MeO·C_6H_4—

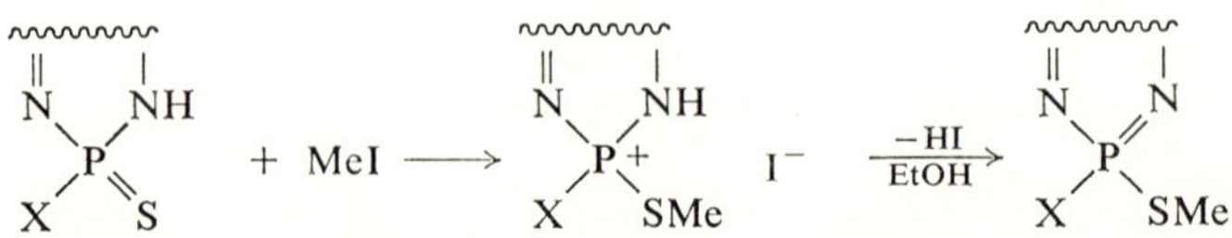

The predominance of the NH-tautomer was established by ^{31}P n.m.r. spectroscopy. This tautomer was converted to the cyclotriphosphazatriene by quaternisation at the sulphur atom and subsequent removal of HI by EtOH or a MeCN–water mixture.

3 Reactions Involving Displacement of Halogen Atoms

A. By Other Halogen Atoms.—A comprehensive study of the fluorination of $N_3P_3Cl_6$ by NaF in nitromethane or in nitrobenzene and of $N_4P_4Cl_8$ by KSO_2F has been reported.[43] Compounds representing all degrees of replacement in the series $N_3P_3Cl_{6-n}F_n$ ($n = 1$—6) and $N_4P_4Cl_{8-n}F_n$ ($n = 1$—8) have been obtained. ^{19}F N.m.r. spectroscopy shows that fluorination in both cases proceeds by a geminal replacement pattern. With $N_4P_4Cl_8$ two alternative geminal routes are available when $n > 2$ and the one taken is as indicated below:

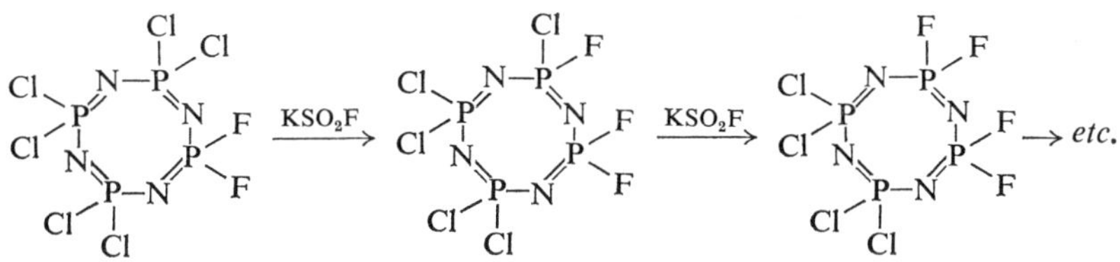

These observations are explicable in terms of a simple electrostatic effect whereby replacement of a chlorine atom at phosphorus by a fluorine atom increases the electrophilicity of the phosphorus atom. Qualitative observations show that the rate of formation of $N_4P_4Cl_6F_2$ is greater than that of $N_3P_3Cl_4F_2$ from the respective monofluoro-derivatives. This behaviour was accounted for in terms of π-inductive effects within the P—N ring system.

Halogen exchange reactions between dimeric phosphazenes $(MeN{=}PCl_3)_2$ and $(MeN{=}PF_3)_2$ (using a 1 : 2 molar ratio in a sealed tube at 110°) lead to the formation of a series of mixed fluoride–chlorides $(MeN{=}P)_2F_nCl_{6-n}$ ($n = 2$—5) which were separated by fractional distillation.[44] The derivatives where $n = 1$ or 2 were obtained by fluorination of $(MeN{=}PCl_3)_2$ by NaF in acetonitrile. The structures of the isomers obtained were not established. In contrast, reaction of $(Cl_3P{=}N{\cdot}CCl_2{\cdot}CCl_2)_2(CH_2)_n$ ($n = 2$—6) with KF in chlorobenzene led to cleavage of the P=N bond and the formation of PF_5 and $(CH_2)_n(CCl_2CN)_2$.[45]

B. Halogen Displacement by Amines.—Interest has recently been shown in the synthesis and properties of amino-derivatives of $N_3P_3F_6$,[46, 47] and, to a

[42] A. Schmidpeter and C. Weingand, *Z. Naturforsch*, 1969, **24b**, 177.
[43] J. Emsley and N. L. Paddock, *J. Chem. Soc.* (*A*), 1968, 2590.
[44] K. Utvary and W. Czysch, *Monatsh.*, 1969, **100**, 681.
[45] V. I. Schevchenko, V. P. Kukhar, and A. V. Kirsanov, *Zhur. obshchei. Khim.* 1967, **37**, 2361 (*Chem. Abs.*, 1968, **68**, 77,683).
[46] O. Glemser, E. Niecke, and H. W. Roesky, *Chem. Comm.*, 1969, 282.
[47] T. Chivers and N. L. Paddock, *Chem. Comm.*, 1969, 337.

lesser extent, in such derivatives of the homologues, $(NPF_2)_n$ (n = 4—6).[46] These amino-derivatives are perhaps best characterised by their ^{19}F n.m.r. spectra which can also give valuable structural information. Aminolysis has been carried out using methylamine,[46] dimethylamine,[46–48] dimethylaminotrimethylsilane,[47, 48] and lithium dimethylamide.[48] The isomers so far characterised in the series, $N_nP_nF_{2n-2}(NMe_2)_2$, are nongeminal.[47, 48] These derivatives are summarised in Table 1 below together with the

Table 1

	X =			
$N_3P_3F_5X$	NHMe[46]	NMe_2[46, 47]	Cl[46, 47]	Br[46, 47]
$N_nP_nF_{2n-1}X$ (n = 4—6)		NMe_2[47]	Cl[47]	Br[47]
$N_3P_3F_4(NMe_2)_2$		NMe_2[47, 48]	(*cis* and *trans*)	
$N_nP_nF_{2n-2}X_2$ (n = 4—6)		NMe_2[47]		
$N_6P_6F_9X_3$		NMe_2[47]		

chloride–fluorides and bromide–fluorides obtained by deaminolysis with HCl or HBr respectively.[46, 47] In view of the relatively high P—N bond orders associated with fluorophosphazenes, it is remarkable that the reaction of $N_3P_3F_6$ with diethylamine is reported to lead to cleavage of the P—N ring system.[46]

Chlorofluorodimethylamino-derivatives may be obtained from the reaction of $N_3P_3Cl_{6-n}(NMe_2)_n$ (n = 2, 3) with SbF_3 or KSO_2F. Fluorinations with these two reagents follow different replacement patterns.[49]

$$N_3P_3Cl_4(NMe_2)_2 \xrightarrow{SbF_3} N_3P_3Cl_2F_2(NMe_2)_2 \xrightarrow{HCl} N_3P_3F_2Cl_4$$

(two geometric isomers with $\equiv$P(F)NMe$_2$ and $\equiv$PCl$_2$ groups)

$$N_3P_3Cl_4(NMe_2)_2 \xrightarrow{KSO_2F} N_3P_3Cl_2F_2(NMe_2)_2$$

(two geometric isomers with $\equiv$PClNMe$_2$ and $\equiv$PF$_2$ groups)

$$N_3P_3Cl_3(NMe_2)_3 \xrightarrow{SbF_3} N_2P_3F_3(NMe_2)_3 \xrightarrow{HCl} N_3P_3Cl_3F_3$$

nongeminal

The course of reactions of $N_3P_3Cl_6$ with a wide range of amines has been studied in recent years but, even now, little is known of the mechanisms of these reactions and their relation to the stereospecific reaction paths observed. Some light is thrown on this matter by the following observations:

48 A. Hawley and G. Nickless, Chemical Society Annual Meeting, Nottingham, 1969, Abstract 7.17.

49 B. Green and D. B. Sowerby, *Chem. Comm.*, 1969, 628.

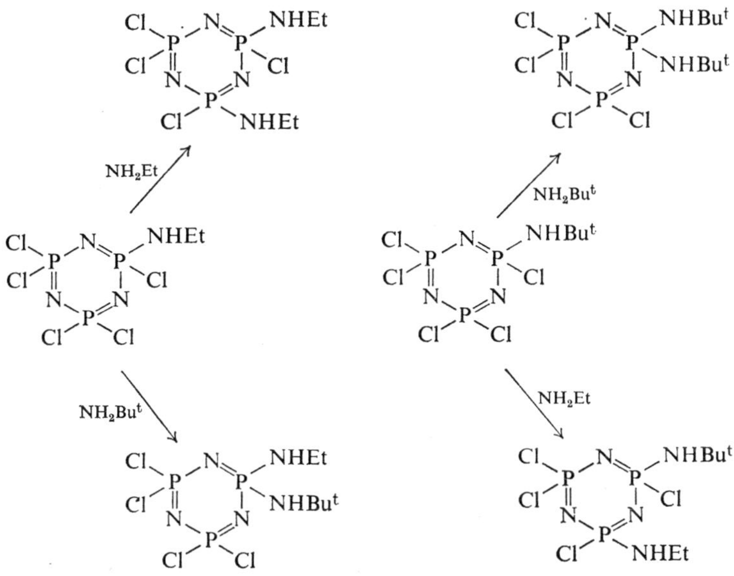

In view of the fact that t-butylamine with $N_3P_3Cl_6$ gives entirely geminal $N_3P_3Cl_4(NHBu^t)_2$, and that ethylamine with $N_3P_3Cl_6$ gives entirely non-geminal $N_3P_3Cl_4(NHEt)_2$, it becomes apparent that at least in some reactions the nature of the nucleophile may be of greater importance than the substituent already present in determining the reaction pattern.[50] It is interesting to note that the reactions of four equivalents of piperidine, and of four equivalents of t-butylamine, with one equivalent of the diphosphatriazines (9),[51a] take a similar route to that observed for the same amines with $N_3P_3Cl_6$.

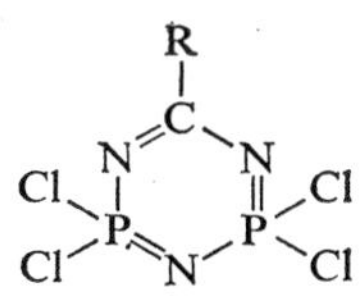

R = Me, Ph

(9)

The former reactions are summarised in the Scheme. ^1H And ^{31}P n.m.r. data were used to show that the geminal route *A* is followed by t-butylamine and that the nongeminal route ***B*** is followed by piperidine. These conclusions were confirmed by the n.m.r. data obtained on the compounds prepared by replacement of the remaining chlorine atoms in (10) and (11)

[50] R. Keat and R. A. Shaw, *Angew. Chem. Internat. Edn.*, 1968, **7**, 212.

[51] [(a)] A. Schmidpeter and N. Schindler, *Chem. Ber.*, 1969, **102**, 856. [(b)] R. Keat and R. A. Shaw, *J. Chem. Soc.* (*A*), 1966, 908.

by methoxy- or dimethylamino-groups. Although it was not experimentally established whether the dipiperidino-derivative (11) had the *cis*- or the *trans*-structure, the latter was postulated on the basis of the '*cis*-effect'.[51b]

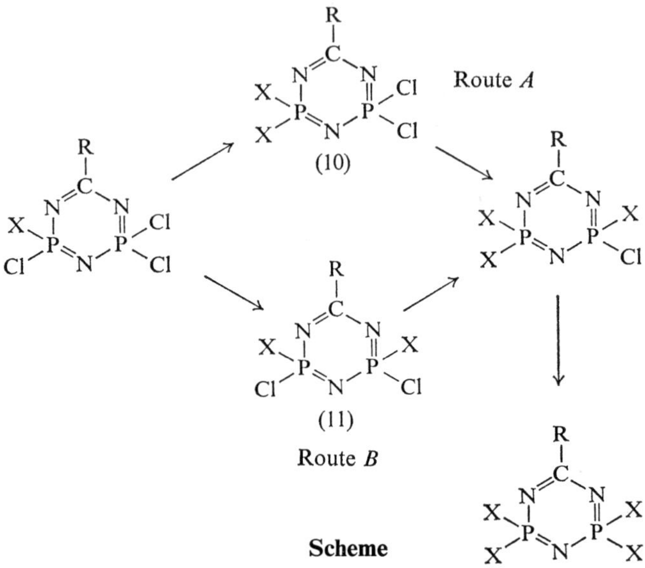

Scheme

A series of mixed pyrrolidine–aziridine derivatives, $N_3P_3(NC_4H_8)_n(NC_2H_4)_{6-n}$ (n = 1—4) have been prepared starting from aziridinochloro-derivatives of known geminal structure and from chloropyrrolidino-derivatives of unknown structure.[52] By a comparison of the isomers obtained, it was possible to establish that pyrrolidine replaces the chlorine atoms in $N_3P_3Cl_6$ by a predominantly nongeminal scheme. Similarly, the preparation of a series of aziridinomorpholino-derivatives of $N_3P_3Cl_6$ was used to establish a nongeminal reaction scheme with morpholine.[53] A series of monoamino-derivatives, $N_3P_3Cl_5R$ (R = $NHCHPhCO_2Me$, $NHNHPr^i$, and NHCHMePh)[54] and a series of α-phenylpyrrolidinyl-derivatives, $N_3P_3Cl_{6-n}$ 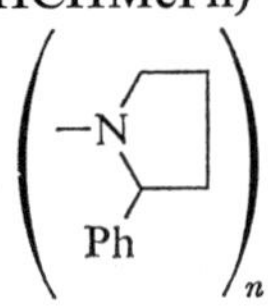

(n = 1—4) (one isomer only in each case)[55] have been described.

[52] A. A. Kropacheva and N. M. Kashnikova, *Zhur. obshchei Khim.*, 1968, **38**, 136; *J. Gen. Chem. U.S.S.R.*, 1968, **38**, 135.

[53] L. E. Mukhina and A. A. Kropacheva, *Zhur. obshchei Khim.*, 1968, **38**, 313; *J. Gen. Chem. U.S.S.R.*, 1968, **38**, 314.

[54] A. A. Kropacheva and N. M. Kashnikova, *Khim. Org. Soedineniya Fosfora, Akad. Nauk S.S.S.R.*, 1967, 186 (*Chem. Abs.*, 1968, **69**, 10,066).

[55] A. A. Kropacheva and N. M. Kashnikova, *Khim. Org. Soedineniya Fosfora, Akad. Nauk S.S.S.R.*, 1967, 188 (*Chem. Abs.*, 1968, **69**, 10,321).

The N—P bonds in $N_3P_3X_6$ (X = F, Cl, Br), $N_4P_4Cl_8$, and $(NPCl_2)_n$ ($n \sim 15{,}000$) undergo a novel form of degradation on reaction with *ortho*-aminophenol.[56] In all cases the same product is obtained which has properties consistent with the five-co-ordinate phosphorus compound (12), rather than with the analogous six-co-ordinate species (13).

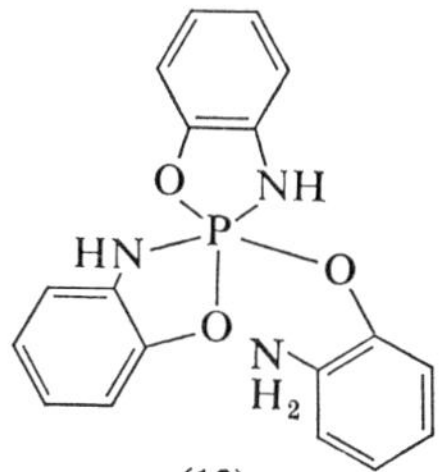

(12)

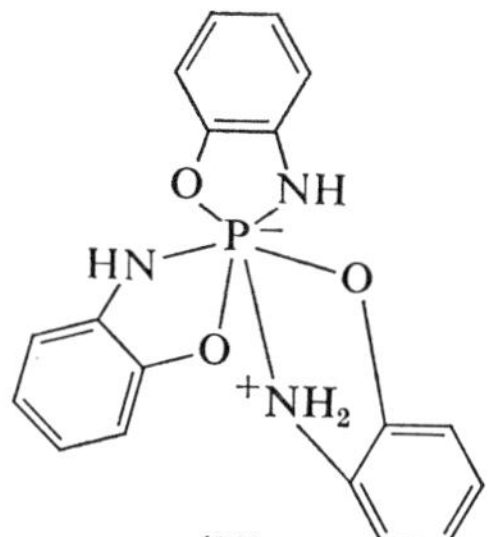

(13)

Preliminary information on the mode of reaction of $N_4P_4Cl_8$ with dimethylamine has appeared and some indication of the complexity of this system is provided by the detection (in one case by t.l.c. chromatography [58]), isolation, and characterisation [57] of five isomeric bisdimethylamino-derivatives, $N_4P_4Cl_6(NMe_2)_2$, one of which was geminal. Of these, the two pairs of nongeminal isomers could be interconverted by pyridine hydrochloride. Only two trisdimethylamino-isomers, $N_4P_4Cl_5(NMe_2)_3$, geminal and nongeminal, were isolated from a possible total of 5.[57] The reaction of $N_4P_4Cl_8$ with an excess of the bulky secondary amine, methylcyclohexylamine, is strongly stereospecific and the nongeminal tetrakis-derivative, $N_4P_4Cl_4(NMe{\cdot}C_6H_{11})_4$, is formed in high yield.[59] The structure of this derivative (14) was established by its reactions with methylamine and phenyl isocyanate, and by ^1H n.m.r. spectroscopy on these materials. It was also suggested that in this case aminolysis proceeds sequentially at the

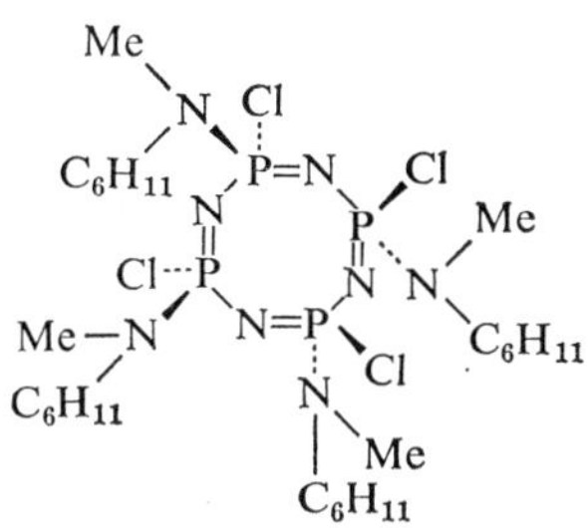

(14)

56 H. R. Allcock and R. L. Kugel, *Chem. Comm.*, 1968, 1606.
57 W. Lehr, *Naturwiss*, 1969, **56**, 214.
58 R. Stahlberg and E. Steger, *J. Inorg. Nuclear Chem.*, 1968, **30**, 737.
59 A. J. Berlin, B. Grushkin, and L. R. Moffett, *Inorg. Chem.*, 1968, **7**, 589.

2,6- and 4,8-positions in the ring, a result only partially observed for dimethylamine.[57]

The reactions of the monomeric phosphazenes $PhN{=}PCl_2R$ [R = NMe_2, NEt_2, $NBu^n{}_2$, and $N(Me)CH_2Ph$] with ammonia give products which are the phosphonium salts $[PhNHP(R)(NH_2)_2]^+Cl^-$, rather than the salts $[PhNH(NH_2)RP{=}N{-}\overset{+}{P}(NH_2)(NHPh)]\overset{-}{Cl}$. The latter type (where R = H) are obtained from ammonia and (15). It was suggested that the observed products might be due to the enhanced stability of $[PhNHPR(NH_2)_2]^+Cl^-$ relative to $[PhNHP(NH_2)_3]^+Cl^-$.[60]

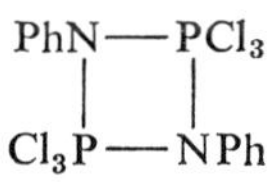
(15)

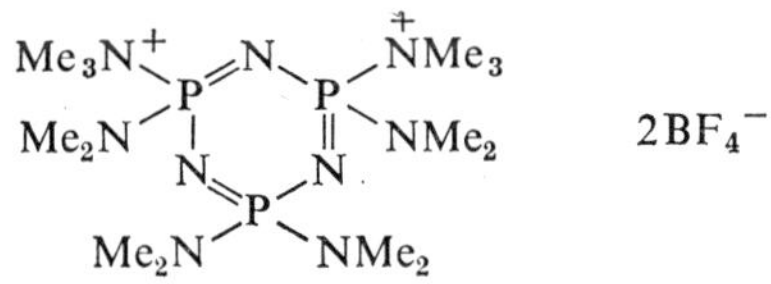
(16)

The Lewis-base properties of aminocyclophosphazenes continue to provide a source of interest. The compound $N_3P_3(NMe_2)_6$ undergoes quaternisation at the exocyclic nitrogen atoms by $Me_3O^+BF_4^-$ to give (16).[61] Evidence for the structure of the cation postulated was obtained by hydrolytic degradation of the salt (16) which gave $Me_2NH_2Cl^-$ and $Me_3\overset{+}{N}H\overset{-}{Cl}$. On the assumption (not necessarily correct) that the ring nitrogen atoms are least hindered to the approach of an electrophile, it was suggested that charge-transfer from the exocyclic nitrogen atoms may not be as extensive as that suggested from protonation studies. On the other hand, *X*-ray crystallographic evidence shows unambiguously that protonation of a ring nitrogen atom occurs in $N_3P_3Cl(NHPr^i)_4$,HCl,[62] and furthermore on that ring nitrogen which has the largest number of $\alpha(NHPr^i)$ substituents, confirming earlier deductions from basicity studies (see section 5 on structural studies). It is likely that the nature of the acid may be of considerable significance in any estimation of the relative basicity of cyclic and exocyclic nitrogen atoms. Steric effects are also important in the consideration of solvent effects on the ^1H n.m.r. spectra of the dimethylamino-derivatives, $N_3P_3X_{6-n}(NMe_2)_n$ (X = Cl, n = 1—4 and 6; X = Br, n = 1—3) which generally show an upfield shift on passing from carbon tetrachloride to benzene solutions.[63] The shift is most pronounced for n = 1. For a given degree of chlorine or bromine replacement the signals from dimethylamino-groups *cis* to one another were shifted upfield to a greater extent than those *trans* to one another. This may reflect a preference of the benzene solvent molecules to approach the phosphazene ring on the side with more dimethylamino-groups than chlorine atoms.

[60] K. Utvary and M. Bermann, *Inorg. Chem.*, 1969, **8**, 1038.
[61] J. N. Rapko and G. R. Feistel, *Chem. Comm.*, 1968, 474.
[62] N. V. Mani and A. J. Wagner, *Chem. Comm.*, 1968, 658.
[63] R. Keat and R. A. Shaw, *J. Chem. Soc.* (*A*), 1968, 703.

The reactions of $N_3P_3(NMe_2)_6$ with HX (X = Cl, Br, or I) in boiling xylene solution have been studied.[64] Adducts of the type $N_3P_3(NMe_2)_6$,HX were initially isolated, but prolonged reactions lead to deaminolysis and the formation of *cis*-$N_3P_3Cl_2(NMe_2)_4$ and *trans*- and *cis*-$N_3P_3X_3(NMe_2)_3$ (X = Cl or Br), although the latter isomer may not have been a primary reaction product. The *cis*-isomer, $N_3P_3Cl_2(NMe_2)_4$, is also obtained from the dimethylaminolysis of $N_3P_3Cl_6$, and its isolation by deaminolysis lends some weight to the suggestion that it is the thermodynamically favoured form. No iododimethylamino-derivatives were isolated. The hydrochlorides, $N_3P_3[NH(CH_2)_nMe]_6$, xHCl ($n = 2$, $x = 2$; $n = 3$, $x = 2$; $n = 4$, $x = 3$) have been obtained from the reaction of the corresponding hexakisamino-derivatives with hydrogen chloride in benzene solution at room temperature.[65]

C. Halogen Displacement by Alcohols.—Relatively little work has recently been reported on the alcoholysis of chlorophosphazenes. Cyclic fluoroalkoxy-oligomers of the type $(NPR_2)_n$ [$n = 3$—7; R = $OCH_2(CH_2)_xY$; $x = 1$—20; Y = H or F] have been prepared from $(NPCl_2)_n$ and the corresponding sodium alkoxides or alcohols. Their chemical and thermal stabilities have been investigated.[66]

The base-catalysed hydrolysis of $N_3P_3(OCH_2CF_3)_6$ proceeds by successive removal of two trifluorethoxy-groups at different phosphorus atoms followed by hydrolytic degradation of the P—N ring:[67]

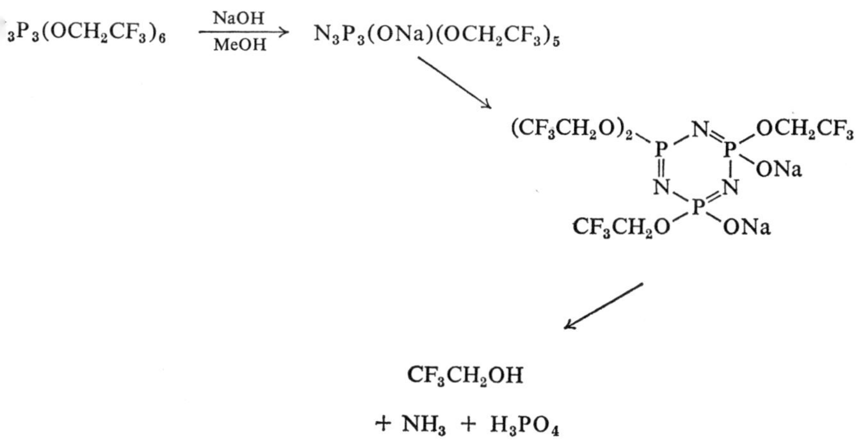

In contrast, the hydrolysis of (17) to give (18) proceeds much more rapidly:[67]

[64] S. N. Nabi, R. A. Shaw, and C. Stratton, *Chem. and Ind.*, 1969, 166.
[65] K. Denny and S. Lanoux, *J. Inorg. Nuclear Chem.*, 1969, **31**, 1531.
[66] Olin Mathieson Chem. Corpn., B.P. 1,104,471/1968 (*Chem. Abs.*, 1968, **69**, 35,419).
[67] H. R. Allcock and E. J. Walsh, *J. Amer. Chem. Soc.*, 1969, **91**, 3102.

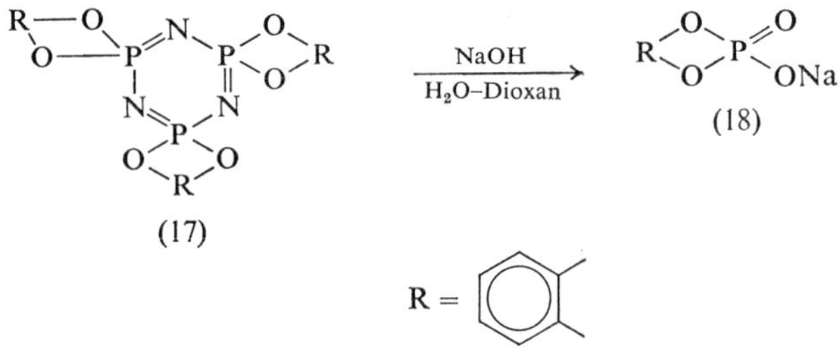

(17) (18)

This behaviour appears to be connected with the strain within the five-membered ring for similar rate enhancement effects have been observed in the hydrolysis of mononuclear phosphorus esters containing a related five-membered ring. The esters $N_3P_3(OPh)_6$ and (19) undergo no detectable hydrolytic attack under similar conditions.

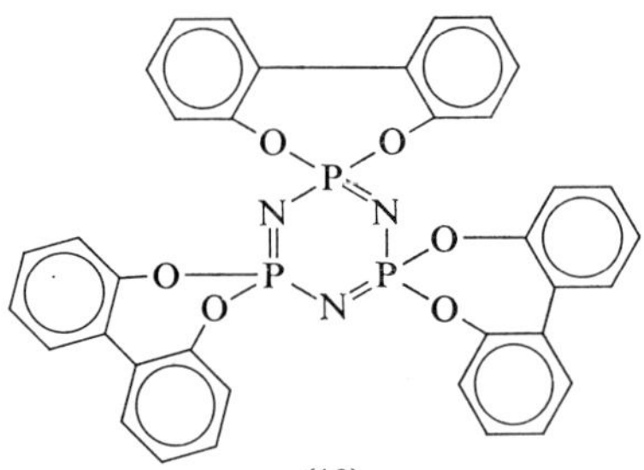

(19)

A new rearrangement of linear phosphazenes has been observed and its mechanism formulated:[68]

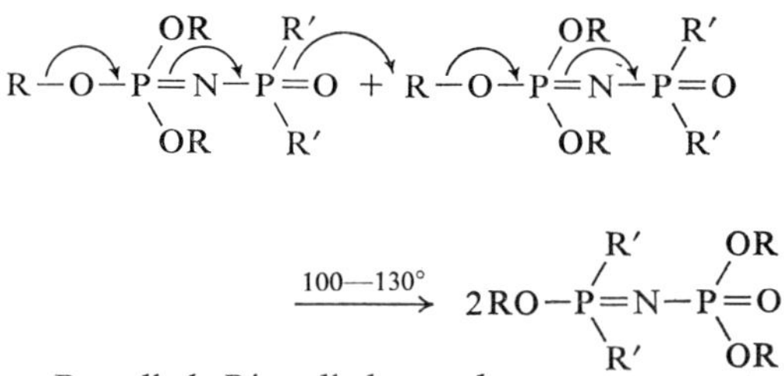

R = alkyl; R′ = alkyl or aryl

The rate of isomerisation decreases with increasing molecular weight of the ester and increases when R′ is changed from alkyl to aryl. Heating also induces the elimination reaction:

$$AlkCCl_2CCl_2N{=}PCl_2(OAlk) \xrightarrow{100^\circ} AlkCCl_2C(Cl){=}NP({:}O)Cl_2 + AlkCl$$

[68] I. M. Filatova, E. L. Zaitseva, A. P. Simonov, and A. Ya. Yakubovich, *Zhur. obshchei Khim.*, 1968, **38**, 1304; *J. Gen. Chem. U.S.S.R.*, 1968, **38**, 1256.

Further alcoholysis of the product occurs at both carbon and phosphorus atoms to give $AlkCCl_2C(OAlk){=}NP({:}O)Cl(OAlk)$.[69]

D. Aryl-derivatives of Cyclophosphazenes.—In contrast to the ring-cleavage reactions commonly observed between $N_3P_3Cl_6$ and organometallic reagents, phenyl-lithium undergoes reaction with $N_3P_3F_6$ in ether solution to give the monophenyl-derivative, $N_3P_3F_5Ph$, and a mixture of diphenyl derivatives $N_3P_3F_4Ph_2$ [*cis*- (20), *trans*- (21), and geminal (22) isomers in the

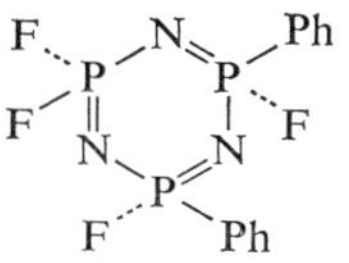
(20)

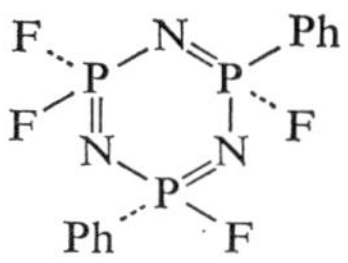
(21)

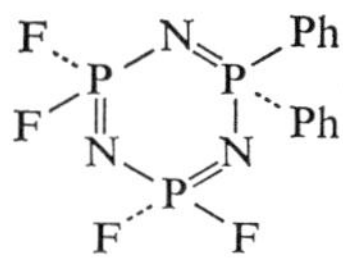
(22)

ratio 3 : 1 : 0·25].[70] With two equivalents of *o*-tolyl-lithium, only the nongeminal *cis*-bis(*o*-tolyl) derivative, $N_3P_3F_4(o\text{-tolyl})_2$, was isolated. Structures were established on the basis of ^{19}F and ^{31}P n.m.r. data and from dipole moment measurements. It was suggested that the previously postulated '*cis*-effect' [51b] should be modified to account for the fact that an exocyclic group which is able to π-bond relatively strongly to phosphorus is able to labilise the phosphorus–halogen bond *cis* to it on the phosphazene ring. This effect can be used to predict the reaction path of dimethylamine and of piperidine with $N_3P_3Cl_6$, in addition to the arylations noted above.

The Friedel–Crafts reactions of $N_3P_3F_6$ with benzene in the presence of aluminium trichloride follow a geminal pattern similar to that observed for $N_3P_3Cl_6$, except that a geminal triphenyl-derivative, $N_3P_3F_3Ph_3$ (23), was

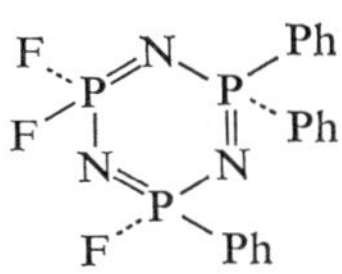
(23)

isolated.[71] No geminal trichloro-derivative, $N_3P_3Cl_3Ph_3$, has been isolated in this way. Structures were again established from ^{19}F and ^{31}P n.m.r. data and correlations were made between this data and the expected π-bonding effects, exocyclic to, and within, the P—N ring.

The ^{19}F n.m.r. spectra of a series of monopentafluorophenyl-derivatives, $N_nP_n(C_6F_5)F_{2n-1}$ (n = 3—8), have been measured and indicate that there is a strong π-withdrawal from the pentafluorophenyl ring in each case (p_π–d_π effect).[72] This effect is dependent on the value of n and varies in a

[69] V. I. Shevchenko, A. A. Koval', and A. V. Kirsanov, *Zhur. obshchei Khim.*, 1968, **38**, 555; *J. Gen. Chem. U.S.S.R.*, 1968, **38**, 541.
[70] C. W. Allen and T. Moeller, *Inorg. Chem.*, 1968, **7**, 2177.
[71] C. W. Allen, F. Y. Tsang, and T. Moeller, *Inorg. Chem.*, 1968, **7**, 2183.
[72] T. Chivers and N. L. Paddock, *Chem. Comm.*, 1968, 704.

similar manner to changes in base strength and ionisation potential in the series $(NPCl_2)_n$.

Some insight has been gained [73] into the reaction of $N_3P_3Cl_6$ with phenylmagnesium bromide, which has previously been shown to lead to the formation of small quantities of $N_3P_3Ph_6$ and relatively large quantities of unidentified materials. The latter have now been shown to consist of a mixture of fully phenylated acyclic phosphazenes which give, *e.g.*, (24) on

$$[Ph_3P{=}N{-}PPh_2{=}N{-}PPh_2{=}\overset{+}{N}H_2]\,X^-$$

$$(X = \text{e.g., } ClO_4)$$

(24)

treatment with Brønsted acids. The compounds $N_3P_3Cl_4Ph_2$ and $N_3P_3Cl_2Ph_4$ undergo reaction with the same reagent more slowly, but it is believed that in the three cases the initial reaction is one involving ring cleavage by phenylmagnesium bromide. With $N_3P_3Cl_6$, the reaction may be represented as shown, although it is possible that two molecules of the organometallic reagent are involved.

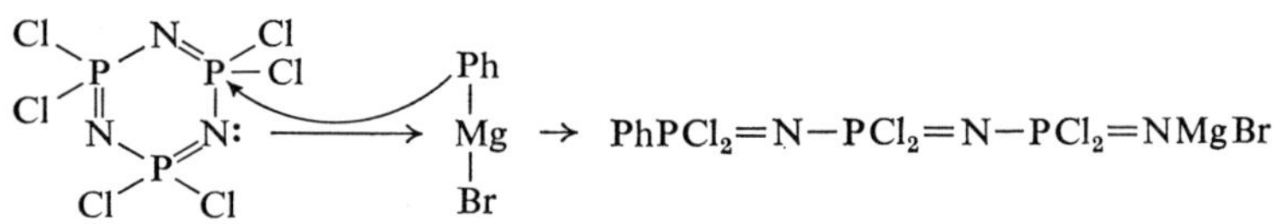

This linear product may then undergo phenylation and subsequent recyclisation:

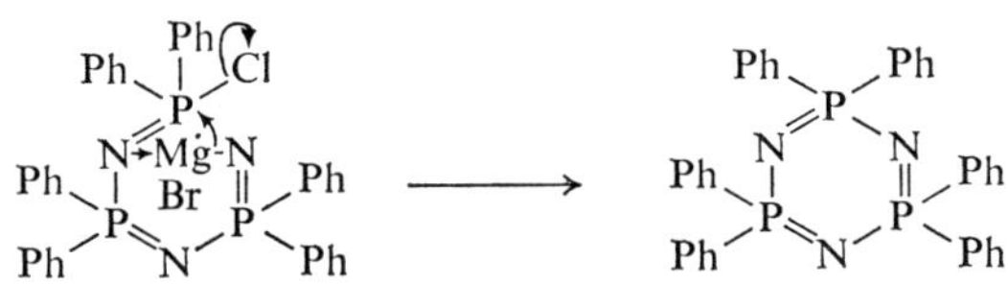

The proportion of acyclic phenylated products is drastically reduced relative to that in the above reactions, and the nature of the cyclic species is completely changed when $N_3P_3Cl_6$ is treated with diphenylmagnesium in 1,4-dioxan.[74, 75] The major product is formulated as (25) and evidence for

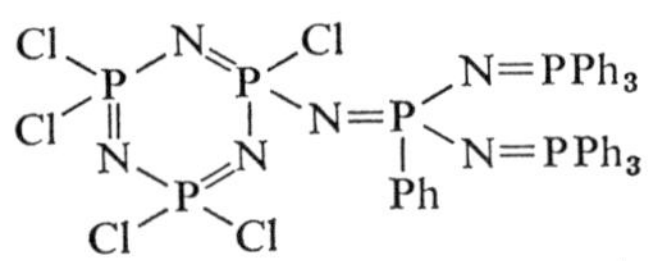

(25)

[73] M. Biddlestone and R. A. Shaw, *J. Chem. Soc.* (*A*), 1969, 178.
[74] M. Biddlestone, R. A. Shaw, and D. Taylor, *Chem. Comm.*, 1969, 320.
[75] M. Biddlestone and R. A. Shaw, *Chem. Comm.*, 1968, 407.

this structure is obtained from degradative hydrolysis, from the ^{31}P spectra obtained on a nuclear-electron double-resonance spectrometer, and from the ^{19}F n.m.r. spectrum of its penta(trifluoroethoxy)-derivative, $N_6P_6Ph_7(OCH_2CF_3)_5$.[74] A minor product has been identified as (26) which displays remarkable stability considering that it contains a P—P bond which is not cleaved by halogens.[75]

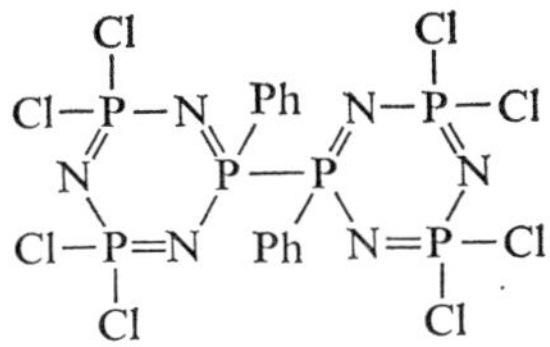

(26)

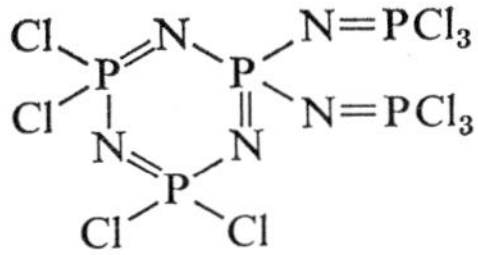

(27)

The arylation of the trichlorophosphazenyl-derivatives, $N_3P_3Cl_5N{=}PCl_3$ and geminal $N_3P_3Cl_4(N{=}PCl_3)_2$ (27), by arylmagnesium halides proceeds by preferential reaction at the terminal phosphorus atom:[76]

$$N_3P_3Cl_5(N{=}PCl_3) \xrightarrow[Et_2O]{ArMgBr} N_3P_3Cl_5(N{=}PAr_3) \quad (Ar = Ph)$$

$$N_3P_3Cl_4(N{=}PCl_3)_2 \xrightarrow[Et_2O]{ArMgBr} N_3P_3Cl_4(N{=}PAr_3)_2 \quad (Ar = Ph,\ p\text{-tolyl})$$

With longer reaction times, ring phenylation was reported to occur geminally to the $—N{=}PPh_3$ group to give $N_3P_3Cl_4Ph(N{=}PPh)_3$. The latter compound and $N_3P_3Cl_4(N{=}PPh_3)_2$ were unreactive to further attempts at arylation. Reaction patterns and ^{31}P n.m.r. evidence were cited in favour of a geminal structure for the derivatives $N_3P_3ClX(N{=}PPh_3)$ (X = Cl, $—NH_2$, Ph, and $—N{=}PPh_3$).

The ring-contracted geminal derivative, $N_3P_3Cl_4Ph(N{=}PPh_3)$, has also been obtained in low yield by the Friedel–Crafts phenylation of $N_4P_4Cl_8$:[77]

$$N_4P_4Cl_8 \xrightarrow[Et_3N]{AlCl_3\text{–}PhH} [X] + N_3P_3Cl_4Ph(N = PPh_3)$$

The second product, X, was not identified, but treatment with dimethylamine gave the tetramer-derivative, $N_4P_4Cl_2Ph_5{\cdot}NMe_2$, which contains a $\equiv PPh{\cdot}NMe_2$ group, but the relative orientations of the remaining phenyl groups were not established.

The hydrolysis of $N_3P_3ClPh_5$:

$$N_3P_3Cl{\cdot}Ph_5 + H_2O + NC_5H_5 \longrightarrow N_3P_3(OH)Ph_5 + C_5H_5NHCl$$

in acetone has been the subject of a kinetic study which suggests that a nucleophilic base-catalysed mechanism may be operative:[78]

76 M. K. Feldt and T. Moeller, *J. Inorg. Nuclear Chem.*, 1968, **30**, 2351.
77 V. B. Desai, R. A. Shaw, and B. C. Smith, *Angew. Chem. Internat. Edn.*, 1968, **7**, 887.
78 C. D. Schmulbach and V. R. Miller, *Inorg. Chem.*, 1968, **7**, 2189.

$$N_3P_3ClPh_5 + NC_5H_5 \underset{k_2}{\overset{k_1}{\rightleftharpoons}} N_3P_3Ph_5\overset{+}{N}C_5H_5 + \overset{-}{Cl}$$

$$N_3P_3Ph_5\overset{+}{N}C_5H_5 + H_2O \xrightarrow{k_3} N_3P_3(OH)Ph_5 + C_5H_5\overset{+}{N}H$$

Supporting evidence for the intermediate, $N_3P_3Ph_5\overset{+}{N}C_5H_5$ was obtained by the isolation of the perchlorate salt, $N_3P_3Ph_5\overset{+}{N}C_5H_5\ \overset{-}{ClO_4}$.

The physical properties and thermal dissociation of the adducts, $N_3P_3Ph_6,3C_2H_2Cl_4$ [79] and $N_3P_3Ph_4Cl_2,C_2H_2Cl_4$ [80] have been studied; they are believed to be clathrates. The reactions of nongeminal $N_4P_4Cl_2Ph_6$ towards water, amines, and caesium fluoride have been studied as summarised:[81]

$$N_4P_4Cl_2Ph_6 \underset{PCl_5-CHCl_3 \text{ or } SOCl_2-CHCl_3}{\overset{\text{pyridine}-H_2O}{\rightleftharpoons}} N_4P_4(OH)_2Ph_6$$

(two isomers of $N_4P_4Cl_2Ph_6$ were isolated from the reaction with $SOCl_2$, but their structures were not established).

$$N_4P_4(OH)_2Ph_6 \xrightarrow{PBr_5} N_4P_4Br_2Ph_6$$

$$N_4P_4Cl_2Ph_6 \xrightarrow[CH_3CN]{CsF} N_4P_4F_2Ph_6$$

$$N_4P_4Cl_2Ph_6 \xrightarrow[CHCl_3]{NHRR'} N_4P_4(NRR')_2Ph_6$$

(R = R′ = H; R = H, R′ = Me; R = R′ = Me; R = R′ = Et;
R = Me, R′ = Ph; RR′ = C_5H_{10})

The preparation of the *cis*- and *trans*-isomers of $N_3P_3Br_3Ph_3$ has been described in detail:[82]

$$PhPBr_4 + NH_4Br \longrightarrow N_3P_3Br_3Ph_3$$

Interconversion of these two isomers was observed in refluxing acetonitrile or bromobenzene solution.[83] It was suggested that isomerisation occurred by a solvent-assisted ionisation mechanism.

4 Formation of Polymeric Phosphazenes

Current work on the synthesis of polymeric phosphazenes is largely reported in the patent literature. Two reviews [84, 85] of recent developments in this aspect of polymer chemistry have appeared as well as a more general, but useful, survey of the technical uses of phosphorus-nitrogen compounds.[86]

[79] R. D. Whitaker, A. J. Barreiro, P. A. Furman, W. C. Guida, and E. S. Stallings, *J. Inorg. Nuclear Chem.*, 1968, **30**, 2921.
[80] R. D. Whitaker and W. C. Guida, *J. Inorg. Nuclear Chem.*, 1969, **31**, 875.
[81] A. J. Bilbo, C. M. Grieve, D. L. Herring, and D. E. Salzbrunn, *Inorg. Chem.*, 1968, **7**, 2670.
[82] P. Nannelli, S.-K. Chu, B. S. Manhas, and T. Moeller, *Inorg. Synth.*, 1968, **11**, 201.
[83] B. S. Manhas, S.-K. Chu, and T. Moeller, *J. Inorg. Nuclear Chem.*, 1968, **30**, 322.
[84] H. R. Allcock, *Chem. Eng. News*, 1968, **46**, 68.
[85] H. Saito, *Kobunshi*, 1968, **17**, 391 (*Chem. Abs.*, 1968, **69**, 59,570).
[86] H.-G. Horn, *Chem. Ztg. Chem. App.*, 1969, **93**, 241.

Interest in the polymerisation of oligomeric cyclophosphazenes to give linear polymers, $(NPCl_2)_n$ continues. It is found[87] that highly purified $N_3P_3Cl_6$ polymerises only slowly at 300°, whereas with relatively crude $N_3P_3Cl_6$ polymerisation was rapid at 270°. The polymerisation process is also accelerated by the presence of oxygen. Acyclic ionic phosphazenes with boron- or aluminium-containing anions[88] have been prepared by the reactions:

$$PCl_5 + (NPCl_2)_n \longrightarrow [Cl(Cl_2P{=}N)_nPCl_3]^+[PCl_6]^- \quad (n = 3, 4)$$

$$\big\downarrow MCl_3$$

$$(M = Al \text{ or } B) \quad [Cl(Cl_2P{=}N)_nPCl_3]^+[MCl_4]^-$$

The resultant fluids were heated and the degree of polymerisation of the cation studied. The pyrolysis of mixtures of $NbNCl_2$ or $NbOCl_3$ and $(NPCl_2)_{10}$ has been studied.[89]

There are several reports of condensation polymerisations involving diols and oligomeric cyclophosphazenes, *e.g.*:

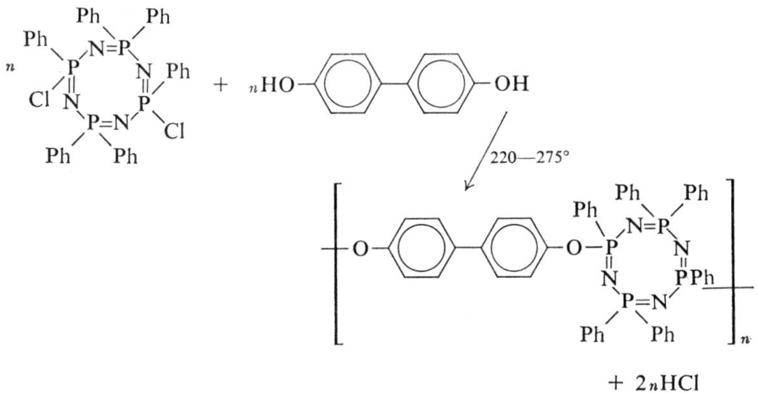

Simple polymer fractionation gave samples with a molecular weight of *ca.* 500,000. Differential thermal analysis and thermogravimetric analysis on this polymer suggested that thermal degradation is initiated in the phosphazene ring.[90] Similar condensation studies were also conducted with analogous polymers using as bridging groups the following difunctional reagents:[90]

87 R. O. Colclough and G. Gee, *J. Polymer Sci., Part C*, 1968, **7**, 3639.
88 E. F. Moran, *J. Inorg. Nuclear Chem.*, 1968, **30**, 1405.
89 U. A. Buslaev, B. V. Levin, S. M. Sinitsyna, M. A. Polikarpova, Z. G. Rumyautseva, and V. V. Mironova, *Izvest Akad Nauk S.S.S.R., Neorg. Materialy*, 1968, **4**, 1706 (*Chem. Abs.*, 1969, **70**, 53,566).
90 A. J. Bilbo, C. M. Douglas, N. R. Fetter, and D. L. Herring, *J. Polymer Sci., Part A*, 1968, **6**, 1671.

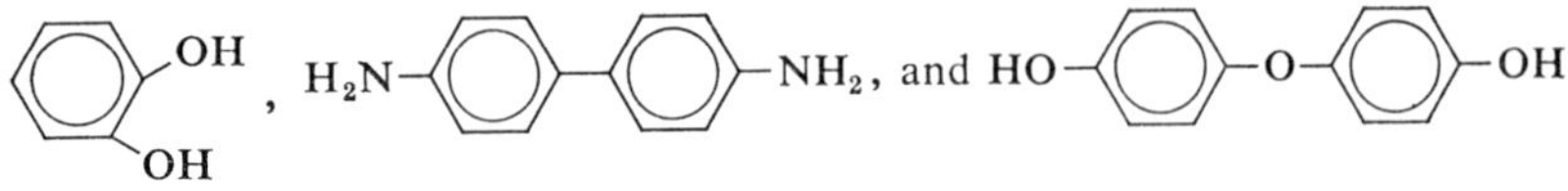

The condensation reactions of $N_3P_3Cl_6$ with 2,6-methylol-*o*-cresol in warm dioxan have been studied.[91] The insoluble polymeric materials obtained from the pyrolysis of $N_3P_3Cl_6$ give, on treatment with ethylene glycol, materials soluble in chloroform, and two hypotheses have been proposed to account for this observation.[92] Various diols and triols of boron have been condensed with $N_3P_3Cl_6$ and $N_3P_3(OBu)_6$ to produce novel polymers,[93] *e.g.*:

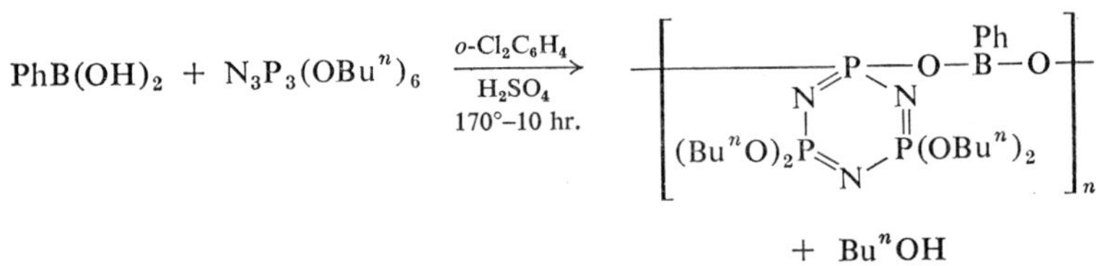

Polymers with useful chemical and thermal stability have been obtained by the reaction of pyrolysed $N_3P_3Cl_6$ with a mixture of sodium fluoroalkoxides, NaOR (R = CH_2CF_3 and $CH_2C_3F_7$).[94] The copolymers obtained have a relatively large proportion of $[NP(OCH_2CF_3)(OCH_2C_3F_7)]$ units. The reader is also referred to the patent literature, where examples of polymeric phosphazenes obtained from reactions involving alcohols,[95–98] amines,[99,100] and various alkyl phosphorus compounds[101,102] are to be found.

5 Structure and Bonding

The output of detailed structural information on the phosphazenes steadily increases. This continues to provide some impetus for theories of bonding within these structures, and a useful review[103] of bonding in cyclic phos-

91 M. Kajiwara and H. Saito, *J. Chem. Soc. Japan, Ind. Chem. Sect.*, 1968, **71**, 1470.
92 M. Pornin, *Bull. Soc. chim. France*, 1968, 759.
93 A. V. Deryabin, S. M. Zhivukhin, V. V. Kireev, and G. S. Kolesnikov, *Plast. Massy.*, 1968, **3**, 29 (*Chem. Abs.*, 1968, **69**, 3200).
94 S. H. Rose, *J. Polymer Sci., Part B, Poly. Letters*, 1968, **6**, 837.
95 M. E. Hull and D. W. Hoch, U.S.P. 3,402,145/1968 (*Chem. Abs.*, 1968, **69**, 103,783).
96 H. R. Allcock and R. L. Kugel, U.S.P. 3,370,020/1968 (*Chem. Abs.*, 1968, **68**, 69,555).
97 S. M. Zhivukhin and S. I. Belykh, U.S.S.R.P. 212,522/1968 (*Chem. Abs.*, 1968, **69**, 19,998).
98 D. J. Jaszka, U.S.P. 3,392,214/1968 (*Chem. Abs.*, 1968, **69**, 59,945).
99 Yu. V. Azhikina, M. Ya. Konoleva, B. M. Maslenikov, and L. Ya. Kulikova, *Izvest. Akad. Nauk S.S.S.R., Neorg. Materialy*, 1968, **4**, 1711 (*Chem. Abs.*, 1969, **70**, 53,572).
100 H. R. Allcock and R. L. Kugel, U.S.P. 3,364,189/1968 (*Chem. Abs.*, 1968, **68**, 50,558).
101 H. H. Sisler, Fr.P. 1,497,470/1967 (*Chem. Abs.*, 1969, **69**, 52648).
102 E. I. Sokolov, V. A. Bartashiv, A. L. Klebanskii, T. I. Saratovkina, T. L. Chernyavskaya, and V. N. Sharov, U.S.S.R.P. 216,710/1968 (*Chem. Abs.*, 1969, **70**, 11,808).
103 K. A. R. Mitchell, *Chem. Rev.*, 1969, **69**, 157.

phazenes has appeared in connection with the more general aspects of 3*d*-orbital bonding. The utilisation of phosphorus 3*d*-orbitals for bonding in the planar ring structures, $N_3P_3F_6$, $N_4P_4F_8$, and $N_3P_3Cl_6$, has been examined in detail[104] and it was suggested that the $3d_{xz}$- and $3d_{x^2-y^2}$-orbitals (for axes see ref. 104) are the most effective in π-bonding. These conclusions are consistent with a bonding scheme that emphasises cyclic electron delocalisation within the two π-systems, symmetric and antisymmetric to reflection in the molecular plane.

Recently reported molecular parameters for cyclic and acyclic phosphazenes are summarised in Table 2. In all molecules where each phosphorus atom has the same pair of substituents the endocyclic P—N bond lengths are equal within experimental error.

There have been a number of attempts to identify the ring conformations of $N_4P_4Cl_8$ in solution by vibrational spectroscopy. The most likely conformation has a point group D_{2d} (saddle shape),[117, 118] but variable symmetry cannot be ruled out.[119] For samples examined in the solid state, agreement is reached with *X*-ray crystallographic evidence on the structure of the $K(S_4)$ and $T(C_{2h})$ forms.[118] These structures have been further confirmed by n.q.r. studies and these results have been tentatively interpreted in terms of differing π-character in the axial and equatorial P—Cl bonds of the *T* form.[120] It has been suggested that $N_4P_4Cl_8$ has a planar ring (D_{4h} symmetry) in the vapour phase.[118]

The vibrational spectra of $N_5P_5Br_{10}$ have been interpreted in terms of a distorted D_{5h} symmetry.[121]

6 Miscellaneous Physical Measurements

The variation of vapour pressures of $N_6P_6Cl_{12}$ and $N_4P_4Br_8$ over a range of temperatures has been measured. These observations give lattice energies of $25{\cdot}3 \pm 0{\cdot}42$ and $29{\cdot}4 \pm 0{\cdot}47$ kcal/mole respectively.[122]

104 K. A. R. Mitchell, *J. Chem. Soc.* (*A*), 1968, 2683.
105 M. I. Davis and J. W. Paul, *Acta Cryst. Supp.*, 1969, **A25**, S116.
106 A. J. Wagner and A. Vos, *Acta Cryst.*, 1968, **B24**, 707.
107 A. W. Schlueter and R. A. Jacobson, *J. Chem. Soc.* (*A*), 1968, 2317.
108 F. R. Ahmed, P. Singh, and W. H. Barnes, *Acta Cryst.*, 1969, **B25**, 316.
109 A. J. Wagner and A. Vos, *Acta Cryst.*, 1968, **B24**, 1423.
110 N. L. Paddock, J. Trotter, and S. H. Whitlow, *J. Chem. Soc.* (*A*), 1968, 2227.
111 G. J. Bullen, P. R. Mallinson, and A. H. Burr, *Chem. Comm.*, 1969, 693.
112 J. Trotter, S. H. Whitlow, and N. L. Paddock, *Chem. Comm.*, 1969, 695.
113 J. C. van de Grampel and A. Vos, *Acta Cryst.*, 1969, **B25**, 651.
114 U. Klement and A. Schmidpeter, *Z. Naturforsch*, 1968, **23b**, 1610.
115 M. L. Ziegler, *Z. anorg. Chem.*, 1968, **362**, 257.
116 J. W. Cox and E. R. Corey, *Chem. Comm.*, 1969, 205.
117 T. R. Manley and D. A. Williams, *Spectrochim. Acta*, 1969, **24**, *A*, 1661.
118 H. C. Hisatsune, *Spectrochim. Acta*, 1969, **25**, *A*, 301.
119 W. P. Griffith and K. J. Rutt, *J. Chem. Soc.* (*A*), 1968, 2331.
120 M. Dixon, H. D. B. Jenkins, J. A. S. Smith, and D. A. Tong, *Trans. Faraday Soc.*, 1967, **63**, 2852.
121 G. E. Coxon and D. B. Sowerby, *Spectrochim. Acta*, 1968, **24**, *A*, 2145.
122 S. Cotson and K. A. Hodd, *J. Inorg. Nuclear Chem.*, 1969, **31**, 245.

Table 2 *Molecular structures of phosphazenes obtained by diffraction methods*

Compound	*Average bond distances* (Å)[a]		*Average bond angles* (°)[a]		*Comments*	*Reference*
	P—N	P—X	$\widehat{NPN}$	$\widehat{XPX}$		
$N_3P_3Cl_6$	1·585 (0·010)	2·006 (0·007)	119·7 (0·3) (average ring angle)	101·8 (1·2)	Preliminary report of electron diffraction (e.d.) study. Slight chair of C_{3V} symmetry, ring puckered by 6°. Accuracy better than previous e.d. studies	105
$N_4P_4Cl_8$ (*T* form)	1·559 (0·012)	1·992 (0·004)	120·5 (0·7)	103·1 (0·2)	*X*-Ray study. Chair-shaped ring with approximately C_{2h} symmetry. Previously studied (*K* form) has boat conformation. Two independent P—N—P angles, 133·6 and 137·6 (0·8°), explained by Cl—Cl interaction. See i.r. studies below	106
$N_3P_3Ph_6$	1·597 (0·006)	1·804 (0·007)	117·8 (0·3)	103·8 (0·3)	*X*-Ray study. Slight chair conformation. Effects of electronegativity on P—N ring parameters discussed for halogenophenylphosphazenes	108

$N_5P_5Cl_{10}$	1·521 (0·013)	1·961 (0·009)	118·4 (0·8)	102·2 (0·4)	*X*-Ray study, nearly planar. Relatively short P—N distance and large P—N—P angle (148·6°) explained in terms of enhanced π' bonding	107
$N_6P_6(NMe_2)_{12}$	1·563 (0·010)	1·669 (0·010)	120·0 (0·5)	103	*X*-Ray study, molecular symmetry $\bar{3}$. Highly puckered ring analogous to boat. Endocyclic P—N bond lengths shorter and $P\widehat{N}P$ [147·5 (0·7°)] longer than in $N_4P_4(NMe_2)_8$. Again explained in terms of π'-bonding	109
$N_8P_8(OMe)_{16}$	1·561 (0·014)	1·576 (0·013)	116·7 (0·7)	99·9 (0·4)	*X*-Ray study. Centrosymmetric. Two approximate planar and parallel six-atom segments joined by a 'step'. Re-entrant N-atoms at each end of 'step'	110
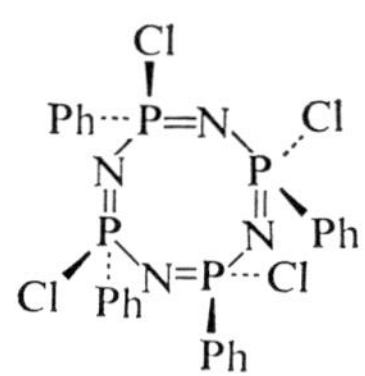	1·57	P—Cl 2·03 P—Ph 1·79	120	104	*X*-Ray study. *β-trans*-Structure confirms earlier assignments. Ring has chair conformation similar to $N_4P_4Cl_8$ (above). Ring symmetry ~ C_{2h}. Estimated deviations ⩽0·1 Å for lengths and <0·6° for angles	111

[a] Estimated deviations in parentheses.

Table 2 (*cont.*)

Compound	*Average bond distances* (Å)[a] P—N	P—X	*Average bond angles* (°)[a] $\widehat{NPN}$	$\widehat{XPX}$	*Comments*	*Reference*
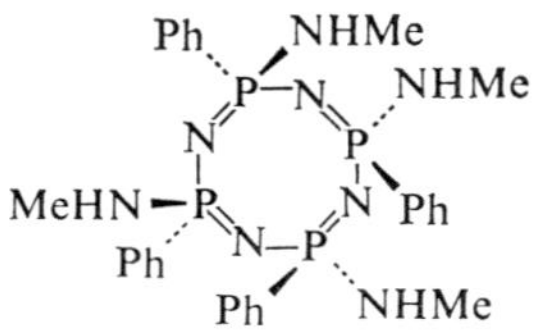	1·59	P—N 1·68 P—Ph 1·81	120	106	*X*-Ray study. *β-trans*-Structure in chair-conformation. Obtained from $N_4P_4Cl_4Ph_4$ above, therefore suggests that net configuration is retained on aminolysis	111
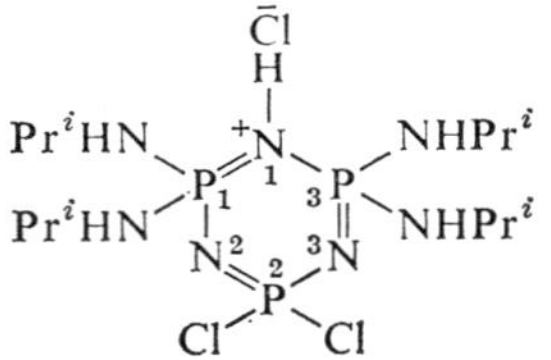	$N_{(1)}P_{(1)}$ 1·67 $P_{(1)}N_{(2)}$ 1·58 $N_{(2)}P_{(2)}$ 1·56 (0·005)	P—N 1·62 (0·005) P—Cl 2·00 (0·002)	$N_{(1)}\widehat{P_{(1)}}N_{(2)}$ 107 (0·4) $N_{(2)}\widehat{P_{(2)}}N_{(3)}$ 120 (0·4)	104	*X*-Ray study. Protonation on ring nitrogen atom. $N_{(1)} \cdots H = 1{\cdot}04\ (0{\cdot}025)$ Å. Ring nonplanar with distorted boat form	62
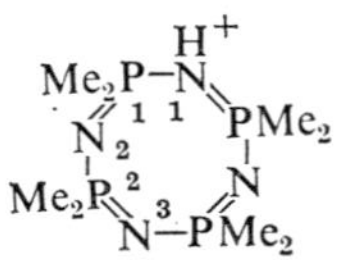 $CoCl_4^-$	$N_{(1)}P_{(1)}$ 1·70 $N_{(2)}P_{(1)}$ 1·54 $N_{(2)}P_{(2)}$ 1·61 $N_{(3)}P_{(2)}$ 1·58 (0·03)				*X*-Ray study. Two types of P—N rings, one in tub and one in saddle conformation	112

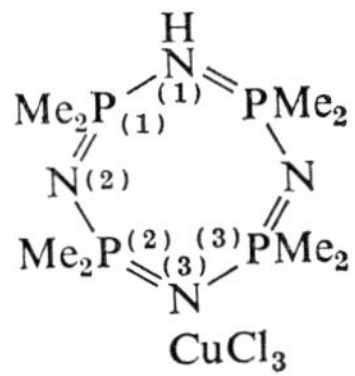	$N_{(1)}P_{(1)}$ 1·67 $N_{(2)}P_{(1)}$ 1·56 $N_{(2)}P_{(2)}$ 1·60 $N_{(3)}P_{(2)}$ 1·63				*X*-Ray study. One type of P—N ring in tub conformation. Proton and $CuCl_3$ group covalently bonded	112
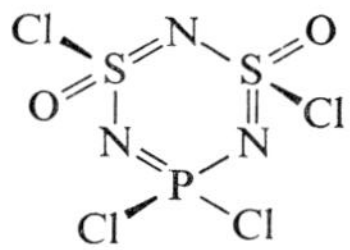	1·585 (0·013)	1·957 (0·006)	115·3 (0·7)	104·4 (0·3)	*X*-Ray study. Ring has distorted chair configuration	113
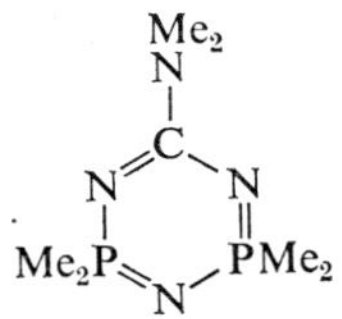					*X*-Ray study. Planar ring. *N*-Methyl groups in plane of ring. Crystal data only	114
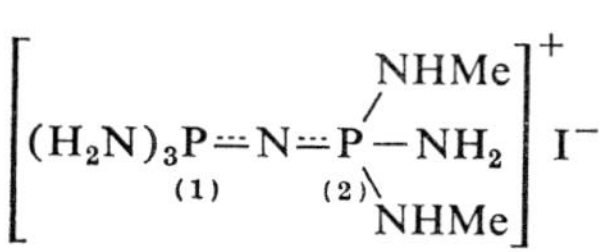	$NP_{(1)}$ 1·583 (0·013) $NP_{(2)}$ 1·539 (0·020)		$P_{(1)}\widehat{N}P_{(2)}$ 129·1		*X*-Ray study. Both methylamino-groups are on same P-atom. One of $P—NH_2$ bonds exceptionally short [1·490 (0·038) Å]	115

[a] Estimated deviations in parentheses.

Table 2 (*cont.*)

Compound	*Average bond distances* (Å)[a] P—N	*Average bond distances* (Å)[a] P—X	*Average bond angles* (°)[a] $\widehat{NPN}$	*Average bond angles* (°)[a] $\widehat{XPX}$	*Comments*	*Reference*
$[Ph_2\underset{(1)}{P}(NH_2)\cdots N\cdots \underset{(2)}{P}(NH_2)Ph_2]^+ Cl^-$	$NP_{(1)}$ 1·58 (0·01) $NP_{(2)}$ 1·57 (0·01)		$P_{(1)}\widehat{N}P_{(2)}$ 136		*X*-Ray study. Bridging P—N bonds equal within experimental error. Terminal P—N bonds 1·64 and 1·66 (0·01) Å	116

[a] Estimated deviations in parentheses.

Basicity measurements have been reported on amino-derivatives of the geminal diphenyl-, $N_3P_3Cl_4Ph_2$, and of the geminal tetraphenyl-cyclotriphosphazatriene, $N_3P_3Cl_2Ph_4$. Amino-substituents used included $-NH_2$, $-NHMe$, $-NHEt$, $-NHC_6H_{11}$, $-NHBu^t$, $-NMe_2$, and $-NC_5H_{10}$. Structures were assigned to positional isomers, $N_3P_3Cl_2Ph_2R_2$ (R = NHMe or $-NC_5H_{10}$) using the deduction that a geminal isomer (28) is generally more basic than the corresponding nongeminal isomer[123] (29). Measure-

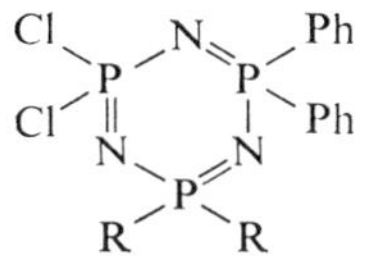

(28)

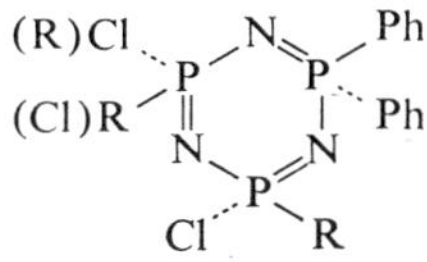

(29)

ments of the effect on basicity of replacing chlorine atoms by other groups in these cyclotriphosphazatrienes has permitted the evaluation of substituent constants α_R and γ_R; these represent the 'increase' in basicity of a ring nitrogen atom, when a chlorine atom on a phosphorus atom α or γ to the nitrogen atom is replaced by the group R. These ideas have been extended to derivatives of $N_3P_3Cl_6$ with mixed substituents including $-NMe_2$, $-OMe$, $-OPh$, $-Ph$, $-SEt$, and $-SPh$.[124] The u.v. spectra and basicities of monophosphazenes of the type $p\text{-}XC_6H_4P(Et_2){=}N\cdot C_6H_4\cdot NO_2\text{-}p$ (X = NMe_2, OMe, Me, H, Cl, and CO_2Me), have also been measured and the basicities correlated with σ *p*-constants for the corresponding X sub-substituent in an attempt to assess the conjugative ability of the P—N bond.[125] The same problem has also been studied by ^1H n.m.r. and u.v. spectroscopy and by dipole moment determination on a similar series of compounds $RR'_2P{=}NC_6H_4NO\text{-}p$ ($R' = C_3H_7$; $R = p\text{-}ZC_6H_4$, Z = H, Cl, Br, NMe_2; or $R = C_3H_7$; $R' = p\text{-}ZC_6H_4$, Z = H, Cl, OMe, NMe_2.)[126]

Attempts have been made to relate the i.r.[127] and u.v. spectra[128] of disubstituted derivatives of $N_3P_3Cl_6$ to the π-electron distribution in cyclotriphosphazatriene rings. In both cases, the importance of a three-centred π-bond system was emphasised.

[123] D. Feakins, S. N. Nabi, R. A. Shaw, and P. Watson, *J. Chem. Soc.* (*A*), 1968, 10.
[124] D. Feakins, W. A. Last, S. N. Nabi, R. A. Shaw, and P. Watson, *J. Chem. Soc.* (*A*), 1969, 196.
[125] T. G. Edel'man and B. I. Stepanov, *Zhur. obshchei Khim.*, 1968, **38**, 195; *J. Gen. Chem. U.S.S.R.*, 1968, **38**, 197.
[126] H. Goetz and D. Probst, *Annalen*, 1968, **715**, 1.
[127] P. Pulay, B. Lakatos, G. Toth, A. Hesz, and Z. Vetessey, *Magyar Kém. Floyóirat*, 1968, **74**, 479.
[128] B. Lakatos, A. Hesz, Z. Vetessey, and G. Horbath, *Magyar Kém. Folyóirat*, 1968 **74**, 468 (*Chem. Abs.*, 1969, **70**, 15,622).

10

Radicals, Photochemistry, and Deoxygenation Reactions

BY R. S. DAVIDSON

1 Photochemistry

Further examples (1—3) of the photoinduced cleavage of P—H bonds to give phosphorus radicals have been reported.[1, 2, 3] When phosphine is flashed (energy *ca.* 1 kjoule) in the presence of cyanogen and nitrogen, two

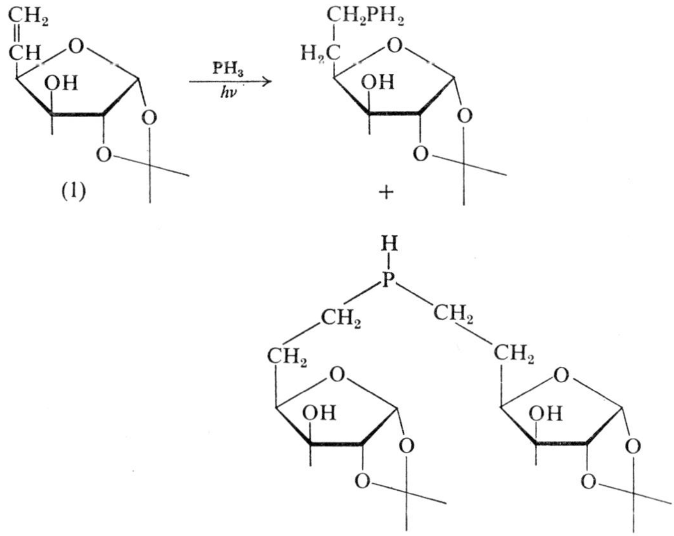

(1)

Ref. 1

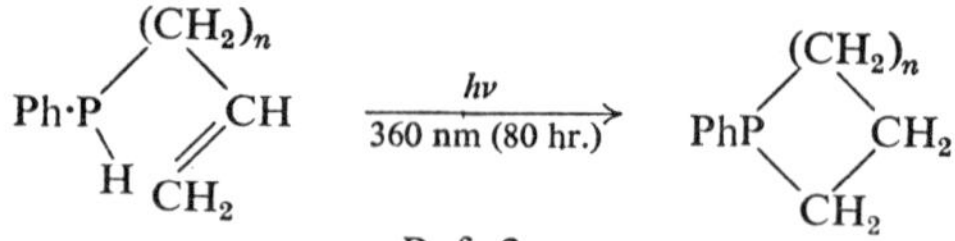

Ref. 2

(2)

n = 2,3,and 4

[1] R. L. Whistler, C.-C. Wang, and S. Inokawa, *J. Org. Chem.*, 1968, **33**, 2495.
[2] Shell Internationale Research Maatschappij N.V., F.P. 1,488,936 (*Chem. Abs.*, 1968, **69**, 36,252X).
[3] Monsanto Co., Bt.P. 1,101,334 (*Chem. Abs.*, 1968, **69**, 3004q).

$$\underset{(3)\quad \text{Ref. 3}}{Ph_2P(:S)H + CH_2{=}CH{-}R} \longrightarrow \underset{R = CN \text{ or } CH_2OH}{Ph_2P(:S)CH_2{\cdot}CH_2R}$$

transients are formed[4] having strong absorptions at 301·4 and 285·7 nm and 336 and 316 nm. The first spectrum was assigned to the PCN radical and this was supported by observation of the same transient when a phosphorus trichloride–cyanogen–nitrogen mixture was flashed. The second absorption was assigned to the HPCN radical. Radicals produced by irradiation of chlorophosphines at low temperatures have been examined by electron spin resonance spectroscopy.[5] Photolysis of phosphorus trichloride gave both the dichlorophosphine (4) and the tetrachlorophosphine (5) radical. The formation of the latter, by reaction of chlorine atoms with phosphorus trichloride, was suppressed by the use of xenon as a diluent.

$$PCl_3 \xrightarrow{h\nu} \underset{(4)}{\dot{P}Cl_2} + Cl_2\cdot$$

$$Cl\cdot + PCl_3 \longrightarrow \underset{(5)}{\dot{P}Cl_4}$$

Photolysis of dichloromethylphosphine gave the chloromethylphosphine radical. When phosphorus trichloride, in the presence of cyclohexane, is energised by radiation from a ^{60}Co source, dichlorocyclohexylphosphine is formed and is the major product.[6] The yield of the product is dependent upon the dosage of the radiation.

Attempts have been made[7] to prepare the phosphonium salt (6) which has been previously postulated as being formed in the photolysis of triphenylphosphine.[8] Reaction of tetraphenylphosphonium bromide with

$$Ph_4\overset{+}{P}\overset{-}{Br} + \overset{+}{K}\overset{-}{P}Ph_2 \longrightarrow \underset{(6)}{Ph_4\overset{+}{P}\overset{-}{P}Ph_2} \longrightarrow 2Ph_3P$$

potassium diphenylphosphide in tetrahydrofuran at the reflux temperature of the solvent gave triphenylphosphine as the sole product. Reaction did not occur at room temperature. The thermal instability of the salt (6) leads one to question its role in the photolysis of triphenylphosphine.

Two radicals (7 and 8) formed by photolysis of triethyl phosphite have been positively identified by e.s.r.[9] These radicals can only have been

$$\underset{(7)}{(EtO)_2P{-}\dot{O}} \qquad \underset{(8)}{(EtO)_2\overset{+\bullet}{P}{-}\bar{O}}$$

[4] N. Basco and K. K. Yee, *Chem. Comm.*, 1968, 152.
[5] G. F. Kokoszka and F. E. Brinckman, *Chem. Comm.*, 1968, 349.
[6] E. I. Babkina and I. V. Vereshchinskii, *Zhur. obshchei Khim.*, 1968, **38**, 1772 (*Chem. Abs.*, 1969, **70**, 4233W).
[7] L. Horner, P. Beck, and R. Luckenbach, *Chem. Ber.*, 1968, **101**, 2899.
[8] L. Horner and J. Dörges, *Tetrahedron Letters*, 1965, 763; M. L. Kaufman and C. E. Griffin, *Tetrahedron Letters*, 1965, 769.
[9] K. Terauchi and H. Sakurai, *Bull. Chem. Soc. Japan*, 1968, **41**, 1736.

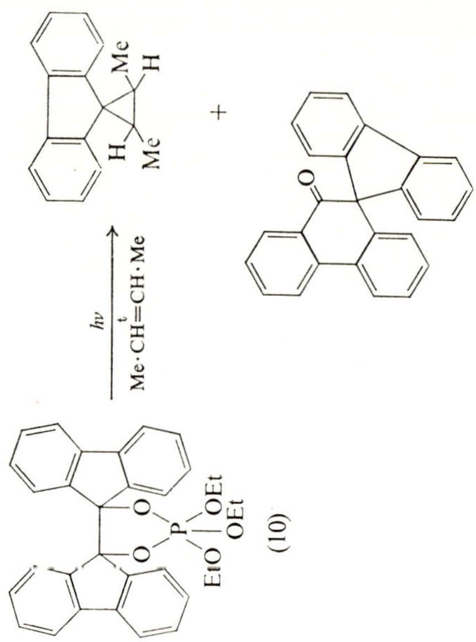

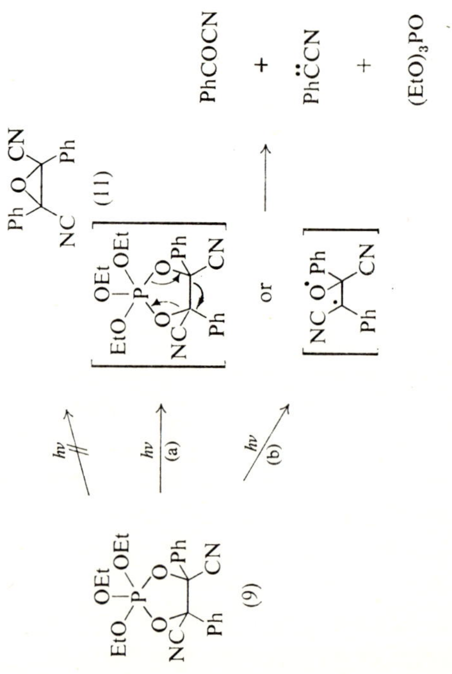

formed by carbon—oxygen bond fission which contrasts with the suggestion made previously[10] to account for the product of the reactions, that phosphorus—oxygen bond fission occurs. Photolysis of the phospholans (9 and 10) to give carbene intermediates has been reported[11] and this undoubtedly occurs by phosphorus—oxygen bond cleavage. Formation of the carbene *via* the epoxide (11) was ruled out but it is still uncertain as to whether the carbene is formed by a concerted (pathway a) or a homolytic (pathway b) route.

The photorearrangement of β-ketoethyl phosphites (12) to dimethyl vinyl phosphates has been reported[12] and suggested as occurring by attack of the excited carbonyl group upon the phosphorus atom. When solvents are used which contain readily abstracted hydrogen atoms, intermolecular reduction of the carbonyl group is an important competing reaction. When cyclohexane was used as solvent, some of the cyclohexyl radicals formed in this way attacked (12) to give phosphate (13). Dimethyl phosphite is also produced in the photoreactions and its yield is relatively unaffected

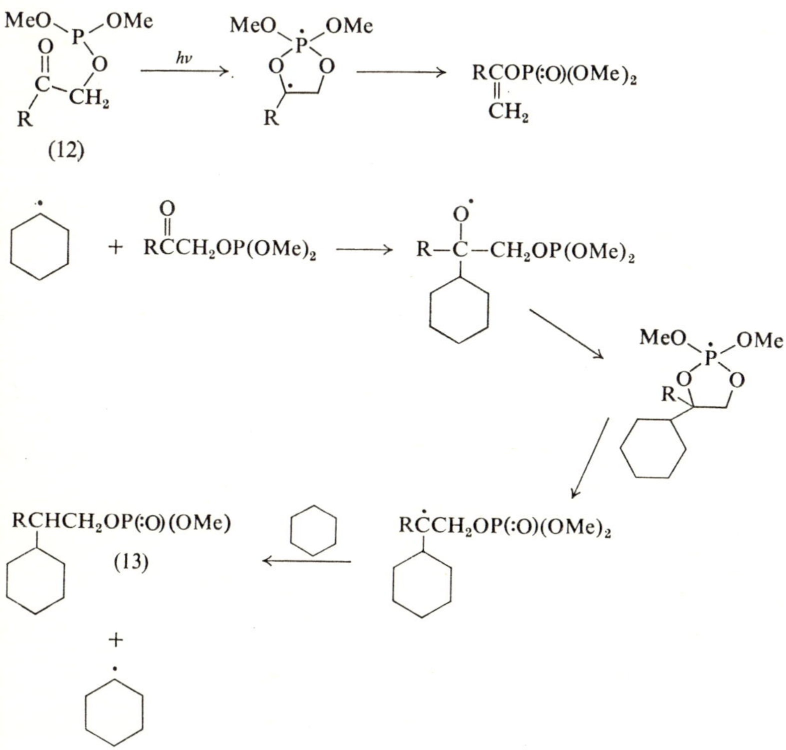

[10] R. B. LaCount and C. E. Griffin, *Tetrahedron Letters*, 1966, 5049.
[11] P. Petrellis and G. W. Griffin, *Chem. Comm.*, 1968, 1099.
[12] C. E. Griffin, W. G. Bentrude, and G. M. Johnson, *Tetrahedron Letters*, 1969, 969.

by changes in solvent which affect the yield of vinyl phosphate. This suggests that the two compounds are formed by different processes. Quenching studies indicate that both are formed *via* the triplet state of the carbonyl group and this was confirmed, in the case of the phosphate, by the observation that the reaction could be sensitised by acetone and benzophenone.

The photoreduction of β-ketoalkyl phosphonates (14) to give the corresponding β-hydroxy compounds or pinacols has been reported.[13] The reaction was suppressed by the addition of triplet quenchers. A bimolecular

$$\underset{(14)}{RCOCH_2P(:O)(OR)_2} \xrightarrow[\text{ether}]{h\nu} RCH(OH)\cdot CH_2P(:O)(OR)_2 \quad \text{or} \quad \begin{array}{l} RC(OH)\cdot CH_2P(:O)(OR)_2 \\ | \\ RC(OH)\cdot CH_2P(:O)(OR)_2 \end{array}$$

hydrogen abstraction rate constant of $6{\cdot}83 \times 10^6$ M^{-1} sec^{-1} was obtained for the reduction of β-keto-n-propylphenylphosphonate by diethyl ether.

Further work on the photoinduced hydrolysis of substituted phenyl phosphates (15) has shown[14] that it occurs by attack of water at the

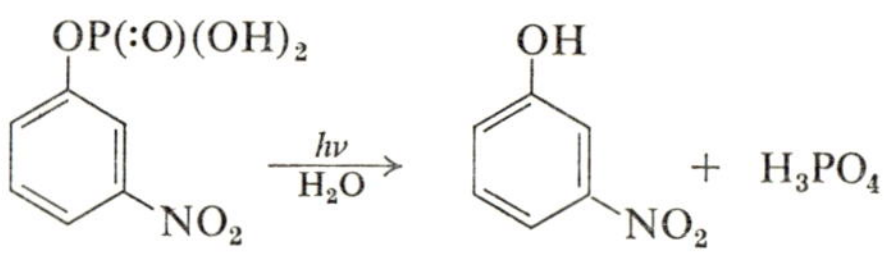

(15)

phosphorus atom which is followed by phosphorus—oxygen bond fission. In strongly alkaline solution hydrolysis occurs by attack of hydroxyl ion at the phosphate-substituted carbon atom and is followed by the loss of phosphate anion. Previous work in this field has been summarised and discussed.[15] A study has been made of the riboflavin-sensitised dephosphorylation of menadiol diphosphate (16) and the conclusion reached[16] was

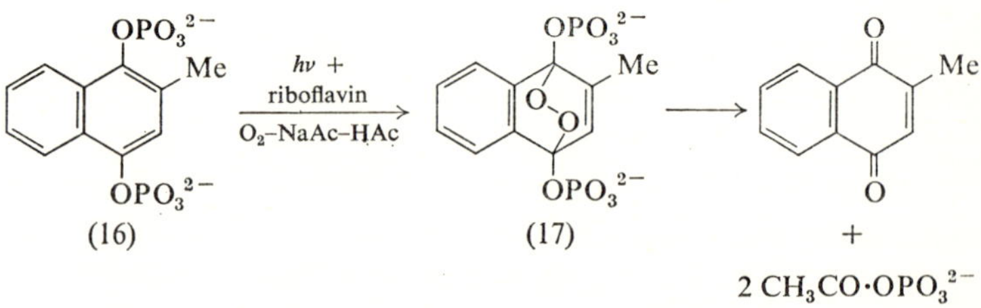

(16) (17) + $2\ CH_3CO{\cdot}OPO_3^{\ 2-}$

[13] H. Tomioka, Y. Izawa, and Y. Ogata, *Tetrahedron*, 1969, **25**, 1501.
[14] R. O. De Jongh and E. Havinga, *Rec. Trav. chim.*, 1968, **87**, 1318.
[15] R. O. De Jongh and E. Havinga, *Rec. Trav. chim.*, 1968, **87**, 1327.
[16] P.-S. Song and T. A. Moore, *J. Amer. Chem. Soc.*, 1968, **90**, 6507.

that triplet sensitiser produces singlet oxygen which adds onto the phosphate to give a peroxide (17) which decomposes to give the observed products.

Photolysis of α-diazophosphine oxides (18) has been found [17] to give products consistent with the initial production of a phosphorus-substituted carbene.

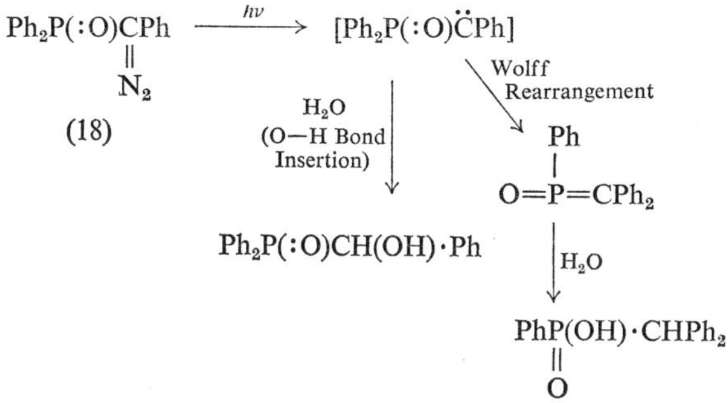

2 Radical Reactions

Several further examples (*e.g.* 19 and 20) [3, 18–20] have been reported of the radical initiated addition of compounds which contain P—H bonds to olefins. It has been reported [21] that at high temperatures and low pressures,

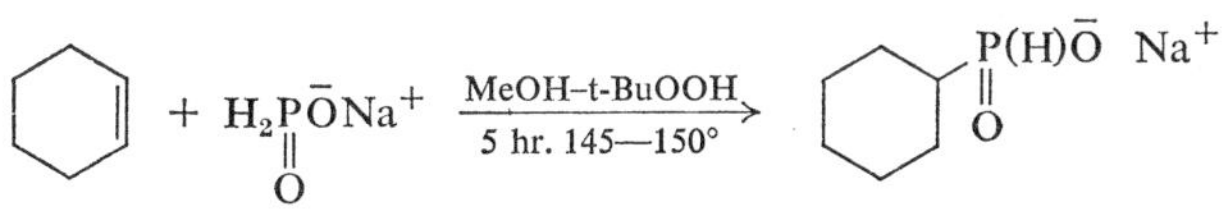

(19)

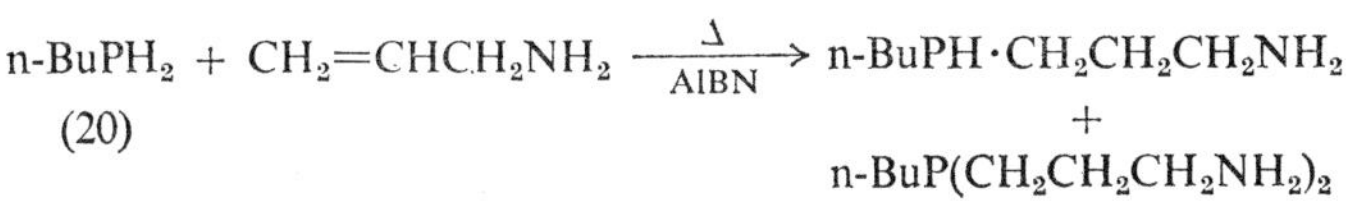

(20)

tetrafluorodiphosphine produces difluorophosphino radicals. Tetraphenyldiphosphine (21) has been shown [22] to react with phenyl-, diphenyl- and triphenyl-acetic acids to give the corresponding hydrocarbons, toluene, diphenylmethane, and triphenylmethane.

[17] M. Regitz, W. Anschütz, W. Bartz, and A. Liedhegener, *Tetrahedron Letters*, 1968, 3171.

[18] K. A. Petrov, T. N. Lysenko, B.-Ya. Libman, and V. V. Pozdnev, *Khim. Org. Soedin, Fosfora, Akad. Nauk S.S.S.R., Otd. obshchei Tekh. Khim.*, 1967, 181 (*Chem. Abs.*, 1968, **69**, 67,487g).

[19] E. E. Nifant'ev and M. P. Koroteev, *Zhur. obshchei Khim.*, 1967, **37**, 1366 (*Chem. Abs.*, 1968, **68**, 39,739b).

[20] K. Issleib, H. Oehme, and E. Leissring, *Chem. Ber.*, 1968, **101**, 4032.

[21] D. Solan and P. L. Timms, *Chem. Comm.*, 1968, 1540.

[22] R. S. Davidson, R. A. Sheldon, and S. Trippett, *J. Chem. Soc. (C)*, 1968, 1700.

$$Ph_2P \cdot PPh_2 + ArCH_2CO_2H \xrightarrow{\Delta} Ph_2POCO \cdot CH_2Ar + Ph_2PH$$
(21)

$$Ph_2POCO \cdot CH_2Ar \longrightarrow Ph_2PO \cdot + ArCH_2\dot{C}O$$

$$ArCH_2\dot{C}O \longrightarrow Ar\dot{C}H_2 + CO$$

$$Ar\dot{C}H_2 + Ph_2PH \longrightarrow ArCH_3 + Ph_2P \cdot$$

Carbon monoxide, and not carbon dioxide, was shown to be a product of the reaction. Hellwinkel has reported[23] that bis-2,2′-biphenylenephosphorane (22) at room temperature and in a nitrogen atmosphere decomposes to a radical (23) which subsequently gives the phosphine (24) and the phosphorane (25). The radical (23) has a *g* value of 2·0025 ± 0·0001 and a

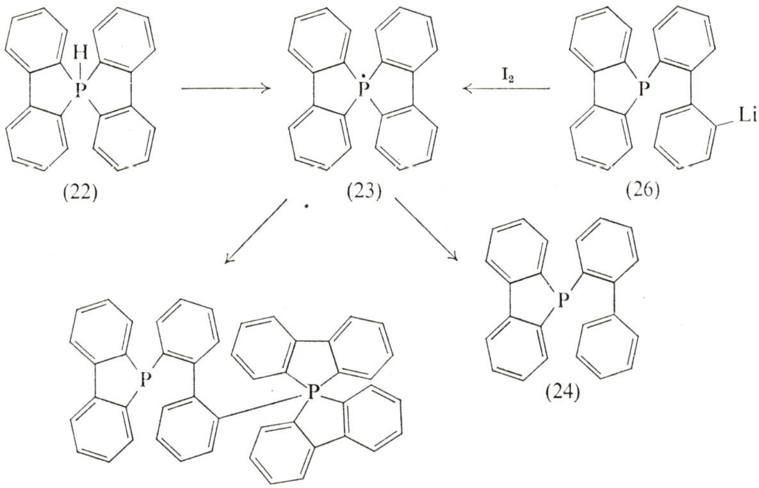

(22) (23) (26) (24) (25)

phosphorus hyperfine coupling constant of 17·9 G. The formation of a radical species in the decomposition was further demonstrated by the observation that the use of good hydrogen donors, such as methanol, as solvents led to the formation of reduced products and that the presence of oxygen gave hydroperoxides. The same radical (23) is formed when the lithium compound (26) reacts with iodine.[24] A mechanism for the formation of the radical (23) was not suggested. Formation of the 1,1-diphenylphosphabenzole (27) by reaction of the phosphabenzole (28) with diphenylmercury has been reported and the radical (29) suggested as an intermediate.[25] Compound (27) was also obtained by reaction of the reduced phosphabenzole (30) with diphenylmercury. The formation of the

[23] D. Hellwinkel, *Chem. Ber.*, 1969, **102**, 528.
[24] D. Hellwinkel, *Chem. Ber.*, 1969, **102**, 548.
[25] G. Märkl and A. Merz, *Tetrahedron Letters*, 1969, 1231.

phenoxyphosphabenzole (31) by reaction of (30) with the 2,4,6-tri-t-butylphenoxy radical was taken as evidence for the feasibility of radical (29)

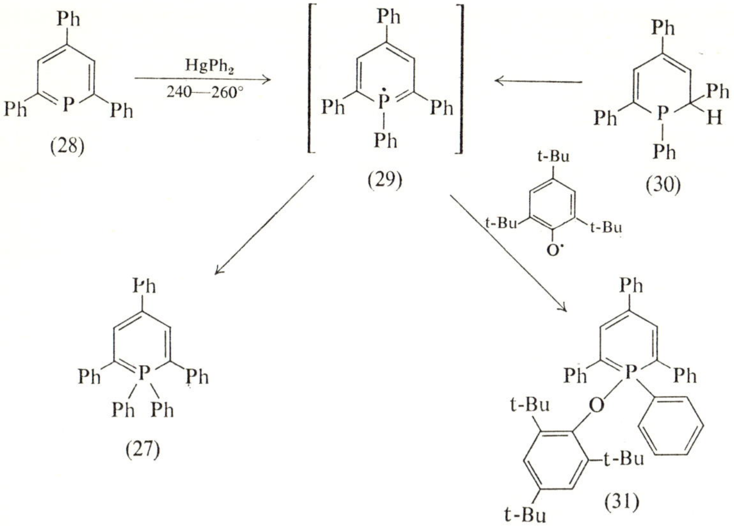

being an intermediate. The reaction of phosphabenzoles with mercuric acetate has been found to give 1,1′-diacetoxyphosphabenzoles (32) which are strongly fluorescent compounds.[26] It was stated that the reaction occurs *via* cation radicals which can be detected by e.s.r. spectroscopy. A preparation of the phenoxy radical (33) has been described and its e.s.r. spectrum analysed.[27] A phosphorus hyperfine coupling constant of 38·56 G was

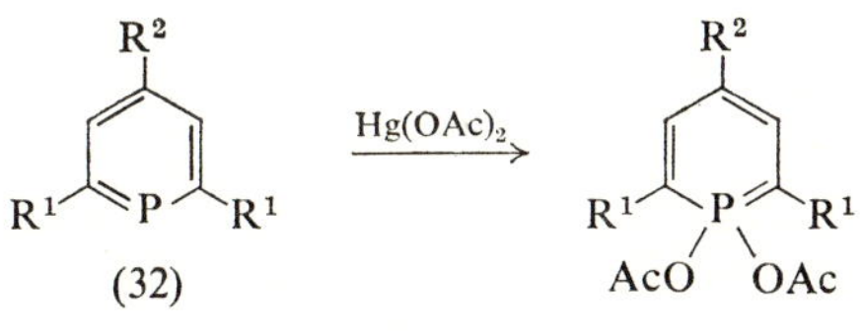

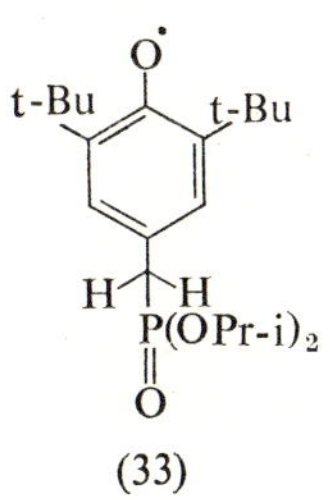

(33)

[26] K. Dimroth and W. Städe, *Angew. Chem. Int. Edn.*, 1968, **7**, 881.
[27] A. Rieker and H. Kessler, *Tetrahedron*, 1968, **24**, 5133.

reported for the radical. The variations of the spectrum with change in temperature were interpreted as being due to variation of the rate of rotation about the carbon—phosphorus bond.

The reactions of a variety of radicals with trivalent phosphorus compounds to give phosphoranyl radicals ($R_4\dot{P}$) have been reported. Dimethylamino radicals react with alkyl diphenylphosphinites to give *NN*-dimethyl diphenylphosphinic amide plus alkyl radicals.[28] In some cases alkyldiphenylphosphine oxides (34) were formed and it was suggested that they

$$R\cdot + Ph_2POR \longrightarrow \underset{(34)}{Ph_2P(:O)R} + R\cdot$$

had been produced by attack of the alkyl radicals upon the phosphinite. That this reaction can indeed occur was confirmed[29] by the finding that phenyl radicals, generated from phenylazotriphenylmethane, react with phosphites to give phenylphosphonates. In the reaction with trimethyl phosphite, 1,1,1-triphenylethane was produced but some doubt was expressed as to its formation by the trapping of methyl radicals by triphenylmethyl radicals. Photolysis of chloromethyl methyl ether (35) in the presence

$$\underset{(35)}{MeCOCH_2Cl} \xrightarrow{h\nu} MeCO\dot{C}H_2 + Cl^{\bullet}$$

$$MeCO\dot{C}H_2 + (EtO)_3P \longrightarrow MeCOCH_2\dot{P}(OEt)_3 \longrightarrow MeCOCH_2P(:O)(OEt)_2 + Et^{\bullet}$$

$$Et^{\bullet} + (EtO)_3P \longrightarrow EtP(:O)(OEt)_2 + Et^{\bullet}$$

$$MeCOCH_2\dot{P}(OEt)_3 + Cl^{\bullet} \longrightarrow MeCOCH_2\overset{+}{P}(OEt)_3\ Cl^- \longrightarrow MeCOCH_2P(:O)(OEt)_2 + EtCl$$

$$MeCOCH_2Cl + (EtO)_3P \longrightarrow MeC(=CH_2)OP(:O)(OEt)_2$$

of triethyl phosphite also gives products which can be visualised as being produced by attack of alkyl radicals upon the phosphite.[30] The photolysis[31]

[28] R. S. Davidson, *Tetrahedron Letters*, 1968, 3029.
[29] W. G. Bentrude, J.-J. L. Fu, and C. E. Griffin, *Tetrahedron Letters*, 1968, 6033.
[30] H. Tomioka, Y. Izawa, and Y. Ogata, *Tetrahedron*, 1968, **24**, 5739.
[31] B. Obrycki and C. E. Griffin, *J. Org. Chem.*, 1968, **33**, 632.

and radiolysis [32] of aryl iodides in the presence of phosphites gives arylphosphonates, presumably by attack of aryl radicals upon the phosphite. The photolytic reaction gave the phosphonates in very high yield. The question of the stereochemistry of phosphoranyl radicals has been investigated. Reaction of tri-t-butyl phosphite with ^{14}C-labelled t-butoxy radicals gave phosphate which contained 75% of the label.[33] This shows that formation of phosphoranyl radicals is irreversible. Reaction of the *cis* and *trans* phosphinites, (36) and (37), with t-butoxy and n-butylthiyl radicals

(36) (37)

gave oxidised products with retention of configuration.[34] Oxidation of optically active phosphines with these radicals gave oxidised products in a stereospecific manner. These results indicate that phosphoranyl radicals which contain more than one type of substituent (*e.g.* $R_3R'\dot{P}$) decompose before they become planar or even pseudorotation takes place. Labelling studies,[33] on the other hand, show that the tetra-t-butoxyphosphoranyl radical must either undergo pseudorotation before decomposition or else loss of the t-butyl group occurs with equal ease from an apical or equatorial substituent. Radicals such as the tetra-t-butoxyphosphoranyl radical, which contains only one type of substituent, would undergo pseudorotation more easily than those which contain two or more different substituents.

3 Electrochemical Generation of Radicals

The electrochemical reduction of triphenylphosphine at a dropping mercury cathode was found [35] to be quasi-reversible when dimethylformamide was used as solvent and irreversible when acetonitrile was used. Reduction in the latter solvent was found to give biphenyl and diphenylphosphinic acid. The formation of the biphenyl is unlikely to occur by dimerisation of two phenyl radicals as suggested, and is probably formed by the collapse of a quinquecovalent phosphorus compound. The observation [36] that quaternary phosphonium amalgams are formed by electrolysis of phosphines at a mercury cathode in dipolar aprotic solvents has been reported and elucidation of the part played by these compounds in the decomposition of the phosphines is awaited with interest.

[32] G. Caspari, H. Drawe, and A. Henglein, *Radiochim. Acta*, 1967, **8**, 102 (*Chem. Abs.*, 1968, **68**, 105,304j).
[33] W. G. Bentrude and R. A. Wielesek, *J. Amer. Chem. Soc.*, 1969, **91**, 2406.
[34] W. G. Bentrude, J. H. Hargis, and P. E. Rusek, *Chem. Comm.*, 1969, 296.
[35] K. S. V. Santhanam and A. J. Bard, *J. Amer. Chem. Soc.*, 1968, **90**, 1118.
[36] W. R. T. Cottrell and R. A. N. Morris, *Chem. Comm.*, 1968, 409.

The cathodic splitting of phosphonium salts ($RPh_3\overset{+}{P}\,\overset{-}{Hal}$) to give phosphine and hydrocarbon has been the subject of a thorough study by Horner and Haufe.[37] The yield of alkane compared with that of benzene was determined for a number of salts. It would appear that alkane formation is favoured in those cases in which the alkyl group is large enough to cause serious steric congestion.

Anodic oxidation of a variety of phosphines at a mercury anode in acetonitrile has been studied[38] and half-wave oxidation and oxidation potentials reported. Mercury complexes of the type (38) were isolated. Triphenylphosphine could not be oxidised at a platinum anode.

$$Ph_3P \xrightarrow{\text{anode}} Ph_3\overset{+}{P}\cdot$$

$$2Ph_3\overset{+}{P}\cdot \xrightarrow{\text{Hg}} (Ph_3P)_2Hg^{++}$$

(38)

4 Phosphinidenes and Related Species

The reactions of the species, which is probably phenylphosphinidene, produced by thermal or photochemical decomposition of pentaphenylcyclopentaphosphine, with disulphides and α-diketones to give (39) and (40)

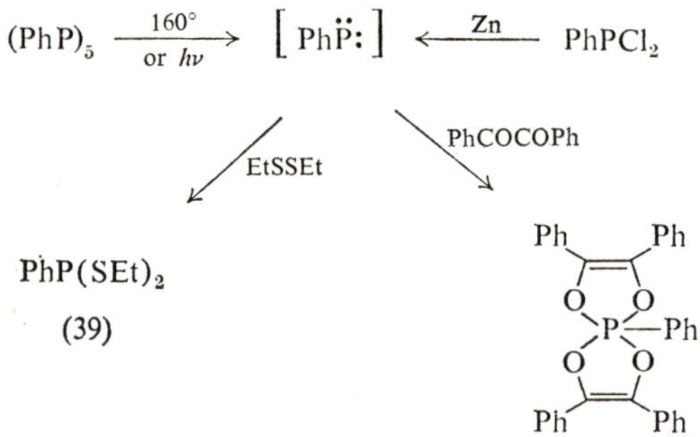

(39)

(40)

respectively have been described.[39] The same products were formed when what is probably the same species was generated by the dechlorination of phenylphosphonous dichloride with zinc.[39, 40] The species, produced in this way, reacts with acrylonitrile to give macromolecular compounds whose composition is dependent upon the temperature of the reaction used.[40] Decomposition of the cyclic phosphine in the presence of dienes gave cyclic compounds of the type (41) and (42)[39] which can be visualised as occurring by attack of the phosphinidene or the diradical (43) upon the olefin. Irradiation of the cyclic phosphine at low temperatures gives a red

[37] L. Horner and J. Haufe, *Chem. Ber.*, 1968, **101**, 2903.
[38] L. Horner and J. Haufe, *Chem. Ber.*, 1968, **101**, 2921.
[39] U. Schmidt, I. Boie, C. Osterroht, R. Schröer, and H.-F. Grützmacher, *Chem. Ber.*, 1968, **101**, 1381.
[40] B. Bloch and Y. Gounelle, *Compt. rend.*, 1968, **266**, *C*, 220.

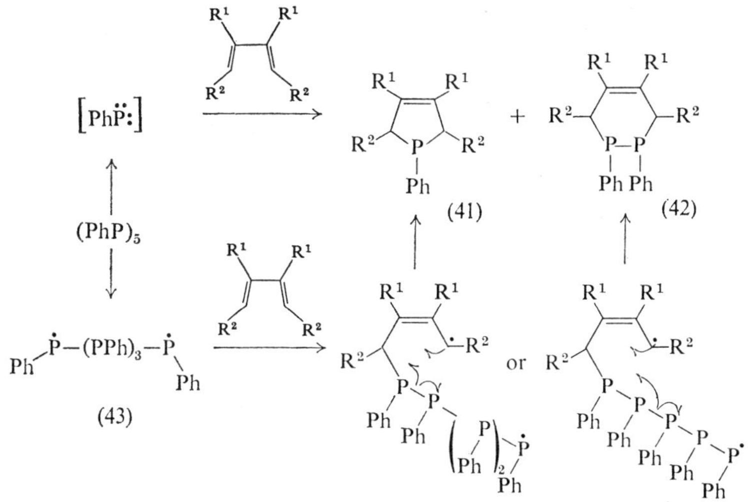

compound which shows an e.s.r. signal ($g = 2{\cdot}0073$). This signal was tentatively assigned to the diradical (43).[39] Peaks attributable to the phosphinidene radical cation have been found[41] in the mass spectrum of phenylphosphine and diphenylphosphine.

The formation of diphosphine disulphides by reaction of thiophosphinyl chloride and of alkyl and arylphosphonothioic dihalides with Grignard reagents has been postulated as occurring by the formation of phosphinidene sulphides (44).[42] Formation of the biphosphines by the halogen

$$R_2P(:S)Cl \xrightarrow{R^1MgX} R_2P(:S)MgX + R^1Cl \xrightarrow{R_2P(:S)Cl} R_2P(=S){-}P(=S)R_2 + MgXCl$$

Scheme 1

$$RP(:S)Hal_2 + R^1MgX \longrightarrow R^1\text{-}Hal + MgXCl + R\ddot{P}S\ (44)$$

$$R\ddot{P}S \xrightarrow{\text{Insertion into P-Hal Bond}} RP(=S)(Hal){-}P(=S)(Hal)R$$

Scheme 2

[41] B. Zeeh and J. B. Thomson, *Tetrahedron Letters*, 1969, 111.
[42] P. C. Crofts and I. S. Fox, *J. Chem. Soc.* (*B*), 1968, 1417.

metal exchange reaction shown in Scheme 1 was ruled out on the grounds that reaction of phosphorus halides with lithium alkyls, which are more likely to undergo such a reaction, gives tertiary phosphine sulphides. The proposed mechanism for the reaction is shown in Scheme 2.

5 Deoxygenation of Peroxides and Desulphurisation of Disulphides

The reaction of peroxides with phosphines has been further investigated. The intermediate, produced in the oxidation reaction was shown [43] (from ^{31}P n.m.r. chemical shifts) to have the covalent structure (45) and not to be a phosphonium alkoxide (46). Ascaridole has been shown to give *p*-cymene

$$R_3P + EtOOEt \longrightarrow R_3P(OEt)_2$$

(45)

$$R_3\overset{+}{P}OEt \quad \overset{-}{O}Et$$

(46)

(47) and not the oxide (48) on heating with triethyl phosphite.[44] It should be noted that at the temperature used for the reaction, ascaridole readily rearranges to give the oxides (49) and (50.)[45] Deoxygenation of cyclic

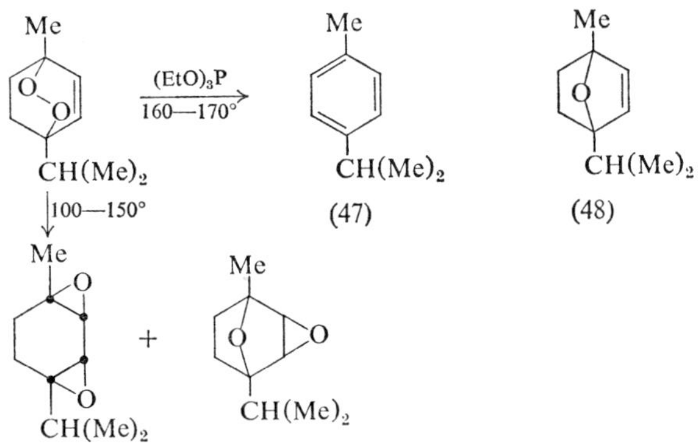

(47) (48)

(49) (50)

peroxyesters (51) by a variety of phosphines and phosphites has been shown to result in fragmentation of the ester (Scheme 3).[46] It was found that increasing the size of the substituent R caused an increase in the ratio of the

[43] D. B. Denney, D. Z. Denney, and L. A. Wilson, *Tetrahedron Letters*, 1968, 85.
[44] T. Kametani and K. Ogasawara, *Chem. and Ind.*, 1968, 1772.
[45] J. Boche and O. Runquist, *J. Org. Chem.*, 1968, **33**, 4285.
[46] W. Adam, R. J. Ramirez, and S.-C. Tsai, *J. Amer. Chem. Soc.*, 1969, **91**, 1254.

amount of keten formed to the amount of decarbonylation. The reverse effect was caused by increasing the polarity of the solvent. Kinetic studies showed that formation of the pentacovalent intermediate (52) is the slow step of the reaction and that its opening up to ionic forms (53) and (54) is

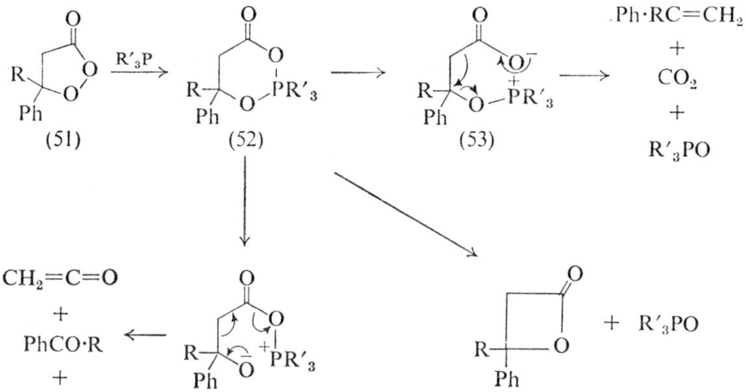

(51) (52) (53) (54)

Scheme 3

fast. Tris-diethylaminophosphine has been found to be particularly effective as a reagent for the desulphurisation of disulphides.[47] Dibenzyl disulphide is smoothly desulphurised to give dibenzyl sulphide. 4-Bromobenzyl benzyl disulphide gave three products: 4,4′-dibromodibenzyl sulphide, dibenzyl sulphide, and 4-bromobenzyl benzyl sulphide. The unexpected formation of the symmetrical sulphides leads to the suggestion that the reaction cannot be occurring *via* such intermediates as (55) and (56). Desulphurisation of (57) and (58) by the aminophosphine occurs

(55) (56)

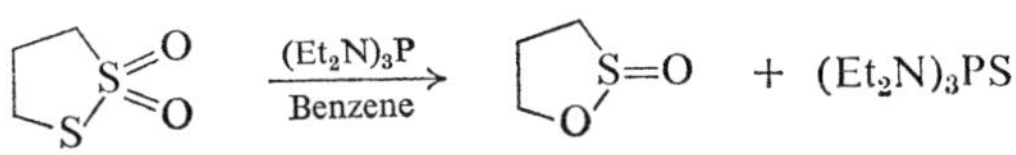

(57)

exothermically to give the products shown.[48] Examples of the synthetical application of the desulphurisation of disulphides by triphenylphosphine have been reported.[49] The photoreaction of alkyl methylphosphinates

[47] D. N. Harpp, J. G. Gleason, and J. P. Snyder, *J. Amer. Chem. Soc.*, 1968, **90**, 4181.
[48] D. N. Harpp and J. G. Gleason, *Tetrahedron Letters*, 1969, 1447.
[49] D. Brewer, R. Rahman, S. Safe, and A. Taylor, *Chem. Comm.*, 1968, 1571; G. M. Blackburn and W. D. Ollis, *Chem. Comm.*, 1968, 1261.

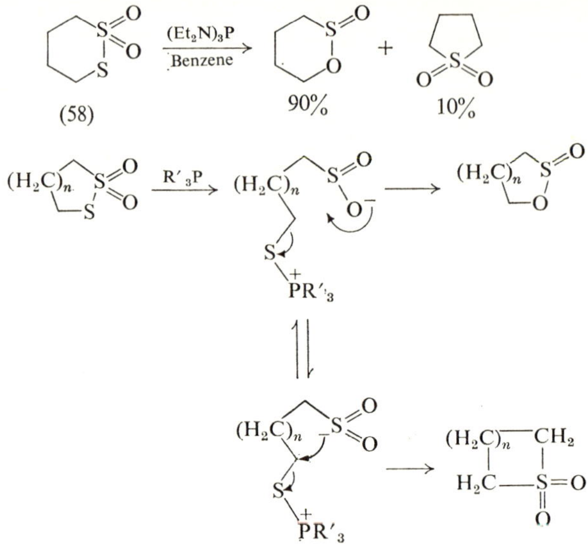

with diphenyl disulphide to give *S*-phenyl alkyl methylphosphinothiolates has been studied, and a radical mechanism proposed.[50] It should be noted that such compounds have been previously shown to react with disulphides in t-butanol solution by an ionic mechanism.[51] Tetra-alkyldiphosphines have been found to react with disulphides under thermal conditions[52] and on irradiation[53] to give *S*-alkyl dialkylphosphinothioite and a radical mechanism was proposed. The desulphurisation of tetrasulphides by triphenylphosphine in carbon tetrachloride solution has been reported to give trisulphides which in turn give disulphides.[54] The rôle of the solvent was not discussed.

6 Deoxygenation of Ozone and Ozonides

Triphenyl phosphite has been found to react with ozone at $-70°$ in methylene chloride solution, to give an adduct which on warming to $-35°$ decomposes to give triphenyl phosphate and singlet oxygen.[55] The usefulness of this adduct as a source of singlet oxygen has been demonstrated by decomposition of the adduct in the presence of cyclohexa-1,3-diene (59) to give norascaridole. The formation of singlet oxygen

[50] W. A. Mosher and R. R. Irino, *J. Amer. Chem. Soc.*, 1969, **91**, 756.
[51] M. Grayson, C. E. Farley, and C. E. Streuli, *Tetrahedron*, 1967, **23**, 1065.
[52] Yu. N. Shlyk, G. M. Bogolyubov, and A. A. Petrov, *Dokl. Akad. Nauk S.S.S.R.*, 1967, **176**, 1327 (*Chem. Abs.*, 1968, **68**, 19,261g).
[53] Yu. N. Shlyk, G. M. Bogolyubov, and A. A. Petrov, *Zhur. obshchei Khim.*, 1968, **38**, 193 (*Chem. Abs.*, 1968, **68**, 67,475b).
[54] J. Tsurugi, T. Horii, T. Nakabayashi, and S. Kawamura, *J. Org. Chem.*, 1968, **33**, 4133.
[55] R. W. Murray and M. L. Kaplan, *J. Amer. Chem. Soc.*, 1968, **90**, 537.

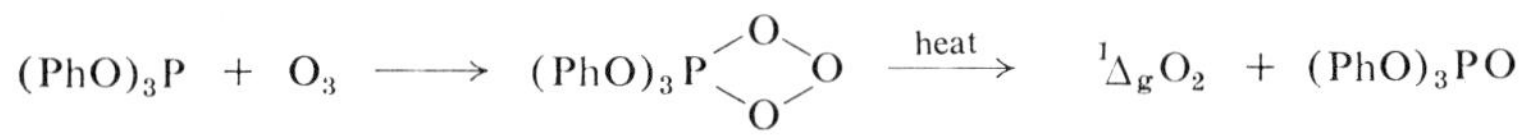

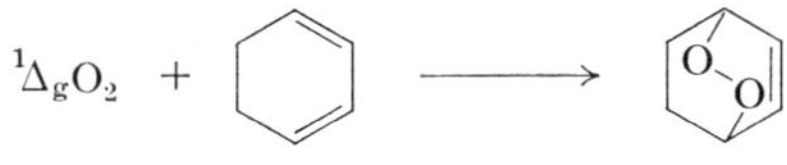

(59)

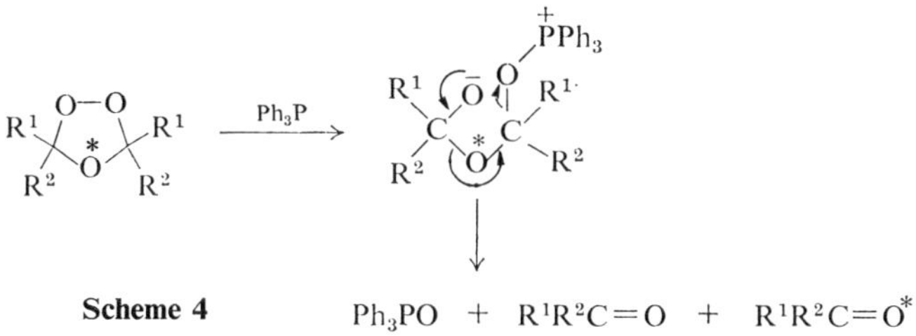

Scheme 4

has also been detected by e.s.r. spectroscopy.[56] By use of ^{18}O-labelled ozonides it has been possible to show that their deoxygenation by phosphines occurs by attack of the phosphine upon an oxygen atom of the peroxide link (Scheme 4).[57]

7 Deoxygenation of Alcohols and Desulphurisation of Thiols

A very thorough study has been made of the reaction of carbon tetrachloride with triethyl phosphite.[58] In the absence of radical initiators and of light, the reaction is ionic. The sigmoid shape of the rate plot was suggested as being due to the change in dielectric constant of the mixture during reaction. In the presence of initiators such as AIBN, at least part of the reaction occurs by a radical mechanism. When the reaction is carried out in the presence of n-butylthiol and a radical scavenger and in the absence of an initiator, the formation of *S*-n-butyl diethylphosphorothioate (60, R = Bu^n) occurs by an ionic mechanism. When an initiator is present and a scavenger is

$$(EtO)_3P + CCl_4 \longrightarrow (EtO)_3\overset{+}{P}CCl_3\ Cl^- \longrightarrow (EtO)_2P(:O)CCl_3 + EtCl$$

$$(EtO)_3\overset{+}{P}CCl_3\ Cl^- \xrightarrow{RSH} (EtO)_3\overset{+}{P}SR\ Cl^- + CHCl_3$$

$$(EtO)_3\overset{+}{P}SR\ Cl^- \longrightarrow (EtO)_2P(:O)SR + EtCl$$

(60)

[56] E. Wasserman, R. W. Murray, M. L. Kaplan, and W. A. Yager, *J. Amer. Chem. Soc.*, 1968, **90**, 4160.
[57] J. Carles and S. Fliszar, *Can. J. Chem.*, 1969, **47**, 1113.
[58] R. E. Atkinson, J. I. G. Cadogan, and J. T. Sharp, *J. Chem. Soc.* (*B*), 1969, 138.

absent, triethyl phosphorothionate is produced by a radical mechanism. In the absence of both an initiator and a scavenger, both mechanisms operate. It was found that in the corresponding reactions with bromotrichloromethane, it was more difficult to assess the role of each mechanism. The synthesis of *OO*-dialkyl-*S*-aryl phosphorothiolates by reaction of phosphites with thiols in the presence of carbon tetrachloride has been investigated.[59] The transformation of alcohols into alkyl chlorides by their reaction with carbon tetrachloride in the presence of triphenylphosphine has been shown to occur with inversion.[60] D-Octan-2-ol was converted into L-2-chlorooctane and the *threo* thiol (61) was converted into the *erythro*-chloro-

$$Ph_3P + CCl_4 \longrightarrow Ph_3\overset{+}{P}CCl_3\ Cl^- \xrightarrow[\text{PhCHD}\cdot\text{CHD}\cdot\text{SH (61)}]{threo} Ph_3\overset{+}{P}SCHD\cdot CHDPh\ Cl^- + CHCl_3$$

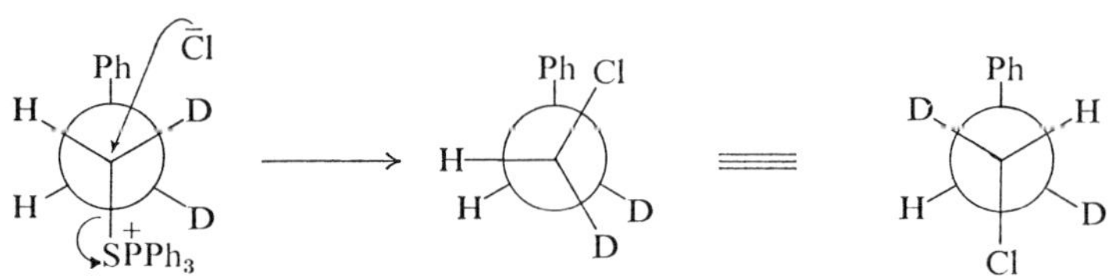

(62)

$$R_3P + R'CX_3 + R''OH \longrightarrow R_3PO + R'CHX_2 + R''X$$

compound (62). Alcohols can also be converted into alkyl halides by reaction with trihalogenomethanes in the presence of triphenylphosphine.[61] Dihalogenomethanes can be used, but they are much less reactive. The ease of reaction with trihalogeno compounds was found to increase with increasing nucleophilic character of the phosphine. Further use of the reaction of alcohols with triphenylphosphine in the presence of *N*-bromosuccinimide to give alkyl bromides has been described.[62] The reaction of the bromodienone (63) with triphenylphosphine in the presence of alcohols to give alkyl bromides has been reported and paramagnetic species in the reaction mixture detected by e.s.r. spectroscopy.[63] This latter observation led to the suggestion that the reaction occurs by way of a stable phenoxy radical (64) and a phosphoranyl radical (65).

[59] L. L. Murdock and T. L. Hopkins, *J. Org. Chem.*, 1968, **33**, 907.
[60] R. G. Weiss and E. I. Snyder, *Chem. Comm.*, 1968, 1358.
[61] I. M. Downie and J. B. Lee, *Tetrahedron Letters*, 1968, 4951.
[62] E. E. Schweizer, W. S. Creasy, K. K. Light, and E. T. Shaffer, *J. Org. Chem.*, 1969, **34**, 212.
[63] M. Tsubota, *Nippon Kagaku Zasshi*, 1968, **89**, 602 (*Chem. Abs.*, 1968, **68**, 106,083u).

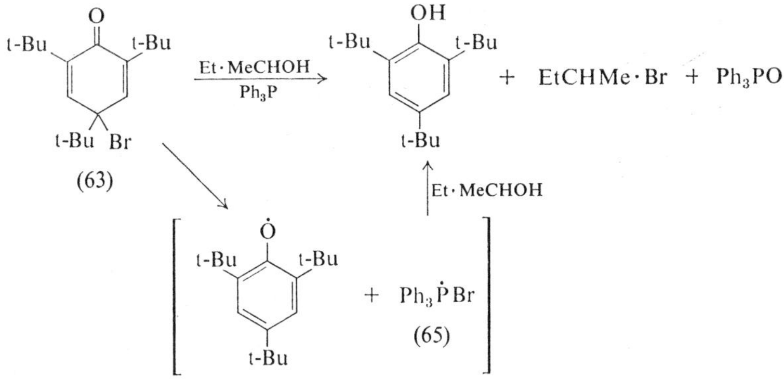

(63) (64) (65)

8 Deoxygenation of Carbonyl Compounds

Tris-dialkylaminophosphines react with *o*-, *m*- ,and *p*-nitrobenzaldehydes to give phospholanes (*e.g.* 66) which are converted at 80° into stilbene oxides.[64] Reaction of the same phosphine with acenaphtha-1,8-quinone

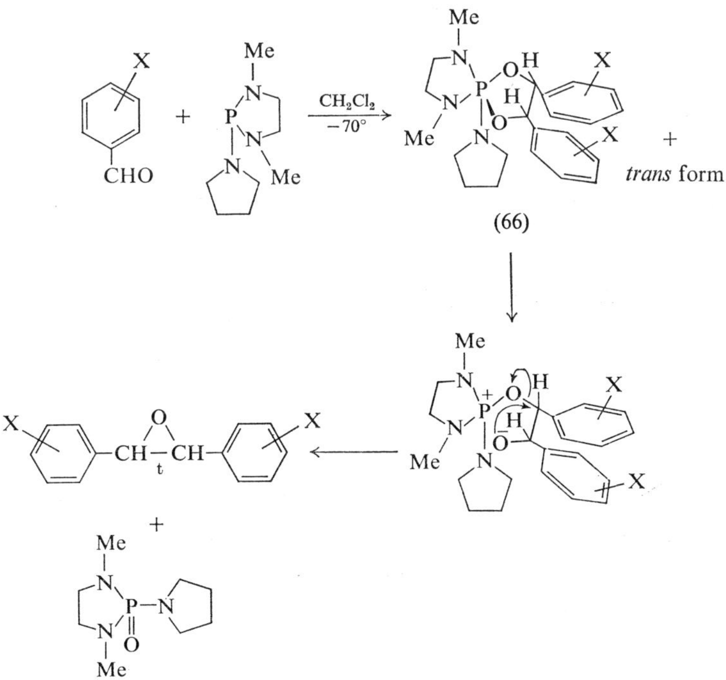

(66)

[64] F. Ramirez, A. S. Gulati, and C. P. Smith, *J. Org. Chem.*, 1968, **33**, 13.

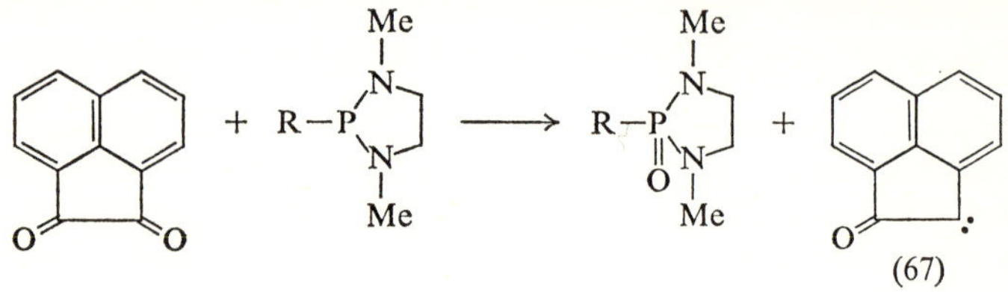
Me
Me
R—P
O
(67)

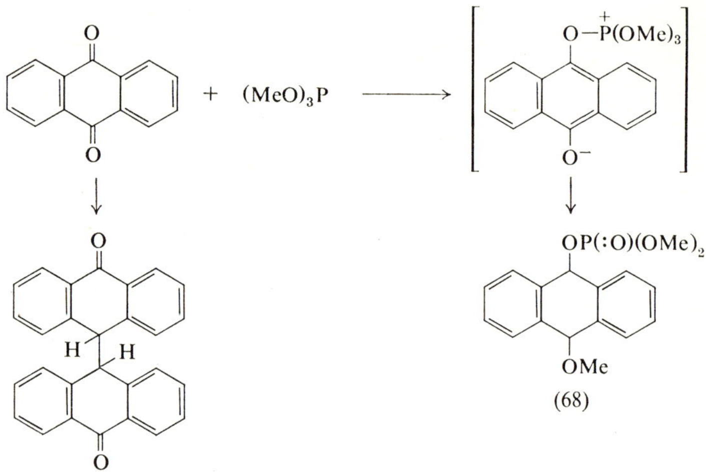
(MeO)3P
O—P(OMe)3
O−
OP(:O)(OMe)2
OMe
(68)

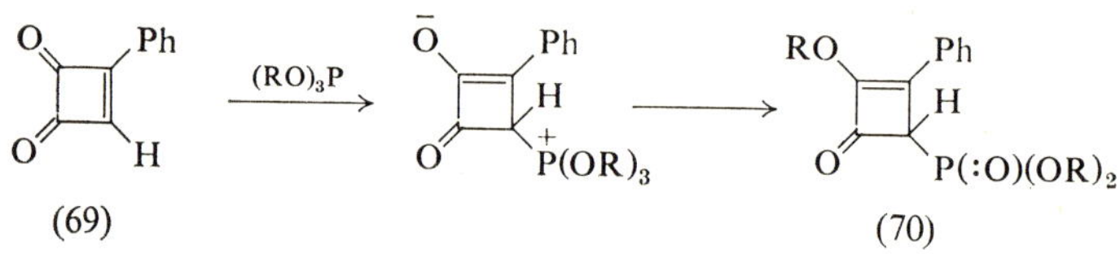
Ph
(RO)3P
P(OR)3
P(:O)(OR)2
(69)
(70)

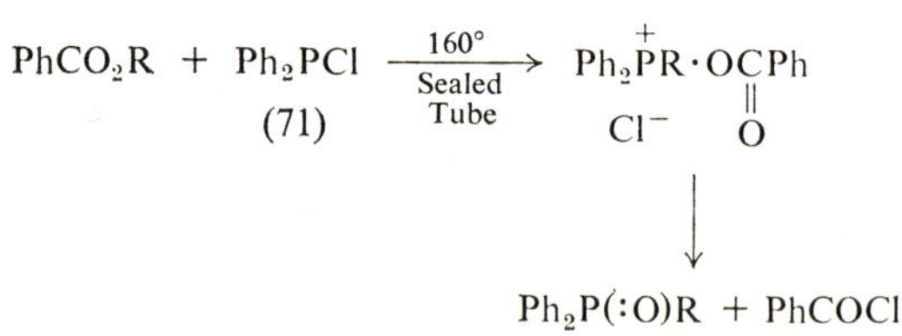
PhCO2R + Ph2PCl
(71)
160°
Sealed
Tube
Ph2PR·OCPh
Cl−
Ph2P(:O)R + PhCOCl

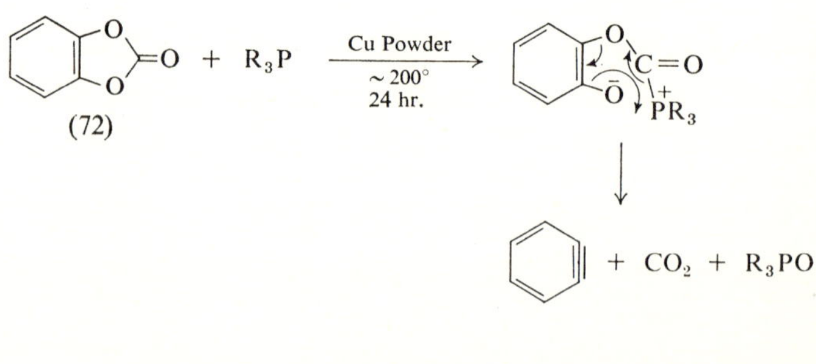
R3P
Cu Powder
~ 200°
24 hr.
(72)
PR3
CO2 + R3PO

gave its phosphine oxide and it was suggested that the carbene (67) is also formed.[65] The fate of the carbene was not enlarged upon. Anthraquinone reacts with trimethyl phosphite to give the phosphate (68) and also some bianthrone.[66, 67] Reaction of cyclobuten-1,2-dione (69) with phosphites gives phosphonates (*e.g.* 70).[68] The reaction of esters with diphenylphosphinous chloride (71) at 160° has been shown [69] to give acid chlorides and is a reaction which may well be of synthetical importance. The deoxygenation of *o*-phenylene carbonate (72) to give benzyne is suggested as occurring by initial attack of the phosphorus compound upon the carbonyl group.[70] Forcing conditions (190—200° for 24 hr in the presence of copper) were required to bring about reaction and low yields of benzyne adducts were obtained. It was shown that the benzyne generated was being trapped by the deoxygenating agent.

9 Deoxygenation of Nitrile Oxides and Nitrones,

Nitrile oxides have been shown [71, 72] to react with ylides such as (73) to give quinquecovalent cyclic adducts of the type (74). These adducts on heating decompose to give azirines and ketimines, and phosphine oxide. Reaction

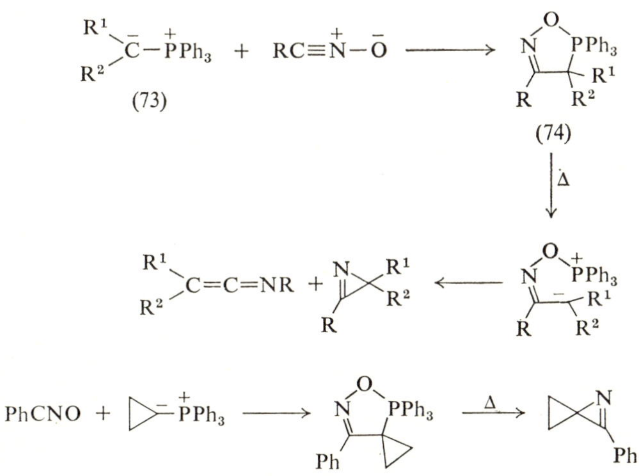

65 F. Ramirez, A. V. Patwardhan, H.-J. Kugler, and C. P. Smith, *Tetrahedron*, 1968, **24**, 2275.
66 F. Ramirez, S. B. Bhatia, A. V. Patwardhan, E. H. Chen, and C. P. Smith, *J. Org. Chem.*, 1968, **33**, 20.
67 J. S. Meek and L. L. Koh, *J. Org. Chem.*, 1968, **33**, 2942.
68 R. C. De Selms, *Tetrahedron Letters*, 1968, 5545.
69 S. T. McNeilly and J. A. Miller, *Chem. Comm.*, 1969, 620.
70 C. E. Griffin and D. C. Wysocki, *J. Org. Chem.*, 1969, **34**, 751.
71 H. J. Bestmann and R. Kunstmann, *Chem. Ber.*, 1969, **102**, 1816.
72 R. Huisgen and J. Wulff, *Chem. Ber.*, 1969, **102**, 1833.

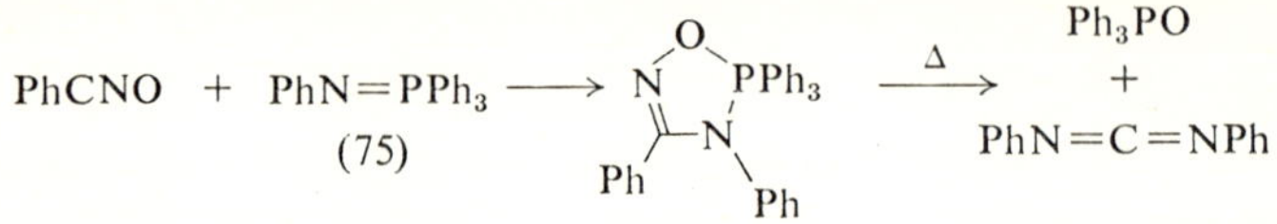

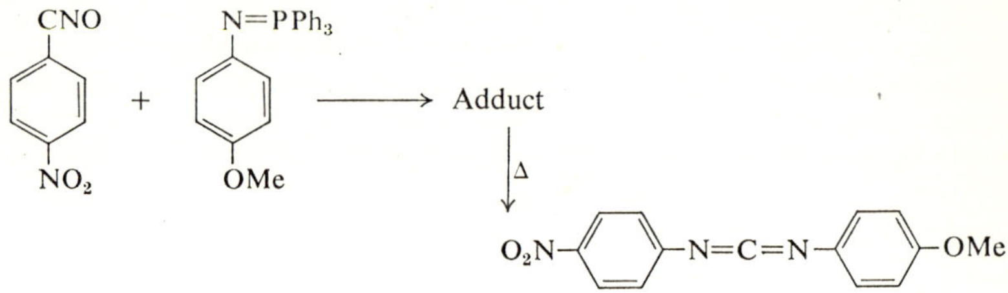

$$(RO)_2PNHPh + PhNCO \longrightarrow (RO)_2P-\overset{+}{N}H\cdot Ph\,(\bar{O}-C{=}NPh) \longrightarrow (RO)_2POH + PhN{=}C{=}NPh$$

(76)

$$(RO)_2POH + PhN{=}C{=}NPh \longrightarrow (RO)_2P(:O)C(NHPh){=}NPh$$

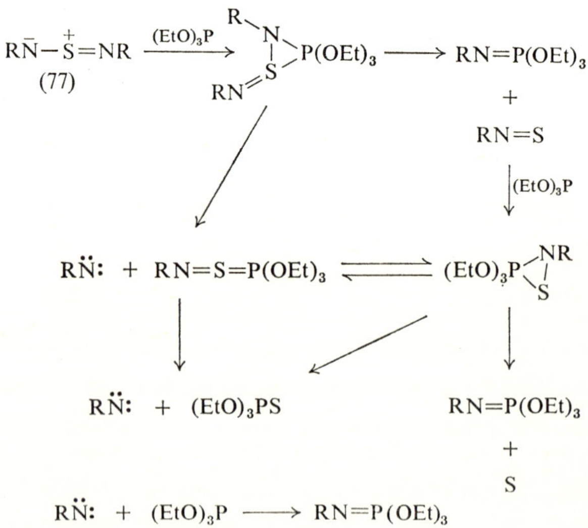

of nitrile oxides [73] with iminophosphoranes (75) gave adducts which can be thermally decomposed to give di-imides. Di-imides are also thought to be formed when phosphoramidites (76) react with isocyanates.[74] Reaction of (77) with triethyl phosphite gives products which suggest that nitrenes are produced in the reaction.[75]

10 Deoxygenation of Nitro- and Nitroso-compounds

The literature on this subject has been reviewed up to early 1968.[76] The deoxygenation of the nitrobenzoxazole (78) to (79) was interpreted as

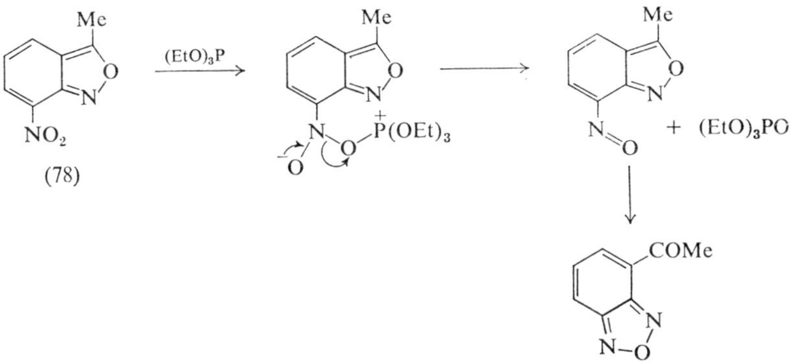

evidence for the formation of an intermediate nitroso-compound.[77] Syntheses of numerous heterocyclic systems by deoxygenation of nitro compounds have been reported and some are shown in Scheme 5. Deoxygenation of (80) [78, 79] to give (81) and (82) has been suggested as occurring by formation of the nitrene followed by insertion into the C—CO_2Et bond. Formation of (83) and (84) was rationalised as occurring by attack of the nitrene upon the carbonyl of the ester group to give initially a diradical or oxazirine. Other examples [80–83] in which an intermediate nitrene attacks a carbonyl group have been reported. The deoxygenation of (85) and (86) to

[73] R. Huisgen and J. Wulff, *Chem. Ber.*, 1969, **102**, 1848.
[74] R. F. Hudson and R. J. G. Searle, *J. Chem. Soc.* (*B*), 1968, 1349.
[75] T. Minami, H. Miki, H. Matsumoto, Y. Ohshiro, and T. Agawa, *Tetrahedron Letters*, 1968, 3049.
[76] J. I. G. Cadogan, *Quart. Rev.*, 1968, **XXII**, 222.
[77] A. J. Boulton, I. J. Fletcher, and A. R. Katritzky, *Chem. Comm.*, 1968, 62.
[78] T. Kametani, T. Yamanaka, and K. Ogasawara, *Chem. Comm.*, 1968, 996.
[79] T. Kametani, K. Ogasawara, and T. Yamanaka, *J. Chem. Soc.* (*C*), 1969, 138.
[80] T. Kametani, T. Yamanaka, and K. Ogasawara, *Chem. Comm.*, 1968, 786.
[81] T. Kametani, T. Yamanaka, and K. Ogasawara, *J. Chem. Soc.* (*C*), 1969, 385.
[82] T. Kametani, K. Nyu, T. Yamanaka, H. Yagi, and K. Ogasawara, *Tetrahedron Letters*, 1969, 1027.
[83] D. G. Saunders, *Chem. Comm.*, 1969, 680.

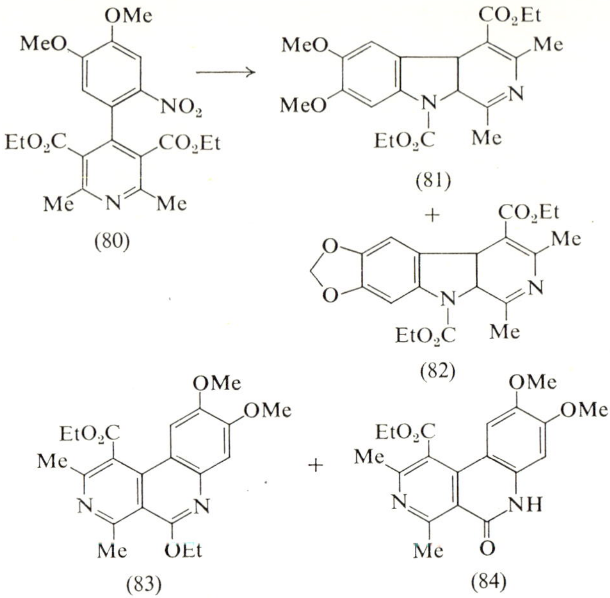

(80) (81) (82) (83) (84)

Ref. 78 and 79

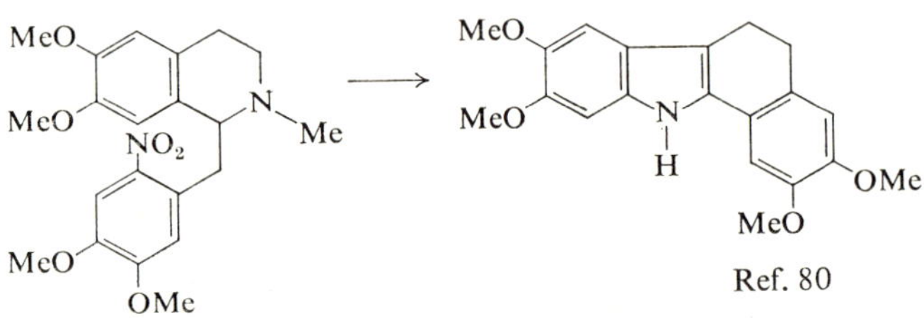

Ref. 80

give the quinoline (87) and the indole (88) shows that steric factors control whether bond insertion or attack upon the carbonyl group occurs. Red phosphorus has been found to deoxygenate 4-substituted nitrobenzenes to give azoxy compounds.[84] The photoinduced deoxygenation of nitro-compounds by phosphites has been studied,[85, 86] and the results indicate that nitrenes are produced in the reaction, *e.g.* irradiation in the presence of diethylamine gives 2-diethylamino-3H-azepines. In the absence of diethylamine, *N*-arylphosphorimidates are produced. Products formed by C—H bond insertion and by hydrogen abstraction were not formed in these reactions. Nitromesitylene, on photoinduced deoxygenation with triethyl

[84] N. S. Kozlov and V. A. Soshin, *Tr. Perm. Gos. Sel'skokhoz Inst.*, 1967, **38**, 95 (*Chem. Abs.*, 1968, **69**, 96,164h).

[85] R. J. Sundberg, W. G. Adams, R. H. Smith, and D. E. Blackburn, *Tetrahedron Letters*, 1968, 777.

[86] R. J. Sundberg, B. P. Das, and R. H. Smith, *J. Amer. Chem. Soc.*, 1969, **91**, 658.

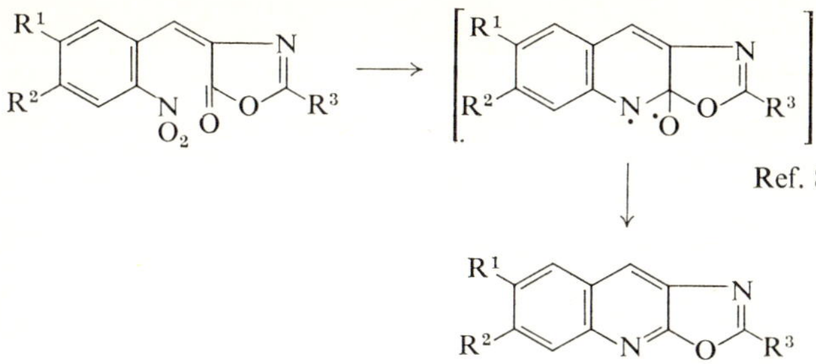

Ref. 81

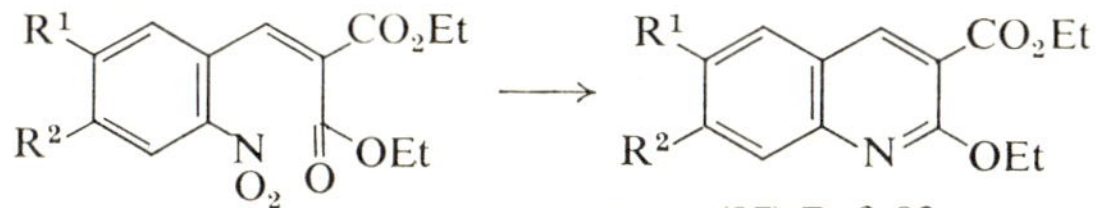

(85)

(87) Ref. 82

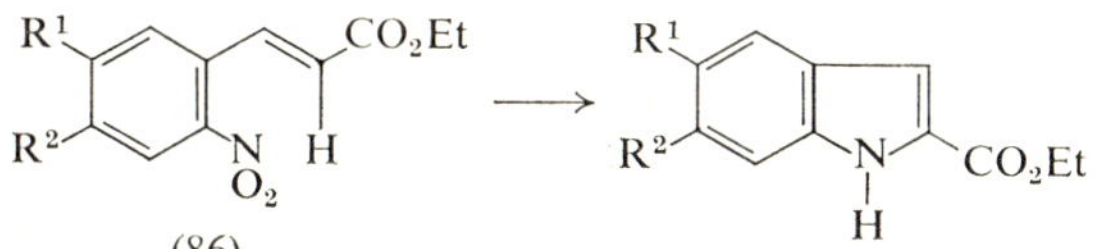

(86)

(88) Ref. 82

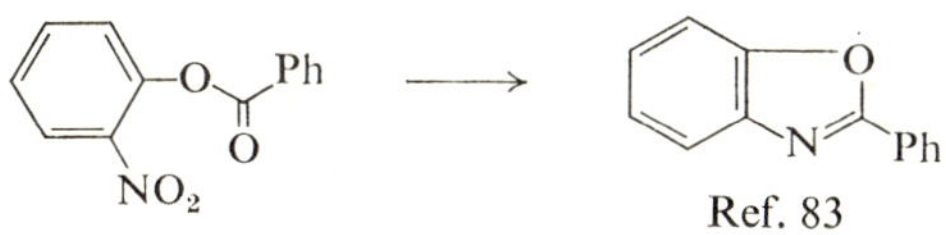

Ref. 83

Scheme 5

phosphite, gave phosphorimidate (89), a pyridine (90), and an azepine (91). It was suggested that the pyridine is formed from the azepine precursor (92). The presence of acetic acid in both the thermal and photoinduced reactions affects the type of products formed.[87] 2-Hydroxyacetanilides (93) and aminophenylphosphonates (94 and 95) are the main products and *N*-arylphosphoramidates are produced in lower yields than those observed in the absence of acetic acid. It was suggested that a substantial fraction of the aryl nitrenes generated are protonated by the acetic acid to give aryl nitrenium ions (96). Deoxygenation of nitroso-compounds is similarly affected by the presence of acetic acid. *N*-Nitrosobenzatriazole is deoxygenated at room temperature by ethyl diphenylphosphinite to give benzyne.[88] High yields of benzyne adducts were obtained. Tris-diethyl-

[87] R. J. Sundberg, R. H. Smith, and J. E. Bloor, *J. Amer. Chem. Soc.*, 1969, **91**, 3392.
[88] J. I. G. Cadogan and J. B. Thompson, *Chem. Comm.*, 1969, 770.

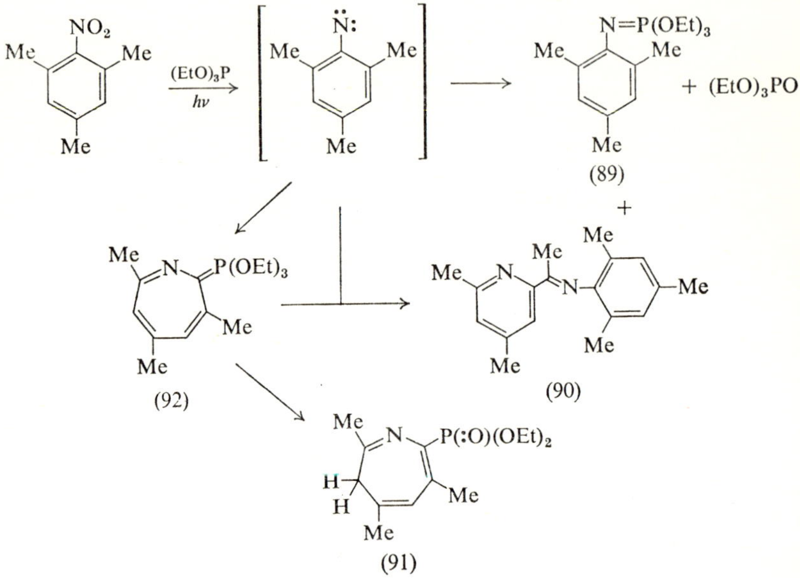

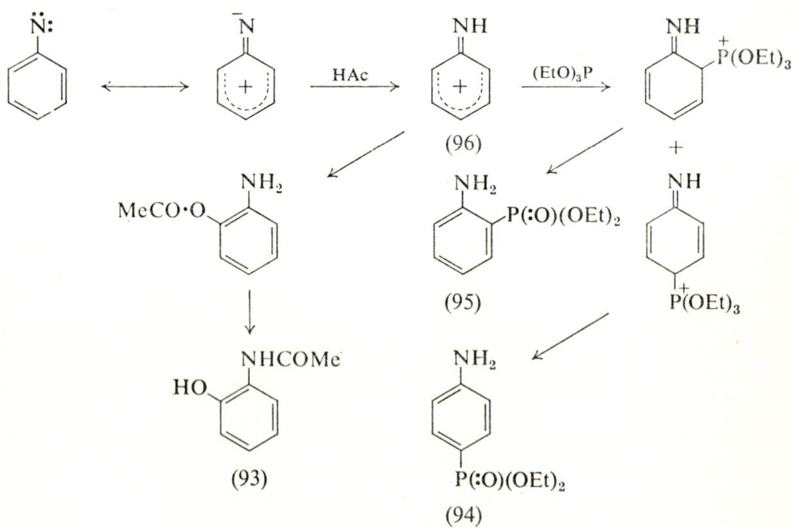

$$(CF_3)_3CNO + Ph_3P \xrightarrow[RT]{Et_2O} Ph_3PO + (CF_3)_3CN{:}PPh_3$$

(97)

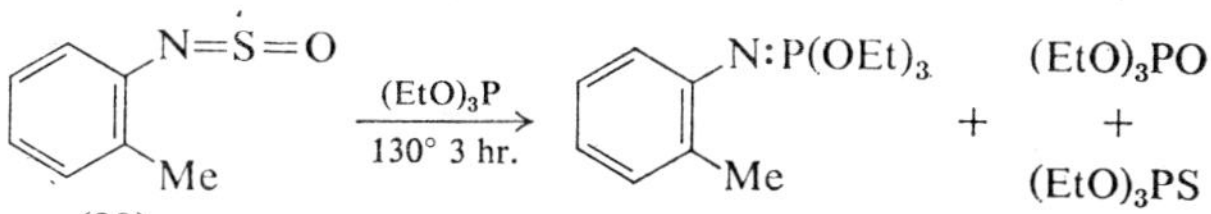

(98)

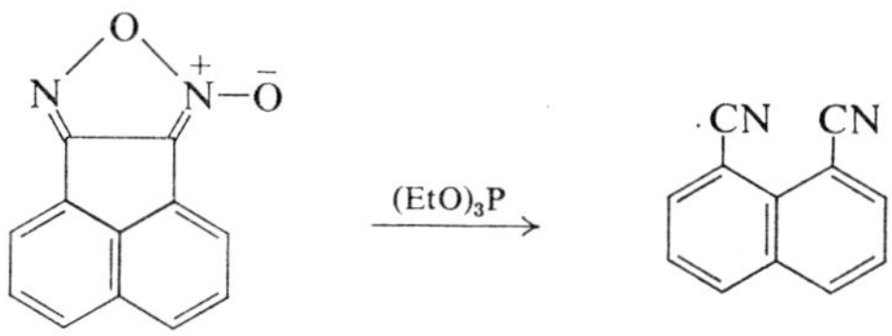

(99)

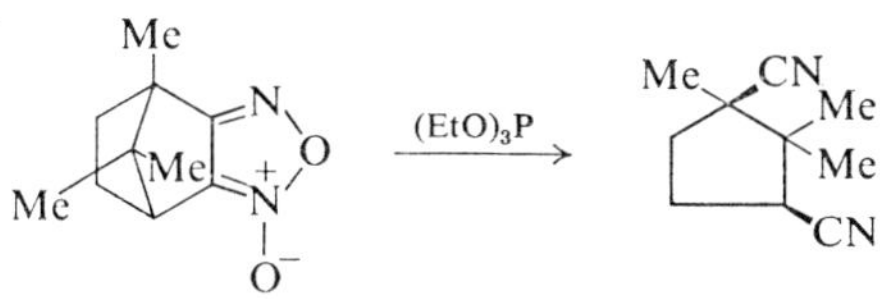

(100)

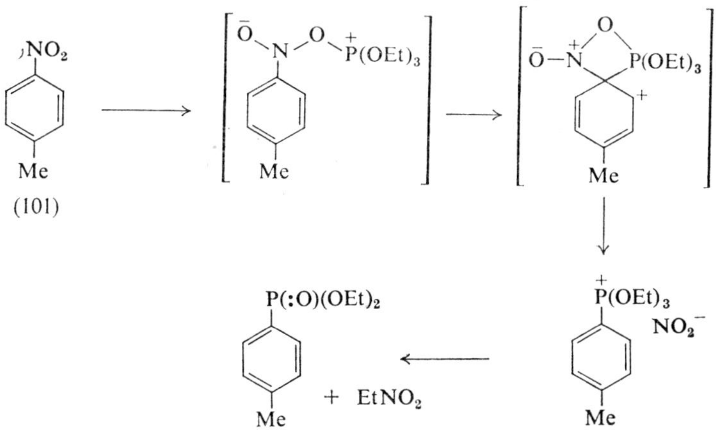

(101)

(103)

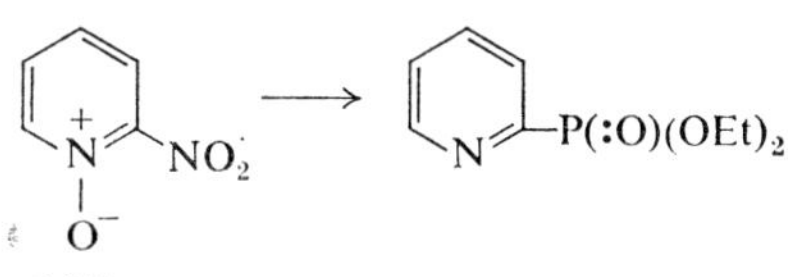

(102)

aminophosphine has been reported to deoxygenate nitrosobenzene to give azoxybenzene.[89] Nitrene formation appears to occur in the deoxygenation of the nitroso-compound (97)[90] and of the *N*-sulphinylamine (98).[91] Deoxygenation of (99) and (100) results in fragmentation to give dinitriles.[92] A number of nitro-compounds (*e.g.* 101 and 102) have been found to react with triethyl phosphite to give arylphosphonates (103) and ethyl nitrite instead of deoxygenation products.[93, 94] It was found that the more nucleophilic the phosphite, the easier the displacement reaction occurred. The isolation of ethyl nitrite is a little surprising in the light of the recent report that nitrites are deoxygenated by phosphines.[95]

89 H. Weingarten and M. G. Miles, *J. Inorg. Nuclear Chem.*, 1968, **30**, 668.
90 B. L. Dyatkin, E. P. Mochalina, Yu. S. Konstantinov, S. R. Sterlin, and I. L. Knunyants, *Izv. Akad. Nauk S.S.S.R. Ser. khim.*, 1967, 2297 (*Chem. Abs.*, 1968, **68**, 77,632).
91 T. Minami, H. Miki, and T. Agawa, *Kogyo Kagaku Zasshi*, 1967, **70**, 1829 (*Chem. Abs.*, 1968, **68**, 38,988v).
92 A-u-Rahman and A. J. Boulton, *Chem. Comm.*, 1968, 73.
93 J. I. G. Cadogan, D. J. Sears, and D. M. Smith, *Chem. Comm.*, 1968, 1107.
94 J. I. G. Cadogan, D. J. Sears, and D. M. Smith, *J. Chem. Soc.* (*C*) 1969, 1314.
95 J. H. Boyer and J. D. Woodyard, *J. Org. Chem.*, 1968, **33**, 3329.

11
Physical Methods

BY J. C. TEBBY

Although as much effort as possible has been made to classify the material into separate sections, inevitably it has been necessary to include some results in inappropriate sections, *e.g.* in the n.m.r. section some chemical shifts and some variable-temperature work will be found under spin–spin coupling constants.

1 Nuclear Magnetic Resonance Spectroscopy

Of all the physical methods available for the study of organophosphorus chemistry, n.m.r. spectroscopy has attracted by far the most interest. The use of this technique is made all the more rewarding for phosphorus compounds since specific information on the bonding and environment of the phosphorus atoms can be obtained directly from the ^{31}P spectra and indirectly from the spectra of other atoms when they are coupled to phosphorus.

A review by Gallagher and Jenkins,[1] on 'Stereochemical Aspects of Phosphorus Chemistry', includes a considerable amount of physical data and covers the literature up to mid-1967. A review [2] on the 'Spectroscopic Properties of Inorganic and Organo-metallic Compounds' covers the 1967 literature and contains sections on fluorophosphorus and phosphorus compounds. A more specific review by van Gorkom and Hall [3] includes a section on equivalence related to phosphorus compounds. In the present report the ^{31}P and ^{1}H chemical shifts are relative to 85% phosphoric acid and tetramethylsilane respectively. The proton shifts are reported in τ (p.p.m.) in order to avoid any confusion which may arise from the quotation of positive δ values (N.B. in most cases δ is negative but the sign is usually omitted).

A. Chemical Shifts and Shielding Effects.—The correlation of ^{31}P chemical shifts with additive group constants has been successfully applied in recent years to phosphines and phosphonium salts. It is found that a simple

1 M. J. Gallagher and I. D. Jenkins, *Topics in Stereochemistry*, 1968, **3**, 1.
2 N. N. Greenwood, J. W. Akitt, W. Errington, T. C. Gibb, and B. P. Straughan, 'Spectroscopic Properties of Inorganic and Organometallic Compounds', The Chemical Society, London, 1968, Vol. 1.
3 M. van Gorkom and G. E. Hall, *Quart. Rev.*, 1968, **22**, 14.

linear equation can also be derived[4] for protonated trialkylphosphines which employs the constants which were derived for tertiary phosphines. Also a theoretical expression has been derived for nuclear shielding of the phosphorus atom.[5]

The ^{31}P chemical shifts of four phosphiranes (1) were at much higher field (+234—+341 p.p.m.) than those of the corresponding tertiary and secondary acyclic phosphines (+46—+99·5 p.p.m.) but the variation within each series showed similar trends.[6]

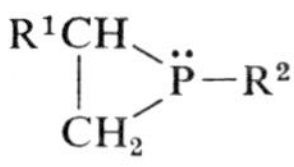
(1)

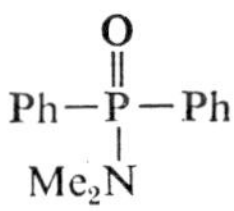
(2)

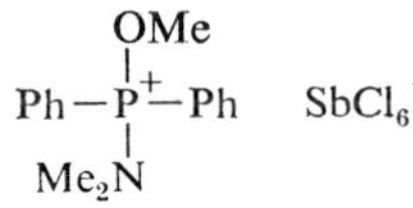
(3)

A series of dimethylamino- and methoxy-phenylphosphine chalcogenides *e.g.* (2), has been converted to quasiphosphonium salts, *e.g.* (3), and their ^{31}P chemical shifts (δ_P) have been compared.[7] Upon methylation, shielding by the chalcogenide decreases, and shielding by the other substituents increases due to changes in their p_π–d_π electron contributions. A comparison[8] of δ_P for a wide range of phenyl compounds with those of the corresponding methyl compounds shows that the general effect of the phenyl group is to increase the shielding of the phosphorus atom, presumably by an Ar—P p_π–d_π interaction. A series of related phosphinimines, (4), (5), and (6), have been prepared. The effects[9] of *S*-methylation and of

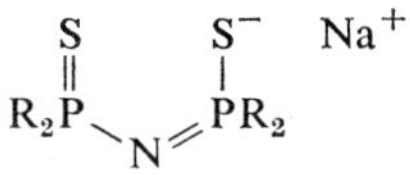
(4)

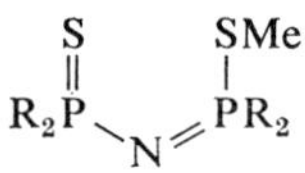
(5)

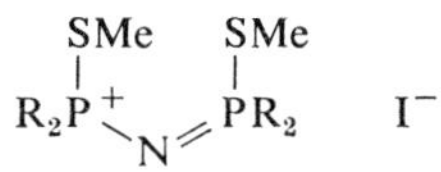
(6)

interchanging methyl and phenyl groups, (R), on the chemical shifts of both phosphorus nuclei are discussed at length. The δ_P values for three series of tetra-co-ordinated compounds of the general structure (7), (8), and (9) have been correlated[10] with the Hammett constants of the variable P-substituent.

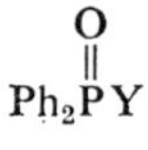
(7)

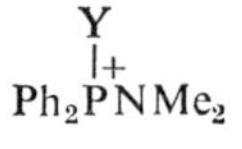
(8)

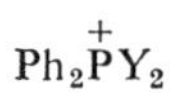
(9)

Y_3P:
(10)

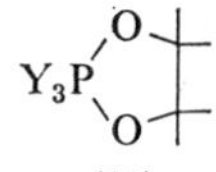
(11)

[4] G. A. Olah and C. W. McFarland, *J. Org. Chem.*, 1969, **34**, 1832.
[5] D. Purdela, *Rev. Roumaine Chim.*, 1968, **13**, 1415 (*Chem. Abs.*, **70**, 72,746).
[6] S. Chan, H. Goldwhite, H. Keyzer, D. G. Rowsell, and R. Tang, *Tetrahedron*, 1969, **25**, 1097.
[7] A. Schmidpeter and H. Brecht, *Z. Naturforsch.*, 1969, **24b**, 179.
[8] A. Schmidpeter and H. Brecht, *Inorg. Nuclear Chem. Letters*, 1968, **4**, 563.
[9] A. Schmidpeter, H. Brecht, and J. Ebeling, *Chem. Ber.*, 1968, **101**, 3902.
[10] A. Schmidpeter and H. Brecht, *Z. Naturforsch.*, 1968, **23b**, 1529.

The ^{31}P chemical shifts (δ_P) of a number of reactive alkylidene-phosphoranes have been measured.[11] The shielding increases upon the introduction of alkyl groups in place of hydrogen at the carbon atom and to a

$Ph_3P{=}CH_2$	$Ph_3P{=}CHMe$	$Ph_3P{=}CHCH_2R$	$Ph_3P{=}CMe_2$	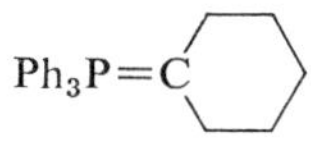
δ_p −20·3	−14·6	−12·6 to −12·2	−11·3	−6·4 ppm

lesser extent on replacing methyl by methylene. An interesting comparison has been made between the δ_P for some related P^{III} and penta-co-ordinate P^V compounds.[12] Upon changing the group Y into the phosphine series (10), the shielding rapidly increases in the order MeO < Me_2N < Ph < Me whereas the same changes have a deshielding effect on the P^V compounds (11).

These observations are in accordance with the work of Rakshys, Taft, and Sheppard,[13] who found that p_π–d_π bonding is more important in penta-co-ordinated compounds. Thus the π-bonding and presumably the shielding, increases, (*a*) with the availability of electrons on Y, and (*b*) with the reduction of the diffuse nature of the phosphorus *d*-orbitals by inductive withdrawal.

The trends were less clear in the alkyl series of (10) and (11). However, in the case of the bisbiphenylene-phosphoranes (12) shielding increases in

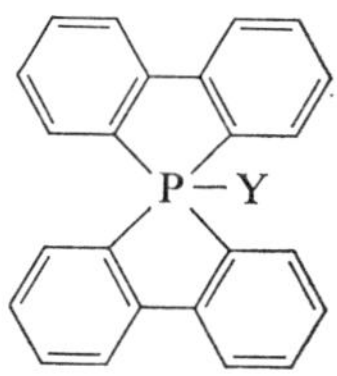

(12)

$R_2BOP(CF_3)_2$

(13)

the order aryl < alkyl < Me < H or D,[14] which is similar to that found in phosphines. The δ_P values for a number of penta-co-ordinate chlorophosphoranes, hexa-co-ordinate phosphates, and phosphine sulphides have been tabulated.[15] The ^{31}P spectra of the boron esters (13) are reported and the phosphorus resonances occur as sharp (1 + 1) septets not appreciably disturbed by the boron quadruple.[16]

The increased shielding of the methyl groups of quasiphosphonium salts (14) with increased electron donor properties of the groups Y has been

[11] S. O. Grim and J. H. Ambrus, *J. Org. Chem.*, 1968, **33**, 2993.
[12] F. Ramirez, C. P. Smith, J. F. Pilot, and A. S. Gulati, *J. Org. Chem.*, 1968, **33**, 3787.
[13] J. W. Rakshys, R. W. Taft, and W. A. Sheppard, *J. Amer. Chem. Soc.*, 1968, **90**, 5236.
[14] D. Hellwinkel, *Chem. Ber.*, 1969, **102**, 528.
[15] H. P. Latscha, *Z. Naturforsch.*, 1968, **23b**, 139.
[16] A. B. Burg and J. S. Basi, *J. Amer. Chem. Soc.*, 1969, **91**, 1937.

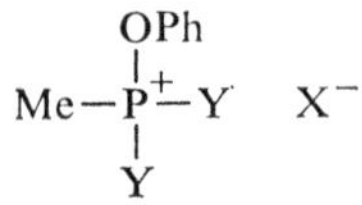

(14)

related[17] to a decrease in the positive charge on the phosphorus atom. A similar conclusion has been arrived at for aminoquasiphosphonium salts.[18]

The carbanionic nature of the α carbon atom of alkylidene-phosphoranes has a considerable shielding effect on the protons attached to this atom, *e.g.* τ 10·8—11·0 for (15)[19] and 10·61 p.p.m. for (16).[20] This shielding effect is considerably reduced by conjugative withdrawal such as that

$Me_3P{=}CHR$ (15) $Ph_3P{=}CH_2$ (16) $Ph_3P{=}CHCOMe$ (17)

$Ph_3P{=}CH{-}\overset{+}{P}Ph_3$ (18)

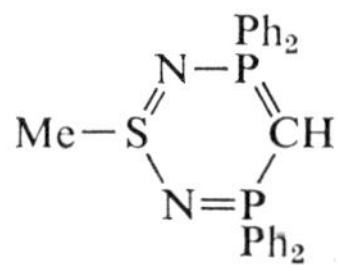

(19)

provided by a carbonyl group, *e.g.* τ 6·32 p.p.m. for (17), but the presence of a second phosphonium centre or a phosphinimine centre has a smaller effect.[21] Thus the compounds (18) and (19) exhibit ^1H resonances at τ 8·08 (8·0)[22] and 8·29 p.p.m. respectively.

A considerable amount of work has been carried out on phosphonitrilic systems[23] and chemical shifts and coupling data have been compiled[24] and systematised[25] for the trimeric compounds.

The electron acceptor capacity of the phosphoryl group, through a p_π–d_π interaction, is a more important determinant of the ^1H parameters of phosphorylated furans and thiophens (20) than for phosphorylated pyrroles.[26]

Although it is well recognised that α methylene or methyl groups of phosphonium salts are shifted *ca.* 1—2 p.p.m. downfield relative to a similar phosphine or carbon compound, the anisotropic and bonding

17 L. V. Nesterov, A. Ya. Kessel, Yu. Yu. Samitov, and A. A. Musina, *Doklady Acad. Nauk S.S.S.R.*, 1968, **180**, 116.
18 H. H. Sisler and S. R. Jain, *Inorg. Chem.*, 1968, **7**, 104.
19 H. Schmidbaur and W. Tronick, *Chem. Ber.*, 1968, **101**, 604.
20 H. J. Bestmann and J. P. Snyder, *J. Amer. Chem. Soc.*, 1967, **89**, 3936.
21 L. Siekmann, H. O. Hoppen, and R. Appel, *Z. Naturforsch.*, 1968, **23b**, 1156.
22 J. S. Driscoll, D. W. Grisley, J. V. Pustinger, J. E. Harris, and C. N. Matthews, *J. Org. Chem.*, 1964, **29**, 2427.
23 C. W. Allen, F. Y. Tsang, and T. Moeller, *Inorg. Chem.*, 1968, **7**, 2183.
24 H. P. Latscha, *Z. anorg. Chem.*, 1967, **362**, 7.
25 M. K. Feldt and T. Moeller, *J. Inorg. Nuclear Chem.*, 1968, **30**, 2351.
26 R. H. Kemp, W. A. Thomas, M. Gordon, and C. E. Griffin, *J. Chem. Soc.* (*B*), 1969, 527.

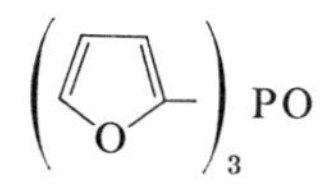

(20) O=O or S

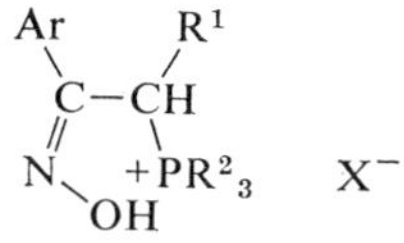

(21)

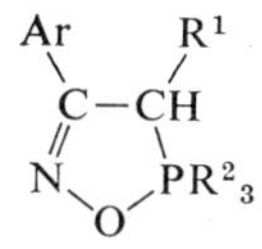

(22)

effects are difficult to separate. A number of closely related oximino-phosphonium salts (21) and phosph(v)oles (22) have been prepared[27] in which the anisotropic differences should be minimal. The phosphonium atom produces a strong deshielding of the α protons which appear 1·3–1·7 p.p.m. to lower field than the corresponding phosph(v)ole. In fact a deshielding effect is exerted through 3, 4, and even 5 bonds.

The 3′-proton of the cyclic phosphate (23) of *arabino*-cytidine resonates 1 p.p.m. downfield of the 3′-proton of *arabino*-cytidine.[28] This dramatic deshielding has been attributed to the anisotropic effect of the phosphate

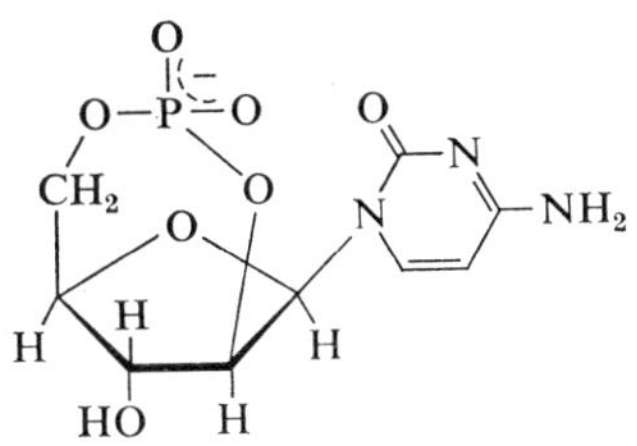

(23)

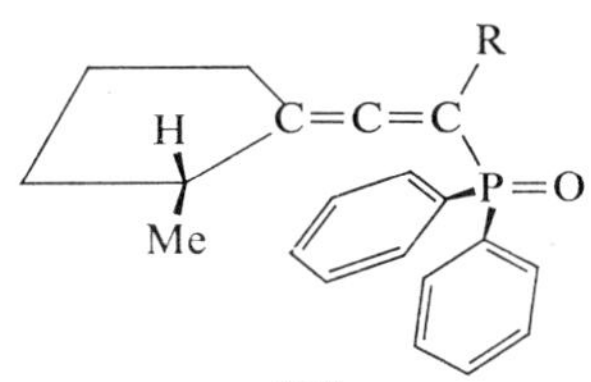

(24)

group. In the allenic phosphine oxides (24) the phosphoryl group is more remote; however, the chemical shift (τ 9·45—9·34 p.p.m.) for the methyl group of the *cis*-isomer (24) is *ca.* 0·5 p.p.m. upfield from that of the *trans*-isomer, presumably due to shielding by the phenyl rings of the Ph_2PO group.[29]

A detailed study of the ^{19}F chemical shifts of the fluorophenyl-phosphonium salts, -phosphine oxides and -ylides (25), indicates that the changes in inductive effects are more important than resonance effects.[30] It was concluded that, (*a*) the highly electronegative character of the phosphorus atom in these compounds is reduced in dimethylsulphoxide (due, possibly, to $Me_2\overset{+}{S}-\overset{-}{O}\cdots\cdots\overset{\delta+}{P}$ complexation), (*b*) stabilisation of the carbanion centre of the ylides increases the positive charge on phosphorus, (*c*) whereas the phosphoryl (P=O) p_π–d_π bonding is fairly constant such bonding in the ylides (P=O) is susceptible to inductive effects of substituents, (*d*) although

27 G. Gaudiano, R. Mondelli, P. P. Ponti, C. Ticozzi, and A. Umani-Ronchi, *J. Org. Chem.*, 1968, **33**, 4431.
28 W. J. Wechter, *J. Org. Chem.*, 1969, **34**, 244.
29 J. P. Battioni and W. Chodkiewicz, *Bull. Soc. chim. France*, 1969, 981.
30 A. W. Johnson and H. L. Jones, *J. Amer. Chem. Soc.*, 1968, **90**, 5232.

inductively withdrawing, the fluorine atom increases the aryl π-donor properties to the 'phosphonium' atom, and (*e*) one substantial conjugative p_π–d_π interaction rules out conjugative interactions with another group, *i.e.* the 3*d* orbital geometry in tetra-co-ordinate phosphorus is such that only one conjugative interaction is possible.

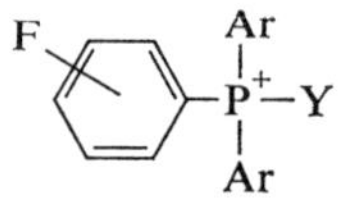

(25) Y = R, O^-, $\bar{C}R_2$

$C_6F_5\ddot{P}F_2$

(26)

$F_2C{=}CF{-}\ddot{P}F_2$

(27)

Also for P^{III} compounds, it is found[31, 32] that the pentafluorophenyl group in (26) has a larger shielding effect on the phosphorus atom (δ_P = −193·4 p.p.m.) than the phenyl group in $C_6H_5PF_2$ (δ_P = −208 p.p.m.), and a similar effect is observed for the perfluorovinylphosphines (33) and for some of the monofluorophenylphosphines (28). However, in the cases of the dimethyl phosphine (28; Y = Me) and the bis(dimethylamino) phosphine (28; Y = Me_2N) the Y_2P groups are weak donors to the aromatic ring.[13] Since $\ddot{Y}$—P: conjugation has only a minor influence on Ar—P: conjugation it seems that different interacting *d* orbitals on phosphorus are involved in these two interactions. These workers also studied

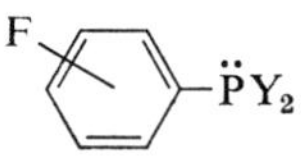

(28)

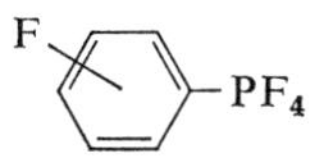

(29)

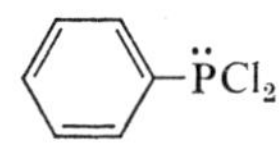

(30)

the tetra-co-ordinate compounds (25; Ar = Ph) and emphasised that the resonance effect of the PPh_2Y group increases with the inductive effect, and therefore the observation of an inductive order for substituent effects does not rule out p_π–d_π acceptor action by P. Also notable are their conclusions from a study of the penta-co-ordinate phosphoranes (29). The Ar—P p_π–d_π interaction was found to be the largest in the whole series. Thus the use of one *d* orbital to form a phosphorane does not inhibit the use of other *d* orbitals for a π-interaction.

Finally, to return to ^{31}P chemical shifts, it is found that δ_P values for a series of fluorophenylphosphines show a linear correlation with the total number of *ortho* fluorine atoms, suggesting that the *ortho*-contributions to δ_P are dominant.[34] Further, the upfield shift on introducing fluorine atoms into the aromatic ring of chlorides such as (30) was greater for the *meta*- than the *para*-orientated fluorine atoms.[35] This provides further evidence

[31] M. Fild and R. Schmutzler, *J. Chem. Soc.* (*A*), 1969, 840.
[32] M. G. Hogben and W. A. G. Graham, *J. Amer. Chem. Soc.*, 1969, **31**, 283.
[33] A. H. Cowley and M. W. Taylor, *J. Amer. Chem. Soc.*, 1969, **91**, 1929.
[34] D. I. Nichols, *J. Chem. Soc.* (*A*), 1969, 1471.
[35] R. De Ketelaere, E. Muylle, W. Vanermen, E. Claeys, and G. P. Van Der Kelen, *Bull. Soc. chim. belges*, 1969, **78**, 219.

No. *ortho* fluorines	0	2	3	4
Phosphine	Ph_3P	$Ph_2C_6F_5P$	2,5-$(C_6H_3F_2)_3P$	$Ph(C_6F_5)_2P$
δ_P (p.p.m.)	8	26·3	34·5	49·7

No. *ortho* fluorines	6	6	6
Phosphine	$(C_6F_5)_3P$	2,3,6-$(C_6H_2F_3)_3P$	2,6-$(C_6H_3F_2)_3P$
δ_P (p.p.m.)	75·5	78·5	78·5

that the main function of the fluorine atom is inductive and enables the aryl ring to back donate to P more efficiently. It was also found that replacement of an aryl group by a chlorine atom resulted in a downfield shift.[35]

B. Relaxation Times.—T_1 has been measured for trimethyl phosphate by a rapid-passage method.[36] T_1 and T_2 have been used to study the interactions of tributyl phosphate and di-isoamyl methylphosphonate and their uranyl nitrate complexes.[37]

C. Studies of Equilibria and Reactions.—The Lewis basicity of tricyclopropylphosphine has been estimated on the basis of n.m.r. peak areas of the free and complexed phosphine.[38] Methyl phosphate at pH 4—10 consists of a rapidly interconverting mixture of mono- and di-anions (31) and (32)

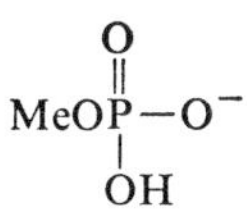
(31)

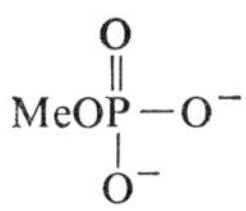
(32)

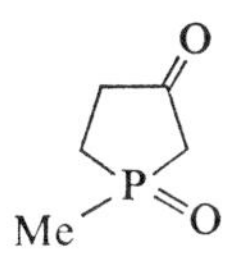
(33)

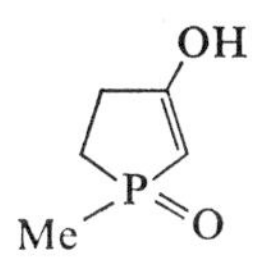
(34)

and only one methyl signal is observed. However, the variation of τ_{CH_3} and J_{POCH} of methyl phosphate with pH is a direct measure of the proportions of the two anions and can therefore be used to estimate the pK value.[39] The chemical shift of water protons has been used[40] to study the equilibrium constants and enthalpy changes for hydrogen bonding of water to tributyl phosphate in the range 10—45° in carbon tetrachloride.

Keto–enol tautomerism, which is strongly concentration dependent, has been identified for the cyclic oxide (33).[41] ^{1}H and ^{31}P Spectra support the i.r. evidence that in high concentration the enol (34) predominates and at low concentration the keto-form (33) predominates. This appears to be the first instance where a significant degree of stabilisation of the enol function is imparted by a phosphoryl group. ^{31}P N.m.r. has also been used to identify

[36] Yu. I. Mitchenko and V. V. Frolov, *Yad. Magn. Resonans Leningrad Gos Univ.*, 1968, 184 (*Chem. Abs.*, **70**, 15,780).
[37] P. M. Borodin, E. N. Sventitskii, and V. I. Chizhik, *Yad Magn. Resonans Leningrad Gos Univ.*, 1968, 69 (*Chem. Abs.*, **70**, 15,834).
[38] A. H. Cowley and J. L. Mills, *J. Amer. Chem. Soc.*, 1969, **91**, 2915.
[39] E. L. Uhlenhopp, J. A. Glasel, and A. I. Krasna, *J. Org. Chem.*, 1969, **34**, 2237.
[40] S. Nishimura, C. H. Ke, and N. C. Li, *J. Amer. Chem. Soc.*, 1968, **90**, 234.
[41] L. D. Quin and J. A. Caputo, *Chem. Comm.*, 1968, 1463.

prototropy in allylphosphonates (35)[42] and to estimate the structure of the alkali-metal derivatives of dimethyl- and diphenyl-phosphine oxides (36).[43]

(35)

(36)

The exchange kinetics for triarylphosphines in pseudotetrahedral complexes of cobalt and nickel halides have been determined using line-widths and chemical shifts of excess of phosphine.[44] The borine and ethylborane derivatives of the aminophosphines (37) and (38) show ^{31}P

$Me_2N\ddot{P}Me_2$ (37) $(Me_2N)_3\ddot{P}$ (38) $Et_2N\ddot{P}MeOPr^i$ (39)

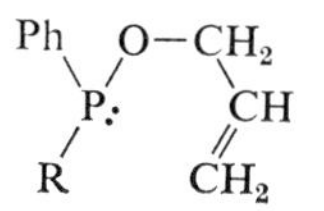

(40)

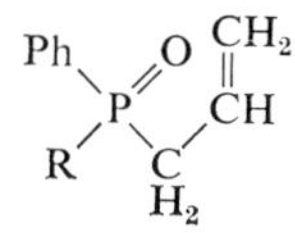

(41)

chemical shifts ($\Delta\delta_P$ 18—22 p.p.m.) downfield of the free phosphine, whereas there is very little change in τ_{CH_3N} ($\Delta\tau_{CH_3}$ 0·2 p.p.m.). These results have been interpreted in favour of a P—B dative bond.[45] Similar conclusions have been drawn for a borane derivative of (39).[46]

^1H N.m.r. has been used to study the reaction kinetics and stereospecificity of the rearrangement of allyl phosphinites (40) to allylphosphine oxides (41),[47] and ^{31}P n.m.r. has been used to follow the reactions of chlorine with phosphines,[48] the alkali hydrolysis of tetrakis(hydroxymethyl)phosphonium chloride,[49] and the reaction of ethyl diphenylphosphinate with phenylmagnesium bromide.[50]

^{31}P and 200 MHz ^1H N.m.r. have been used to study the ligand exchange rates and isomerisation of the rhodium complex (42), which is used for the homogeneous hydrogenation of olefins. The results were rationalised in terms of the formation of a loose complex (43) which indicates that there is, in addition to a reactive site on the metal, a means of accumulating molecules in a 'second co-ordination sphere' ready for reaction.[51]

$$\underset{(42)}{Rh(Ph_3P)_3Cl + Ph_3P} \rightleftharpoons \underset{(43)}{[Rh(Ph_3P)_3Cl]Ph_3P}$$

[42] G. Mavel, R. Mankowzki-Favelier, G. Lavielle, and G. Sturtz, *J. Chim. Phys.*, 1967, **64**, 1698.
[43] K. Issleib, B. Walther, and E. Fluck, *Z. Chem.*, 1968, **8**, 67.
[44] L. H. Pignolet and W. D. Horrocks jun., *J. Amer. Chem. Soc.*, 1968, **90**, 922.
[45] J. P. Laurent, G. Jugie, and G. Commenges, *J. Inorg. Nuclear Chem.*, 1969, **31**, 1353.
[46] J. P. Laurent, G. Jugie, and R. Wolf, *J. Chim. Phys. Physicochim. Biol.*, 1969, **66**, 409.
[47] A. W. Herriott and K. Mislow, *Tetrahedron Letters*, 1968, 3013.
[48] D. B. Denney, D. Z. Denney, and B. C. Chang, *J. Amer. Chem. Soc.*, 1968, **90**, 6332.
[49] W. J. Vullo, *J. Org. Chem.*, 1968, **33**, 3665.
[50] H. R. Hays, *J. Amer. Chem. Soc.*, 1969, **91**, 2736.
[51] D. R. Eaton and S. R. Suart, *J. Amer. Chem. Soc.*, 1968, **90**, 4170.

D. Pseudorotation.—Pseudorotation is an intramolecular process by which the trigonal bipyramidal orientation of groups arranged about a penta-coordinate central atom may be changed without the cleavage of bonds. The process has recently received detailed attention.[52, 53] Calculations based on n.m.r., i.r., and Raman data confirm that a tetragonal bipyramid provides the lowest energy path leading to a vibrational exchange of orientation.[54]

A review by Westheimer[52] on 'Pseudorotation in the Hydrolysis of Phosphate Esters' includes a discussion of the variable-temperature n.m.r. evidence for the pseudorotation of oxyphosphoranes. This subject is taken further in a review by Ramirez[55] on isolable oxyphosphoranes which includes a discussion of the application of solvent effects, δ_P, and *X*-ray crystallography to the problem of the structure and equilibria of oxyphosphoranes.

Chemical evidence has indicated that the presence of small rings (four- or five-membered) and an electronegative substituent restricts alkyl and aryl groups of oxyphosphoranes to an equatorial orientation.[56] Variable-temperature n.m.r. studies on the oxyphosphoranes (44), (45), and (46) support this conclusion. Two stereoisomeric forms (44a) and (44b) have

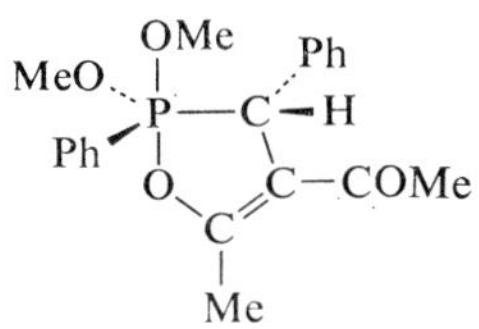

(44a)

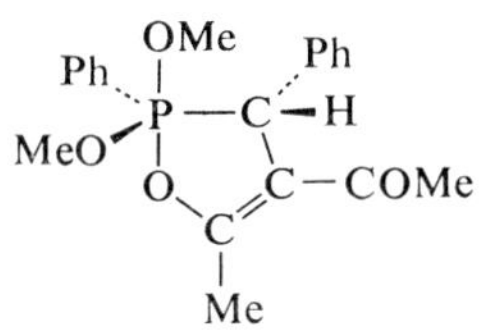

(44b)

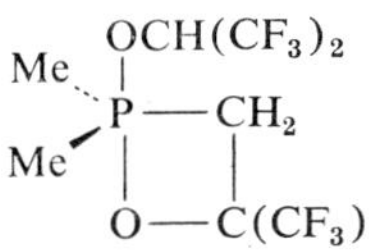

(45)

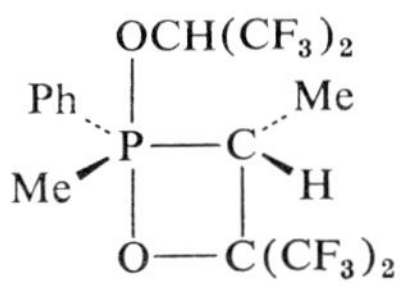

(46a)

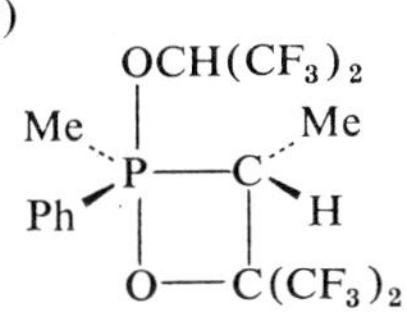

(46b)

been identified[57] in which the *P*-phenyl group retains an equatorial orientation even when the *P*-alkoxy-groups are undergoing rapid apical–equatorial exchange relative to the n.m.r. time scale. At −20° the spectrum is that of the two separate isomers, at 0—5° interconversion of apical and equatorial methoxy-groups becomes apparent, and at 52° positional exchange of the

52 F. H. Westheimer, *Accounts Chem. Res.*, 1968, **1**, 70.
53 P. C. Lauterbur and F. Ramirez, *J. Amer. Chem. Soc.*, 1968, **90**, 6722.
54 R. R. Holmes sen. and R. M. Deiters, *J. Amer. Chem. Soc.*, 1968, **90**, 5021.
55 F. Ramirez, *Accounts Chem. Res.*, 1968, **1**, 168.
56 W. Hawes and S. Trippett, *Chem. Comm.*, 1968, 295; *J. Chem. Soc.* (*C*), 1969, 1465; S. E. Cremer, R. J. Chorvat, and B. C. Trivedi, *Chem. Comm.*, 1969, 769; J. R. Corfield, J. R. Shutt, and S. Trippett, *ibid.*, 1969, 789.
57 F. Ramirez, J. F. Pilot, O. P. Madan, and C. P. Smith, *J. Amer. Chem. Soc.*, 1968, **90**, 1275.

phenyl groups becomes important as indicated by the coalescence of the methoxyl doublets and acetyl singlets.

The ^1H n.m.r. spectrum of (45) did not change in the range -70 to $+30°$ which was interpreted in terms of lack of pseudorotation due to the restricting influence of the four-membered ring. The phosphetane (46) was isolated as a mixture of the two diasteriomers, (46a) and (46b).[58] The rate of stereomutation was appreciable at 120°; however, pyrolysis to phosphonite and olefin began to compete at elevated temperatures.

A revealing variable-temperature study has been carried out[59] on the spiro-oxyphosphorane (47) in perdeuteriotoluene solution. In the range -70 to 0° the ring methylene protons appeared as part of an AA′BB′X system and the methyl groups appeared as two sharp signals (7·5 Hz separation due to chemical shift). This means that the methylene protons and methyl groups labelled 1 and 2 are equivalent, as are those labelled 3 and 4. Therefore the interconversion of (47a) and (47b) is fast even at low

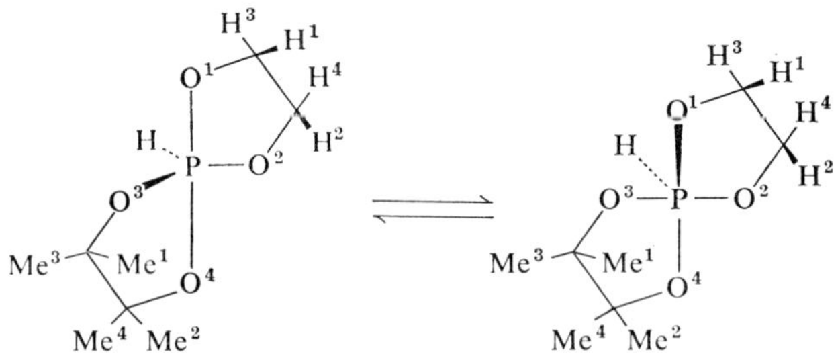

(47a) (47b)

temperature. At 37° the two methyl peaks coalesce and become a sharp line at high temperature whilst the ring methylene protons change to an A_4X system. The P—H group appeared as a doublet ($^1J_{PH}$ 808 Hz) which was independent of temperature. The higher temperature process renders equivalent all the substituents of one ring and can be explained by the operation of a pseudo-rotatory process involving a strained trigonal bipyramidal structure with an apical hydrogen, *e.g.* (47c) for the interconversion of (47a) and (47d).

These results are in accordance with the rapid exchange observed for the fluorophosphorane (48) compared with the corresponding six-membered ring compound,[60] and are contrary to the belief that small rings increase the barrier to pseudorotation. Indeed it can be argued that they lower the energy barrier between structures which maintain apical–equatorial bridging by the small ring(s). Thus not only should the small ring(s) stabilise the trigonal bipyramidal intermediates but also the tetragonal

[58] F. Ramirez, C. P. Smith, and J. F. Pilot, *J. Amer. Chem. Soc.*, 1968, **90**, 6726.
[59] D. Houalla, R. Wolf, D. Gagnaire, and J. B. Robert, *Chem. Comm.*, 1969, 443.
[60] W. Mahler and E. L. Muetterties, *Inorg. Chem.*, 1965, **4**, 1520.

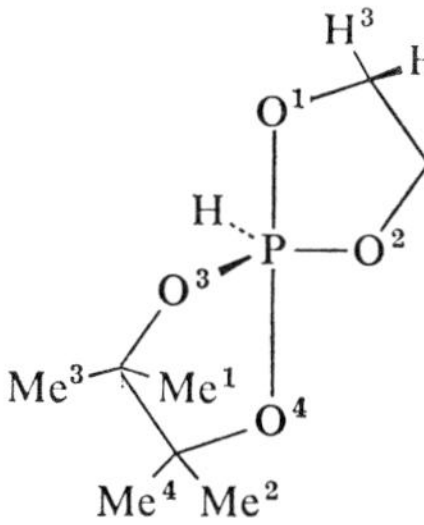

(47a)

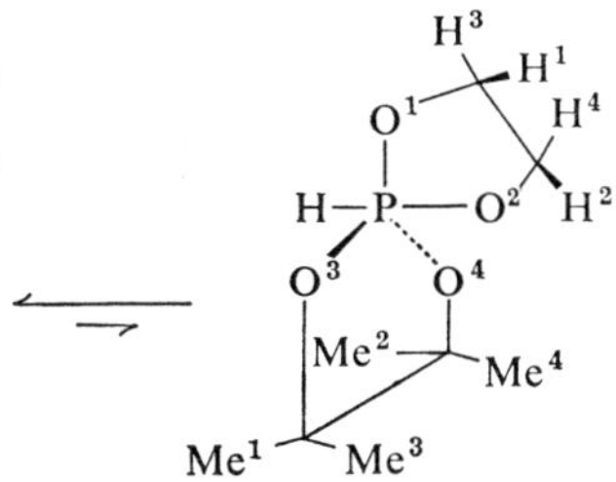

(47c)

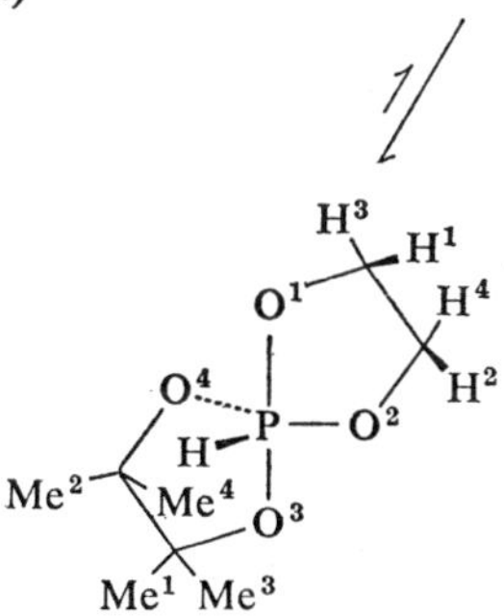

(47d)

(48)

bipyramidal transition state, and moreover they should reduce nonbonding interactions in the latter to a greater degree. Furthermore, for small rings with identical atoms bound to phosphorus, the trigonal bipyramidal intermediate is destabilised by the forced occupation of an electropositive atom in an apical position or an electronegative atom in an equatorial position.

However, the small rings do raise considerably the energy of some trigonal bipyramidal structures, *e.g.* (47c).* Also recent calculations based on molecular orbital theory indicate that a small ring lowers the occupation of the 3*d* orbitals of phosphorus,[61] a factor which has been put forward to account for the slow pseudorotation of the perfluorovinylphosphoranes (49) compared with that of the saturated analogues (50).[62]

[61] D. B. Boyd, *J. Amer. Chem. Soc.*, 1969, **91**, 1200.
[62] A. H. Cowley and M. W. Taylor, *J. Amer. Chem. Soc.*, 1969, **91**, 1934.

* The isolation of stable optically active bisbiphenylenephosphoranes of the type (48) bears evidence of the barrier raised by such unstable trigonal bipyramidal structures. The rapid pseudorotation between (48a) and (48b) which may occur when R is pivot does not lead to loss of chirality in this case.

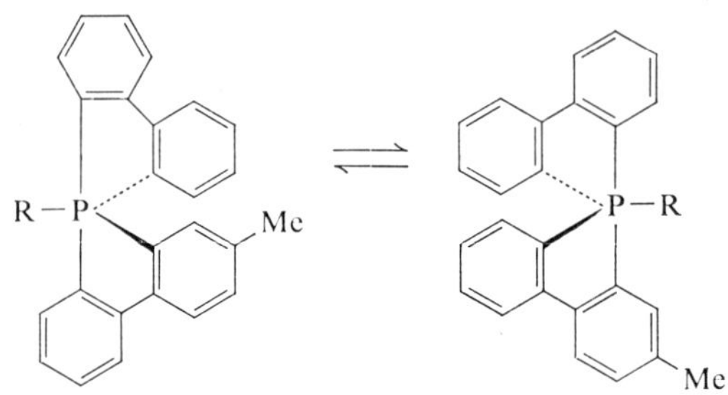

(48a) (48b)

Upon cooling (49) to $-20°$ the ^{19}F spectrum showed two F(P) environments in 2 : 1 abundance in accordance with equatorial vinyl groups as shown. The spectrum of (51) showed spectroscopic equivalence of the

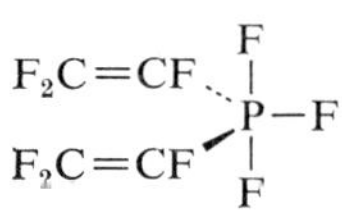

(49)

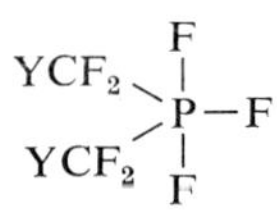

(50) Y = F or CF_3

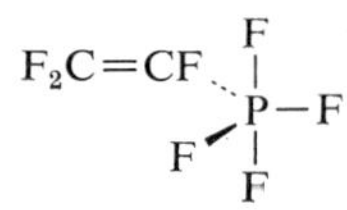

(51)

phosphorane fluorine atoms, F(P), down to $-100°$. The stereochemistry of (51) was estimated by comparing its δ_F values with those of (49). The average chemical shift of the F(P) atoms should be 67·9, *i.e.* $[\frac{3}{4}(76{\cdot}8) + \frac{1}{4}(41{\cdot}3)]$ if the perfluorovinyl group is predominantly apical but 59·1 if it is equatorial. Since the observed shift is 54·9 p.p.m. a predominantly equatorial vinyl group is favoured.

The spectra of the aminotetrafluorophosphoranes are consistent with an equatorial amino-group,[63–65] whereas the spectrum of phenyltetrafluorophosphorane $PhPF_4$, is characteristic of rapid exchange [31] as is the spectrum of the cyclic diphosphorane (52).[65] This appears to be another example of ready pseudorotation induced by the presence of a small ring. The spectra of ethyl- and methyl-trifluorophosphorane (53; R = Et or Me) [66] and a number of phenoxy- and mercapto-difluoro- and trifluoro-phosphoranes [67] and other trifluorophosphoranes [68] are all consistent with the fluorine atoms occupying apical orientations.

It is difficult to rationalise the factors which cause slow pseudorotation of fluorophosphoranes. It is possible that a slow process occurs only when one

[63] F. N. Tebbe and E. L. Muetterties, *Inorg. Chem.*, 1968, **7**, 172.

[64] M. A. Landau, V. V. Sheluchenko, G. I. Drozd, S. S. Dubov, and S. Z. Ivin, *Zhur. struct. Khim.*, 1967, **8**, 1097 (*Chem. Abs.*, **69**, 35,214).

[65] J. J. Harris and B. Rudner, *J. Org. Chem.*, 1968, **33**, 1392.

[66] R. A. Goodrich and P. M. Treichel, *Inorg. Chem.*, 1968, **7**, 694.

[67] S. C. Peake and R. Schmutzler, *Chem. Comm.*, 1968, 665.

[68] G. I. Drozd, S. Z. Ivin, V. V. Sheluckenko, B. I. Tetel'baum, and A. D. Varshavzkii, *Zhur. obshchei Khim.*, 1968, **38**, 567 (*Chem. Abs.*, **69**, 43,972).

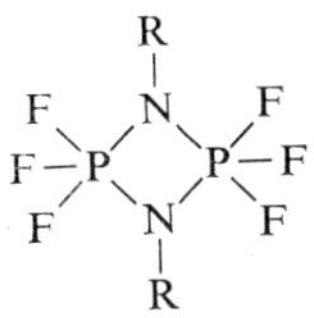

(52)

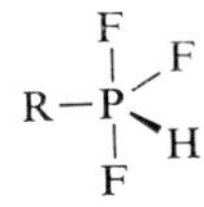

(53)

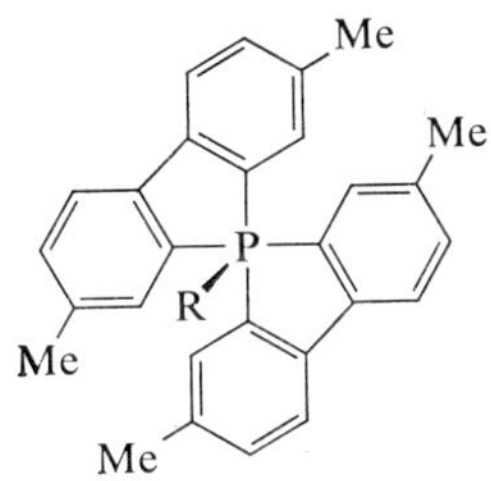

(54)

equatorial substituent (*e.g.* F, OR, NR_2) can strongly p_π–d_π bond to the central atom and when the other two equatorial substituents (*e.g.* H, R, NR_2) do not vigorously compete for apical orientations. To explain the rapid exchange which occurs for (50) it would be necessary to classify the trifluoromethyl and pentafluoroethyl groups as strongly electronegative and as providing strong competition for the apical positions.

Finally, some further work has been carried out on the bis(biphenylene)-phosphoranes (54).[69] When R is aromatic the only major change is in the steric bulk of the group and, as had already been established, pseudorotation is made more difficult as R increases in size.

This result implies that the equatorial ligands approach the pivot ligand (V) to a greater extent than the apical ligands move away from V and therefore the transition state (55) is favoured over the alternative structure (56).

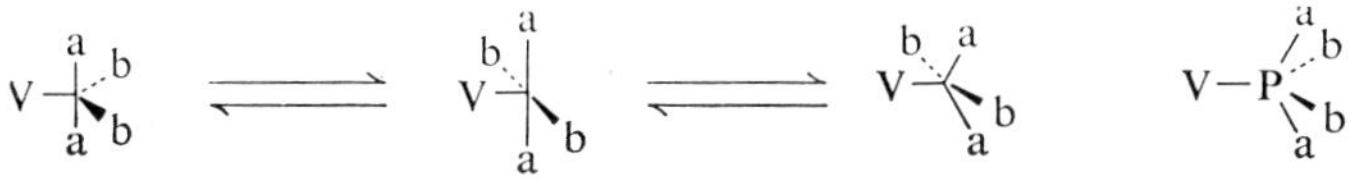

(55) (56)

It is interesting to find that when R is methyl, ethyl, or benzyl [69] in (54), pseudorotation is slower than in the β-naphthyl and phenyl examples. It seems that either π-bonding is restricted to the biphenylene rings or that a second factor comes into play which more than cancels the loss in π-bonding when R is alkyl.

[69] D. Hellwinkel, *Chimia (Switz.)*, 1968, **22**, 488.

E. Restricted Rotation.—The ^{1}H n.m.r. spectrum of the bis(biphenylene)-phosphorane (57)[69] exhibits four different methyl resonances indicating restricted rotation about the P–biphenylene bond. A similar observation has been made for the mercapto-fluorophosphoranes (58) and (59).[70]

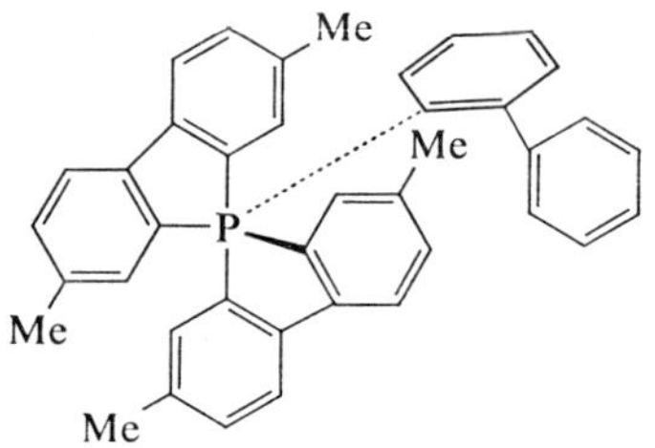

(57)

(58)

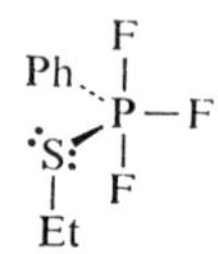

(59)

Upon cooling, the ^{19}F spectrum of (58) shows two nonequivalent apical fluorine atoms and two equivalent equatorial atoms, as a complicated 32-line pattern. The spectrum of (59) at $-60°$ shows three nonequivalent fluorine atoms.

A large number of aminophosphines of the general structure (60) show variable-temperature spectra,[71] which have been interpreted in terms of

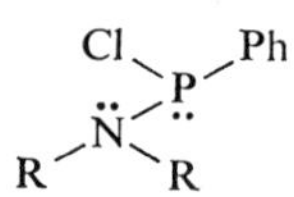

(60)

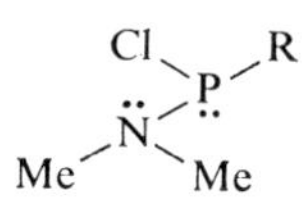

(61)

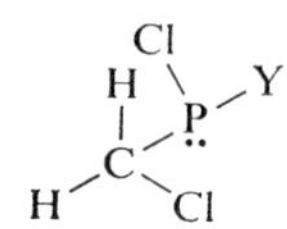

(62)

rapid Cl exchange with P-inversion above 80°, slow P-inversion at room temperature, and slow rotation about the P—N bond at $-80°$. The rotational barrier was dependent on the size of the substituents. It was suggested that p_{π}–d_{π} bonding may be partially responsible for the relatively high barrier to rotation compared with derivatives of ethane and hydrazine. The spectrum of (60; R = Me) has also been interpreted by other workers.[72] The ^{1}H spectrum of this compound contains two clearly defined *N*-methyl doublets below $-60°$ ($\Delta\tau_{Me}$ 0·53 p.p.m.) which have also been attributed to restricted rotation. The di-isopropylamino-compound (60; R = Pr^{i}) showed four nonequivalent methyl groups at $-50°$, coalescence occurring at $-10°$. The barrier to rotation was attributed to a lone pair–lone pair interaction and not p_{π}–d_{π} interaction since the latter was not expected to limit the amino and phosphine groupings to a specific orientation. Slow nitrogen inversion was ruled out by the increase in coalescence temperature

[70] S. C. Peake and R. Schmutzler, *Chem. Comm.*, 1968, 1662.
[71] D. Imbery and H. Friebolin, *Z. Naturforsch.*, 1968, **23b**, 759.
[72] A. H. Cowley, M. J. S. Dewar, and W. R. Jackson, *J. Amer. Chem. Soc.*, 1968, **90**, 4185.

which occurs when the bulk of the *N*-substituents are increased. Fresh evidence in favour of restriction due to p_π–d_π bonding has been presented.[73] Compounds with groups of widely differing electronegativities attached to P would be expected to alter the lone pair electron distribution and so change the rotational barrier. However, the phosphines (61; R = CF_2CHCl_2) and (61; R = Me_2CH) possess coalescence temperatures similar to that of (61; R = phenyl). Further, nonequivalence is not observed for the corresponding phosphine oxides, or the phosphines $ClCH_2PFNMe_2$, $Me_2CHPPhNMe_2$, and MeC≡$CPPhNMe_2$. It is suggested that these results are in accordance with restricted rotation due to p_π–d_π bonding and that an electronegative substituent such as Cl is required in order to promote *d*-orbital contraction, but that substituents such as O^- and F, which themselves form strong d_π–p_π bonds, lead to a low multiple bond order for the P—N bond and a low rotational barrier. [For a discussion of i.r. evidence, see this chapter (3B).]

N.m.r. confirms earlier reports, based on i.r. and Raman spectra,[74] of restricted rotation about phosphorus carbon bonds in compounds of the type (62). The ^1H spectrum of the chloride (62; Y = NMe_2) shows [75] in addition to the expected nonequivalence of the methylene protons, geminal PCH coupling constants with opposite signs (±26 and ∓3 Hz). This has been explained as due to a strong conformational preference such as that shown. Addition of triethylammonium chloride to chloroform solutions led to the collapse of the ABX pattern to an A_2X pattern indicating the occurrence of a rapid chloride-induced exchange reaction at P, with inversion of configuration. The ^1H spectra of (63; R = isopropyl or pinacolyl; Ar = *p*-nitrophenyl or *o*-chloro-*p*-nitrophenyl) show doubling of the alkyl ester protons.[76]

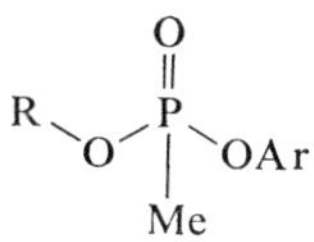

(63a)

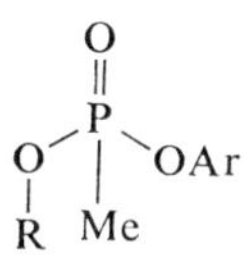

(63b)

The nonequivalence is not maintained at 150° and in the *o*-chloro-*p*-nitrophenyl esters the doubled peaks were in a 2 : 3 intensity ratio. This observation has been attributed to restricted rotation about the RO—P bond giving the conformations (63a) and (63b). The restricted rotation may, presumably, be due to preferred p_π–d_π bonding by the alkyl ether oxygen. Addition of uranyl nitrate removed the doubling as well as shifting the resonances downfield in accordance with the fixation of the molecules in one of the conformations.

[73] H. Goldwhite and D. G. Rowsell, *Chem. Comm.*, 1969, 713.
[74] E. Steger, J. Rehak, and H. Faltus, *Z. Physik. Chem.*, 1965, **229**, 110.
[75] H. Goldwhite and D. G. Rowsell, *Chem. Comm.*, 1968, 1665.
[76] R. V. Jardine, A. H. Gray, and J. B. Reesor, *Canad. J. Chem.*, 1969, **47**, 35.

Less surprising is the restricted rotation about the P–aryl bond of trimesitylphosphine (64).[77]

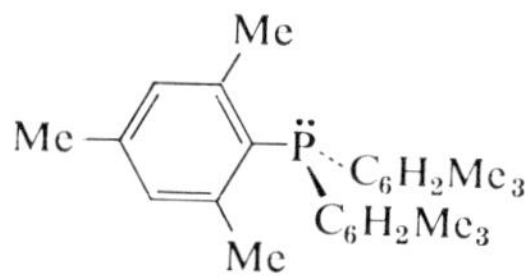

(64)

Further work on the carbalkoxymethylenephosphorane system[78] shows that the major rotamer is *transoid* for (65; Y = Me, R = Et) but *cisoid* for (65; Y = H or Ph, R = Et). The ease of rotation about the C—C bond

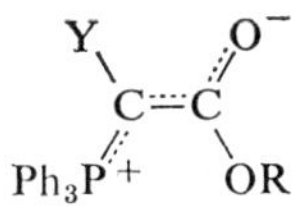

(65) *transoid*

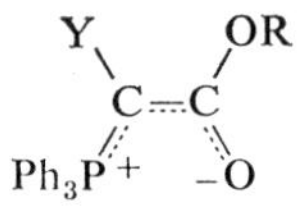

(65) *cisoid*

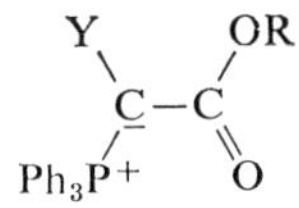

(66)

increases in the order Me < H < Ph indicating an increased contribution of the phosphorane form (66) to the resonance hybrid.[78a]

The same trend is found for the methyl esters.[78b] The preference for *transoid* stereochemistry increases with the bulk of Y when it is an alkyl group, but remains constant (80 : 20 in favour of the *cisoid* rotamer) for the aryl compounds (65; Y = C_4H_4Z, R = Me) upon variation of the *para*-substituent (Z). There is conflicting evidence concerning the effect of acid on the rotational barriers. A trace of acid is reported to decrease the coalescence temperature (T_c) of the methyl esters by *ca.* 30° [rotation occurring *via* the conjugate acid (67)][78b] whereas the T_c value of the ethyl

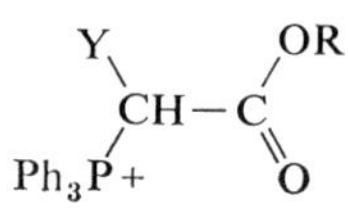

(67)

ester (65; Y = H, R = Et) is reported[79] to be unaffected by the presence of its conjugate acid. In fact, the addition of D_2O to (65; Y = H, R = Et) raises T_c by nearly 60°.

Although different compounds were studied, the different observations may also be related to the nature of the medium (and therefore the quantity and type of 'impurities' present). If this is so one can envisage rapid proton

[77] A. Rieker and H. Kessler, *Tetrahedron Letters*, 1969, 1227.
[78] (*a*) D. M. Crouse, A. J. Wehman, and E. E. Schweizer, *Chem. Comm.*, 1968, 866.
(*b*) H. I. Zelinger, J. P. Snyder, and H. J. Bestmann, *Tetrahedron Letters*, 1969, 2199.
[79] P. Crews, *J. Amer. Chem. Soc.*, 1968, **90**, 2961.

exchange being important in polar media, betaine solvation predominating in less polar media.

When two ester groups are conjugated α and γ to a methylene-phosphorane grouping as in (68) only one shows slow rotation.[80] The absence of variable-temperature spectra for the acylmethylenephosphoranes (69) and

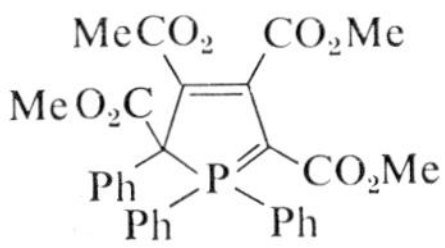

(68)

$Ph_3P{=}CHCOR$

(69)

$R_2S{=}CHCOR$

(70)

corresponding sulphur ylides (70) has been attributed to the absence of restricted rotation, since treatment of the latter with $Me_3\overset{+}{O}\,\overset{-}{B}F_4$ gave a mixture of *cis*- and *trans*-*O*-methylation products.[81]

This is contrary to expectation since the more powerful electron acceptor properties of the acyl groups should increase the C—C double bond character relative to the same bond in compounds containing carbalkoxy-groups, *e.g.* (65).

F. Other Temperature and Medium Effects.—The nonequivalence ($\Delta\delta$) of the isopropyl methyl signals of the phosphine sulphide (71) is due to the

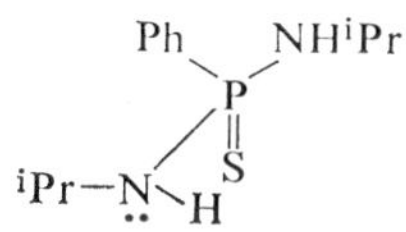

(71)

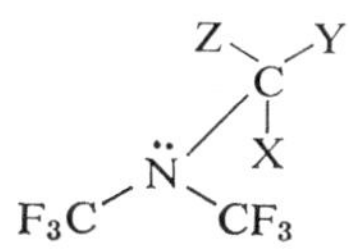

(72)

intrinsic asymmetry of the phosphorus atom. In this case the magnitude is temperature dependent—coalescence occurring at 50°. This observation was first attributed to inversion at N since the addition of acid also caused coalescence.[82] However, a reinvestigation showed that the nonequivalence diminishes smoothly with increased temperature.[83] and the singlets remain sharp just before and at superposition, indicating that the coalescence does not result from a critical rate of site exchange. It has been suggested therefore that the change with temperature is due to changes in conformer populations. The change in chemical shift on the addition of acid (in fact $\Delta\delta$ was found to increase with small addition of acid) was attributed to the

[80] N. E. Waite, J. C. Tebby, R. S. Ward, and D. H. Williams, *J. Chem. Soc.* (*C*), 1969, 1100.

[81] S. H. Smallcombe, R. J. Holland, R. H. Fish, and M. C. Caserio, *Tetrahedron Letters*, 1968, 5987.

[82] R. Keat, W. Sim, and D. S. Payne, *Chem. Comm.*, 1968, 191.

[83] A. H. Cowley, M. J. S. Dewar, W. B. Jennings, and W. R. Jackson, *Chem. Comm.*, 1969, 482.

sum of the nonequivalences in the protonated and non-protonated species weighted to the respective mole fractions. It appears that in the original study the HCl concentration was by chance that where $\Delta\delta = 0$.

It is interesting to note that slow nitrogen inversion, associated with steric hindrance, is favoured as the major factor causing nonequivalence of the two trifluoromethyl groups in the halogenoamines (72) at low temperatures.[84] However, Gorkom and Hall, on p. 22 of their review,[3] have pointed out that slow nitrogen inversion is not necessary to explain nonequivalence when there is an adjacent chiral centre.

The solvent effects (carbon tetrachloride–benzene) on the spectra of a wide range of dimethylamino-phosphine oxides, sulphides, and cyclotriphosphazatrienes have been reported.[85] The proton signals were shifted upfield in benzene solution in accordance with the formation of a weak complex in which one or more benzene molecules are orientated towards the positive end of the dipole in the phosphorus compound.

Although weak complex formation is a popular explanation for aromatic solvent effects, an increasing number of workers attribute solvent effects to changes in conformer populations. The nonequivalence of the methylene protons in the benzyl phosphonium bromides (73)[86] and the methyl groups in (74)[87] have been compared at different temperatures, in several solvents and

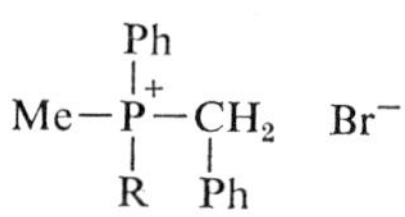

(73)

(74)

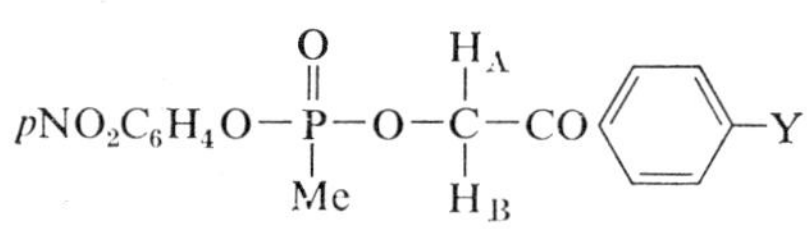

(75)

also for analogues of (74) where PhP is replaced by PhN, S, and Se. The nonequivalence is greatest at low temperatures and in the less polar solvents. This indicates that intrinsic asymmetry is not the sole source of of nonequivalence and that changing conformer populations may be important. The extent of conformer preferences of the phosphonates (75) has been examined[88] by considering the variation of J_{PH_A}, J_{PH_B}, and $J_{H_AH_B}$ with solvent and temperature. The degree of conformer preference appears to follow an inverse dependence with the dielectric constant of the solvent.

[84] M. G. Barlow and K. W. Cheung, *Chem. Comm.*, 1969, 870.
[85] R. Keat and R. A. Shaw, *J. Chem. Soc.* (*A*), 1968, 703.
[86] F. Caesar and W. D. Balzer, *Chem. Ber.*, 1969, **102**, 1665.
[87] W. McFarlane and J. A. Nash, *Chem. Comm.*, 1969, 524.
[88] L. S. Frankel, H. Klapper, and J. Cargioli, *J. Phys. Chem.*, 1969, **73**, 91.

The lower the dielectric constant the greater the repulsion energy and hence the greater the conformer preference.

Contrary to earlier reports the spectra of phosphonium salts can show anion dependence especially in less polar solvents such as chloroform. The cyclopropenyl salts (76) [89] show changes in the methine chemical shift and coupling constants whereas it is the extent of the nonequivalence of the

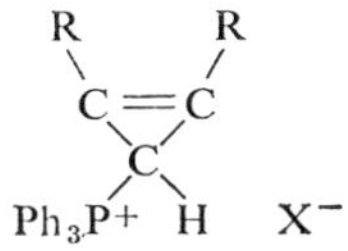

(76)

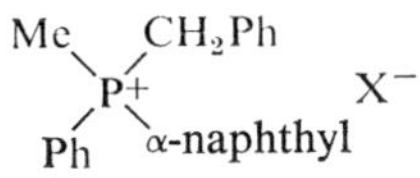

(77)

methylene protons which varies in the spectra of chiral benzylphosphonium salts (77).[90] Although mixed salts gave averaged signals showing a short life for the ion pairs, the anion dependence shown by the spectra of the benzylphosphonium salts is believed to be caused by association since the measured molecular weight is greater than that calculated for the salt. Aggregation in solution has also been investigated by paramagnetic effects [see this chapter (1I)].

Variable-temperature studies have been used to study the molecular motions of phospholipids [91] and egg-yolk lecithin.[92]

Temperature and solvent effects relating directly to inversion and spin–spin coupling are discussed in the appropriate sections.

G. Inversion and Configuration.—The study of the stereochemical lability of diphosphines [93] by use of variable temperature n.m.r. has been taken further. Evidence is presented which discounts dissociation as an inversion pathway, and rotational isomerism was ruled out by the lack of doubling for Me_2PPMe_2 and Me_2PPh_2 and from the fact that diphosphines have been shown to prefer one conformation only (*trans*). Consequently the doubling of methyl and phenyl protons shown by (78) below 180° is confirmed as

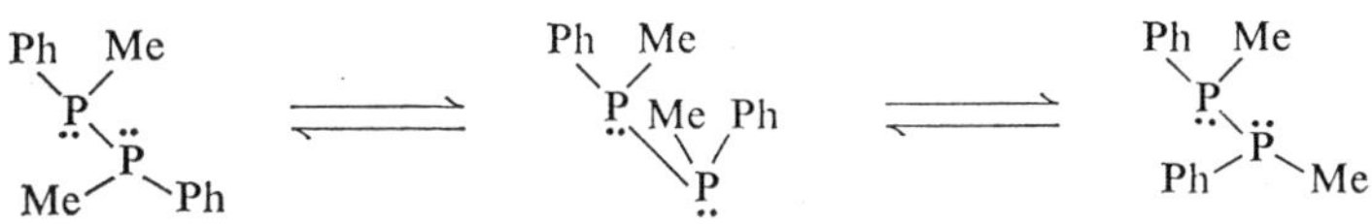

(78) *meso* (78) (±)

being due to the interconversion of *meso*- and (±)-forms through inversion at phosphorus. The low Arrhenius activation energy (26 ± 2 kcal mole^{-1})

[89] D. T. Longone and E. S. Alexander, *Tetrahedron Letters*, 1968, 5815.
[90] G. P. Schiemenz and H. Rast, *Tetrahedron Letters*, 1969, 2165.
[91] N. J. Salsbury and D. Chapman, *Biochim. Biophys. Acta*, 1968, **163**, 314.
[92] S. A. Penkett, A. G. Flook, and D. Chapman, *Chem. and Phys. Lipids*, 1968, **2**, 273.
[93] J. B. Lambert, G. F. Jackson, and D. C. Mueller, *J. Amer. Chem. Soc.*, 1968, **90**, 6401.

compared with that (30 kcal mole^{-1}) of monophosphines is believed to be due to stabilisation of the transition state for inversion by p_π–d_π bonding.

The ^{1}H spectra of (−) menthol esters of n-alkylphenylphosphinic acids are diagnostic of the configuration at phosphorus.[94] There is an upfield shift of one of the methyl groups of the isopropyl group for the $(S)_P$ diasteriomer due to shielding by the phenyl ring. The separation of the methyl group signals is sufficient to allow quantitative estimations.

2-Phenyl-2-methoxyethyl bromide reacts quantitatively and stereospecifically with aliphatic and aromatic tertiary phosphines[95] to give diasteriomeric salts from which optical purities may be estimated.

H. Spin–Spin Coupling.—(*i*) J_{PP}. The coupling between directly bound atoms has been reviewed by McFarlane.[96] The magnitude and sign of $^1J_{PP}$, which may vary from −400 to +600 Hz, has been rationalised in terms of the resonance integral $-\beta$ between the outershell *s* electrons of the two nuclei.[97] $^1J_{PP}$ has also been calculated[98] for the *cis*, *trans*, and *gauche* conformers of P_2F_2 and P_2H_2. An improved method of calculating one- and two-bond constants has been published.[99] The properties of symmetrical spin systems for compounds with two or more phosphorus nuclei have been analysed[100] and methods of determining J_{PP} directly from the spectra are discussed. Four techniques are listed for clarifying deceptively simple spectra.

The ^{31}P and the ^{19}F spectra of the aminobis(difluorophosphine) (79) show unusually large geminal coupling constants (J_{PNP} = 370—446 Hz) which decrease in the order R = Me ~ Et > Ph ~ *m*-ClC_6H_4.[101] It seems likely that the p_π–d_π bonding and the *s*-character of the phosphorus orbitals are

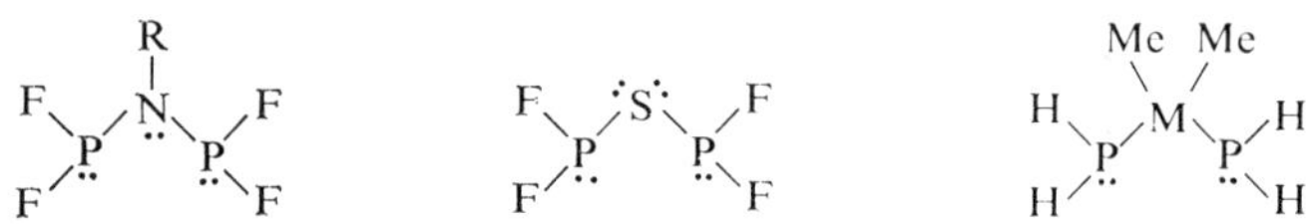

(79) (80) (81) M = Si or Ge

important factors. The coupling constant of the bisphosphine (80) is also very large whereas those of the dimethylgermane and silane bisphosphines (81) are an order of magnitude smaller (*ca.* 1 and 8 Hz respectively).[102] Further, the coupling constants of (80) and (81) are strongly temperature dependent; in each case the constant changes by over 25% over a temperature range of *ca.* 100°.

[94] R. A. Lewis, O. Korpiun, and K. Mislow, *J. Amer. Chem. Soc.*, 1968, **90**, 4847.
[95] J. P. Casey, R. A. Lewis, and K. Mislow, *J. Amer. Chem. Soc.*, 1969, **91**, 2789.
[96] W. McFarlane, *Quart. Rev.*, 1969, **23**, 187.
[97] E. G. Finer and R. K. Harris, *Chem. Comm.*, 1968, 110.
[98] A. H. Cowley, W. D. White, and M. C. Damasco, *J. Amer. Chem. Soc.*, 1969, **91**, 1922.
[99] A. H. Cowley and W. D. White, *J. Amer. Chem. Soc.*, 1969, **92**, 1917.
[100] R. K. Harris and E. G. Finer, *Bull. Soc. chim. France*, 1968, 2805.
[101] J. F. Nixon, *J. Chem. Soc. (A)*, 1969, 1087.
[102] R. A. Newmark, A. D. Norman, and R. W. Rudolph, *Chem. Comm.*, 1969, 893.

It is possible to establish the solution geometry of bisphosphine metal complexes by n.m.r. since $^2J_{PMP}$ is usually small when the phosphines are *cis*-orientated but large when they are *trans*. In the case of methyl phosphines, *e.g.* Ph_2MeP and $PhMe_2P$, advantage may be taken of the 'virtual coupling' effect. The methyl resonance appears as a triplet, rather than a doublet, when the phosphorus nuclei are strongly coupled, *e.g.* the osmium tetrachloride complex (82).[103] However, in some cases the doublet or triplet

$(PhMe_2P)_2OsCl_4$ (82) $(Ph_2MeP)_2PdX_2$ (83)

may be collapsed to a singlet. The causes of collapse have been discussed at length and theoretical evidence has been presented[104] which suggests that one might expect to find chemical situations, *e.g.* (83; X = Cl) and (83; $X_2 = S_2CO$), where the proton n.m.r. pattern is a doublet, a 'virtual coupled' triplet, or a singlet depending on the magnitude of J_{PMP} and the rate of ligand exchange.

$PhPMe_2{\cdot}(PhO)_3P{\cdot}PdI_2$ (84) $Bu_3P{\cdot}(PhO)_3P{\cdot}PdI_2$ (85) $PhPMe_2{\cdot}Bu_3PPdI_2$ (86)

Several studies have been made to evaluate $^2J_{PMP}$. The coupling constant has been observed directly from ^{31}P spectra by using two different phosphines. The coupling constants of the *trans*-palladium iodide complexes, (84), (85), and (86), were 829, 758, and 551 Hz respectively[105] and in agreement with these values, (87) had J_{PPdP} 8 (*cis*) and 565 Hz (*trans*).[106]

$(Me_3P)_2PdI_2$ (87)

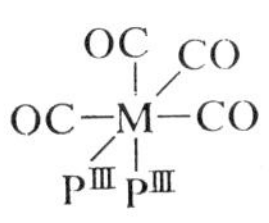

(88)

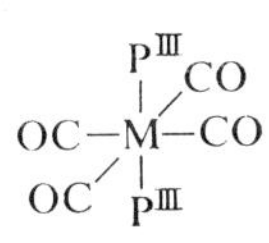

(89)

However, the *cis*- and *trans*-coupling constants are much closer for a number of carbonyl complexes. For example, a mixed phosphine molybdenum complex showed *cis* and *trans* constants of 21 and 49 Hz respectively.[107a] Also triplets, assigned to virtual coupling, have been observed for the *cis*- and *trans*-carbonyl complexes of chromium, molybdenum, tungsten,[107b] and manganese (88).[108] The geometries were assigned on the basis of i.r. frequencies and chemical shifts. The ^{19}F and ^{31}P spectra of a

[103] J. Chatt, G. J. Leigh, D. M. P. Mingos, E. W. Randall, and D. Shaw, *Chem. Comm.*, 1968, 419.
[104] J. P. Fackler jun., J. A. Fetchin, J. Mayhew, W. C. Seidel, T. J. Swift, and M. Weeks, *J. Amer. Chem. Soc.*, 1969, **91**, 1941.
[105] A. Pidcock, *Chem. Comm.*, 1968, 92.
[106] R. G. Goodfellow, *Chem. Comm.*, 1968, 114.
[107] (*a*) S. O. Grim, D. A. Wheatland, and P. R. McAllister, *Inorg. Chem.*, 1968, **7**, 161. (*b*) W. E. Stanclift and D. G. Hendricker, *Inorg. Chem.*, 1968, **7**, 1242.
[108] P. K. Maples and C. S. Kraihanzel, *Chem. Comm.*, 1968, 922.

series of *cis*-molybdenum carbonyl complexes show that ${}^2J_{PMoP}$ increases (from 38 to 55 Hz) with the electron-withdrawing power of the phosphorus substituents.[109] These results are in accordance with an increase of both the *s* character of the P-hybrid orbital which is forming the σ-bond to the metal and the amplitude of the P *s*-orbital at the nucleus.

The signs of J_{PMoP} have been estimated for some trimethylphosphite molybdenum carbonyl complexes using double resonance of ${}^{13}C$ satellites.[110] It was concluded that the coupling constant in *cis*- (88) and *trans*- (89) complexes is negative ($-40{\cdot}5 \pm 1$ Hz) and positive ($+162 \pm 5$ Hz) respectively. A *cis*-coupling constant of 40 ± 5 Hz was also observed for the nickel chloride complex of the quadridentate ligand (90).[111] The PMP

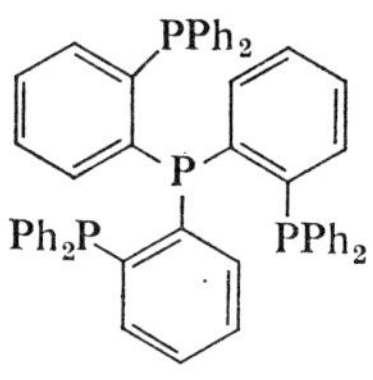

(90)

$(p\text{-tolyl}_3P)_2Ru(C_3H_5)_2$ (91) $(PhPMe_2)_3Fe(C_4H_7)_2$ (92)

coupling constants of the ruthenium and iron allyl complexes (91) and (92) were estimated to be 21 ± 1 and 14 ± 1 by comparing calculated and observed spectra.[112]

(ii) $J_{P^{13}C}$. Evidence has been presented that ${}^1J_{P^{13}C}$ of P^{III}, P^{IV} and P^{V} compounds depend on the energies of the *s*-electrons of the coupled nuclei[113] and parallel ${}^2J_{PCH}$ and the predicted ${}^1J_{PP}$.[97]

(iii) J_{PF}. ${}^1J_{PF}$ is found to decrease with increasing electronegativity of the carbon atoms joined to P for the phosphine oxides (93).[114] The opposite trend is found for the fluorophosphines (94)[115] which are in the range

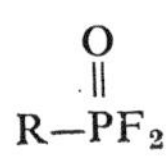

(93)

$R\ddot{P}FX$

(94)

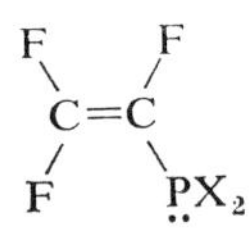

(95)

[109] C. G. Barlow, J. F. Nixon, and J. R. Swain, *J. Chem. Soc.* (*A*), 1969, 1082.
[110] R. D. Bertrand, F. Ogilvie, and J. G. Verkade, *Chem. Comm.*, 1969, 756.
[111] J. W. Dawson and L. M. Venanzi, *J. Amer. Chem. Soc.*, 1968, **90**, 7229.
[112] G. A. Fuchs and C. A. Reilly, International Symposium on N.M.R., Birmingham, July, 1969.
[113] G. Mavel and M. J. Green, *Chem. Comm.*, 1968, 742.
[114] L. N. Mashlyakovzkii, B. I. Ionin, V. B. Lebedev, A. A. Petrov, and I. S. Okhrimenko, *Khim. Org. Soedineniya Fosfora Akad. S.S.S.R., Otdel., obshchei Tekh. Khim.*, 1967, 238 (*Chem. Abs.*, **68**, 114,704).
[115] G. I. Drozd, S. Z. Ivin, V. N. Kulakova, and V. V. Sheluchenko, *Zhur. obshchei Khim.*, 1968, **38**, 576 (*Chem. Abs.*, **69**, 43,973); C. G. Barlow, R. Jefferson, and H. F. Nixon, *J. Chem. Soc.* (*A*), 1968, 2692.

($^1J_{PF}$ 900—1191 Hz). Whereas $^3J_{PF}$ for the pentafluorophenylphosphines are in the range 30—60 Hz (*e.g.* 44 Hz for $C_6F_5PF_2$), the corresponding oxides show vicinal coupling constants of a much smaller magnitude, *e.g.* 1 Hz for $C_6F_5POF_2$.[32] The ^{19}F spectra of the perfluorovinylphosphines (95) indicate that J_{PCF} values are of opposite sign for (95: X = F) and (95: X = Cl).[116]

(*iv*) J_{PH} and J_{HH}. The ^{31}P spectra of a series of P^{III} compounds dissolved in 100% sulphuric acid [117] and fluorosulphonic acid [118] have been measured. A large range of values for $^1J_{PH}$ was observed which could be related to the amount of *s*-character of the P—H phosphorus orbital for the phosphines only.[118] A direct relationship between $^1J_{PH}$ and $^3J_{HPCH}$ is also present. For a wider range of P^{III} compounds [117] it was necessary to take into account the change in effective nuclear charge (Z_P) caused by the inductive effects of the P-substituents. There was a very good linear relationship between $^1J_{PH}$ and the corresponding value of $^1J_{P^{183}W}$ for the tungsten carbonyl complexes.[119] This indicates the latter is also dominated by changes in Z_P [see this chapter (3C)].

The P^V—H group in the bis(biphenylene)phosphorane (96) [14] possesses a smaller coupling constant ($^1J_{PH}$, 482 Hz) and a chemical shift (τ 0·67 p.p.m.) at lower field than a P^V—H group in oxy- and fluoro-phosphoranes which are in the ranges $^1J_{PH}$ 730—1084 Hz and τ 3·6—1·38 p.p.m.

The P^{III}—H group in phosphines shows smaller coupling constants. The phosphirane (97) shows $^1J_{PH}$, 158 and 159 Hz, for its *cis*- and *trans*-isomers

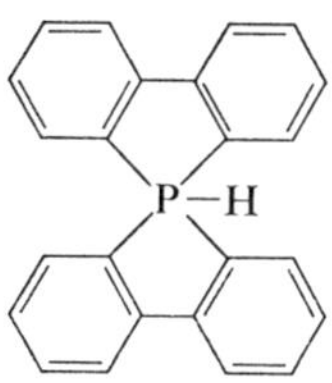

(96)

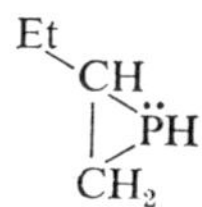

(97)

$ClCH_2\ddot{P}H_2$

(98)

(half the doublet was hidden by the ring protons at τ 7·5—9·9 p.p.m. and the other half appeared at τ 11·4—12·4 p.p.m.);[6] the chloromethylphosphine (98) shows $^1J_{PH}$ +199·4 Hz (geminal and vicinal coupling constants were also positive).[120]

Geminal PCH coupling constants for a wide range of acyclic compounds have been reviewed by Gallagher.[121] A comparison of $^2J_{PCH}$ between the oximinophosphonium salts (21) and the corresponding phosphorane (22)

[116] A. H. Cowley and M. W. Taylor, *J. Amer. Chem. Soc.*, 1969, **91**, 1026.
[117] W. McFarlane and R. F. M. White, *Chem. Comm.*, 1969, 744.
[118] G. A. Olah and C. W. McFarland, *J. Org. Chem.*, 1969, **34**, 1832.
[119] For correlation of J_{WP} and ν_{CO} see S. O. Grim, P. R. McAllister, and R. M. Singer *Chem. Comm.*, 1969, 38.
[120] H. Goldwhite and D. G. Rowsell, *J. Phys. Chem.*, 1968, **72**, 2666.
[121] M. J. Gallagher, *Austral. J. Chem.*, 1968, **21**, 1197.

[see this chapter (1A)] shows a fall in magnitude for the latter series.[27] Assuming $^2J_{PCH}$ is negative for the salts (21), the change is a positive one in agreement with the low *s*-character of the bonding orbitals of the P^V atom. A positive shift for the negative geminal coupling constant is also predicted for increases in the *s*-character of the bonding orbitals of the carbon atom. In agreement, a series of 9-phosphofluorene salts (99) showed that upon

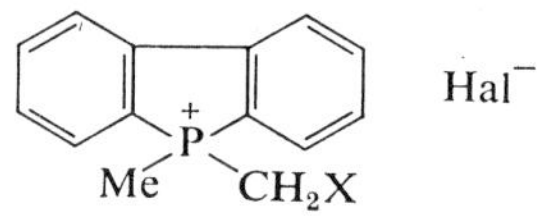

(99)

increasing the electronegativity of X the magnitude of J_{PCH_2} decreased whereas that of J_{P-Me} remained constant.[122] A similar trend was shown by phosphine oxides. These relatively small increases in the *s*-character of the bonding orbitals of carbon were sufficient to reduce the negative $^2J_{PH}$ to zero.

Larger increases in the *s*-character of the bonding orbitals of carbon are obtained by considering an sp^2 hybridised carbon atom; in such tetra-co-ordinated phosphorus compounds $^2J_{PCH}$ is almost certainly positive.[123–125] It is generally in the region 5—23 Hz for vinyl 'phosphine' oxides (100),[123, 124, 126] decreasing with increase in electronegativity of Z, but

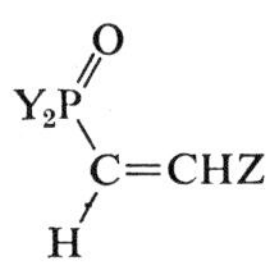

(100)

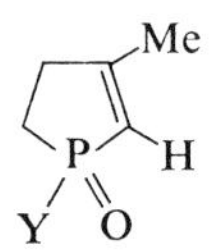

(101)

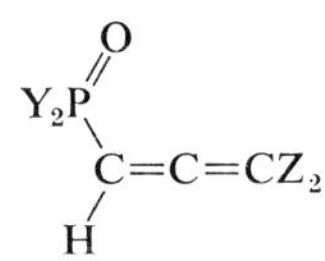

(102)

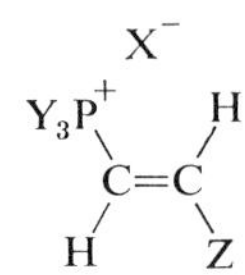

(103)

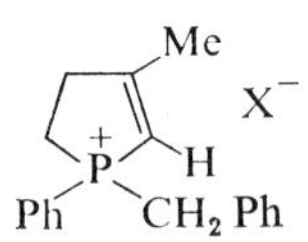

(104)

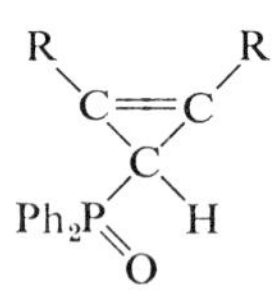

(105)

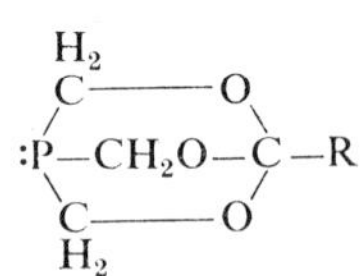

(106)

[122] D. W. Allen, I. T. Millar, and J. C. Tebby, *Tetrahedron Letters*, 1968, 745.
[123] T. N. Timofeeva, B. I. Onin, Yu. L. Kleiman, N. V. Morkovin, and A. A. Petrov, *Zhur. obshchei Khim.*, 1968, **38**, 1255 (*Chem. Abs.*, **69**, 77,353); J. E. Lancaster, *Spectrochim. Acta*, 1967, **23**, *A*, 1449.
[124] W. Hagens, H. J. T. Bos, W. Voskuil, and J. F. Arens, *Rec. Trav. chim.*, 1969, **88**, 71.
[125] M. P. Williamson, S. Castellano, and C. E. Griffin, *J. Phys. Chem.*, 1968, **72**, 175.
[126] R. M. Lequan and M. P. Simonnin, *Compt. rend.*, 1969, **268**, *C*, 1400.

24—26 Hz for the cyclic oxides (101);[127] $^2J_{PCH}$ is 0—8 Hz for allenic 'phosphine' oxides (102) [128] but much larger (16—28 Hz) for the chlorides (102; Y = Cl); and $^2J_{PCH}$ is 17—23 Hz for vinylphosphonium salts (103) [129] but 28·7 Hz for the cyclic salt (104).[127]

An exceedingly large geminal coupling constant (40·8 Hz) is observed [89] for the cyclopropenylphosphine oxide (105) and even larger constants (47—52 Hz) for the cyclopropenylphosphonium salts (76) [see this chapter (1F)]. The authors of this report predict that these constants will also be positive. The sign and magnitude of $^2J_{PH}$ for the phosphine (106), its oxide, sulphide, BH_3 complexes, and metal carbonyl complexes have been discussed.[130] The results confirm a change in the sign of $^2J_{PCH}$.

The assignment of the stereochemistry about a double bond by n.m.r. does not present any difficulties provided the vinyl protons are not obscured or equivalent. The large values of *trans*-PC=CH coupling constants (usually 28—51 Hz) for tetra-co-ordinated phosphorus derivatives of ethylene are typified by $^3J_{PH}$ = *ca.* 40 Hz for the phosphole- and dihydrophosphole-oxides and sulphides (107) [131] and (108).[127] The corresponding

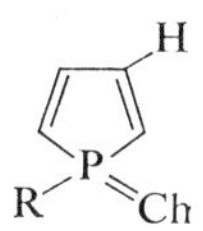

(107)

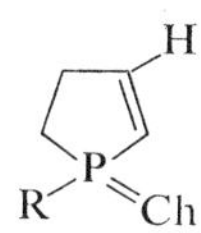

(108)

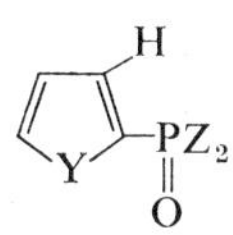

(109)

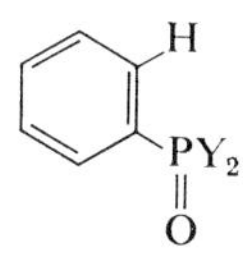

(110)

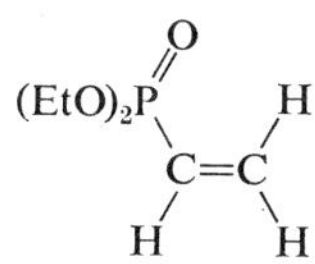

(111)

cis-PC=CH coupling constants are much smaller (10—20 Hz), but the inclusion of the double bond as part of a heteroaromatic ring (109; Y = O, S or N) reduces the constants to the range 2—8 Hz.[26] On the other hand, $^3J_{PH}$ is 11·8 Hz for triphenylphosphine oxide (110; Y = Ph).[132] It is interesting that once again, *viz.* (102),[128] the presence of the $POCl_2$ grouping raises the coupling constant and $^3J_{PH}$ is 17·0 Hz for (110; Y = Cl).[132] In the case of the vinylphosphonate (111) [125] the vinyl protons are almost equivalent and it would not have been possible to analyse the spectrum

[127] L. D. Quin, J. P. Gratz, and T. P. Barket, *J. Org. Chem.*, 1968, **33**, 1034.
[128] M. P. Simonnin and C. Charrier, *Org. Magn. Resonance*, 1969, **1**, 27.
[129] G. Pattenden and B. J. Walker, *J. Chem. Soc.* (*C*), 1969, 531.
[130] E. J. Boros, R. D. Compton, and J. G. Verkade, *Inorg. Chem.*, 1968, **7**, 165.
[131] G. Markl and R. Potthast, *Tetrahedron Letters*, 1968, 1755.
[132] R. Keat, *Chem. and Ind.*, 1968, 1362.

without making use of the coupling to phosphorus. The relative signs of J_{PH} for vinyl- and allyl-phosphonates, which in general parallels H—H couplings in the same compounds, have been reviewed.[133] The configurations of the vinylphosphine oxides (110) were assigned [124] by a comparison of the observed $^3J_{HH}$ with the calculated values for the *cis*- and *trans*-isomers. The constants $^3J_{PH}$ as well as $^2J_{PH}$ decrease for both isomers with increase in the electronegativity of Z. It is also of note that for the series of *trans*-vinyl 'phosphine' oxides (100; Z = OEt), $^3J_{PCCH}$ varies with the Taft constant σ_i of the phosphorus substituent (Y).[126]

The coupling constants of tri-co-ordinated phosphorus compounds are complicated by the presence of the phosphorus lone pair. It has been argued [134] that the very high *s*-character of the lone pair of electrons of phosphines will give these electrons little or no directional properties, and on this assumption the phosphorinane (112) has been assigned the con-

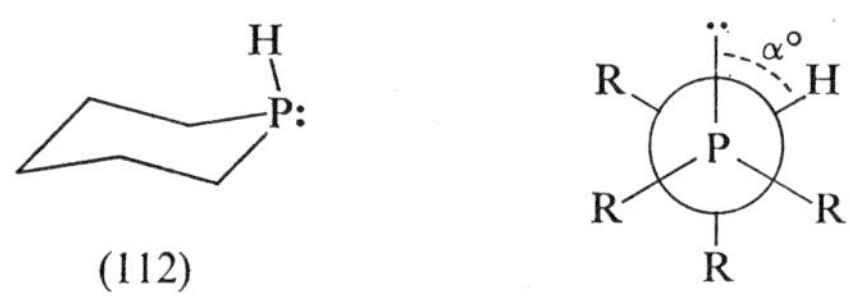

(112)

formation with an axial P—H group as shown. This conclusion follows from the appearance of the P—H proton as a triplet of triplets (J_{HPCH} 12 and 2·5 Hz), the larger coupling constant being associated with axial–axial coupling.

However, there is considerable evidence [135] that the geminal and vicinal coupling constants to the phosphorus atom of P^{III} compounds, *i.e.* $^2J_{PH}$ and $^3J_{PH}$, are strongly dependent on the dihedral angle subtended by the C—H bond and the lone pair of electrons on phosphorus.

The signs and magnitudes of the geminal coupling constant $^2J_{PH}$ for a number of cyclic phosphines bearing sp^3 hybridised carbon atoms have been examined.[136] A graph has been drawn which indicates J_{PCH} to be at maximum (*ca.* +27 Hz) when the proton and lone pair of electrons are closest ($\alpha = 0°$) and at a minimum (*ca.* −5 Hz) when the dihedral angle α is 120°. These observations support the existence of two coupling terms of opposite signs.[135]

An example (113) of $^2J_{PCH}$ involving an sp^2 hybridised carbon atom has been reported.[127] The magnitude (42 Hz) is much larger than those above and is in accordance with a positive 'through space' contribution now being reinforced by the normal bonding contribution which has become positive in sign due to the increased *s*-character of the bonding orbitals of carbon.

[133] G. Mavel and R. Mankowski-Favelıer, *J. Chim. phys.*, 1967, **64**, 1808.
[134] J. B. Lambert, W. L. Oliver jun., and G. F. Jackson, *Tetrahedron Letters*, 1969, 2027.
[135] G. Mavel, *J. Chim. Phys. Physicochim. Biol.*, 1968, **65**, 1692.
[136] J. P. Albrand, D. Gagnaire, and J. B. Robert, *Chem. Comm.*, 1968, 1469; J. P. Albrand, D. Gagnaire, J. Martin, and J. B. Robert, *Bull. Soc. chim. France*, 1969, 40.

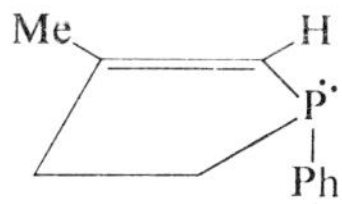

(113)

The influence of the orientation of the lone pair of electrons on vicinal coupling constants to phosphorus has also been the subject of several papers. The ^1H spectrum of the ethylenephosphite (114) was in agreement [137]

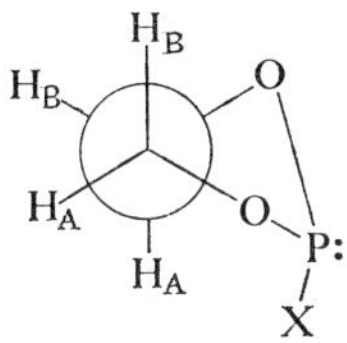

(114)

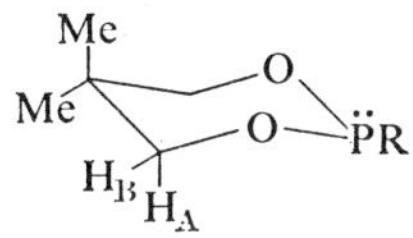

(115)

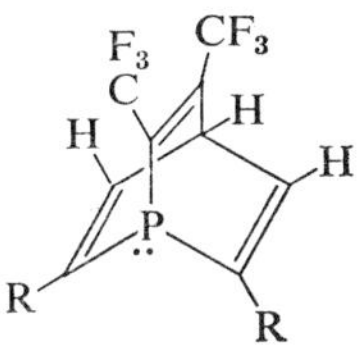

(116)

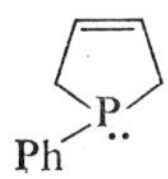

(117)

with the twist envelope conformation shown, possessing $^3J_{PH_A}$ = *ca.* 2 Hz and $^3J_{PH_B}$ = *ca.* 9 Hz. No change occurred in the spectrum up to 100°. Also no inversion was observed for the cyclic phosphonite (115) in the temperature range −50 to 160°.[138] The spatial arrangement of the phosphorus lone pair is considered to be important, the coupling constant being smaller for the axial proton H_A ($^3J_{PH}$ = 3—4 Hz) than for the equatorial proton H_B ($^3J_{PH}$ = 11—20 Hz). The existence of *syn*- and *anti*-isomers for similar compounds had been deduced from δ_P considerations.[139] Note that in the case of vicinal coupling both the 'through space' and the 'through bond' coupling are varying with the stereochemistry of the molecule.

Several examples of vicinal coupling involving sp^2 hybridised β carbon atoms in cyclic compounds have been reported. Thus $^3J_{PH}$ is 7 Hz for the phosphabarrelene (116; R = Ph or CMe_3), 14 Hz for (116; R = Me),[140] and 8 Hz for (117).[131] These values are considerably smaller than the corresponding constants in trivinyl phosphine ($^3J_{PH}$*trans* +30·2 Hz;

137 P. Haake, J. P. McNeal, and E. J. Goldsmith, *J. Amer. Chem. Soc.*, 1968, **90**, 715.
138 D. Gagnaire, J. B. Robert, and J. Verrier, *Bull Soc. chim. France*, 1968, 2392.
139 A. V. Bogatskii, A. A. Kolesnik, Yu. Yu. Samitov, and T. D. Butova, *Zhur. obshchei Khim.*, 1967, **37**, 1105 (*Chem. Abs.*, **68**, 95,896).
140 G. Markl and F. Lieb, *Angew. Chem. Internat. Edn.*, 1968, **7**, 733.

${}^3J_{PH}$ *cis* + 13·6 Hz). The *cis* coupling constants, involving the *ortho*-proton in 2-thienyl phosphines [141] and aryl phosphines,[132] are of the order 6—8 Hz and the P—H coupling constants of tri-3-thienylphosphine are all smaller than the equivalent H—H constants,[142] and all have the same sign, presumably positive.

From these results it appears that the influence of the orientation of the lone pair of electrons is significant for vicinal atoms especially when coupling occurs through a multiple bond.

The normal dependency of vicinal coupling constants on the dihedral angle has been used to establish the configuration and conformation of a phosphoniaethane sulphonate.[143] The *threo*-form (118) was identified from its very small PCCH coupling constant. The cyclic phosphate (119)

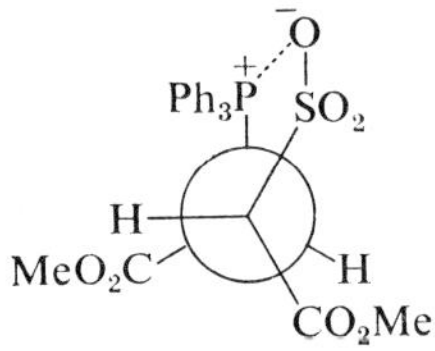

(118)

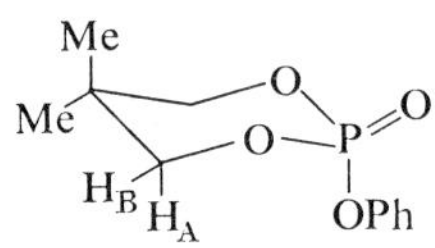

(119)

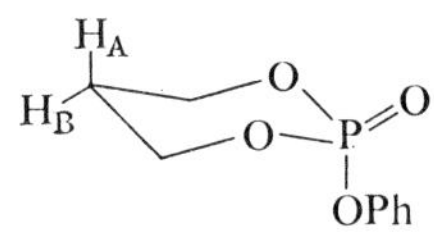

(120)

possessed J_{POCH_A} 3 Hz and J_{POCH_B} 21 Hz. The variation of stereochemistry is also important through four bonds.[144] By double irradiation of the CH_2O protons in (120), ${}^4J_{PH_A}$ and ${}^4J_{PH_B}$ were found to be 1 and 2·6 Hz respectively. The constants ${}^3J_{PH}$ and ${}^4J_{PH}$ normally fall in the ranges 7—13 Hz and 0·3—1·3 Hz, respectively, for the ethyl esters of phosphonic acids.[145]

Proton exchange has been identified as the cause of the variable-temperature geminal coupling constants for the alkylidenephosphoranes (121), (122), and (123). The exchange arises from a transylidation process for the

$(Me)_3P{=}CH_2$	$Ph_3P{=}CHCOR$	$Ph_3P{=}CHCO_2R$
(121)	(122)	(123)

reactive methylenephosphorane (121),[146] but arises from the presence of traces of conjugate phosphonium salts for the acyl and carboalkoxymethylenephosphoranes (122) and (123).[147] Thus J_{PCH} is 24·5 Hz for very pure (122; R = Ph) but is collapsed to a broad singlet at 50° in the presence of 1% $Ph_3\overset{+}{P}CH_2COPh\overset{-}{X}$.

141 M. J. Bulman, *Tetrahedron*, 1969, **25**, 1433.
142 H. J. Jakobsen and J. Aa Nielson, *Acta Chem. Scand.*, 1969, **23**, 1070.
143 M. A. Shaw, J. C. Tebby, R. S. Ward, and D. H. Williams, *J. Chem. Soc.* (*C*), 1968, 2795.
144 L. D. Hall and R. B. Malcolm, *Chem. and Ind.*, 1968, 92.
145 M. P. Williamson and C. E. Griffin, *J. Phys. Chem.*, 1968, **72**, 4043.
146 H. Schmidbaur and W. Tronich, *Chem. Ber.*, 1968, **101**, 604.
147 F. J. Randall and A. W. Johnson, *Tetrahedron Letters*, 1968, 2841; H. J. Bestmann, H. G. Liberda, and J. P. Snyder, *J. Amer. Chem. Soc.*, 1968, **90**, 2964.

Large vicinal coupling constants (34—40 Hz) have been reported for a series of phosphorins (124)[148] and diphospha-*s*-triazines (125; Y = Ph, X = Ph or Me).[149] For (125; Y = Me) $^4J_{PH}$ was 2·5—3·3 Hz and for (125; Y = NMe_2) $^5J_{PH}$ was 0·45 Hz.

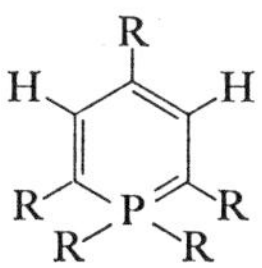

(124)

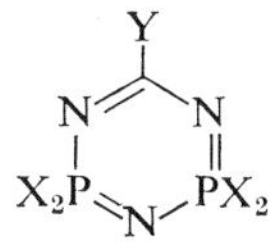

(125)

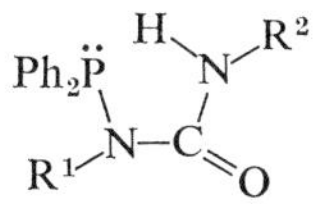

(126)

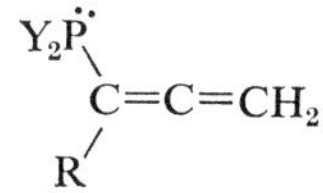

(127) R = H, Me

The origin of the exceptionally large four-bond coupling constant ($^4J_{PH}$ 16 Hz) observed for the 'phosphine' (126)[150] may be due to the close proximity of the phosphorus atom and the N—H proton. The N—H doublet becomes a singlet and moves downfield on forming the 'phosphine' oxide.

A linear relationship has been found between the coupling constants of the allenic 'phosphines' (127) and the sum of the inductive and resonance effects of the P substituents;[151] $^4J_{PH}$ (0·5—5 Hz) was negative in most cases but may be positive when the P substituent is a strong donor, *e.g.* Y = NMe_2. The magnitude of $^2J_{PH}$ and $^4J_{PH}$ for the 'phosphines' (127) and their oxides varied with temperature and solvent.[127]

The multiplicity of the proton H_A in the phosphine (128) was greater than that expected for the vinyl grouping. The extra multiplicity was

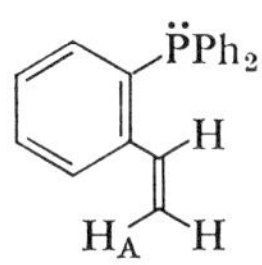

(128)

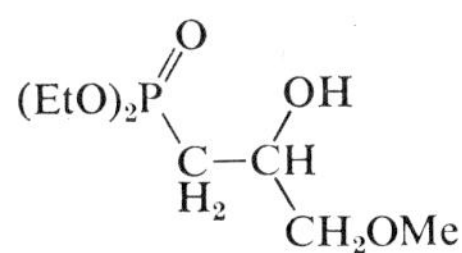

(129)

removed by ^{31}P irradiation and is therefore attributed to long-range coupling ($^5J_{PH}$1·1 Hz).[152] A coupling constant ($^4J_{PH}$) of 1·2 Hz is reported for an interaction through three sp^3-hybrid carbon atoms for the phosphonate (129).[153]

I. Paramagnetic Effects.—The observation of large proton chemical shifts for diamagnetic cations in the presence of certain paramagnetic anions is a very sensitive technique for detecting ion pairing or aggregation in solution. The n.m.r. spectra of the unsymmetrical diamagnetic cation $Ph_3\overset{+}{P}Bu$ in

[148] G. Markl and A. Merz, *Tetrahedron Letters*, 1968, 3611; *ibid.*, 1969, 1231.
[149] A. Schmidpeter and J. Ebeling, *Chem. Ber.*, 1968, **101**, 3883.
[150] R. F. Hudson and R. J. G. Searle, *J. Chem. Soc.* (*B*), 1968, 1349.
[151] M. P. Simonnin and M. C. Charrier, *Compt. rend.*, 1968, **267**, *C*, 550.
[152] A. G. Moritz, J. D. Saxby, and S. Sternhell, *Austral. J. Chem.*, 1968, **21**, 2565.
[153] C. E. Griffin and S. K. Kundu, *J. Org. Chem.*, 1969, **34**, 1532.

the presence of paramagnetic anions such as $Ph_3PNiBr_3^-$ or $Ph_3PCoI_3^-$ have been used to estimate inter-ionic distances.[154] Comparisons of calculated and observed linewidths and dipolar shifts are consistent with the approach of the cation along the C-3 axis of the anion on the opposite side to the bulky phosphine ligand (130).

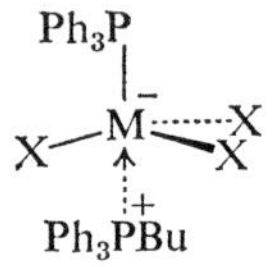

(130)

Low-field dynamic nuclear polarisation results for phosphorus in different environments are reported.[155] The order of enhancement is found to be:

$$(EtO)_3P > (EtO)_2\overset{O}{\overset{\|}{P}}H > (EtO)_3P{=}O$$

which supports an exchange–polarisation interpretation of the molecular encounter.

The Overhauser effects on phosphorus oxides, sulphides, and selenides have been measured in CS_2 in the presence of tributylphenoxyl radicals and large positive enhancements have been observed for the P^{III} compounds.[156] This has enabled the estimation of $^1J_{PSe}$ (263 ± 10 Hz) from the analysis of the ^{77}Se satellites of P_4Se_3.

2 Electron Spin Resonance Spectroscopy

E.s.r. has been used to investigate the mechanism of the photoisomerisation of triethylphosphite (131) to diethyl ethylphononate (132).[157] Two signals were observed below 100° corresponding to $>P{-}\dot{O}$ and ethyl radicals.

$$\underset{(131)}{(EtO)_3P:} \xrightarrow{h\nu} \underset{(132)}{(EtO)_2\overset{O}{\overset{\|}{P}}Et}$$

At 100° the $>P{-}\dot{O}$ spectrum was replaced by a $>\dot{P}{-}O$ spectrum and at 110° this latter spectrum disappeared leaving the spectrum due to ethyl radicals, with a reduced intensity.

[154] R. H. Fischer and W. D. Horrocks jun., *Inorg. Chem.*, 1968, **7**, 2659.
[155] J. A. Potenza, P. J. Caplan, and E. H. Poindexter, *J. Chem. Phys.*, 1968, **49**, 2461.
[156] R. A. Dwek, R. E. Richards, D. Taylor, G. J. Penney, and G. M. Sheldrick, *J. Chem. Soc.* (*A*), 1969, 935.
[157] K. Terauchi and H. Sakurai, *Bull. Chem. Soc. Japan*, 1968, **41**, 1736.

The reaction of alkyl methylphosphinates (133) with α-methylbenzylamine and sulphur has been shown to proceed *via* the stable free radical (134) [158] (*g* value 2·0289) to give the methylthiophosphonate (135).

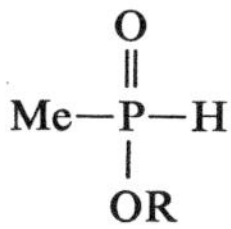
(133)

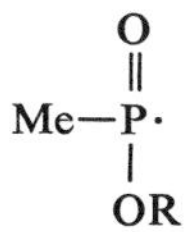
(134)

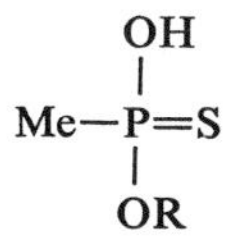
(135)

The reaction of t-butoxy and t-butylthiyl radicals with phosphites and phosphines has also been followed by e.s.r.[159] Very clean reactions were observed which can be rationalised in terms of the formation of a four-co-ordinate phosphorus radical, *e.g.* (136), which fragments to give the

$Y_3\dot{P}$—OBu^t
(136)

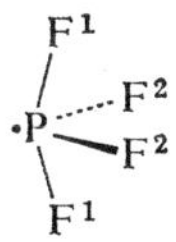
(137)

$\overset{+}{P}H_4\ I^-$
(138)

most stable radical. Whereas trialkylphosphites react with t-butoxy and t-butylthiyl radicals to give a t-butyl radical, triallylphosphite gives allyl radicals,* and triphenylphosphite gives phenoxy radicals.* On the other hand, phosphines give a t-butyl radical only with t-butylthiyl radicals and with t-butoxy radicals the sole radicals produced come from the phosphine, *e.g.* an ethyl radical from triethylphosphine.

By comparing MO calculated and observed e.s.r. isotropic hyperfine splittings, the stereochemistry of the tetrafluorophosphoranyl radical (137) has been estimated to be similar to that of a SF_4 molecule.[160] The F^1PF^1 angle is $174 \pm 5°$ and the F^2PF^2 angle is $109 \pm 9°$.

The structure of phosphonium iodide (138) has been determined at liquid nitrogen temperature from the e.s.r. spectrum of I_2^- centres produced by *X*-ray irradiation.[161]

3 Vibrational Spectroscopy

In addition to the normal extensive use for structural elucidation, vibrational spectroscopy has been applied to stereochemical and conformational problems and the estimation of bonding interactions. Frequencies are expressed in S.I. units mm^{-1} ($100\ mm^{-1} \equiv 1{,}000\ cm^{-1}$).

158 W. A. Mosher and R. R. Irino, *J. Amer. Chem. Soc.*, 1969, **91**, 756.
159 J. K. Kochi and P. J. Krusic, *J. Amer. Chem. Soc.*, 1969, **91**, 3944.
160 J. Higuchi, *J. Chem. Phys.*, 1969, **50**, 1001.
161 C. L. Marquardt, *J. Chem. Phys.*, 1967, **48**, 994.

* The reaction with butoxy-radicals only was reported.

A. Band Assignments and Structural Elucidation.—Analyses of the i.r. and Raman spectra of [$^2H_9$]trimethylphosphine,[162] phosphiranes,[6, 163] triethylphosphine sulphide, triethylphosphine selenide,[164] and a series of quasi-phosphonium chlorides and antimony hexachlorides [165] have been reported. I.r. spectra are frequently used to establish the presence of *P*-phenyl groups [166] and the *P*-phenyl stretching and bending vibrations in the 53·0—10·0 mm^{-1} region have now been assigned.[167] The i.r. spectra of aromatic [168] and unsaturated alkyl [169] phosphonates have been studied. It is found that compounds which possess hydroxyl- or amino-groups on a side-chain, *e.g.* (139),[158] or aromatic ring, *e.g.* (140),[168] will H-bond, inter-

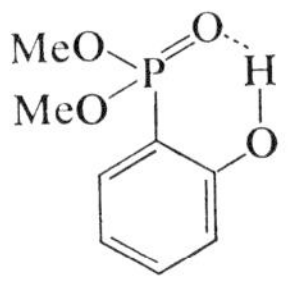
(139)

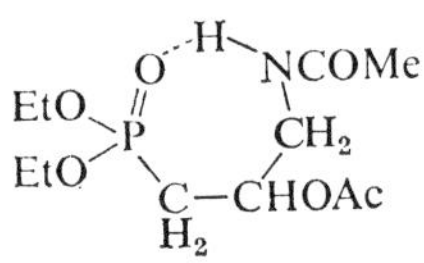
(140)

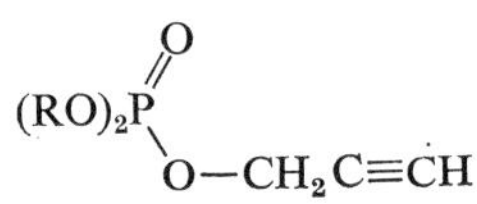
(141)

or intra-molecularly, to the phosphonyl group causing ν_{PO} to shift from 124·2—125·8 mm^{-1} down to 120·3—123·0 mm^{-1}. Hydrogen-bonding is also reported for the enol (34) [41] [see this chapter (1C)]. However H-bonding involving acetylenic hydrogen (*e.g.* 141) did not alter $\nu_{P=O}$.[170] Further evidence has been presented [80] which confirms the very low wavenumbers (150—154 mm^{-1}) of the i.r. bands of carboxylic ester groups bonded to the γ position of an allylidenephosphorane such as (68) [see this chapter (1E)]. The i.r. spectra of a series of phosphinimines (142) has been examined [171] and $\nu_{P=N}$ has been identified using ^{15}N isotope effects. Its variation with mass and bonding effects (especially those connected with R^1) is discussed. The azophosphinimines (143) [172] and related compounds [173] have also been studied in detail. I.r. solution spectra of the major product from the reaction of *o*-aminophenol with hexachlorocyclotriphosphazine $(NPCl_2)_3$ was in accordance with (144).[174] However, in the solid phase the presence of a broad

162 P. J. D. Park and P. J. Hendra, *Spectrochim Acta*, 1968, **24**, *A*, 2081.
163 R. W. Mitchell, L. J. Kuzma, R. J. Pirkle, and J. A. Merritt, *Spectrochimica Acta*, 1969, **25**, *A*, 819.
164 J. R. Durig, J. S. Di Yorio, and D. W. Wertz, *J. Mol. Spectroscopy*, 1968, **28**, 444.
165 A. Schmidt, *Z. anorg. Chem.*, 1968, **362**, 129.
166 A. M. Aguiar and M. G. R. Nair, *J. Org. Chem.*, 1968, **33**, 579.
167 K. Shobatake, C. Postmas, J. R. Ferraro, and K. Nakamoto, *Appl. Spectroscopy*, 1969, **23**, 12.
168 R. Obrycki and C. E. Griffin, *J. Org. Chem.*, 1968, **33**, 632.
169 L. P. Raskina, B. S. El'tsefon, G. S. Radovskii, and A. A. Berlin, *Zhur. Priklad. Spektroscopii*, 1968, **9**, 691 (*Chem. Abs.*, **70**, 62,574).
170 V. V. Tarasov, Ya. S. Arbisman, N. S. Rylyakova, and Yu. A. Kondrat'ev, *Zhur. fiz. Khim.*, 1968, **42**, 2720 (*Chem. Abs.*, **70**, 77,083).
171 W. Wiegrabe and H. Boch, *Chem. Ber.*, 1968, **101**, 1414.
172 H. Boch, M. Schnoellr, and H. tom Dieck, *Chem. Ber.*, 1969, **102**, 1363.
173 H. Boch and M. Schnoellr, *Angew Chem. Internat. Edn.*, 1968, **7**, 636.
174 H. R. Allcock and R. L. Kugel, *Chem. Comm.*, 1968, 1606.

$R_3P{-}NR^1$ (143) $R_3P{=}N{-}N{=}CR^1_2$ (142)

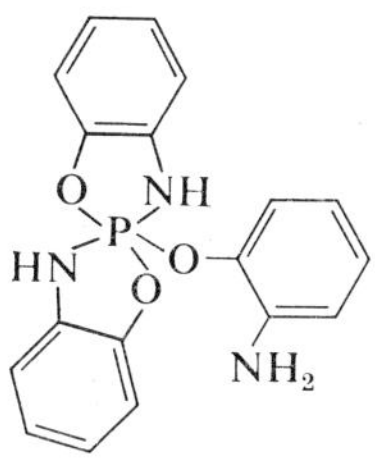

(144)

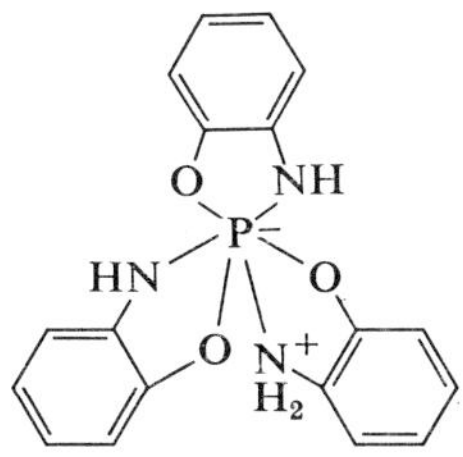

(145)

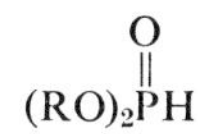

(146)

band at 321·5 mm^{-1} and the absence of the NH_2 deformation band at 164·0 mm^{-1} indicated that (145) may be present. Deuterium studies of (146) [175] indicate that both δ_{PH} and γ_{PH} occur in the region 95—101 mm^{-1}.

B. Stereochemical Aspects.—I.r. and Raman spectra are commonly used to establish the stereochemistry of fluorophosphoranes. In contrast to microwave spectra [see this chapter (4)] the i.r. and Raman spectra of (147)

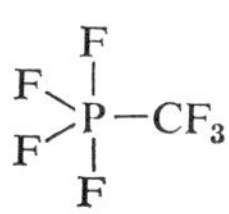

(147)

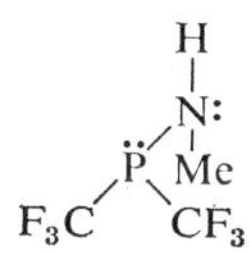

(148a)

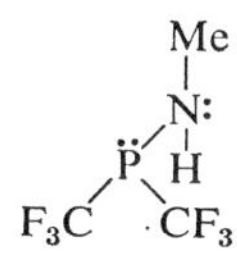

(148b)

favour an equatorial CF_3 grouping.[176] Considered together, these results may indicate the operation of an exchange process. The i.r. spectrum of the aminophosphine (148) has a well-resolved unsymmetrical doublet (346·5 and 343·0 mm^{-1}) in the N—H stretching region in accordance with the presence of unequal amounts of two rotational isomers in thermal equilibrium.[177] Overtones were identified at 689·2 and 684·1 mm^{-1} and the spectrum was the same in gas and liquid phases and benzene solution, which rules out doubling due to hydrogen-bonding. Larger groups in place of methyl caused the conformer absorbing at the lower frequency to become more predominant; however, there was no evidence of N—P π-bonding and the results were in accordance with a steric barrier to rotation. ^{1}H N.m.r. confirmed the presence of two rotamers [see this chapter (1E)]. If rapid N inversion occurs, the two conformers could possess the averaged structures (148a) and (148b).

175 R. A. Nyquist, *Spectrochim. Acta*, 1969, **25**, *A*, 47.
176 J. E. Griffiths, *J. Chem. Phys.*, 1968, **49**, 1307.
177 N. N. Greenwood, B. H. Robinson, and B. P. Straughan, *J. Chem. Soc.* (*A*), 1968, 230.

The i.r. spectra of some acyl phosphines (149) also exhibit two bands, in this case in the 160—180 mm^{-1} region.[178] The band at higher frequency

R_2PCOR^1 (149) $\quad$ $(MeO)_2P(=S)X$ (150) $\quad$ $H-C\equiv C-CH_2OP(=S)Cl_2$ (151)

is attributed to the *cis*-conformer. When the group R^1 is larger than methyl or phenyl the *cis*-conformer is not observed presumably due to steric reasons.

There have been further reports of doubling of phosphoryl and thiophosphoryl bands. Restricted rotation about the P—O bond is believed to account for the doubling in the spectra of phosphorochloridothioates, *e.g.* (150; X = Cl) and (151).[179] For trimethyl thiophosphate (150; X = MeO) one of the lines of the Raman $\nu_{P=S}$ doublet disappears on cooling[180] and in (152) one of the two ν_{S-H} bands is intramolecularly hydrogen-bonded.[175] In another case, triethyl sulphide, the higher band of a proposed P=S doublet has been assigned to a skeletal stretching vibration.[164]

$(RO)_2P(=S)SH$
(152)

The bands due to $\nu_{P=O}$ are reported to be a doublet for a series of aromatic phosphates.[181] The high frequency part of the doublet increases in intensity with temperature due to changes in conformer populations.

Contrary to previous reports $\nu_{P=O}$ in 1,3,2-dioxaphosphorinanes (153) is found to be dependent upon conformation.[182] Carbon tetrachloride

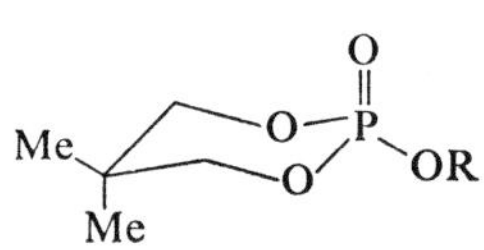

(153a)

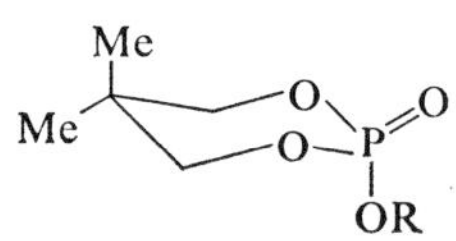

(153b)

solutions show two bands in the regions 126·7—129·8 and 130·8—132·2 mm^{-1} which shift to lower frequencies on the addition of phenol. These bands are assigned to a conformational equilibrium of (153a) and (153b) in which the latter, absorbing at higher frequency, predominates. Phosphoryl groups have been shown, by *X*-ray diffraction, to occupy an equatorial orientation in the solid state.

[178] R. G. Kostyanovskii, V. V. Yakshin, S. L. Zimont, and I. I. Chervin, *Izvest. Akad. Nauk S.S.S.R., Ser. khim.*, 1968, 190, 188 (*Chem. Abs.*, **69**, 43,987, 43,988).
[179] R. A. Nyquist and W. W. Muelder, *J. Mol. Structure*, 1968, 2465.
[180] J. R. Durig, J. S. DiYorio, *J. Mol. Structure*, 1969, 3179.
[181] A. M. Noskov and V. V. Moskovskikh, *Zhur. priklad Spektroskopii*, 1968, **9**, 235 (*Chem. Abs.*, **70**, 15,549).
[182] M. Kainosho, T. Morofushi, and A. Nakamura, *Bull. Chem. Soc. Japan*, 1969, **42**, 845.

Although sodium trimetaphosphate, $Na_3P_3O_9$, and triphosphoramidate, $Na_3(PO_2NH)_3$, are known to possess C_{3v} symmetry in the solid phase, it is

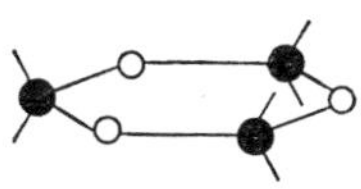

D_{3h} symmetry

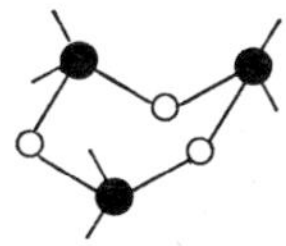

C_{3v} symmetry

found that, in aqueous solution, the vibrational spectra are in accordance with planar geometry (D_{3h} symmetry) similar to phosphonitrilic chlorides $(PHCl_2)_3$.[183]

C. Studies of Bonding.—The i.r. spectra of triaziridophosphine (154) and corresponding mono- and di-compounds (both P^{III} and P^{IV}) indicate a lack

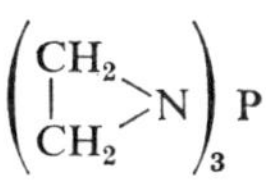

(154)

(155)

(156)

of conjugation other than donation of the N lone pair to P.[184] The P—O frequencies of trimethylphosphine oxide and a number of its complexes with metal halides have been compared with those of amine oxides.[185] The shift of ν_{P-O} to lower frequencies upon co-ordination is attributed to a reduction in P—O π-bonding. The i.r. spectra of a series of phosphinic and carboxylic acid halides (155) and (156) have been compared.[186] Although the polarity of the P=O and P—X bonds is little affected compared to that of the C=O bond, an inductive effect raises ν_{PO} and reduces the basicity of the oxygen atom.

A careful study has been carried out on the force constants relating mainly to P—S bonds. In general the force constants of bonds to phosphorus increase with the percentage *s*-character of the bond in a magnitude similar to those of C—H bonds.[187] In the Table the force constants are presented as a percentage of that of the *sp*³ hybrid. It was also found that the force constants increased (*a*) with increased electronegativity of the P-substituents, and (*b*) on transition from negatively charged ions to neutral molecules. Although ν_{PO} increased regularly with the force

[183] W. P. Griffith and K. J. Rutt, *J. Chem. Soc.* (*A*), 1968, 2331.
[184] R. R. Shagidullin and N. P. Grechkin, *Zhur. obshchei Khim.*, 1968, **38**, 150 (*Chem. Abs.*, **69**, 66,727).
[185] S. H. Hunter, V. M. Langford, G. A. Rodley, and C. J. Wilkins, *J. Chem. Soc.* (*A*), 1968, 305.
[186] A. A. Neimysheva and I. L. Knunyants, *Doklady Akad. Nauk S.S.S.R.*, 1968, **181**, 888 (*Chem. Abs.*, **69**, 91,515).
[187] J. Goubeau, *Angew. Chem. Internat. Edn.*, 1969, **8**, 328.

constant (k_{PO}), ν_{PS} did not. This was attributed to strong coupling of the SP stretching vibration with the other stretching vibrations of the molecule

Table

Bond	*Apical* P^V	P^{III}	P^{IV}	*Axial* P^V
Hybridisation	*pd*	(*s*)*p*	sp^3	sp^2
Approx. % *s*	0	10	25	33
Force constant	61—71	76—94	100	101—109

(strong coupling is also reported for triethylphosphine sulphide).[164] The coupling was largely compensated in the calculation of the force constant k_{PS}, and a plot of k_{PS} *vs.* the sum of the electronegativities of the three substituents gave two straight lines, one for compounds with light substituent atoms (F, O, N, C) and one for the heavier substituent atoms (Cl, Br, S). It was concluded that the heavier atoms permitted a larger degree of p_π–d_π bonding in the P=S bond relative to small atoms with the same electronegativity. The P=S bond orders (1·9—1·3) calculated from k were significantly less than the P=O bond orders (2·4—1·9). The P=S and P=Se stretching frequencies of the tricyclopropyl phosphine chalcogenides (157) and corresponding tri-isopropyl compounds (158) are

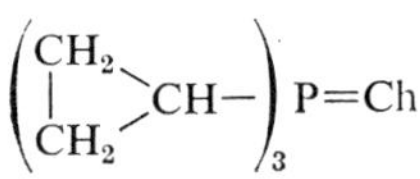

(157)

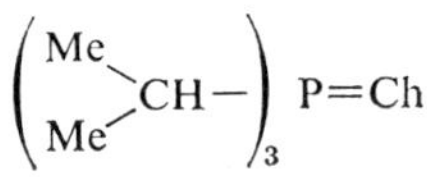

(158)

all higher than the values previously reported for tertiary alkyl or aryl phosphine chalcogenides. These observations support a shorter P=Ch bond and could be due to electron withdrawal by the cyclopropyl groups and a steric effect by the isopropyl groups.[38]

The carbonyl stretching frequencies of the mono-tertiary P^{III} complexes[119] and mono- and di-cyclopropylphosphine complexes,[38] (159) and (160),

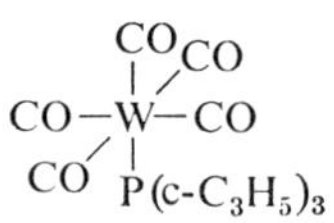

(159)

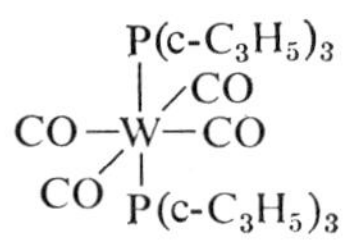

(160)

$Ph_2P—C≡C—PPh_2$

(161)

fit linear relationships with $^1J_{PW}$. The least basic P^{III} compound, triphenylphosphite, shows the largest values of ν_{CO} (195·9 mm^{-1}) and $^1J_{PH}$ (411 Hz) and the most basic, tributylphosphine, shows the smallest ν_{CO} (193·4 mm^{-1}) and $^1J_{PH}$ (200 Hz). The most likely reasons are that the P → W σ-bond, which transmits the coupling effect, is strengthened by a synergic π-interaction, which is at a maximum for the less basic phosphorus ligands,[119] and that the shift of ν_{CO} to higher frequencies is caused by an increase in electron-withdrawing power of W resulting from decreased back donation from the metal to the carbonyl ligands. The complex of the tricyclopropyl phosphine possessed ν_{CO} and J_{PW} similar to those of diphenylalkylphosphines indicating the former to be a good π acceptor.[38] N.m.r. studies have indicated that the nuclear charge on P is a dominant factor and therefore the correlation between J_{PW} and the CO stretching frequency can be interpreted [117] by a mechanism involving σ-bonds only. Two reviews have been published [188] which include discussions of the information which may be gained from ν_{CO}.

The stretching frequency $\nu_{C\equiv C}$ of the bis(diphenylphosphino)acetylene (161) appears at 209·7 mm^{-1} in the Raman spectrum.[189] The low frequency is caused, at least in part, by p_π–d_π bonding. This π donation to the phosphorus atom is reduced upon complexation or formation of the dioxide or disulphide, as shown by an increase in $\nu_{C\equiv C}$, in accordance with a reduction in the phosphine–acetylene interaction. The change in $\nu_{C\equiv C}$ has been used to estimate the π-bonding ability of a wide variety of metals to the phosphine (161) which acts as a bidentate bridging ligand. The i.r. spectrum of the acetylenic phosphine oxide (162) indicates the presence of a C≡C—P interaction but no C≡C—P=O interaction. The intensity of $\nu_{C\equiv C}$ increased with the electronegative character of P substituents.[190]

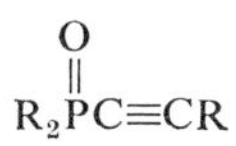

(162)

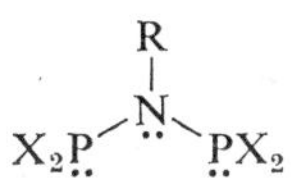

(163)

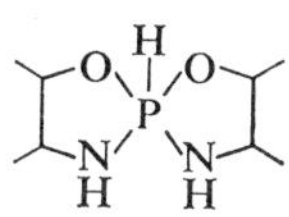

(164)

Although weakly basic, due to p_π–d_π bonding, the diphosphinamine (163) possesses a band at 276—282 mm^{-1} typical of the group $\rangle\ddot{N}$—R in which the lone pair is not used in bonding.[191] However, the variation of ν_{NH} with solvent for spirophosphoranes of the type (164) shows a marked resemblance to pyrrole. This can be interpreted as a result of sp^2 hybridisation of the N atom and p_π–d_π N—P conjugation.[192] It was also found that the variation of ν_{PH} paralleled $^1J_{PH}$.

4 Microwave Spectroscopy and Dipole Moments

The microwave spectrum of trifluoromethyltetrafluorophosphorane (147) is consistent [193] with a C_{3v} structure in agreement with the n.m.r. spectrum. However, since the i.r. spectrum indicates a C_{2v} structure it seems reason-

[188] L. Vaska, *Accounts Chem. Res.*, 1968, **1**, 335; E. W. Abel and F. G. A. Stone, *Quart. Rev.*, 1969, **23**, 325.
[189] A. J. Carty and A. Efraty, *Chem. Comm.*, 1968, 1559.
[190] V. V. Tarasov, Ya. S. Arbisman, Yu. A. Kondrat'ev, and S. Z. Ivin, *Zhur. obshchei Khim.*, 1968, **38**, 130 (*Chem. Abs.*, **69**, 58,713).
[191] J. F. Nixon, *J. Chem Soc.* (*A*), 1968, 2689.
[192] R. Mathis, R. Burgada, and M. Sanchez, *Spectrochim. Acta*, 1969, **25**, *A*, 1201.
[193] E. A. Cohen and C. D. Cornwell, *Inorg. Chem.*, 1968, **7**, 398.

able to conclude that an exchange process occurs in which the CF_3 group and fluorine atoms all participate. The stereochemistry of phosphirane (165) has been evaluated from its microwave spectrum.[194] Of particular

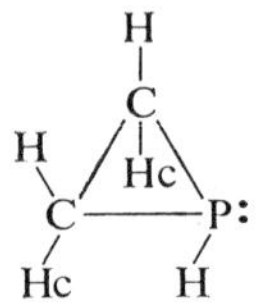

(165)

(166)

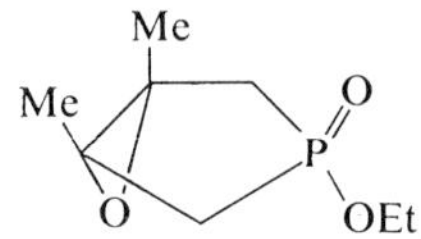

(167)

note is the repulsion between the *cis*-hydrogens, H_C, and the P—H group. The bond angles indicate that the bonds to P are nearly pure *p* and the lone pair of electrons nearly pure 3*s*. The dipole moment was also measured. The microwave spectrum of difluorophosphine oxide (166) has been recorded. Its dipole moment was $2{\cdot}65 \pm 0{\cdot}03$ D [195] which is only 0·08 D smaller than the dipole moment calculated by use of the bond moments of PF_3, PH_3, and F_3PO.

The dipole moments of the P—S bond decrease in the order SPR_3 (4·9) > $SP(SR)_3$ (3·8) > $SPCl_3$ (2·0) > $SP(OR)_3$ (1·4), which supports the overriding importance of the inductive effect on the polarity of the P=S bond.[196]

Dipole moments have been used to establish the stereochemistry of several cyclic compounds. Epoxidation of a phospholene oxide was shown to occur *trans* to the P=O group to give (167),[197] and the cyclic phosphates (168) and the corresponding amino-derivatives (169) have been found to

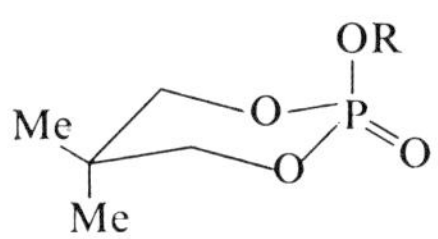

(168)

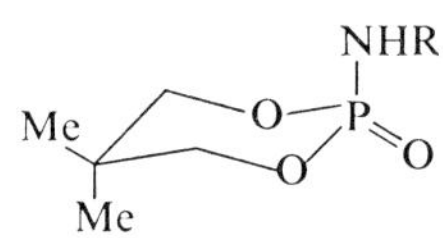

(169)

possess equatorial P=O groups;[198] also see this chapter (1H and 3B). In the case of the cyclic phosphites (170) advantage was taken of the large dipoles induced by the co-ordination of BH_3.[199] The data indicated the methoxy-groups to be axial. Together with variable-temperature n.m.r. spectra it was concluded that the phosphite exists in two stereomeric forms (170a) and (170b) in the ratio of 1 : 3·3 at 21° and 1 : 6·9 at $-80°$. Restricted rotation

[194] M. T. Bowers, R. A. Beaudet, H. Goldwhite, and R. Tang, *J. Amer. Chem. Soc.*, 1969, **91**, 17.
[195] L. F. Centofanti and R. L. Kuczkowski, *Inorg. Chem.*, 1968, **7**, 2582.
[196] J. P. Fayet, P. Mauret, M. C. Labarre, and J. F. Labarre, *J. Chim. Phys. Physicochim. Biol.*, 1968, **65**, 722.
[197] B. A. Arbuzov, A. P. Anastas'eva, A. N. Vereshchagin, A. O. Vizel, and A. P. Rakov, *Izvest. Akad. Nauk S.S.S.R., Ser. khim.*, 1968, 1729 (*Chem. Abs.*, **70**, 29,008).
[198] M. Kainosho and T. Shimozawa, *Tetrahedron Letters*, 1969, 865.
[199] D. W. White, G. K. McEwen, and J. G. Verkade, *Tetrahedron Letters*, 1968, 5369.

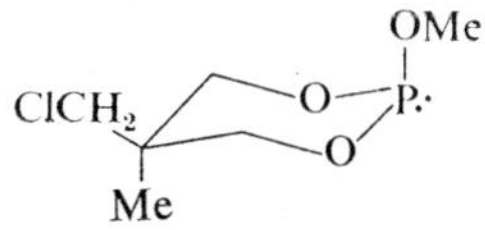

(170a)

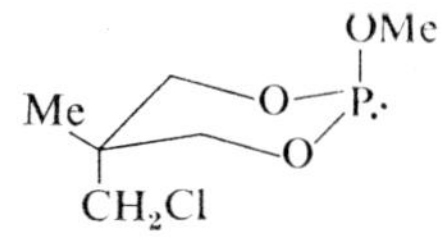

(170b)

(171)

has been invoked to explain the dipole moments of certain diethyl alkylphosphonates.[200] Dipole moments as well as u.v. spectra have been used to study the merocyanine type of mesomerism shown by the phosphinimines (171).[201]

5 Electronic Spectroscopy

The single maximum above 220 nm in the u.v. spectra of vinylphosphines, *e.g.* λ_{max} 245 nm (log ε 2·36) for (172), has been interpreted in terms of promotion of a nonbonded electron on P to the empty π^* orbital of the vinyl group.[202] In this case the bathochromic shift for *trans*-1,2-vinylene diphosphine (173) (λ_{max}, 263; log ε, 6·6) may be attributed to stabilisation

(172)

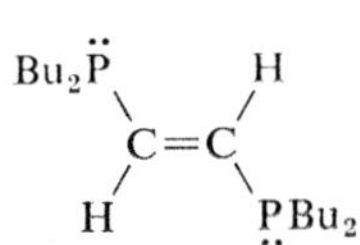

(173)

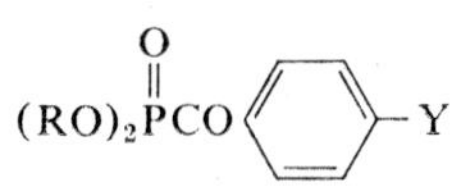

(174)

of the excited state by p_π–d_π conjugation with the P^{III} atom. The further bathochromic shift of the monoxide of (173) (λ_{max} 273, log ε 5·15) and the lack of any maximum above 220 nm for the dioxide of (173) add weight to this explanation.

Aroyl phosphonates (174) show bathochromic shifts compared with the corresponding aromatic aldehyde for both the $\pi \rightarrow \pi^*$ and $n \rightarrow \pi^*$ transitions of the carbonyl group.[203] The interaction (overlap effect), which may involve the phosphorus *d* orbitals appears to be far less sensitive to rotation around the bond axis than is the p_π–p_π bonding of α-diketones. An interesting linear correlation between the wavelengths of the intense $n \rightarrow \pi^*$ bands and the Hammett substituent constant of Y may be drawn.

200 K. S. Mingaleva, B. I. Ionin, and A. A. Petrov, *Zhur. obshchei Khim.*, 1968, **38**, 560 (*Chem. Abs.*, **69**, 76,478).
H. Goetz and D. Probst, *Annalen*, 1968, **715**, 1.
M. A. Weiner and G. Pasternack, *J. Org. Chem.*, 1969, **34**, 1130.
K. Terauchi and H. Sakurai, *Bull. Chem. Soc. Japan*, 1969, **42**, 821.

The *p*-tolylphosphonium salts (175) show two bands in the 265—275 nm region which are attributed to lack of conjugation between rings (after Jaffé),

$p\text{-tolyl}_3\overset{+}{P}R\ Hal^-$

(175)

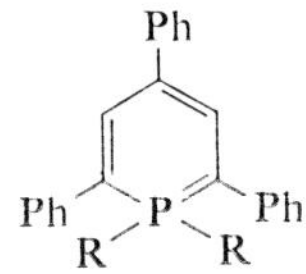

(176)

and a band at 233—234 nm which, from its high intensity, may be due to charge transfer from the halide ion.[204]

The spectra of a number of reactive alkylidenephosphoranes have been determined.[11] In addition to the phenyl band at 265 nm, methylenephosphorane ($Ph_3P{=}CH_2$) absorbs at 341 nm and this maximum is increased by 35 ± 2 nm on replacing one α-hydrogen by an alkyl group and by 45 nm on replacing both hydrogens by alkyl groups. Although a $\pi \rightarrow \pi^*$ transition was favoured over a $\pi \rightarrow d$ transition, substitution by alkyl groups would raise the π-bond energy and thereby lower the transition energy in both cases. The electronic spectra of a number of phosphorins (176) have been measured and in addition to a band in the 335—345 region, bands are observed at 394—400 and 513—525 nm.[148]

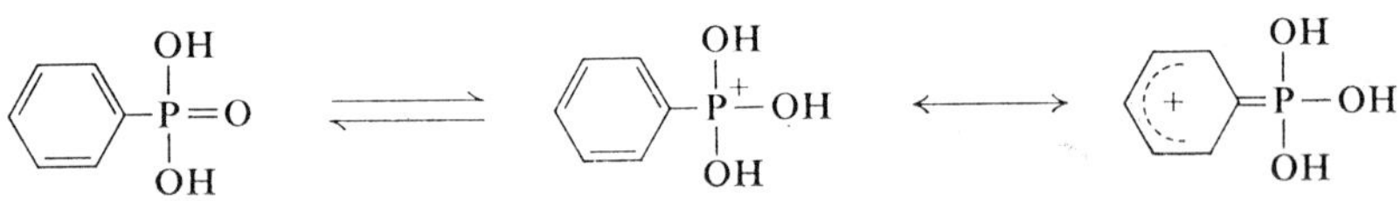

(177) (178)

Phenylphosphonic acid (177) possesses a band at 270 nm which increases in λ_{max} and intensity with concentration of H_2SO_4 and $HClO_4$. Benzylphosphonic acid did not show this effect indicating that upon protonation of (177) to give (178) there is increased π-donation from the phenyl ring towards P.[205]

The u.v. spectra of disubstituted phosphazines have been interpreted as exhibiting no absorption bands characteristic of aromatic systems.[206] The

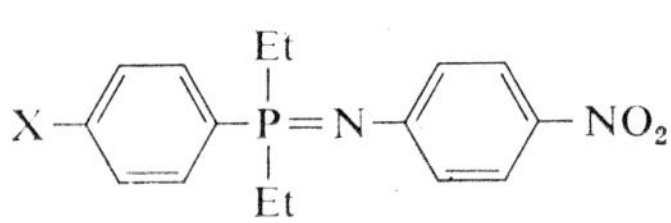

(179)

two main u.v. bands of the phosphinimines (179) corresponded to bands of *p*-nitroaniline and diethylarylphosphine indicating that the P=N acts as an

[204] M. A. A. Beg and Samiuzzaman, *Tetrahedron*, 1968, **24**, 191.
[205] A. Modro, T. Jasinski, and T. A. Modro, *Chem. and Ind.*, 1969, 381.
[206] B. Lakatos, A. Hesz, Z. Vetessy, and G. Horvath, *Magar kem. Folyóirat*, 1968, **74**, 468 (*Chem. Abs.*, **70**, 15,622).

interrupting link. The pK_a values in nitromethane confirmed this interpretation.[207]

6 Magnetic Rotation and Susceptibilities

Further work by French workers [208] on magnetic rotation (the Faraday effect) has been directed towards the tetra-alkyldiphosphines (180) and the

R_2P-PR_2

(180)

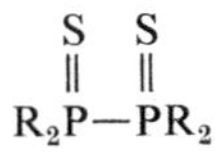

(181)

corresponding disulphides. The rotations due to the P—P and P=S bonds have been calculated. The latter rotations are considerably higher than those of monosulphides ($R_3P=S$) and indicate that the two P=S bonds in (181) are conjugated.[208a] Work has also been carried out on phosphine–boron complexes [208b] including magnetic susceptibility determinations. The magnetic moment of a phosphite cobalt(II) complex is also reported.[209]

7 Circular Dichroism and Refraction

The c.d. curves of methyl phosphine oxides (182) have been compared with those of sulphoxides (183) possessing a similar configuration. They

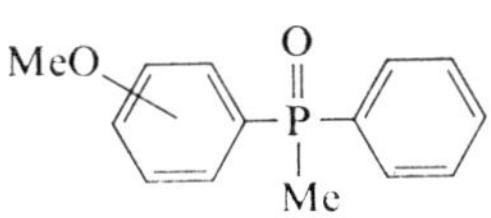

(182)

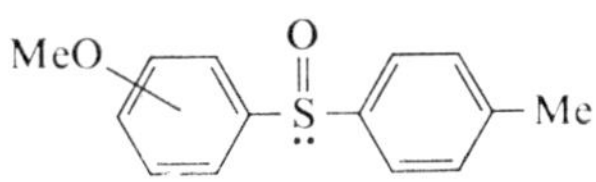

(183)

display a remarkable correspondence of Cotton effects. This intersystem matching provides striking evidence that the lone pair on sulphur in the anisyl *p*-tolylsulphoxides (183) is not significantly implied in the generation of the optical active transitions.[210]

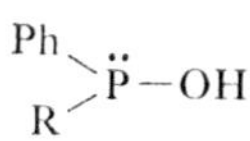

(184)

The molar refraction of phosphinous acids (184; R = Et or $CH=CH_2$) has been measured.[211] The mean bond refraction of the $P^{III}-C_{phenyl}$ bond in esters of this type is 4·404, similar to that of triphenylphosphine.

207 T. G. Edel'man and B. I. Stepanov, *Zhur. obshchei Khim.*, 1968, **38**, 195 (*Chem. Abs.*, **69**, 58,748).

208 (*a*) D. Voigt, R. Turpin, and M. C. Labarre, *Bull. Soc. chim. France*, 1968, 3561; (*b*) J. P. Laurent and G. Jugie, *Bull. Soc. chim. France*, 1969, 26.

209 E. A. Zgadzai, G. P. Naumova, and A. D. Troitskaya, *Tr. Kazan Khim-Technol. Inst.*, 1967, 96 (*Chem. Abs.*, **69**, 100,661).

210 F. D. Saeva, D. R. Raynor, and K. Mislow, *J. Amer. Chem. Soc.*, 1968, **90**, 4176.

211 F. M. Kharrasova and G. Kamai, *Zhur. obshchei Khim.*, 1968, **38**, 359, 617 (*Chem. Abs.*, **69**, 43,974, 52,215).

8 Diffraction

X-Ray diffraction has shown that the crystal structure of the reactive alkylidenephosphoranes (185; R = H or Me) has a phosphorus atom with a distorted tetrahedral co-ordination.[212] The phenyl groups, which are arranged like a propeller, have unequal rotations about the P—C bonds. The P=CH$_2$ bond length (166 pm) is among the shortest observed so far and is close to the sum of the double bond radii of phosphorus and carbon. The implication of these results on the extent of p_π–d_π bonding is fully discussed.

$$Ph_3P{=}CR_2$$
(185)

The stereochemistry of the phosphorins (186) [213] and (187) [214] has been determined. Planarity of the phosphorin rings is achieved by expansion of

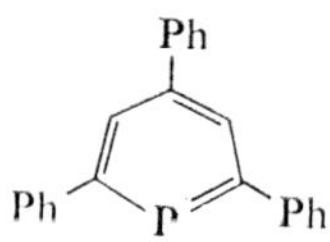

(186)

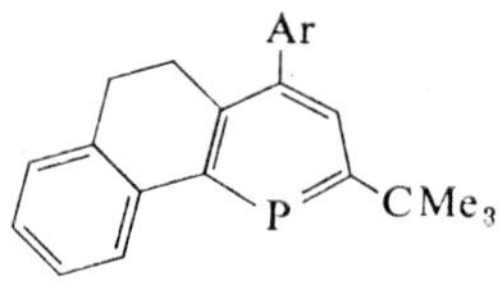

(187)

the PCC and CCC angles to 122—126° in order to accommodate the CPC bond angle of 103°. The P—C bond lengths (173—175 pm) are intermediate between those of triphenylphosphine (183 pm) and methylene-phosphoranes (165—168 pm) and compare with those (175—176 pm) of the phosphoranylidenephosphetane (188). The planarity of the phosphorin rings and equality of P—C bond lengths are in character with an aromatic

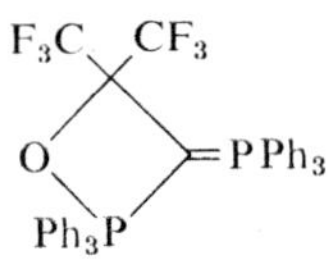

(188)

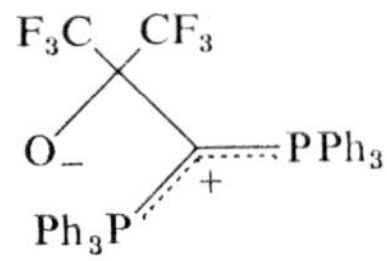

(189)

ring. The essentially equal P—C bond lengths [215] and long P—O bond (201 pm) in (188) indicate that (189) makes an important contribution to the resonance hybrid. *X*-Ray diffraction of two other phosphetanes (190) [216] and (191) [217] has been reported. There was no mention of nonplanarity for (190), but in (191) the ring is 24° out of plane with the larger groupings

[212] J. C. J. Bart, *Angew. Chem. Internat. Edn.*, 1968, **7**, 730; *J. Chem. Soc.* (*B*), 1969, 350.
[213] J. C. J. Bart and J. J. Daly, *Angew. Chem. Internat. Edn.*, 1968, **7**, 811.
[214] W. Fischer, E. Hellner, A. Chatzidakis, and K. Dimroth, *Tetrahedron Letters*, 1968, 6227.
[215] G. Chioccola and J. J. Daly, *J. Chem. Soc.* (*A*), 1968, 568.
[216] D. D. Swank and C. N. Caughlan, *Chem. Comm.*, 1968, 1051.
[217] C. Moret and L. M. Trefonas, *J. Amer. Chem. Soc.*, 1969, **91**, 2255.

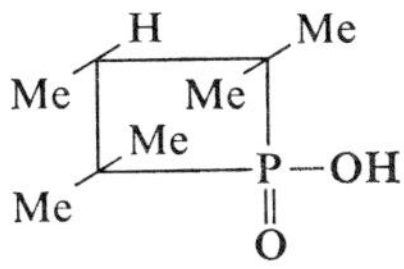

(190)

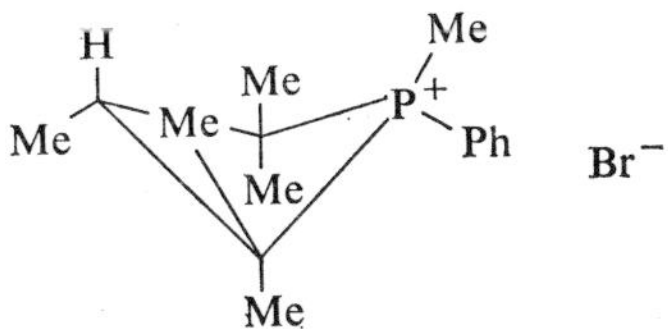

(191)

occupying the equatorial orientation as shown. Such stereochemical preferences probably play a very important role in the reactions of these compounds.[56] The CPC bond angle in the ring was very similar (82·5°) for (190) and (191), but it is difficult to believe that the ring bond lengths differ by *ca.* 10 pm. An unusually short hydrogen bond (249 pm) from the proton of the phosphinic acid grouping to the water of hydration is also reported.

Although the highfield chemical shift of the phosphorus atoms (+3·5 p.p.m.) of the hexaphenyl analogue of (192) is thought to be due to delocalisation, the hexa-alkyl compound (192; X = Me, Y = Et) possesses P—C and C=C bond lengths of 182 and 130 pm respectively, indicating a diene structure. As expected the ring is planar.[218]

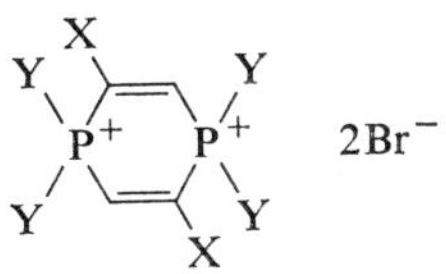

(192)

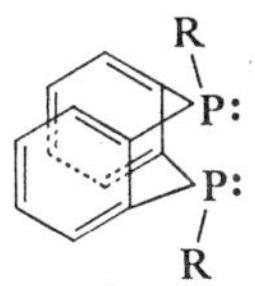

(193)

5,10-Dihydro-5,10-diethylphosphenanthrene is known in two isomeric forms, assigned *cis*- and *trans*-geometries. It is now found that the crystal structure of an arsenic analogue has a *cis* 'butterfly' conformation which adds weight to the earlier conclusion that the major isomer of the phosphorus compound has the *cis*-configuration (193).[219]

The crystal structure of a copper derivative of the triphenylmethylphosphonium cation (194) showed the cation to be a nearly perfect tetrahedron[220] with the phenyl rings in the expected propeller arrangement.

X-Ray diffraction shows that the aminocyclophosphazine hydrochloride (195) has a distorted boat conformation[221] and confirms that the ring nitrogen atoms are the stronger basic centres. The stereochemistries of the non-geminally substituted phosphazines (196) and the corresponding

[218] R. Majeste and L. M. Trefonas, *J. Heterocyclic Chem.*, 1969, **6**, 269.
[219] O. Kennard, F. G. Mann, D. G. Watson, J. K. Fawcett, and K. A. Kerr, *Chem. Comm.*, 1968, 269.
[220] R. M. Wing, *J. Amer. Chem. Soc.*, 1968, **90**, 4828.
[221] N. V. Mani and A. J. Wagner, *Chem. Comm.*, 1968, 658.

$Ph_3\overset{+}{P}Me$

(194)

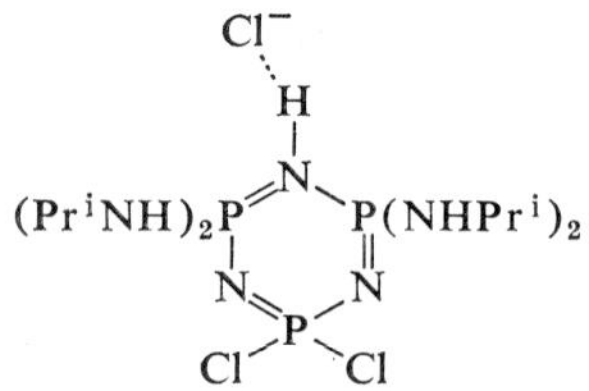

(195)

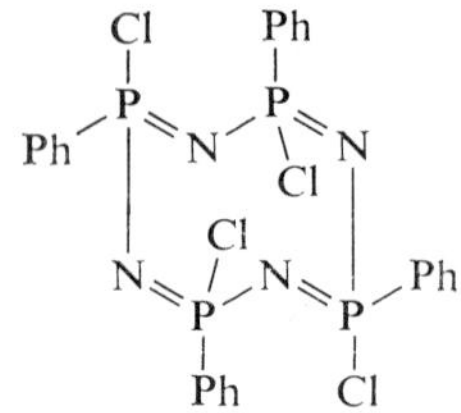

(196)

tetramethylamino-compound have been determined and show that substitution of Cl by NHMe proceeds with retention of configuration and causes a considerable reduction in the P—N—P bond angle.[222]

The ring system of the cyclo-octaphosphazine, $[NP(OMe)_2]_8$, consists of two approximately planar and parallel six-atom segments joined by a step as shown in (197).[223] The molecular structure as a whole supports earlier conclusions as to the importance of the π system involving orbitals with axes perpendicular to the ring.

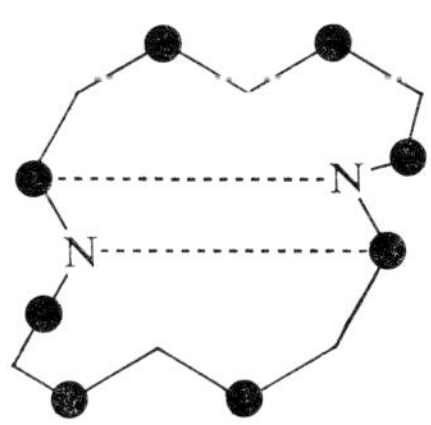

(197) ● = P

The crystal structure of the cobalt(II) chloride and copper(II) chloride complexes of the octamethylcyclophosphazine cation $(NPMe_2)_4H^+$ shows an alternation of P—N bond lengths which is independent of the conformation of the ring. This is interpreted in terms of the approximate equality of the components of a dual π-system.[224]

Electron diffraction indicates a pyramidal structure for disilylphosphine with an SiPSi angle of 96—97°.[225] This is in contrast to earlier evidence for a planar heavy-atom skeleton.

9 Electrochemical Studies

The half-wave potentials of phosphonium salts have been determined by rapid polarography.[226] The more positive waves corresponded to the formation of P^{III} compounds and the more negative waves corresponded

222 G. J. Bullen, P. R. Mallinson, and A. H. Burr, *Chem. Comm.*, 1969, 691.
223 N. L. Paddock, J. Trotter, and S. H. Whitlow, *J. Chem. Soc.* (*A*), 1968, 2227.
224 J. Trotter, S. H. Whitlow, and N. L. Paddock, *Chem. Comm.*, 1969, 695.
225 B. Beagley, A. G. Robiette, and G. M. Sheldrick, *J. Chem. Soc.* (*A*), 1968, 3002.
226 L. Horner and J. Haufe, *J. Electroanalyt. Chem. Interfacial. Electrochem.*, 1969, **20**, 245.

to the degradation of tertiary and secondary compounds. The position of the half-wave potentials were influenced by the measuring conditions. Polarography, cyclic voltammetry, and coulometry have been used[227] to study the electroreduction of triphenylphosphine and its oxide.

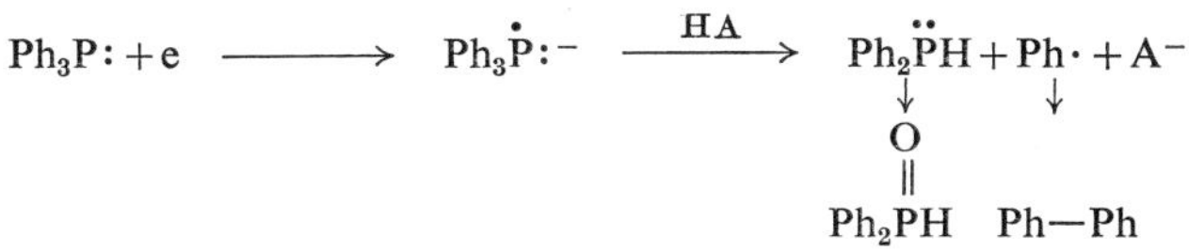

The conductivity of tungsten chloride[228] and rhodium chloride[229] biphosphine–carbonyl complexes in nitromethane and acetonitrile respectively, confirmed their ionic character and showed them to be 1 : 1 electrolytes. However, chloroform solutions of the former were non-conducting. The semiconductor properties of the adducts of *OO*-dialkyl dithiophosphates and amines were confirmed by their increase in conductivity with temperature.[230]

In contrast to earlier experiments but in confirmation of the ^{31}P n.m.r., dichlorotriphenylphosphorane (198) dissociates as shown. It does not ionise to $Ph_3\overset{+}{P}Cl$ and $Ph_3\overset{-}{P}Cl_3$ since the addition of tetraethylammonium chloride did not produce an inflection in the conductance.[231]

$$\underset{(198)}{Ph_3PCl_2} \rightleftharpoons Ph_3\overset{+}{P}Cl + \overset{-}{Cl}$$

The paper electrophoresis of phosphorus oxyacids, phenyl compounds, and hexose or inosine phosphate esters has been examined and a method of estimating dissociation constants and molecular weights has been devised.[232]

10 Mass Spectrometry

The mass spectra of phenyl, diphenyl, triphenylphosphines,[233] and the cyclopentaphosphine (199)[234] all contain an ion at *m/e* 108 attributed to the phenylphosphinidene ion. This ion is also present in the spectra of most phenyl-substituted phosphine oxides, phosphonium salts, and ylides,[235, 236] but are not observed for the halogen compounds, *e.g.*

227 K. S. V. Santhanam and A. J. Bard, *J. Amer. Chem. Soc.*, 1968, **90**, 1118.
228 P. M. Boorman, N. N. Greenwood, and M. A. Hildon, *J. Chem. Soc.* (*A*), 1968, 2466.
229 J. T. Mague and J. P. Mitchener, *Inorg. Chem.*, 1969, **8**, 119.
230 L. Almasi, A. Hantz, E. Weissmann, and E. Hamburg, *Rev. Roumaine Chim.*, 1967, **12**, 1269 (*Chem. Abs.*, **69**, 2399).
231 G. S. Harris and M. F. Ali, *Tetrahedron Letters*, 1968, 37.
232 Y. Kiso, M. Kobayashi, Y. Kitaoka, K. Kawamoto, and J. Takada, *J. Chromatog.*, 1968, **36**, 215.
233 B. Zeeh and J. B. Thomson, *Tetrahedron Letters*, 1969, 111.
234 U. Schmidt, I. Boie, C. Osterroht, R. Schroer, and H. F. Grutzmacher, *Chem. Ber.*, 1968, **101**, 1381.
235 D. H. Williams, R. S. Ward, and R. G. Cooks, *J. Amer. Chem. Soc.*, 1968, **90**, 966.
236 R. G. Cooks, R. S. Ward, D. H. Williams, M. A. Shaw, and J. C. Tebby, *Tetrahedron*, 1968, **24**, 3289.

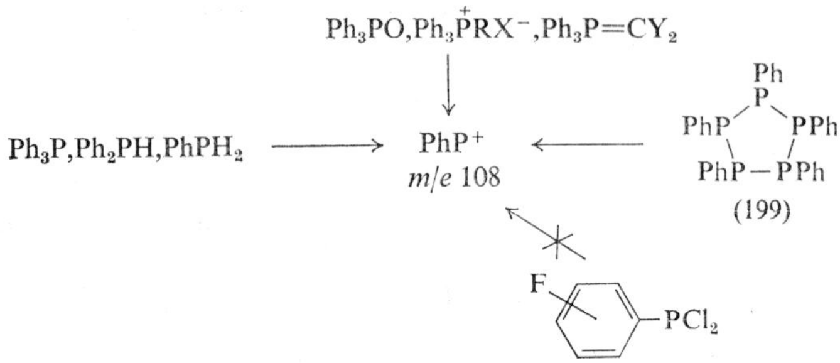

(200)

$PhPCl_2$.[233] Loss of chlorine radicals occurs more readily and in the case of the fluorophenyl compound (200) loss of the aryl ring is also ready.[35]

A wide variety of compounds, possessing at least two phenyl rings attached to phosphorus, give molecular ions (201) which undergo a bond-forming rearrangement and produce abundant $M-1$ ions due to the stable phosphafluorenyl ion (202). It has been shown[235] for (201; Y = Ph,

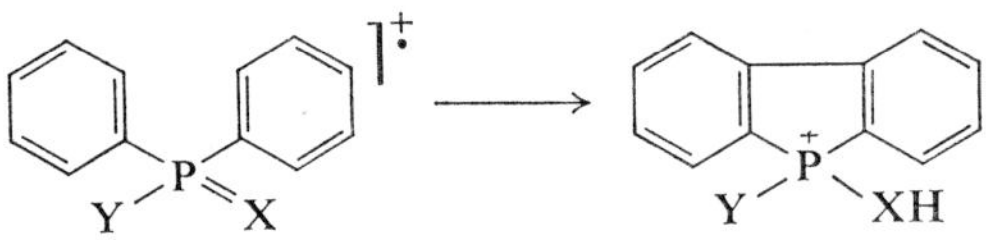

(201) (202)

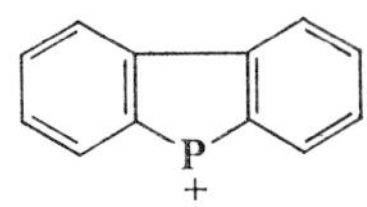

(203)

X = O, S, CH_2) that the cyclisation occurs without scrambling of the phenyl protons. However, for triphenylphosphine, the major fragmentation pathway involves loss of a phenyl group with scrambling of the phenyl protons with the eventual formation of (203), an ion common to the spectra of most triphenylphosphorus compounds. This ion is not prominent in the spectra of the diphenylphosphinate (201; Y = HO or RO, X = O). In this case the major fragmentation pathway involves the loss of water or alcohol from the corresponding phosphafluorenyl ion (202) to give the phosphinylium ion (204).[237] The cyclic phosphinate (205) also loses water or alcohol to give an analogous ion but the keto-derivative (206) tends to form heteroaromatic ions such as (207).

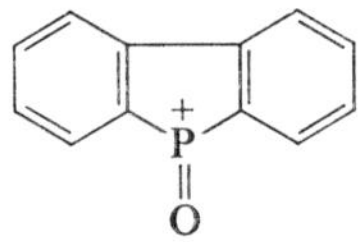

(204)

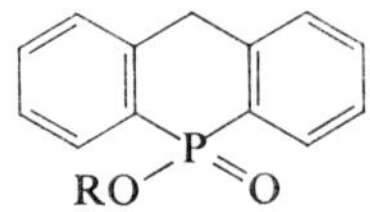

(205)

[237] P. Haake, M. J. Frearson, and C. E. Diebert, *J. Org. Chem.*, 1969, **34**, 788.

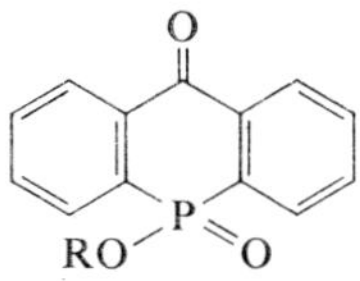

(206)

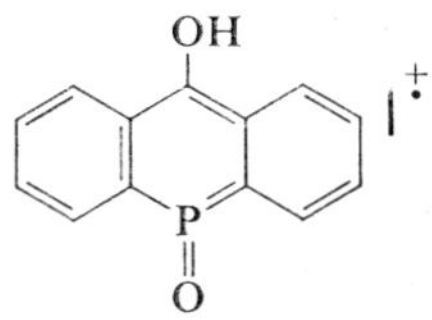

(207)

The mass spectra of alkylidenephosphoranes stabilised by CN, CO_2R, and COR groups have been studied.[236] For the acyl methylenephosphoranes (208) loss of an alkyl or aryl radical was always predominant, but for the carboalkoxy methylenephosphoranes (209; Z = CN or CO_2Me) cyclisation to (210) competed with the loss of methoxy- and carbomethoxy-groups.

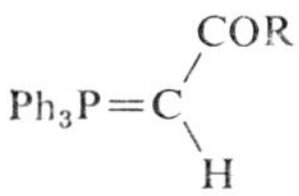

(208)

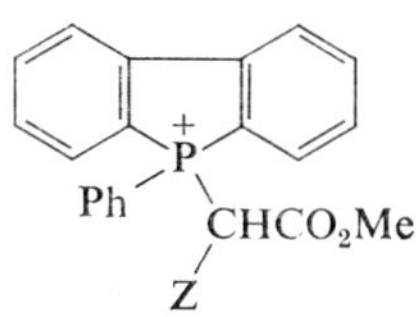

(210)

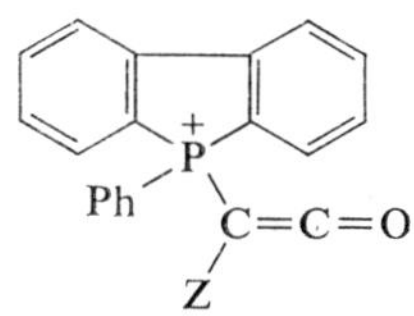

(211)

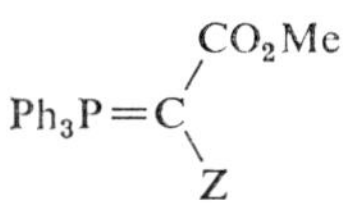

(209)

The transfer of an aromatic hydrogen to a more labile environment is confirmed by the ready loss of methanol from (210) to give (211). When the α-carbon atom already possesses a hydrogen atom loss of methanol can also occur before cyclisation. In the case of the ethyl esters (212) competition arises from a process in which the α-carbon atom gains a hydrogen atom *via* the cyclic loss of CO_2 and C_2H_4; further fragmentation follows a pathway characteristic of $Ph_3P{=}CHZ$.

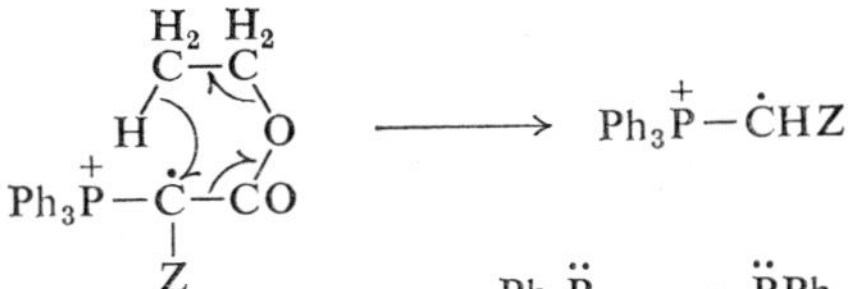

(212)

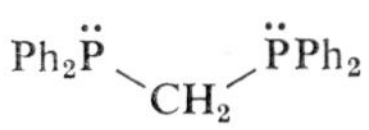

(213)

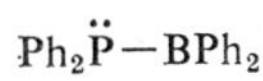

(214)

The bis(diphenylphosphino)methane (213) is reported[238] to show the migration of phenyl groups to give Ph_3P^+ cations which fragment in the usual way. Low-abundance ions corresponding to Ph_3P^+, Ph_3B^+,

[238] R. Colton and Q. N. Porter, *Austral. J. Chem.*, 1968, **21**, 2215.

Ph_4P^+, and Ph_4BP^+ appear in the spectrum of (214). The dominant pathway involves fission of the P—B bond.[239] The mass spectra of the compounds (215) have been cited as evidence for their oxyphosphorane structures.[27] In contrast to the corresponding oximinophosphonium salts, see (21), this chapter (1A), the spectra of the oxyphosphoranes showed no fragment ions above Bu_3PO^+ indicating a cyclic structure with a preformed P—O bond.

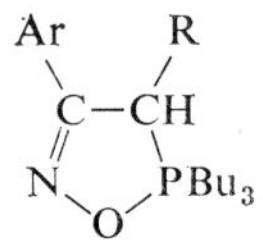
(215)

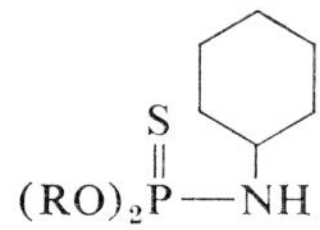
(216)

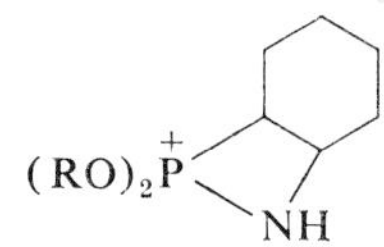
(217)

The molecular ion of phosphoramidothioates such as (216) showed loss of SH with cyclisation to form ions, *e.g.* (217).[240] This type of fragmentation was less important in the chloridothioates (218), loss of Cl, RO, and other radicals providing an easier pathway. The diphenylphosphorothioates (219) showed the expected P=S → P=O isomerisation but aryl migration to sulphur predominated over alkyl migration.

S ‖ $(RO)_2PCl$	S ‖ $ROP(OPh)_2$
(218)	(219)

Initial fragmentation by loss of CF_3 and in some cases CF_4 is reported for the trifluoromethyl compounds (220—223). Extensive rearrangement

S ‖ $(CF_3)_2PH$	$(CF_3)_2PP(CF_3)_2$	$(CF_3)_2PX$	CF_3PX_2
(220)	(221)	(222)	(223)

to species containing P—F bonds was also observed[241] for the P^{III} compounds.

Electron bombardment of phosphazines causes similar fragmentations to those produced by pyrolysis.[242] The most abundant ions in the spectra of $P_3N_3Cl_6$ and $P_4N_4Cl_8$ occur from loss of Cl. The dominance of even-electron ions over odd-electron ions is striking.[243] Chloro-, fluoro-,[244] and

[239] E. W. Abel, R. A. N. McLean, and I. H. Sabherwal, *J. Chem. Soc.* (*A*), 1968, 2371.
[240] R. G. Cooks and A. F. Gerrard, *J. Chem. Soc.* (*B*), 1968, 1327.
[241] R. G. Cavell and R. C. Dobbie, *Inorg. Chem.*, 1968, **7**, 101, 690; R. C. Dobbie, L. F. Doty, and R. G. Cavell, *J. Amer. Chem. Soc.*, 1968, **90**, 2015.
[242] B. Zeeh and R. Beutler, *Org. Mass Spectrometry*, 1968, **1**, 791.
[243] C. D. Schmulback, A. G. Cook, and V. R. Miller, *Inorg. Chem.*, 1968, **7**, 2463.
[244] C. E. Brion and N. L. Paddock, *J. Chem. Soc.* (*A*), 1968, 388, 392.

mixed chlorobromo-phosphonitriles [245] are reported to fragment to cyclic and linear ions.

Appearance potentials have been used to estimate the energy of P—P bonds of diphosphine [246] and diphosphine tetrachloride and tetraiodide.[247]

11 p*K* and Reaction Rate Studies

The majority of kinetic studies has been directed towards the investigation of reaction mechanism.

The mechanism of the acid and base hydrolyses of *p*-nitrophenyl phosphates,[248] with or without the presence of cationic micelles, has been studied. Also the rates of acid hydrolysis of methylphosphonates have been compared with the ageing of phosphorylated cholinesterase.[249] The factors which lead to reactivity differences for acyclic and cyclic phosphoramidites [250] and acyclic, mono- and bi-cyclic phosphinates,[251] have been investigated by comparing rate constants and enthalpies.

The effect of free radical inhibitors and initiators on the rate of attack of triethylphosphite on carbon tetrachloride has been used to show that the reaction may proceed *via* heterolytic or homolytic mechanisms.[252]

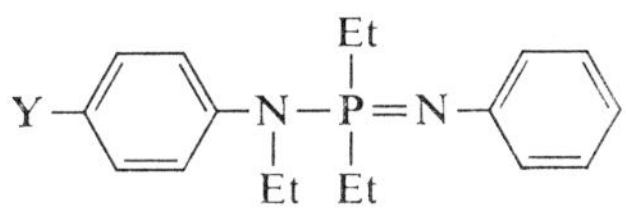

(224)

The transmission of polar effects in phosphinimines (224) has been studied by correlating alkylation rate constants and pK_a values with Hammett constants.[253] The best correlation is obtained with the nucleophilic substituent constant σ^-. Also a successful correlation of ionisation constants and rate constants of benzylations has been achieved using the extended Hammett equation and the σ_I and σ_R substituent constants defined for carbon substituents.[254]

245 G. E. Coxon, T. F. Palmer, and D. B. Sowerby, *J. Chem. Soc.* (*A*), 1969, 358.
246 N. N. Grishin, G. M. Bogolyubov, and A. A. Petrov, *Zhur. obshchei Khim.*, 1968, **38**, 2683.
247 A. Finch, A. Hameed, P. J. Gardner, and N. Paul, *Chem. Comm.*, 1969, 391.
248 C. A. Bunton and S. J. Farber, *J. Org. Chem.*, 1969, **34**, 767; C. A. Bunton and L. Robinson, *J. Org. Chem.*, 1969, **34**, 773, and references therein.
249 J. I. G. Cadogan, D. Eastlick, F. Hampson, and R. K. Mackie, *J. Chem. Soc.* (*B*), 1969, 144.
250 R. F. Hudson and R. J. G. Searle, *J. Chem. Soc.* (*B*), 1968, 1349; R. F. Hudson and A. Mancuso, *Chem. Comm.*, 1969, 522, and references therein.
251 R. Kluger and F. H. Westheimer, *J. Amer. Chem. Soc.*, 1969, **91**, 4143.
252 R. E. Atkinson, J. I. G. Cadogan, and J. T. Sharp, *J. Chem. Soc.* (*B*), 1969, 138.
253 V. A. Gilyarov, A. M. Maksudov, B. A. Korolev, B. I. Stepanov, and M. I. Kabachnik, *Izvest. Akad. Nauk S.S.S.R., Ser. khim.*, 1968, **7**, 1656 (*Chem. Abs.*, **69**, 95,815).
254 M. Charton, *J. Org. Chem.*, 1969, **34**, 1877; see also M. Charton, and B. I. Charton, *J. Org. Chem.*, 1969, **34**, 1882.

The pK_a values of a number of phosphines[255] and phosphine oxides[256] have been measured for solutions in nitromethane by potentiometric titrations.

12 Cryoscopic Studies

Pyridine solutions of copper(I) hydride and trialkyl or triarylphosphines have been studied by freezing-point depression. Cryoscopic titration curves for the more basic phosphines show a molality minimum which corresponds to the formation of $R_3P(CuH)_2$.[257]

[255] B. A. Korolev and B. I. Stepanov, *Izvest. V.U.Z., Khim. i khim. Tekhnol.*, 1968, **11**, 1193.
[256] B. I. Stepanov, B. A. Korolev, and A. I. Bokanov, *Zhur. obshchei Khim.*, 1969, **39**, 316, 321.
[257] J. A. Dilts and D. F. Shriver, *J. Amer. Chem. Soc.*, 1969, **91**, 4088.

Author Index